RF MEASUREMENTS FOR CELLULAR PHONES AND WIRELESS DATA SYSTEMS

RF MEASUREMENTS FOR CELLULAR PHONES AND WIRELESS DATA SYSTEMS

ALLAN W. SCOTT

REX FROBENIUS

Published by John Wiley & Sons, Inc., Hoboken, New Jersey
Published simultaneously in Canada

For general information on our other products and services or for technical support, please contact our Customer Care Department within the United States at (800) 762-2974, outside the United States at (317) 572-3993 or fax (317) 572-4002.

Wiley also publishes its books in a variety of electronic formats. Some content that appears in print may not be available in electronic formats. For more information about Wiley products, visit our web site at www.wiley.com.

Library of Congress Cataloging-in-Publication Data:

Scott, Allan W.
RF measurements for cellular phones and wireless data systems/Allan W. Scott, Rex Frobenius.
p. cm.
ISBN 978-0-470-12948-7 (cloth)
1. Radio frequency integrated circuits—Testing. 2. Wireless communication systems—Equipment and supplies—Design and construction. 3. Cellular telephones—Equipment and supplies—Design and construction. I. Frobenius, Rex. II. Title.
TK7874.S36 2008
621.3845′6—dc22

2008004929

Printed in the United States of America
10 9 8 7 6 5 4 3 2 1

CONTENTS

FOREWORD

In the late 1960s as part of a technical seminar team, I traveled with HP's first-generation Automatic Network Analyzer (ANA), discussing and demonstrating new measurements and s-parameter design techniques. One of our stops was at a U.S. East Coast based defense organization, where a large group gathered to hear our talk. At the conclusion, a man with a skeptical expression on his face indicated that he had two questions. In a somewhat hostile manner he asked, "Are you telling me that with this new equipment I can reliably characterize active devices at microwave frequencies?" We assured him that for small-signal applications it was true. Then he went on, "If I turn off the equipment today, and repeat my measurement tomorrow, will I get the same data?" Again, we replied that after proper calibration, he will have the same results. Shaking his head in disbelief he said, "I cannot believe such b.s.," and stormed out of the lecture hall.

It is hard to understand such a reaction today. However, until the introduction of the network analyzer, obtaining reliable and repeatable y-parameter component characterization with its predecessor, the General Radio RF Bridge, was not possible. Without accurate data or component models, microwave circuit design was more of an art than science.

Even after the spectrum analyzer, network analyzer, and modern power meters became available, relatively simple gain, impedance, power, harmonic and two-tone intermodulation measurements represented a large percentage of microwave testing. This is in sharp contrast to what test engineers and technicians face today, working on products using a wide range of mixed-mode signal processing. In addition, they have to understand and measure parameters, Bit Error Rate (BER), constellation and eye diagrams, Adjacent Channel Power (ACP), just to mention a few. They also have to be familiar with various digital modulation systems, including analog concepts. Last but not least, in the globally competitive marketplace, measurements must be performed rapidly and inexpensively.

The authors of this book based the contents on their extensive experience teaching continuing education courses to practicing professionals of the RF and microwave industries. Their course material is fine-tuned with the feedback provided by course participants and constantly updated to keep up with changes in technology. Measurements described in the book range from basic to advanced types, in addition to reviewing the necessary technical background of cellular and wireless communication systems. I am not aware of any other textbook having such a wealth of information, written in a simple, easily understandable style, without constant use of complex mathematics. Learning the techniques described in the book will elevate the value of anyone working in the field.

LES BESSER

Besser Associates
Mountain View, CA

ACKNOWLEDGMENTS

We would like to thank all the organizations and people who helped make this book possible.

The idea began when the Test and Measurement Division of Hewlett-Packard offered Besser Associates the loan of a suite of RF test equipment for demonstration in their RF classes. Besser Associates is the worldwide leader in RT training, having trained over 40,000 engineers, technicians and managers in RF topics. We serve as instructors for Besser Associates. So beginning in 1998, we began a demonstration class on RF Measurements using the HP equipment.

When HP split off their Test and Measurement Equipment division to become Agilent Technologies, Agilent continued the loan of RF test equipment ot Besser. As the cellular phone and wireless data industry grew and became more technically complicated, so did the test equipment. We are grateful to Susan Owen of Agilent for continuing to support this ongoing relationship, which always allowed us access to the latest models of their equipment. We would also like to thank Ben Zarlingo, who offered extensive support and insight into the operation of the Vector Signal Analyzer.

Many other companies, like Anritsu, Rhode and Schwartz, Aeroflex, and Keithley, to name a few, also make excellent test equipment. Often they demonstrated their test equipment in our course. Along these lines we enjoyed a great deal of support from David Vondran of Anritsu corporation, who provided detailed background on the Scorpion Network Analyzer and noise figure measurements. We did not try to make comparisons between test equipments, but continued to conduct a course using the Agilent equipment, because of their generosity in always loaning us the latest and best they had. This kept us busy learning how to operate their continually evolving equipment, as the cell phone industry itself evolved.

The Besser RF Measurements course has continued to grow in popularity. Initially we taught the five-day course a few times a year, then several times a year. Last year (2007) we taught it 7 times. About a year ago, we decided to write this book, based on the RF Measurements course.

We would also like to thank everyone at Besser Associates, Founder Dr. Les Besser, President Jeff Lange, VP of Sales Annie Wong, and all the administrative staff and instructors who helped and encouraged us to write this book. We would also like to thank Allen Podell for providing numerous technical insights as well as practical tips on keeping our fragile lab components in good repair.

Finally, we would like to thank all the engineers, technicians, and managers who have taken our RF measurements course and made valuable suggestions on how to make it better.

We hope you will enjoy our book and find it useful. We hope that it will improve your understanding of RF measurements by at least 7 dB.

Al Scott

Rex Frobenius

CHAPTER 1

INTRODUCTION

1.1 THE MARKET FOR CELLULAR PHONES AND WIRELESS DATA TRANSMISSION EQUIPMENT

The market for cellular phones and wireless data transmission equipment has changed dramatically since the late 1970s when cellular phones were first introduced and the late 1980s when wireless data equipment became available. As would be expected, during this time RF test requirements and RF test equipment has changed dramatically.

The original cellular phones, which were introduced in North America in the 1970s, were FM analog voice phones with a limited data capability of less than 10 kbps. These analog phones are now called first generation (1G). Cellular phones were digitized in the early 1980s to provide for an increased number of user channels in a given RF frequency band. These digital phones are now called second generation (2G).

During the 1990s the use of 2G cell phones increased dramatically throughout the world, growing to over 2 billion handsets worldwide by 2005. Eighty percent of 2G phones are Global System for Mobile Communications (GSM), using digital FM modulation. The reasons for the expansive growth of GSM phones was (1) the excellent voice quality of the digital signal, which could accurately digitize any language, and (2) an effective worldwide management and billing system for all of its customers.

During the growth of GSM phone capacity worldwide, the North American cellular industry was divided between proponents of using a Time Division Multiple Access (TDMA) system similar to GSM, but carefully designed to be backward

RF Measurements for Cellular Phones and Wireless Data Systems. By A. W. Scott and R. Frobenius

compatible with the RF part of the original analog system, and a new Code Division Multiple Access (CDMA) concept advocated by Qualcomm, which provided greater user capacity in a given RF bandwidth. After extensive field trials conducted throughout the United States in the late 1980s, the CDMA system demonstrated an approximate doubling of voice capacity compared to TDMA.

As digital voice cell phone usage grew in the 1980s, equipment manufacturers began the development of third generation (3G) phones that, in addition to providing high-quality wireless voice service, could also provide a wide range of data related services including the following:

- Data rate transfer exceeding 1 Mbps at any location within the cell where voice phones worked
- Wireless connected photographic cameras
- Wireless connected video cameras
- GPS information
- Windows operating system with Word, Excel, and PowerPoint
- Internet access

In order to accomplish this high data rate capacity, the greater bit rate transfer capability of CDMA systems is required. It is predicted that all cellular phone systems will be converted to CDMA by 2012.

The 20% of system operators who had originally opted for CDMA voice phones are already providing data transfer capability up to 1 Mbps, even though voice service still accounted for 90% of their business in 2005.

The GSM service providers, who provide 80% of worldwide cellular voice service, face an economic problem because of the vast amount of installed GSM base station equipment. However, the GSM community now has a worldwide evolution plan to grow from the limited 100 kbps data capability of GSM phones to a data capability of several gigabits per second using Wideband Code Division Multiple Access (WCDMA)/High-Speed Downlink Packet Access (HSDPA). However, the implementation of this high data rate equipment by the GSM community will lag that of the current CDMA carriers by about 3 years.

The importance of these facts is that the measurement equipment needs for cellular phone equipment are stabilized for the next 5 years, until fourth generation (4G) phones replace the 3G phones.

In a similar way, the requirements for short-range, high data rate equipment like Wi-Fi (802.11a, b, g, and n) are stabilized. These systems achieve data rates up to 200 Mbps because their ranges are short. Consequently, the received power is high and complex modulation schemes like 64-quadrature amplitude modulation (64QAM), which transmits 6 digital bits in every Hertz of bandwidth, can be used.

A significant change in RF test equipment occurred in the early 2000s in order to meet the needs of testing the evolving cell phone and wireless local area network (LAN) equipment. Extensive digital processing was added to conventional RF signal generators, vector network analyzers (VNAs), and spectrum analyzers to improve their measurement uncertainty and increase their capability.

For example, many of the newest VNAs now use a Windows operating system instead of a proprietary operating system. This change gives increased capacity for data processing and allows measurements to be easily transferred to laptops or other computers for further analysis and archiving. Electronic calibration of the VNAs is now available to reduce the uncertainty of their measurements that is due to handling damage of the calibration standards and operator error. Measurement of absolute power in decibels relative to 1 mW (dBm) in a VNA is about ± 1 dBm. Provision is now available to calibrate the VNA with a power meter and achieve power measurements within an uncertainty of only ± 0.2 dBm.

Hardware and software options can now be added to the latest generation of spectrum analyzers to permit them to make the specialized signal analysis measurements required for cell phone and wireless LAN. These upgrades to the spectrum analyzers include the following:

- Measurement of phase noise and noise figure
- Measurement of the spectral regrowth of digitally modulated RF carriers
- Ability to function as a vector signal analyzer (VSA)
- Measurement of the key specifications for any cell phone or wireless LAN system

The life cycle of RF measurement equipment (with hardware and software upgrades) is about 15 years, so the latest versions of RF measurement equipment will cover RF measurement needs throughout the lifetime of the current cell phone and wireless LAN evolutions.

1.2 ORGANIZATION OF THE BOOK

RF Measurements for Cellular Phone and Wireless Data Equipment is organized as follows:

Part I (Chapters 2–4) provides a review of basic RF principles. Many of the users of this book already have knowledge of basic RF terminology, but many do not. For those users who do not, Part I will provide this knowledge and should be studied first. For those users who have this knowledge already, Part I will provide a good review.

Part II (Chapters 5–14) describes RF measurement equipment, including signal generators, power meters, frequency meters, VNAs, spectrum analyzers, VSAs, and other equipment.

Part III (Chapters 15–28) describes the RF devices that are used in cellular phones and wireless data transmission equipment: how they work, what their critical performance parameters are, how they are tested, and what typical test results are.

Part IV (Chapters 29–36) describes the testing of RF devices and systems that use digitally modulated signals to represent the voice, video, or data that the RF wave is carrying. The same RF device will have different performance, depending on the data modulation being used.

1.3 PART I: RF PRINCIPLES

Chapters 2–4 in Part I describe RF principles.

1.4 SUMMARY OF CHAPTER 2: CHARACTERISTICS OF RF SIGNALS

Chapter 2 describes the characteristics of RF signals, which include frequency and wavelength, power (dB and dB relative to 1 mW), and phase.

The range of RF power that must be measured in cellular phones and wireless data transmission equipment varies from hundreds of watts in base station transmitters to picowatts in receivers.

For calculations to be made, all powers must be expressed in the same power units, which is usually milliwatts. A transmitter power of 100 W is therefore expressed as 100,000 mW. A received power level of 1 pW is therefore expressed as 0.000000001 mW. Making power calculations using decimal arithmetic is therefore complicated. To solve this problem, the dBm system is used, which is fully explained in Chapter 2. Figure 1.1 shows the range of RF power and its value in watts and dBm.

1.5 SUMMARY OF CHAPTER 3: MISMATCHES

Chapter 3 describes mismatches, including definition of mismatches: return loss, standing wave ratio (SWR), and reflection coefficient; conversion between units; matching; and use of the Smith Chart for matching design.

Figure 1.2 illustrates the mismatch problem. Figure 1.3 shows how to minimize the mismatch by adding a matching component using a Smith Chart design.

1.6 SUMMARY OF CHAPTER 4: DIGITAL MODULATION

The purpose of a wireless communication system is to transmit voice, video, or data signals wirelessly from one location to another using the least amount of RF bandwidth. The various types of digital modulation [frequency shift keying (FSK), phase shift keying (PSK), and QAM] are explained in this chapter. Trade-offs between capacity and complexity of modulation are presented.

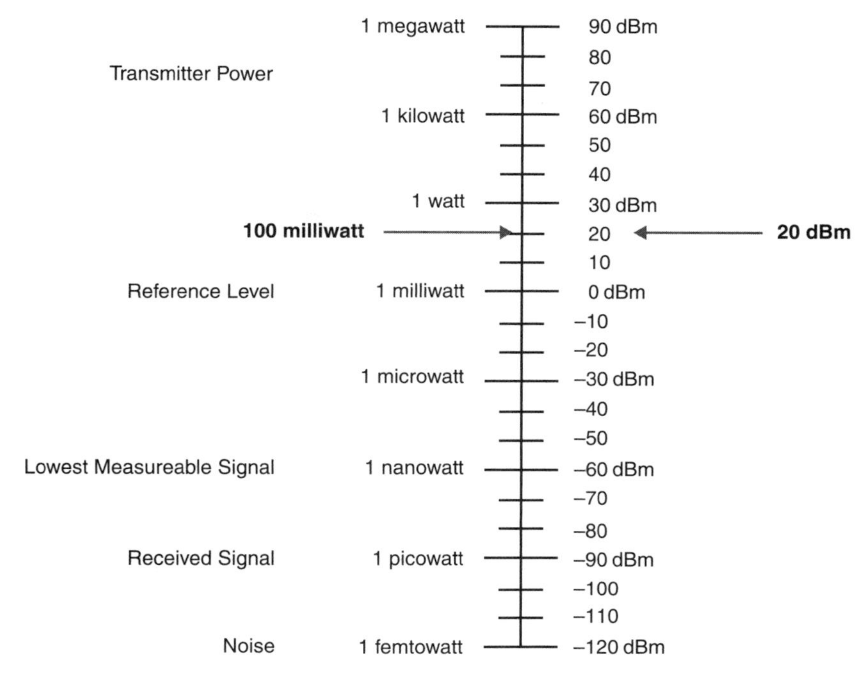

Figure 1.1 Range of RF power in watts and dBm.

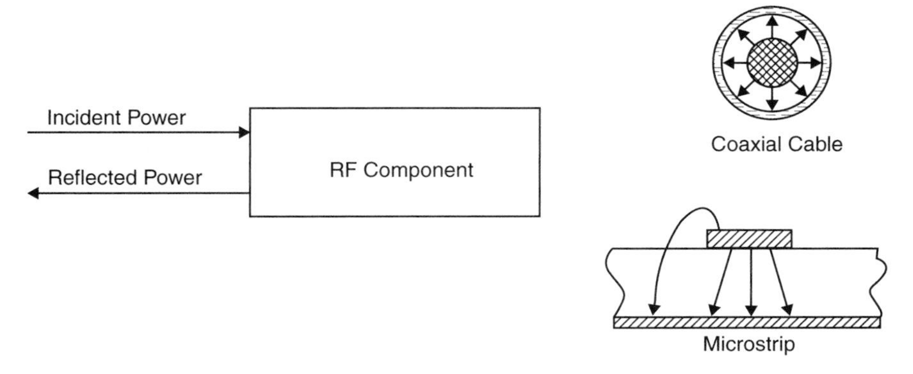

Figure 1.2 Mismatches. Some RF power is reflected as it tries to enter a component because the RF fields do not match. The mismatch is expressed as the percentage of reflected power, return loss, SWR, and reflection coefficient.

Figure 1.4 shows types of digital modulations on an RF wave. The RF wave is called the carrier, because it is carrying digital information by its modulation. The upper curve shows a bipolar digital data stream that is to be transmitted. In a wired communications system, this digital signal is simply transmitted as a voltage

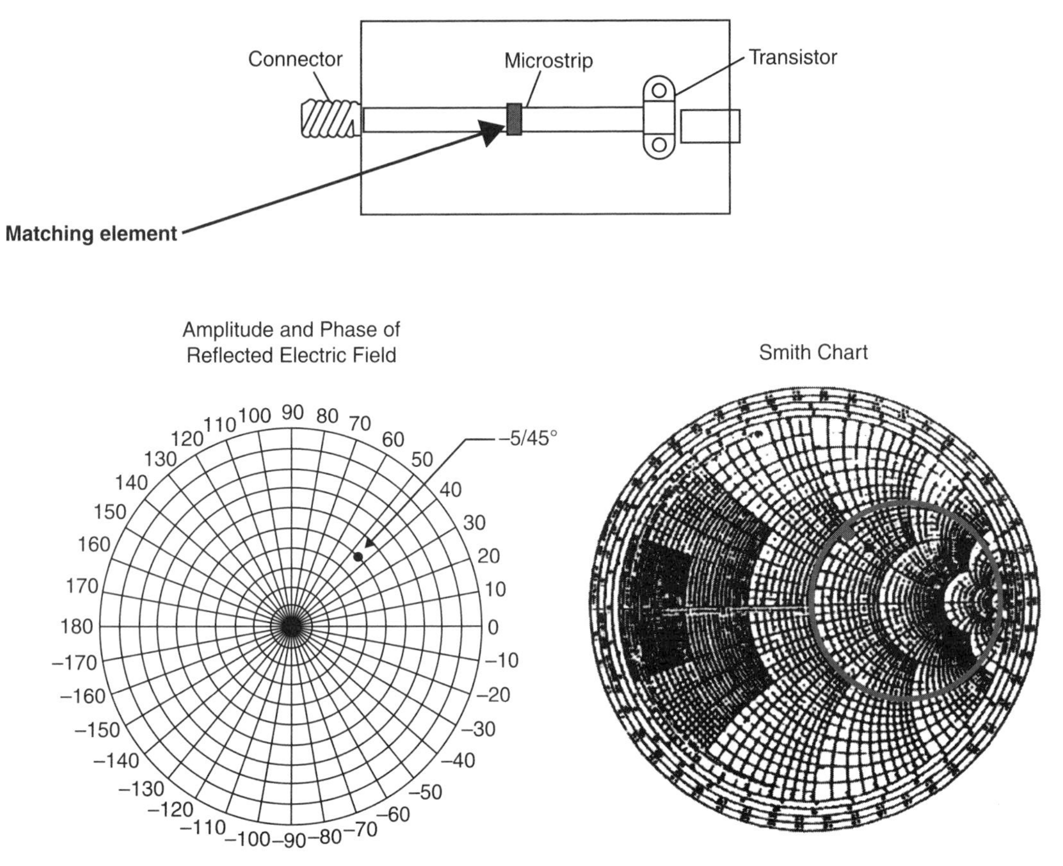

Figure 1.3 Matching with the Smith Chart.

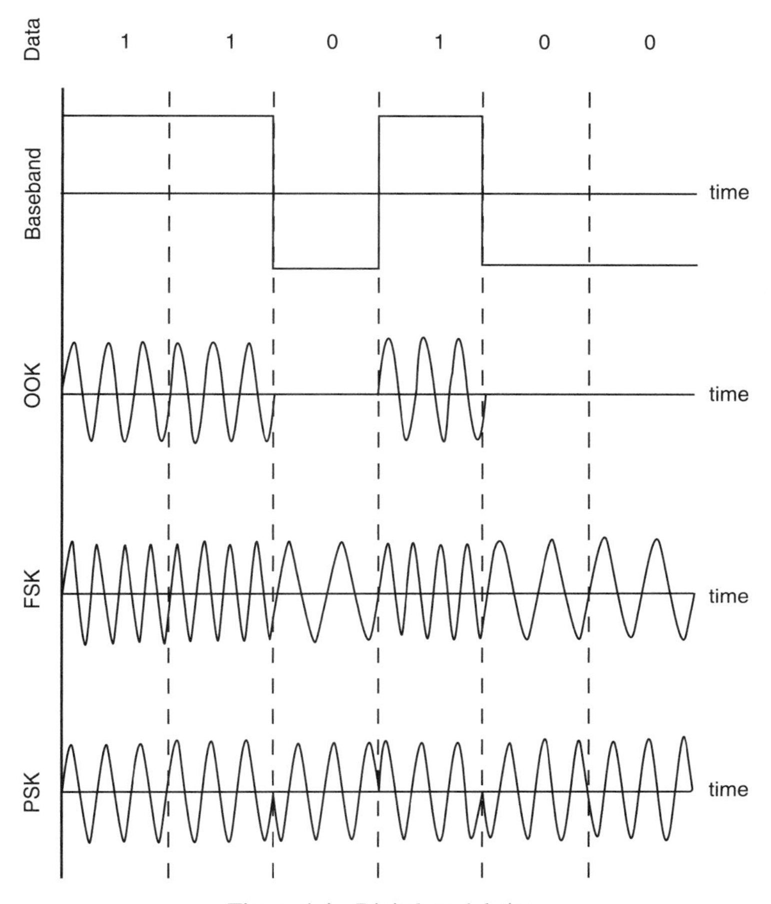

Figure 1.4 Digital modulation.

through wire, coaxial cable, or optical fiber. The data stream in this example is 110100.

The second curve in Figure 1.4 shows the amplitude modulation of a wireless carrier. There are various types of amplitude modulation. The simplest type shown here is on–off keying (OOK). When the RF signal is on, the data is a 1; when the RF signal is off, the data is a 0.

The third curve shows digital FSK. The amplitude (or power level) of the FSK modulated wave is constant, but the frequency is changed to represent the digital information. When the frequency is low, the data is a 0. When the frequency is high, the data is a 1.

The fourth curve shows phase modulation. The amplitude and the frequency of the wave are constant, but the phase is changed to represent information. When the phase is 0°, the data is a 0. When the phase is 180°, the data is a 1. Actually there is no way of telling from the RF wave whether the phase is 0° or 180°. The change can be detected, but not the absolute value. Therefore, a second phase reference signal must be transmitted along with the phase modulated wave. Alternately, a special technique called "differential" PSK (DPSK) can be used, where a change in phase represents a digital 1 and no change in phase represents a digital 0.

Note that the amplitude of the RF wavelets is constant when phase modulation is used. However, the RF amplitude varies during the phase transition between data pulses, and this amplitude change creates difficult design problems for the power amplifier that amplifies the digitally modulated RF signal before transmission.

Each time the amplitude, frequency, or phase of the RF carrier is changed, approximately 1 Hz of bandwidth is used. Therefore, if the data rate is 1 Mbps, the required RF bandwidth to transmit the information is about 1 MHz.

To reduce the RF bandwidth requirements for transmission of a given data rate signal, multiple levels of amplitude, frequency, or phase are used and sometimes two types are modulation are used simultaneously. The number of bits that can then be transmitted in a 1 Hz bandwidth is increased, and this increase is called the "spectral efficiency" of the modulation system.

Constellation diagrams of various multiple level modulation systems are provided in Figure 1.5. These diagrams show the phase of the RF signal in the angular direction and the amplitude of the signal in the radial direction.

The upper left-hand drawing in Figure 1.5 shows the simplest modulation scheme, binary (two level) PSK (BPSK), which transmits 1 bit for every 180° of phase change of the carrier.

The upper center drawing in Figure 1.5 shows QPSK modulation with four phase positions of 45°, 135°, 225°, and 315°, which represent bits 00, 01, 10, 11, respectively. As stated earlier, note that with any phase shift modulation, either a second

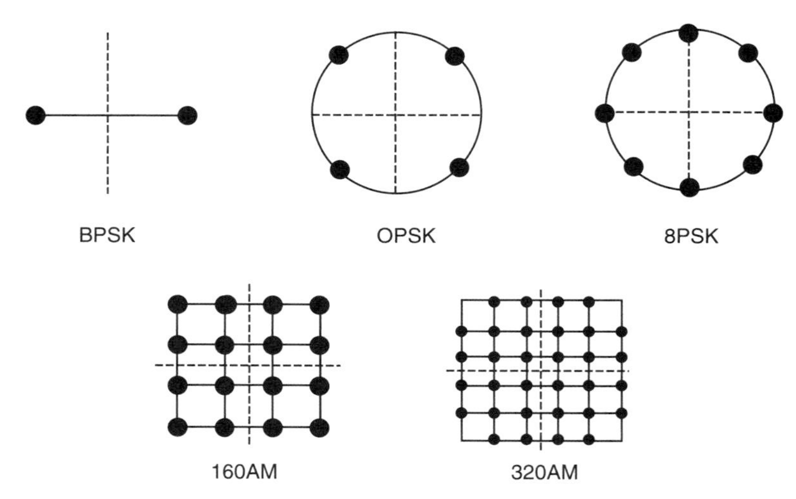

Figure 1.5 Constellation diagrams.

unmodulated carrier must be transmitted or a differential phase shift (DQPSK) modulation must be used. In either case the constellation is the same.

The upper right-hand drawing shows 8PSK modulation, where the constellation points are 45° apart. With 8PSK, 3 bits are transmitted every time I Hz of bandwidth is used.

The lower two drawings in Figure 1.5 show constellation diagrams in which the amplitude and phase are changed simultaneously. These modulation schemes are called QAM. In the lower left-hand drawing there are 16 possible phase and amplitude positions, so that 4 bits are transmitted every time 1 Hz of bandwidth is used. In the lower right-hand drawing, there are 32 possible phase and amplitude positions, so that 5 bits are transmitted every time 1 Hz of bandwidth is used.

1.7 PART II: RF MEASUREMENT EQUIPMENT

Part II describes RF measurement equipment. Chapters 4–12 describe RF measurement equipment and techniques in the following order:

Chapter	Measurement Equipment
5	RF signal generators
6	Power meters
7	Frequency meters
8	VNAs
9	Spectrum analyzers
10	VSAs
11	Noise figure meters
12	Coaxial cables and connectors
13	Measurement uncertainty
14	Measurement on components without coax connectors

1.8 SUMMARY OF CHAPTER 5: RF SIGNAL GENERATORS

To test any RF device or system, an RF signal is required, which is provided by an RF signal generator.

The signal generator provides a single RF signal with characteristics selected by the user, which remain constant until the user changes them. Typical characteristics that can be adjusted are as follows:

Frequency
Power
Type of AM or FM modulation
Type of digital modulation

Digital modulation types can be specified for particular cell phone systems
New modulation techniques can be programmed into the signal generator

The signal generator can also be adjusted to supply a fixed set of multiple signals. Bit error rate (BER) testing on systems can also be done by the signal generator. Figure 1.6 shows an RF signal generator that provides this performance.

1.9 SUMMARY OF CHAPTER 6: POWER METERS

Power meters provide the most accurate measurement of RF power of any of the types of RF measurement equipment. Power meters can be stand-alone instruments, or they can be built into other instruments like signal generators, spectrum analyzers, and VSAs. Some power meters can display RF power as a function of time.

RF power meters provide absolutely *no* information about the frequency distribution of the RF power. The indicated RF power is the total power incident at the power meter. If the signal is a single frequency, the power meter displays its power. However, if multiple signals are present at different frequencies, the power meter displays the total RF power of all of the signals together. Figure 1.7 shows an RF power meter with its power sensor.

1.10 SUMMARY OF CHAPTER 7: FREQUENCY COUNTERS

RF frequency counters measure the frequency of a *single* RF signal. If more than one frequency is present, the power meter turns off its display.

At RF frequencies up to about 500 MHz, frequency counters simply count the cycles of the single frequency RF wave with a digital counter. Accuracy can be as good as 1 part in 1 million.

Digital counting circuits do not work above about 500 MHz. Thus, for counting higher RF frequencies, some type of downconversion is used.

One type of downconversion is "prescaling." Prescaling involves simple division of the input frequency by an integer N to reduce the frequency to a value that can be counted by a digital counter. Typically, N ranges from 2 to 16. The counted prescaled value is then multiplied in a signal processing circuit by the integer N and displayed. This technique allows counting to about 1.5 GHz.

For counting to higher RF frequencies, a heterodyne converter is used. The counter contains a signal generator, a mixer, and a lower frequency digital counter. The RF signal to be counted is mixed down to a lower frequency that can be counted, and the displayed signal is the sum of the frequency of the lower frequency signal generator and the difference frequency of the mixer. Accuracy is determined by the frequency accuracy of the internal RF signal generator. Figure 1.8 shows an RF frequency counter.

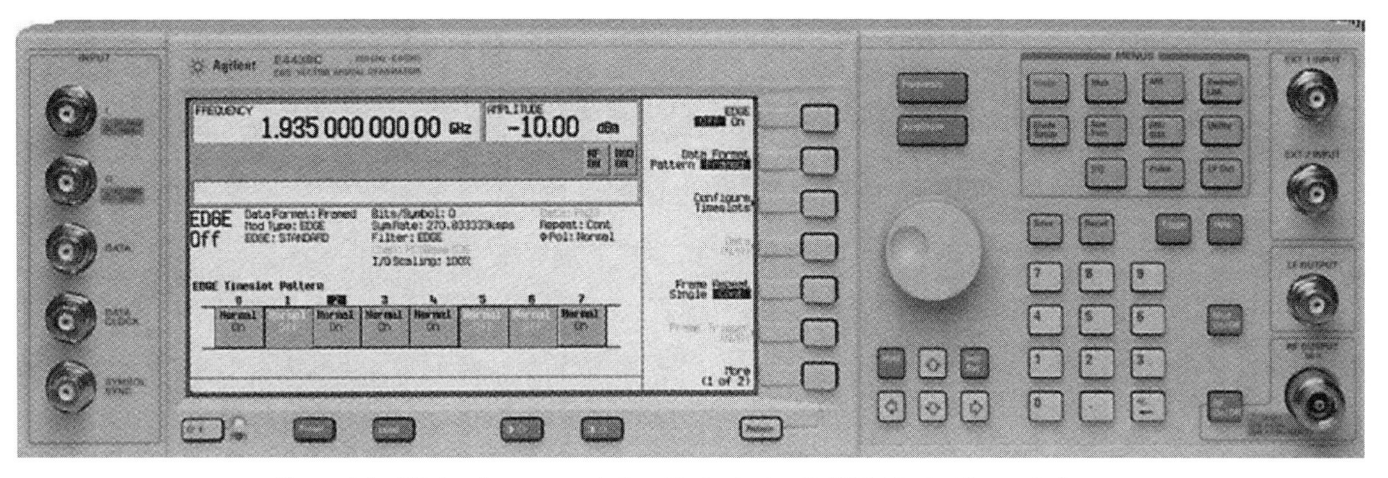

Figure 1.6 RF signal generator. Agilent Technologies © 2008. Used with permission.

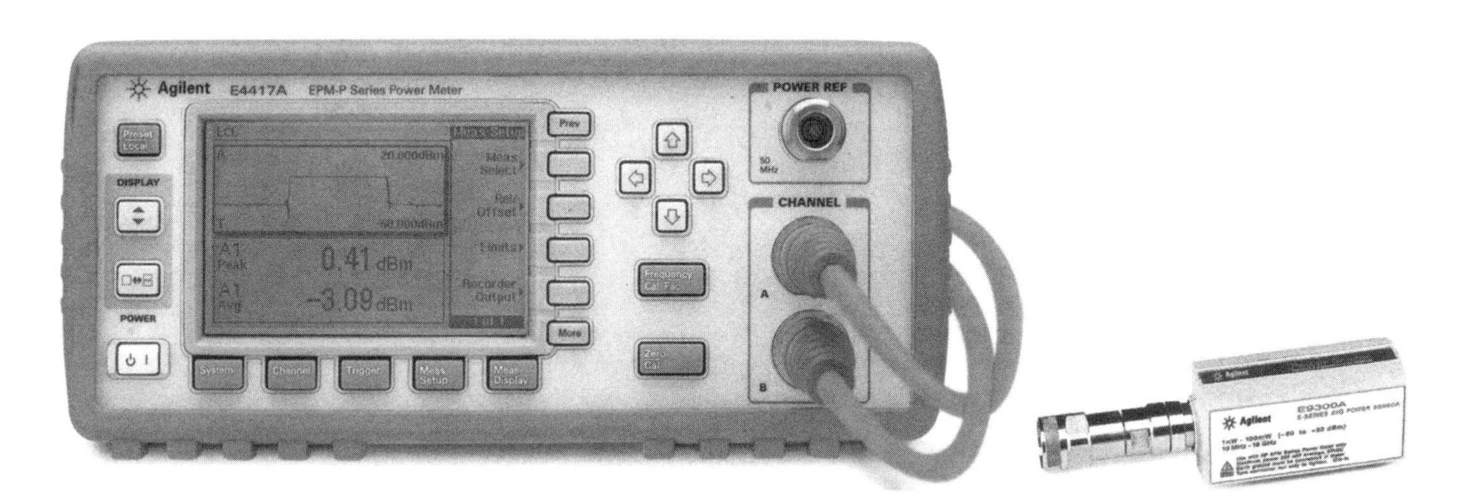

Figure 1.7 Power meter. Agilent Technologies © 2008. Used with permission.

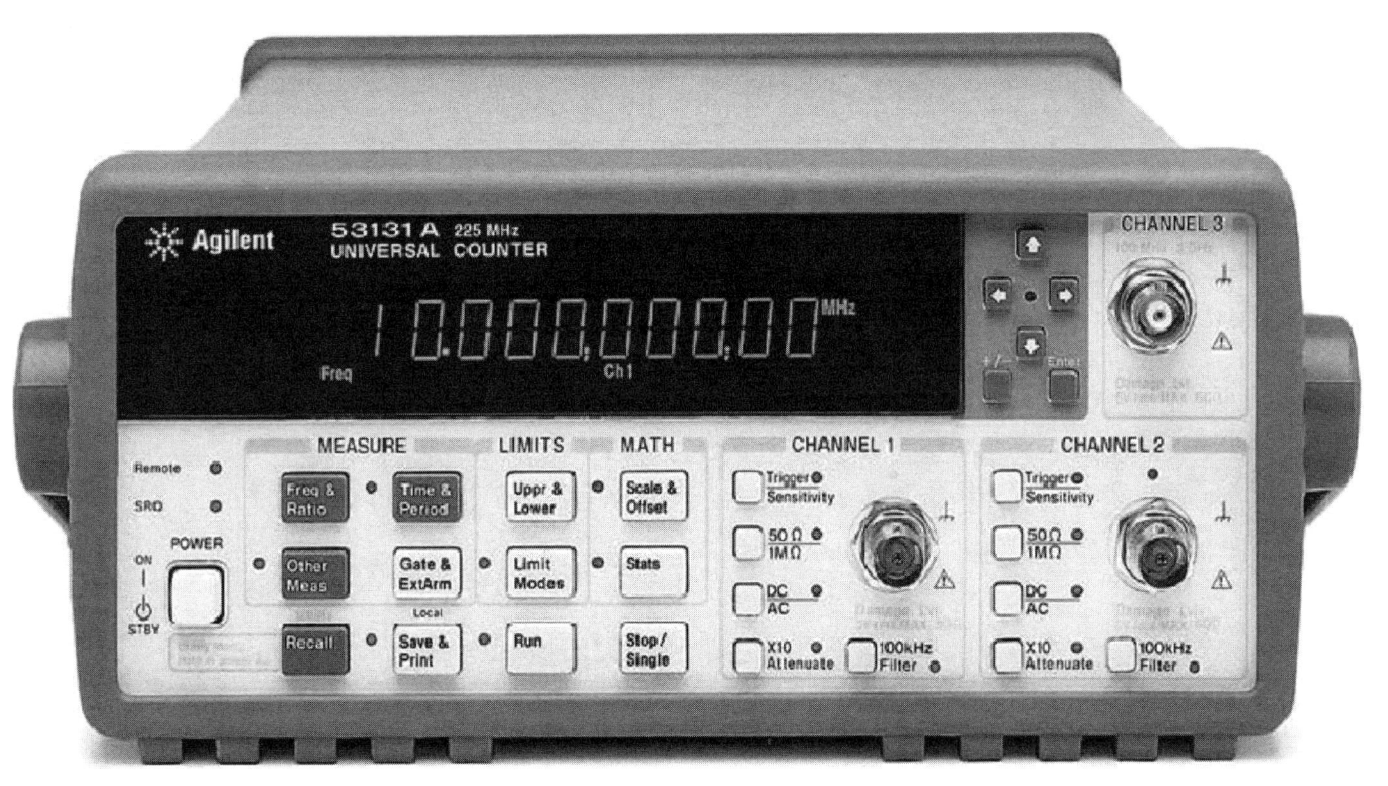

Figure 1.8 Frequency counter. Agilent Technologies © 2008. Used with permission.

1.11 SUMMARY OF CHAPTER 8: VNAs

An RF VNA measures the response of both RF devices and networks (which is a group of devices) as a function of the frequency of an applied continuous, nonmodulated, RF signal. The VNA measures the response of the network one frequency at a time, but it varies the measurement frequency over the user adjusted RF bandwidth very rapidly, making hundreds of measurements in 1 s.

The term vector designates the fact that the VNA measures both the amplitude and *phase* of the RF signal. Figure 1.9 shows a VNA.

The VNA measures the incident test signal, the reflected test signal, and the transmitted signal from the RF device. Then it automatically reverses the connections to measure the same quantities looking into the device from the opposite direction. The VNA can display these measured quantities as a function of frequency. However, it usually processes the information first to display derived quantities such as return loss, insertion loss, scattering parameters (S-parameters) in amplitude and phase, Smith Charts, group delay, and other performance characteristics.

The frequency range over which the VNA sweeps can be adjusted by the user. Alternately, the frequency can be fixed at a constant value and the power level can be swept so that the display shows the device or network performance as a function of power at a fixed frequency.

For ratio measurements, such as return loss or insertion loss, where two power levels are being compared to each other, the VNA's measurement accuracy can be improved to 0.1 dB or better by first calibrating the VNA to a set of standards, usually a short, open, load, and through (SOLT). This calibration can be done manually by the operator. It can also be done electronically using an add-on device that contains the standards and electronically operated relays to perform the calibration automatically. This electronic calibration eliminates operator error and also protects the standards from handling damage.

The accuracy of the VNA when it is used for absolute power measurements can be improved to ± 0.2 dBm by automatically calibrating the VNA with a power meter.

1.12 SUMMARY OF CHAPTER 9: SPECTRUM ANALYZERS

Spectrum analyzers can measure all of the individual frequencies that exist in any particular RF signal and display the power level of each frequency separately. They accomplish what the power meter and the frequency meter can only measure separately. Figure 1.10 displays an RF spectrum analyzer.

Note the difference between a spectrum analyzer and a VNA. The VNA analyzes the performance of a single RF device or combination of devices, either of which is called a "network." It measures the performance of the network one frequency at a time. The spectrum analyzer analyzes a signal to describe the power of each of the frequencies that make up the signal. The spectrum analyzer may be used to measure the distortion that the RF device creates on the different frequency components of the signal passing through it.

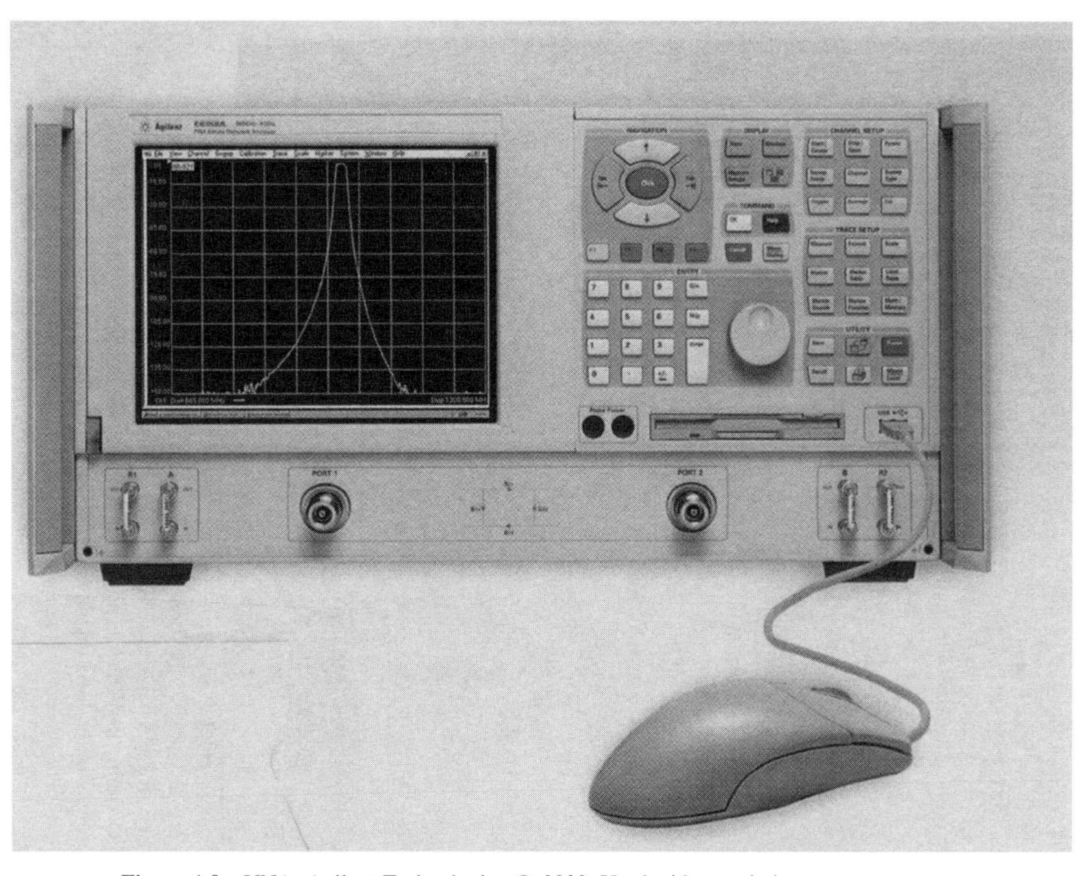

Figure 1.9 VNA. Agilent Technologies © 2008. Used with permission.

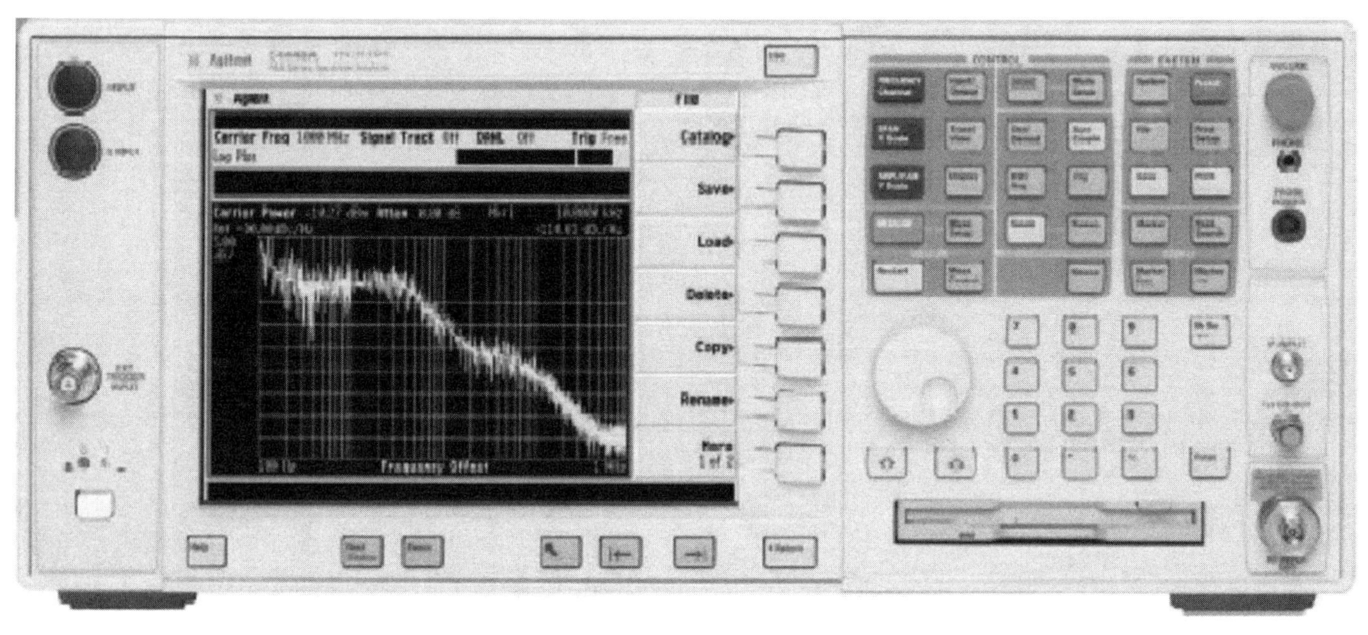

Figure 1.10 Spectrum analyzer. Agilent Technologies © 2008. Used with permission.

The spectrum analyzer not only displays the various frequencies and their power levels that make up an RF signal, but it can also analyze the display to provide specific facts about the signal. For example, it can determine the frequency range of 95% of the power in a given signal or how much of the power of a signal is spread into an adjacent channel frequency band that is assigned to another user. It can also measure the effect of device distortion on the specification requirements of different communication systems. With extra software and hardware added, it can measure complicated performance characteristics like oscillator phase noise and spectral regrowth. With additional hardware added, it can also demodulate an RF signal to permit it to be analyzed by VSA software, as described in the next section.

1.13 SUMMARY OF CHAPTER 10: VSAs

Like the spectrum analyzer, the VSA measures the characteristics of an RF signal, but it displays the signal characteristics in a different way. The VSA can be a stand-alone instrument, but it is most often implemented with a spectrum analyzer that provides the RF demodulating circuits and a software disk and laptop computer that converts the demodulated signal into the displays. Figure 1.11 shows this setup. The device under test (DUT) is shown in the center. The spectrum analyzer described in the previous section is used to demodulate the signal to be analyzed. Notice that its display is blank. The demodulated waveform is sent to a laptop computer, where the complex modulation is analyzed and displayed.

Figure 1.12 shows the display of a VSA when it is measuring an RF signal that has been modulated with $\pi/4$DQPSK modulation. The upper left-hand display is a vector diagram, which shows the amplitude and phase of the RF signal during the transition between measurement points. The upper right-hand display is an eye diagram. It is more complicated than the eye diagram of wired digital transmission systems, because of the multilevel value of the modulation. The lower right-hand display is a constellation diagram. These three displays give insight into the cause of the distortion, but they do not quantify it. Quantization is given by the error vector magnitude (EVM) shown in the lower left-hand display. These displays are explained in detail in Chapter 33.

1.14 SUMMARY OF CHAPTER 11: NOISE FIGURE METERS

Most modern spectrum analyzers can be equipped with special hardware and software to measure noise figure and gain of low noise receiver components such as low noise amplifiers (LNAs), input filters, cabling, and mixers.

A soft key switches the spectrum analyzer back and forth between its function as a spectrum analyzer and a noise figure meter. When it is in its noise figure meter mode, all hard keys except the numeric keypad are deactivated, and all control is by soft keys. A noise figure measurement setup using a spectrum analyzer is provided in Figure 1.13.

A LNA in the noise figure meter hardware is automatically connected between the spectrum analyzer input port and the spectrum analyzer mixer. This reduces the noise figure and increases the gain of the spectrum analyzer. The use of this amplifier

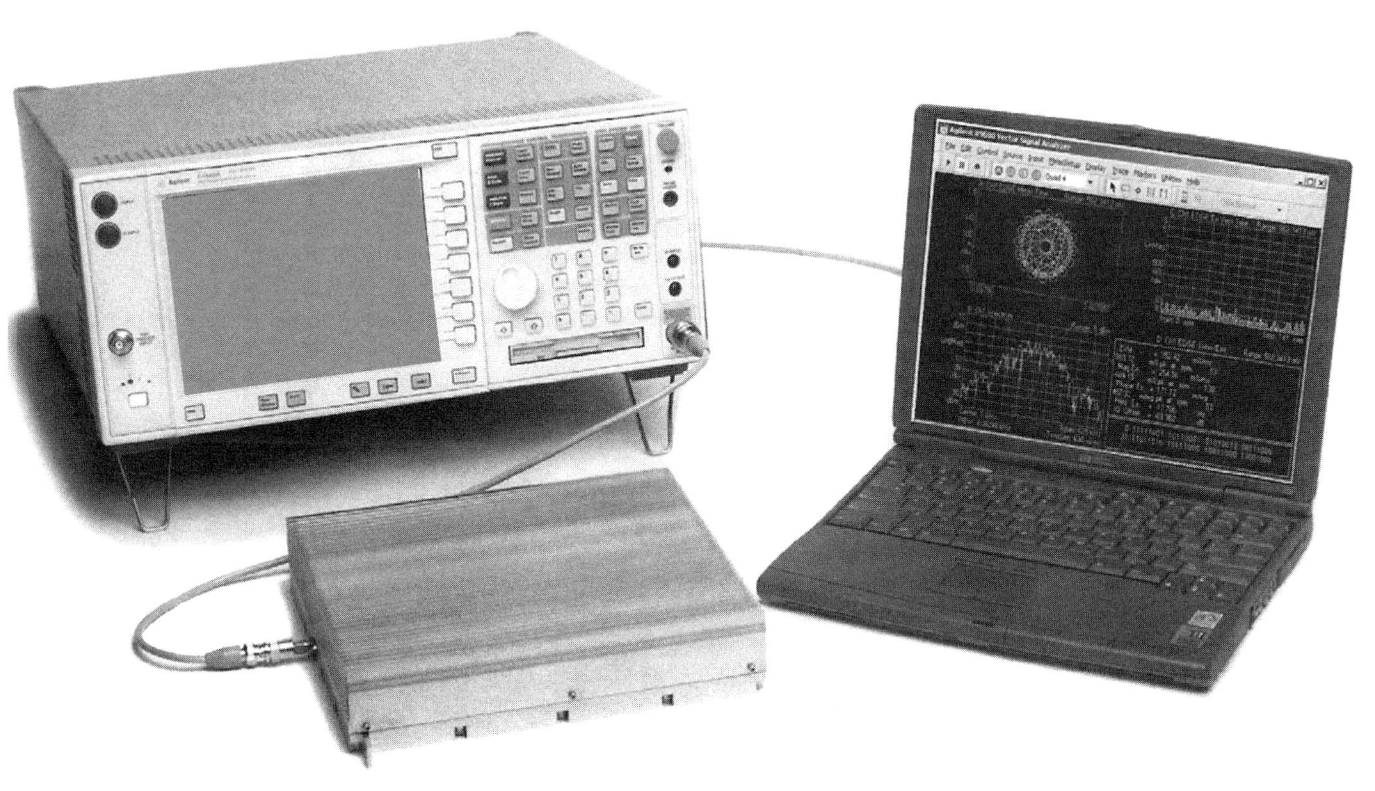

Figure 1.11 VSA. Agilent Technologies © 2008. Used with permission.

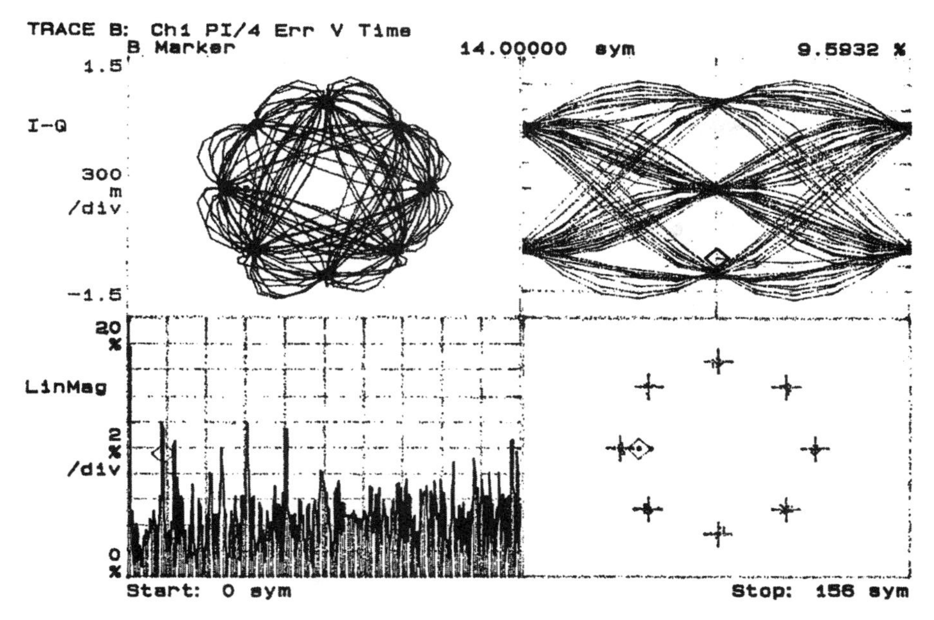

Figure 1.12 Various ways of analyzing modulation distortion.

reduces the effect of the spectrum analyzer noise figure on the noise figure measurements of the DUT.

To make the noise figure measurements, a known signal and a known noise must be sequentially connected to the input of the DUT. These signals are supplied by the noise source. The output signal must be then measured for each condition. The input signal/noise ratio (S/N) will therefore be known, and the output S/N can be calculated. Their ratio is the noise figure of the DUT.

1.15 SUMMARY OF CHAPTER 12: COAXIAL CABLES AND CONNECTORS

Chapter 12 describes the various coaxial transmission lines and connectors used to connect the RF test equipment with the RF devices under test. Recommended practices to insure measurement accuracy are also described.

1.16 SUMMARY OF CHAPTER 13: MEASUREMENT UNCERTAINTIES

The uncertainty of RF measurements and steps that can be taken to minimize them are explained in this chapter. The uncertainty of all RF measurements is affected by source and sensor mismatches. Steps to minimize this type of uncertainty are explained. Additional uncertainties are specific to each RF measurement type.

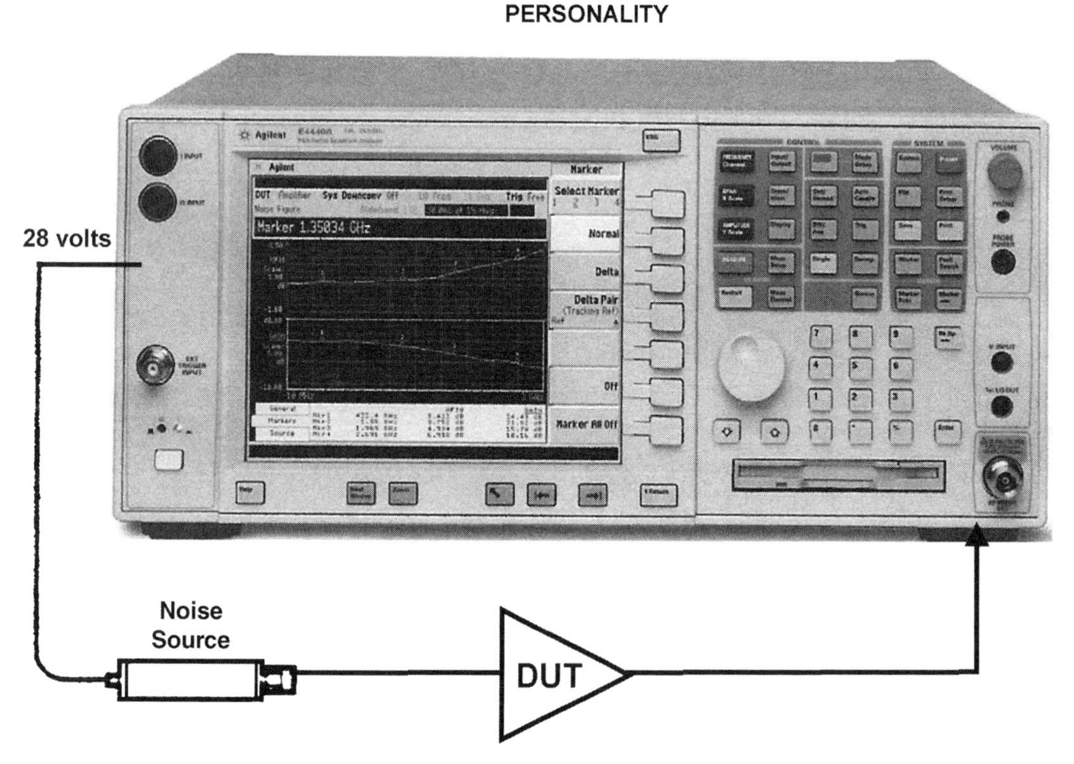

Figure 1.13 Noise figure test setup. Photo Agilent Technologies © 2008. Used with permission.

If great care is taken, RF power can be measured within an accuracy of ± 0.2 dB with a power meter. Ratioed measurements, such as the comparison of output power to input power of a device, can be measured to ± 0.05 dB with a VNA, if the VNA is calibrated with standards. The calibration can be done manually or electronically. The VNA can measure absolute power to only ± 1 dBm. However, it can be calibrated with a power meter to achieve power meter accuracy, with the disadvantage that the power meter calibration takes several minutes.

Spectrum analyzer power measurements have been greatly improved in the latest available models by including built-in power meter calibration. Power measurements with the spectrum analyzer can now be made with about ± 0.5 dBm uncertainty. Frequency can be measured with a spectrum analyzer to an uncertainty of about $\pm 3\%$ of the span.

1.17 SUMMARY OF CHAPTER 14: MEASUREMENT OF COMPONENTS WITHOUT COAXIAL CONNECTORS

All of the test equipment types that are discussed have coaxial fittings to which the device to be tested must be connected. However, many DUTs do not have coaxial input and output connectors. In order to make measurements on devices without coaxial connectors, the device has to be mounted in a test fixture with transitions between the device connections and coaxial connectors that can connect to the test equipment.

With this arrangement the test equipment will measure the device plus the test fixture. The measurements must then be corrected to give the characteristics of the device alone. The four methods of achieving this are discussed in this chapter.

1.18 PART III: MEASUREMENT OF INDIVIDUAL RF COMPONENTS

Part III describes RF measurement of individual RF devices. Chapter 15 shows a generic RF communication system block diagram. Chapters 16–24 describe the individual RF devices that make up the block diagram, and the measurements made on them, in the following order:

Chapter	RF Device
16	Signal control components
17	Phase locked oscillators (PLOs)
18	Upconverters
19	Power amplifiers
20	Antennas
21	RF receiver requirements
22	Filters
23	LNAs
24	Mixers

The measurement of noise figure and intermodulation products, which are common to most of the receiver components, are described in Chapters 25 and 26.

Overall receiver performance is calculated from the measurements of the individual RF parts in Chapter 27. RF integrated circuits (RFICs) and systems on a chip (SOC), in which several RF parts are fabricated on a single RF chip, are described in Chapter 28.

1.19 SUMMARY OF CHAPTER 15: RF COMMUNICATIONS SYSTEM BLOCK DIAGRAM

Figure 1.14 is a block diagram of an RF communication system. The block diagram is generic. It applies to any type of wireless RF communications system: cellular phone, wireless LAN, satellite communications system, and even a deep space probe. Any RF communications system must contain all of the devices shown. Of course, the performance requirements of each device vary from system to system.

PLO: generates the RF carrier at the correct frequency

Modulator: varies the frequency, amplitude, or phase of an intermediate frequency (IF) carrier to put information onto it

Upconverter: shifts the modulated IF signal to RF

Power amplifier: increases the power level of the modulated RF carrier

TX antenna: transmits the RF carrier in the direction of the receiver

RX antenna: collects the transmitted RF signal at the receiver

RF filter: allows only a specified range of RF frequencies to pass, and blocks all other frequencies

LNA: amplifies the weak received RF carrier

Mixer and IF amplifier: shifts the RF carrier to a lower frequency below the RF band, and amplifies it to a level where it can be demodulated

Demodulator: removes the information from the low frequency carrier

1.20 SUMMARY OF CHAPTER 16: SIGNAL CONTROL COMPONENTS

RF signal control components vary the frequency, power, and other characteristics of the RF signal. Because many of these control components use semiconductor devices for their operation, these devices will be discussed. Then, PIN diode attenuators will be explained.

1.21 SUMMARY OF CHAPTER 17: PLOs

The function of the PLO in a wireless communication system is to generate the RF signal that will carry the digital information, in the form of modulation of

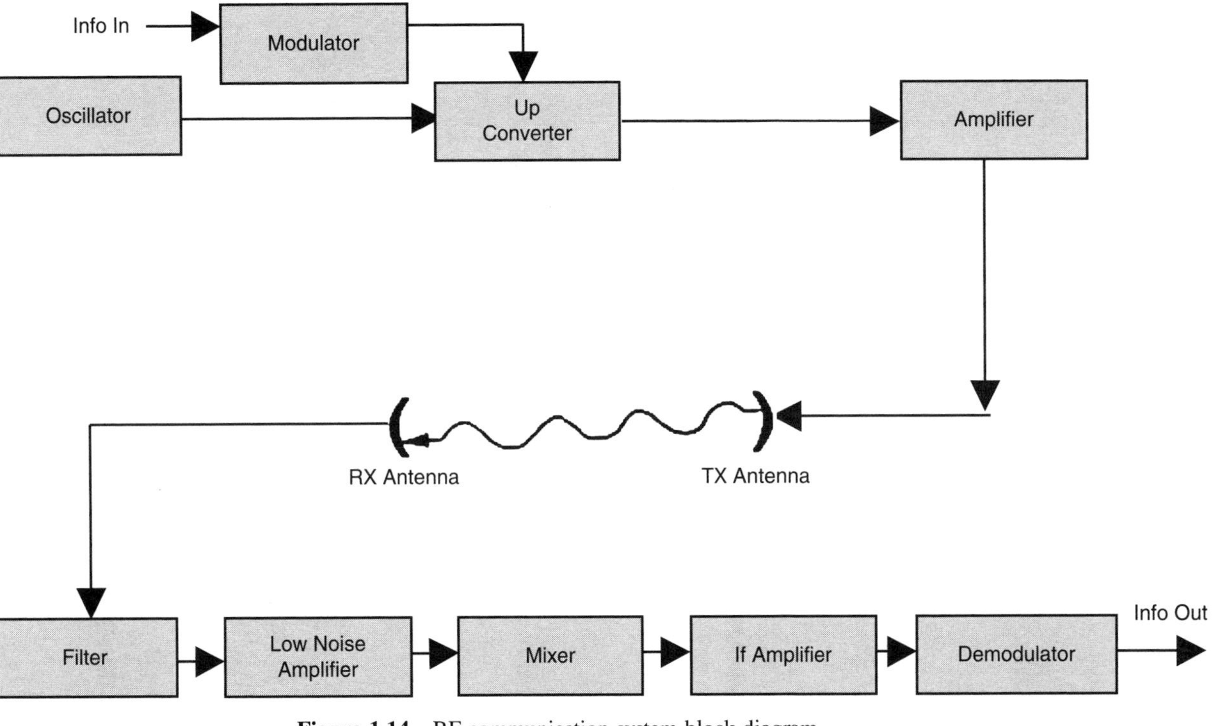

Figure 1.14 RF communication system block diagram.

amplitude, frequency, or phase, wirelessly from the transmitter location to the receiver location.

The most important characteristics of a PLO are its RF frequency stability, its capability of being rapidly tuned from one RF frequency to another, and phase noise.

The frequency stability requirements are 10^{-6} or better, which would be a stability of 1 kHz for a 1 GHz RF signal.

At frequencies below the RF band, the transmitter can be stabilized by using a quartz crystal as its resonant circuit. Unfortunately, quartz crystals do not have resonant frequencies in the RF band. Therefore, the RF PLO must divide its frequency to a value below RF, where its divided down frequency can be compared to the reference frequency of a quartz crystal.

Figure 1.15 shows a block diagram of an RF PLO. The PLO consists of two parts: a voltage controlled oscillator (VCO), shown in the shaded box, and a phased locked loop (PLL). The characteristics of the PLO that need to be measured are its frequency, output power, tuning sensitivity, and phase noise.

1.22 SUMMARY OF CHAPTER 18: UPCONVERTERS

Chapter 18 describes upconverters. The complicated modulations that are used in cellular phones and wireless data transmission systems are difficult to generate at RF frequencies. Thus, in most RF communication systems, the modulated signals are generated at low frequency with digital processing chips that do not work at RF and then upshifted to the desired RF frequency in an upconverter.

Figure 1.16 shows a block diagram of an upconverter. A stable frequency below the RF range, for example, 100 MHz, is generated with a simple quartz crystal oscillator. Digital information is then modulated onto this low frequency carrier in a digital IC. An RF PLO, such as that discussed in Chapter 17, is then used to generate an RF signal at 900 MHz, which is 100 MHz below the desired transmitted frequency. The modulated low frequency signal and the RF signal are then added together to form a sum frequency of 1000 MHz that is now carrying the digitally modulated signal. The upconverter also produces a difference frequency at 800 MHz, and this must be removed by a bandpass filter.

The details of how the upconverters work and the design and measurement of the upconverter are discussed in Chapter 18.

1.23 SUMMARY OF CHAPTER 19: POWER AMPLIFIERS

As the RF signal exits the upconverter, it is at the correct RF frequency, which is controlled by the PLO, and it is carrying the digital information to be transmitted that was applied to the RF carrier by the upconverter. However, the RF power level is only a few milliwatts; and when it is attenuated by the 90 dB or more of path loss, it would be lost in the noise of the RF receiver. Therefore, before the

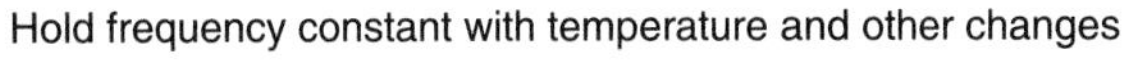

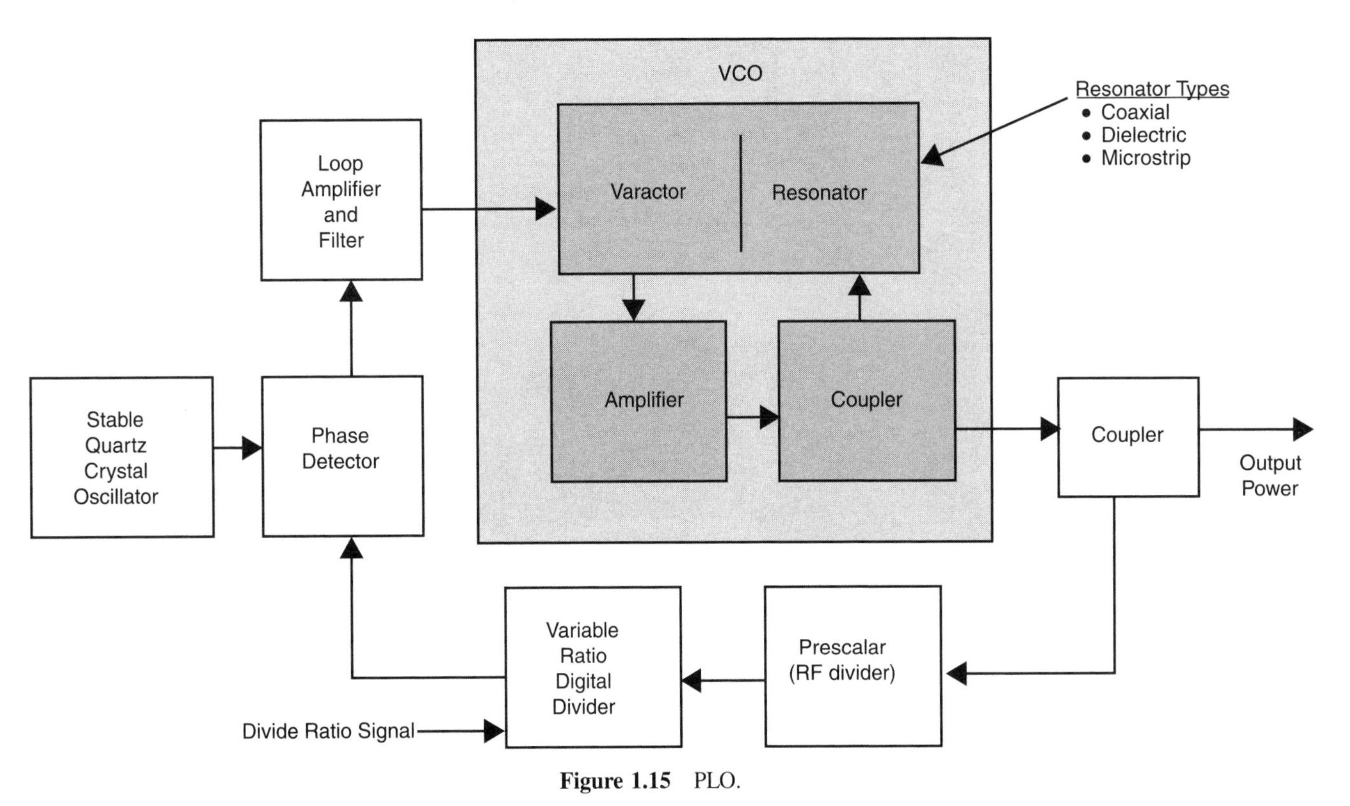

Figure 1.15 PLO.

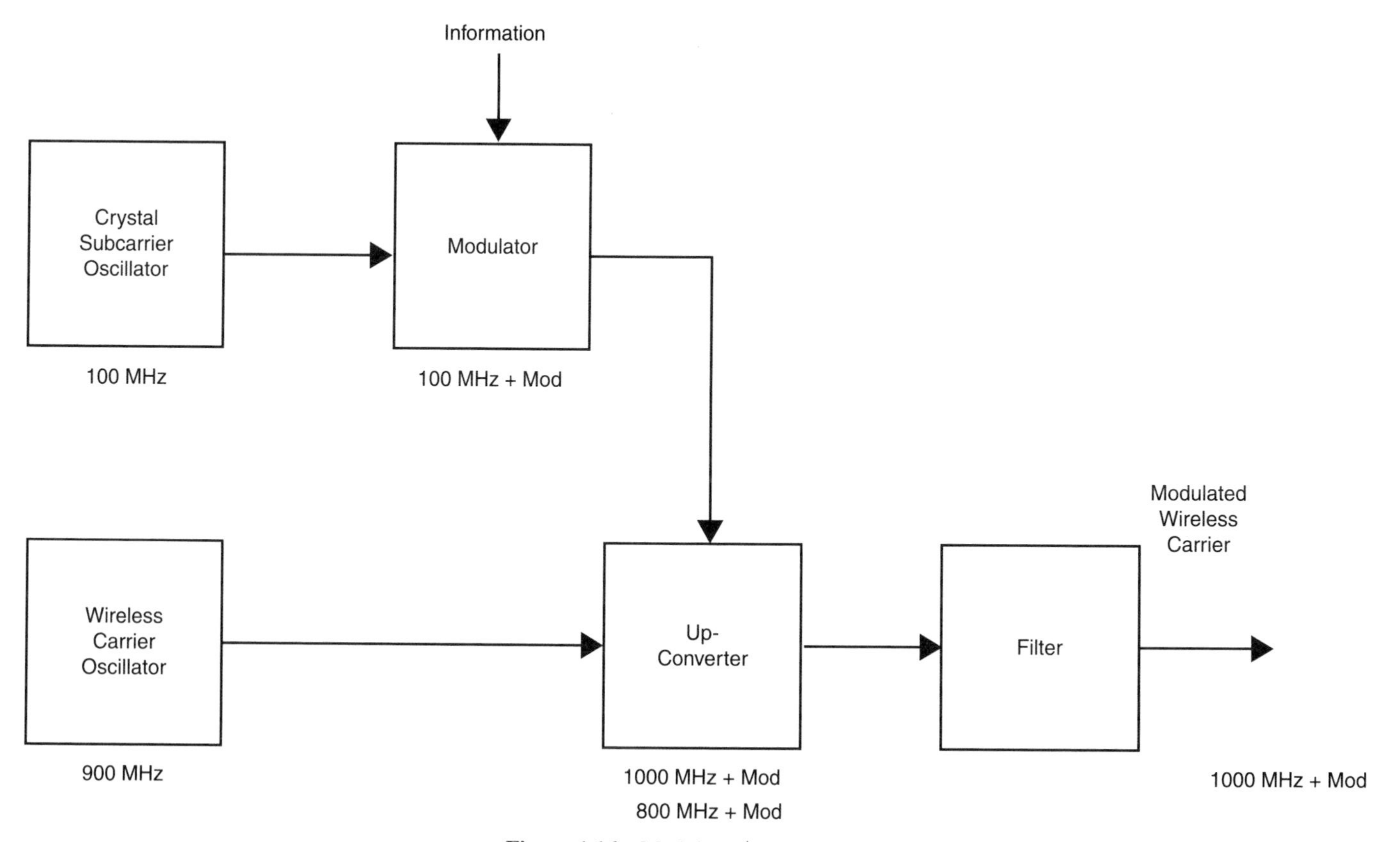

Figure 1.16 Modulator/upconverter.

modulated RF signal can be transmitted, it must be amplified in an RF power amplifier. The major performance requirements of the power amplifier are the following:

1. to amplify the RF power from the upconverter to 30–40 dBm,
2. to generate this power with high efficiency, and
3. to not distort the digital modulation during the amplifying process.

RF power amplifiers are either bipolar transistors or field effect transistors (FETs), but they are different from their low frequency counterparts because of "transit time" effects. Transit time effects occur because the electrons travel through the semiconductor material of the RF transistor at approximately 1/3000 of the velocity of light or 10^5 m/s. This is not a problem with low frequency transistors, but it is definitely a problem with RF transistors. To understand how critical the transit time effect is, realize that at 1 GHz, one RF cycle is 1 ns (10^{-9} s). In one cycle at 1 GHz, the electrons will travel 10^5 m/s $\times$ 10^{-9} s = 100 microns (μm). For reasonable performance the electrons must move through the transistor in less than one-tenth of a cycle, which means the spacing between doping regions in the transistor must be less than 10 μm. At 10 GHz, the spacings must be less than 1 μm.

RF power amplifiers use two techniques to achieve the required transit times: reduced spacings between transistor elements and use of semiconductor materials like gallium arsenide (GaAs) and silicon germanium (SiGe), in which electrons move faster than they do in silicon.

Figure 1.17 shows a measurement of a typical RF power amplifier made with a VNA. For this measurement, the VNA is adjusted to measure and display gain and output power as a function of RF input power at a single frequency of 2.45 GHz. The VNA is calibrated with a power meter, so the output power measurement has only ±0.2 dBm uncertainty. The gain measurement is calibrated with the VNA standards, and so it has an uncertainty of only ±0.05 dB. The input power is swept over the power range from −7 to +13 dBm, so each horizontal scale division is 2 dBm. The left-hand graph shows the gain in decibels and the right-hand graph shows the output power in decibels relative to 1 mW. At the left-hand edge of each graph, the amplifier is operating in the linear range, where the RF output power is exactly proportional to the RF input power. At the right-hand edge of each graph, the amplifier is operating approximately at saturation. The markers on both graphs are set at the 1 dB compression point, where the gain has dropped from its linear value by 1 dB.

Figure 1.17 shows characteristics that are common to all RF amplifiers. Every RF amplifier has a nonlinear output power versus input power curve because the amplifier cannot generate more power that its battery supplies. Typical amplifier efficiency is about 50% at saturation, which is its maximum power output point. Most RF transistors draw the same power from their battery, regardless of whether they are operated at full power or at an input level that provides very little output power. At small output power levels, the RF output power is proportional to the RF input power. This is called the linear range. Operation near saturation causes distortion. Operation in the linear range causes low efficiency. The usual compromise is to use a 2× higher

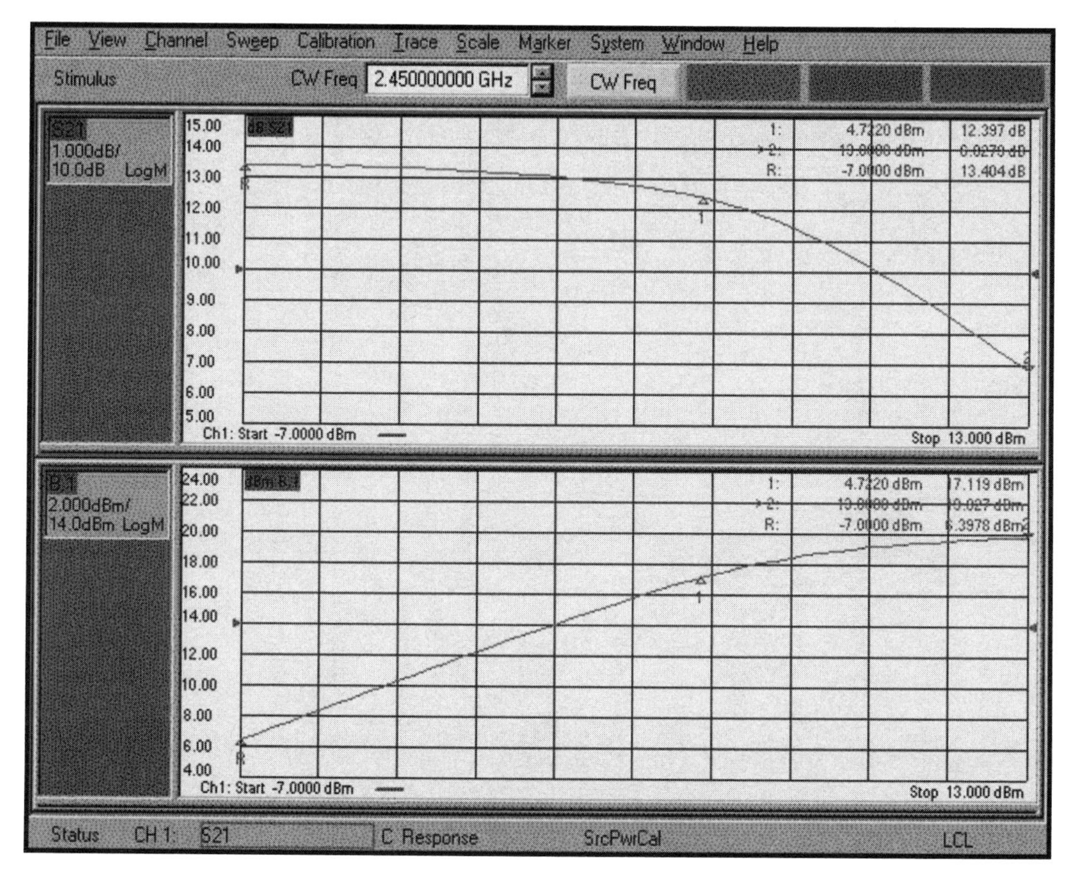

Figure 1.17 Power amplifier swept gain and output power.

power transistor at its 1 dB gain compression point (where the power has dropped by about half), where the distortions of the signal defined by spectral regrowth and modulation distortions are satisfactory.

1.24 SUMMARY OF CHAPTER 20: ANTENNAS

As Figure 1.14 demonstrates, an antenna must be used on both the transmitter and receiver end of any wireless system. These antennas may be the same or different on the two ends of the system, but in either case they serve different functions. The transmitter antenna launches the power in the direction of the receiver and concentrates it in this direction. The receiver antenna simply collects the power from the transmitter.

Specifications for the transmitter antenna are the following:

Gain
Beamwidth
Pattern
Polarization
Impedance match

Gain is a measure in dB of how well the antenna concentrates the power in the direction of the receiver, relative to an isotropic antenna. An isotropic antenna is defined as an idealized antenna that radiates power equally in all directions. Beamwidth is the angular width of the beam generated by the antenna. Gain and beamwidth are related. To achieve more gain, the width of the beam must be decreased.

The antenna pattern defines radiation in undesired directions that may jam other systems. Polarization defines the direction of the electric field of the radiation, whether directed vertically or horizontally to the Earth's surface. The impedance of free space, which is the ratio of the electric field to the magnetic field of the RF wave, is a physical constant equal to 377 Ω. Every antenna serves as an impedance transformer, transforming the impedance of the antenna at its RF connector to 50 or 75Ω.

The pattern, polarization, and impedance match are the same for the receiving antenna as for the transmitting antenna. However, for the receiving antenna the gain and beamwidth are replaced by the area of the antenna, which determines how much of the incident signal is received.

Figure 1.18 shows common RF antennas. The upper left drawing shows a half-wave dipole. This antenna is used on most mobile units, and it also serves as a building block for higher gain antennas. It consists of a feed line, shown as parallel wires, which are bent at right angles to the feed line in the antenna region. The total length of the antenna is 0.5 wavelength at the operating frequency. It has an almost isotropic pattern, except that it does not radiate along the antenna wires. Consequently, it

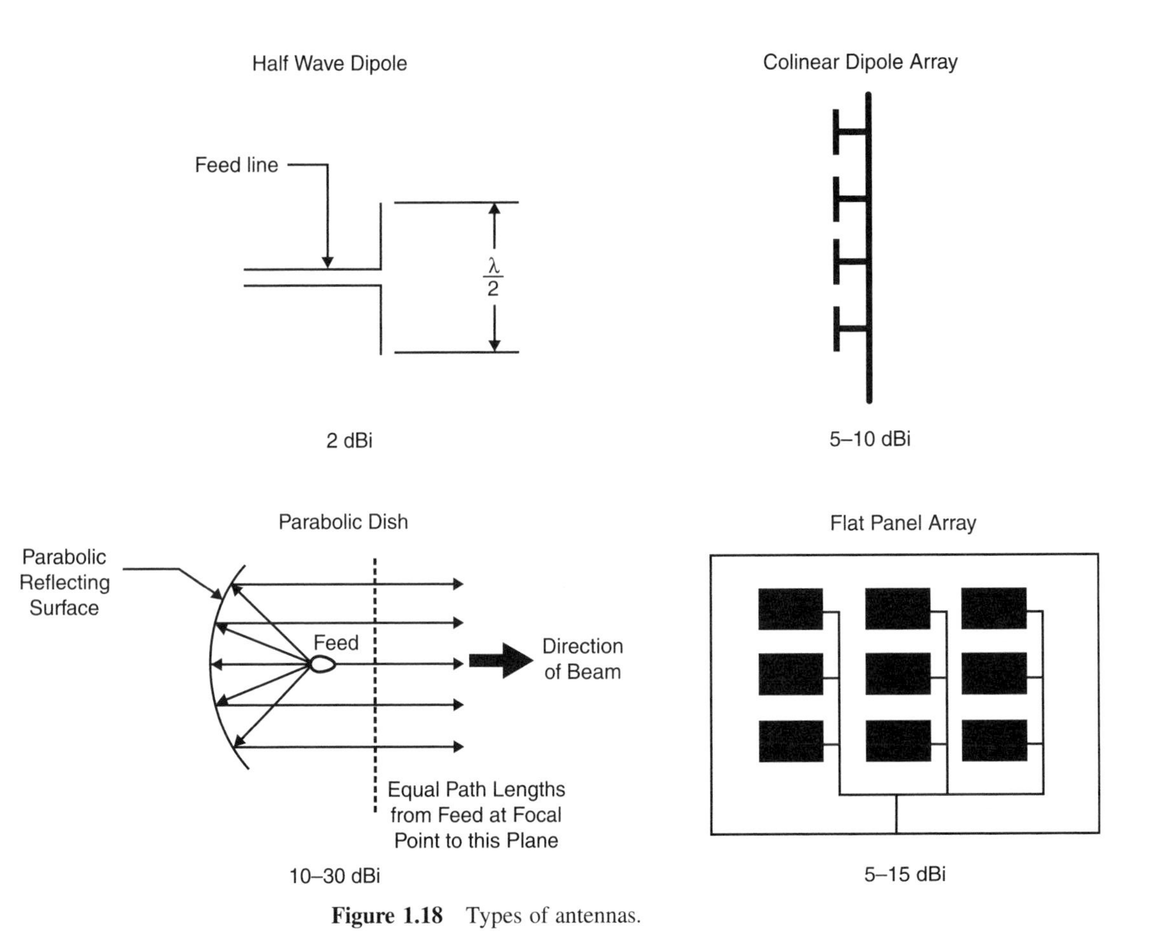

Figure 1.18 Types of antennas.

has a net gain in the direction at right angles to the antenna wires of 2.1 decibels isotropic (dBi). Its total length of 0.5 wavelength controls its impedance transformation from the 377 Ω value of free space to the value at its feed wires. If its length is 0.5 wavelength, the impedance is transformed to 62.5 Ω, which is easy to match to 50 Ω for RF applications or 75 Ω for television applications.

The half-wave dipole antenna also serves as a building block for higher gain antennas. The upper right drawing of Figure 1.18 shows a colinear dipole array, which is formed by several half-wave dipoles stacked vertically. The half-wave dipoles are 1 wavelength apart, so the signals radiating from all of them are in phase. This configuration is used for most base station antennas, although the details cannot be seen because the antenna is covered by a plastic tube to protect it from the environment. Stacking the antennas vertically reduces the radiation above the antenna and directly below the antenna where there are no users, and thus increases its gain into the cell site by the ratio of the number of antennas.

The lower left-hand drawing of Figure 1.18 shows a parabolic dish antenna. The parabolic dish is fed from the focal point of the parabolic reflecting surface. The parabolic shape has the characteristics that all rays from the feed point travel the same distance to a plane perpendicular to the axis of the dish, so that all of the reflections have the same phase, thus adding up in this direction. The gain of the parabolic antenna is approximately proportional to the square of its diameter in wavelengths. A 1 ft diameter parabolic dish has a gain of about 10 dB at 1 GHz, and it will have 100 times (20 dB) more gain at 10 GHz where the wavelength is one-tenth as great.

A simple rectangular patch that is about 0.5 wavelengths in length and that can be photoetched on a microstrip board has the same radiation characteristics as a half-wave dipole. This patch antenna, mounted on a high dielectric constant ceramic, provides a very small antenna that is popular with cellular phones and wireless data transmission mobile units. As the lower right-hand drawing of Figure 1.18 shows, multielement patch antennas, whose performance approaches that of the parabolic dish, can be fabricated for much less cost than the parabolic dish.

The properties of the antennas described earlier can be measured easily using a VNA, which can measure the input power to the antenna and the output power received from the antenna about 10 wavelengths away in the far field region and then calculate the gain. If the antenna is mounted on a rotating platform, its antenna pattern can be measured. The one requirement for all of these measurements is that they be made in an anechoic test room. Figure 1.19 shows such a room, which has absorbing material mounted on its walls, floors, and ceiling so that no reflected signal degrades the measurement results.

1.25 SUMMARY OF CHAPTER 21: RF RECEIVER REQUIREMENTS

The receiver in a wireless mobile unit must operate under a variety of conditions. When the mobile unit is at the edge of the coverage area, the receiver must have a low noise figure to receive the very low signal from the base station transmitter

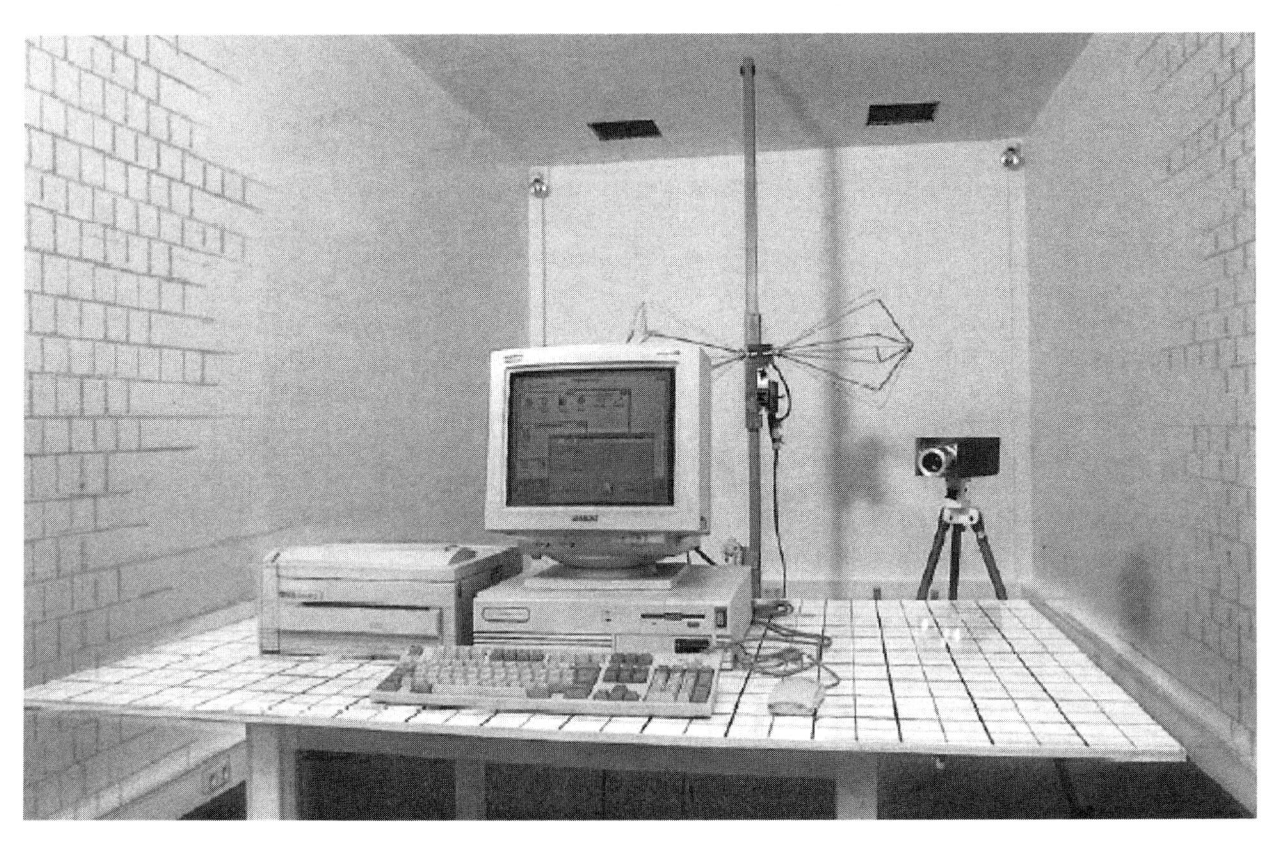

Figure 1.19 Antenna test setup in the anechoic chamber.

with a satisfactory S/N for achieving the required BER. When the mobile unit is close to the base station, the received signals intended for all the mobile units in the cell are very large, and two signals may mix in the mobile receiver to form a signal that jams the signal intended for the mobile. This jamming signal is called an "intermodulation product."

The received RF signal varies over a 90 dB dynamic range, depending on the transmitter–receiver separation and multipath fading. The receiver must provide adjustable gain, depending on the received signal level, to provide a constant output level of about 0 dBm for demodulation.

Every RF receiver is made up of four basic parts, whose performances are as follows:

1. RF filter: allows only a specified range of RF frequencies to pass and blocks all other frequencies
2. LNA: amplifies the weak received RF carrier
3. Mixer: shifts the RF frequency to a lower frequency where it can be more easily amplified and filtered
4. IF amplifier: amplifies the IF frequency to a power level where it can be demodulated

All components contribute to the gain or loss of the RF signal as it passes through the receiver. The filter and the RF amplifier contribute to the noise figure of the receiver. The LNA and the mixer contribute to the intermodulation products.

1.26 SUMMARY OF CHAPTER 22: RF FILTERS

Filters are used for different reasons in RF communications systems:

1. System filter: in the RF receiver to block out the signals from other systems
2. Upconverter filter: in the RF transmitter to remove the unwanted sideband of the upconverter
3. Harmonic filter: in the RF transmitter to reduce the harmonic content of the transmitting signal
4. Image noise filter: in the RF receiver to filter image noise generated by the receiver mixing process
5. User channel filter: just before the IF amplifier, to select individual user's signal

The characteristics of all filters include the following:

Passband frequency
Filter attenuation outside the passband

Attenuation in the passband

Attenuation in the skirt, which is the frequency range between full blocking and full passing of the signal

Figure 1.20 shows the attenuation versus frequency characteristic of the system filter in the RF receiver, which is used to block the RF signals from other systems. These measurements were made with the VNA shown in Figure 1.9. The filter was designed to pass signals in the 2.40–2.48 GHz frequency band used for wireless LANs. Within this range, the insertion loss, as shown by the markers, is less than 1.25 dB. Outside this band at ±280 MHz, the attenuation is greater than 20 dB.

Other characteristics that need to be measured on filters are the mismatches measured at both the input and the output of the filter, and the phase shift through the filter.

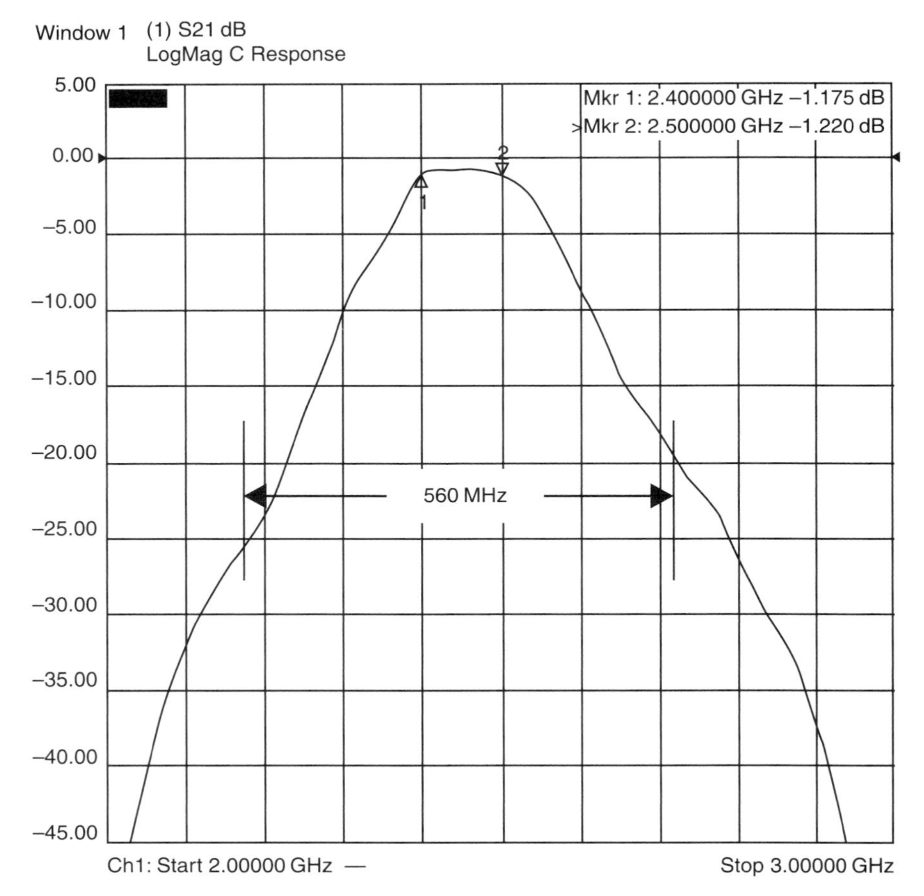

Figure 1.20 Insertion loss of the RF filter versus frequency.

1.27 SUMMARY OF CHAPTER 23: LNAs

The function of the LNA is to boost the RF power level of the incoming RF signal without adding additional noise to it. Figure 1.21 shows a block diagram of an LNA. The use of representative power levels will help explain the characteristics of the LNA. The input signal, which is received from the antenna and passed through the RF system filter, is −110 dBm, which is typical of the received cell phone signal when the cell phone is located at the edge of the cell during a multipath fade. The LNA has a gain of 30 dB, so the output power of the LNA is −80 dBm, as shown.

Noise from the environment also enters the LNA through the antenna. The received noise is −114 dBm/1 MHz of channel bandwidth if the antenna is pointed along the earth, as is the case with most cell phones and wireless data transmission systems. If the bandwidth of a single user channel is 100 kHz (0.1 MHz), the received noise is −124 dBm, so the S/N at the input to the LNA is 14 dB, as shown. The output noise would be expected to be −124 dBm + 30 dB = −94 dBm. Actually, the output noise is −90 dBm, which is 4 dB higher. This is because the LNA not only amplifies the noise at its input but also adds additional noise of its own. The difference between the S/N going into the LNA and the S/N coming out is called the "noise figure" of the LNA.

The noise figure of a passive device, like a filter or a length of transmission line, is equal to its attenuation. A significant part of the overall system noise figure is caused by the system filter that precedes the LNA.

To measure the noise figure of the LNA, a noise source is used. This noise source can be seen as part of the noise figure measurement setup shown in Figure 1.13. The noise source contains a special PN diode. When no voltage is applied to the diode, room temperature noise of −114 dBm/1 MHz is generated. When 28 V is applied to the noise diode, the noise output increases by about 14 dB to −100 dBm/1 MHz. The exact value of the noise when the diode is turned on, above the thermal noise value when the diode is turned off, is called the excess noise ratio (ENR).

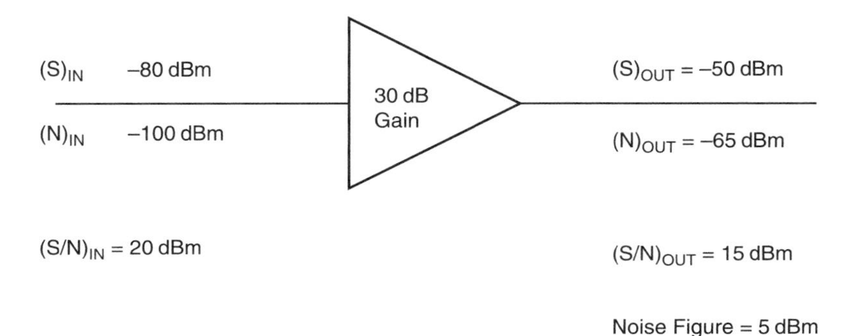

Figure 1.21 Low noise amplifiers. "Noise figure" is the ratio of the S/N going into a device compared to the S/N coming out.

A sensitive spectrum analyzer can measure the output of the LNA with the noise diode on or off and determine the output S/N. It can calculate the noise figure with its internal software.

In a typical wireless communications system, the mobile unit is continually moving relative to the base station. The base station is sending out a constant RF power, so that the received signal at the mobile unit is continually changing over a 90 dB range as the mobile moves. The noise figure of the LNA is the critical factor controlling system performance when the mobile unit is at the edge of the coverage cell. However, as the mobile unit moves closer to the base station, the received signal is well above the noise level. When the mobile unit is midway between the edge of the cell and the base station, the S/N is 60 dB. However, as the mobile unit moves closer to the base station, the LNA faces another problem: intermodulation products. Intermodulation products and their measurement are discussed in detail in Chapter 26.

1.28 SUMMARY OF CHAPTER 24: MIXERS

The purpose of the mixer is to convert the modulated RF signal that is carrying information to a lower frequency where it can be more easily amplified up to a level of about 0 dBm, where the individual voice and/or data channels can be separated. All RF communications systems are licensed to use 25 MHz or more of the RF spectrum to serve hundreds of users. All of the users' signals in the system come through the RF filter and through the LNA. There is no practical way to filter an individual voice or data signal at RF frequencies. The selection of the individual user channels is accomplished in the mixer. Figure 1.22 shows how a mixer does this.

The numerical values shown are similar to those of an IS-136 cellular phone operating in the band from 869 to 894 MHz in North America. IS-136 is a TDMA system, where three voice channels use a frequency channel in different time slots. The channel frequency width is 30 kHz, so 832 TDMA channels can be fitted into the 25 MHz RF bandwidth, as shown in the lower left-hand sketch of Figure 1.22. All of these 832 channels pass through the RF filter and the LNA. The desired 30 kHz wide channel is selected by the mixer.

Assume initially that the lowest channel at 869 MHz is to be selected. The local oscillator is set to 783 MHz by the system, via proper setting of the digital divider in a PLO, which serves as the mixer local oscillator (LO) in the receiver. The desired 30 kHz wide TDMA channel is shifted to the difference frequency of 86 MHz (right-hand sketch, Fig. 1.22), and it passes through the IF filter which has a bandwidth of 30 kHz centered at 86 MHz. All of the other 831 IF TDMA channels are shifted down to the lower frequency range, but they cannot pass through the 30 kHz filter to the IF amplifier. Assume that the next call is assigned the highest frequency in the RF bandwidth at 894 MHz. The local oscillator is set at 808 MHz and shifts the 894 MHz RF signal down to 86 MHz, so it can pass through the IF filter.

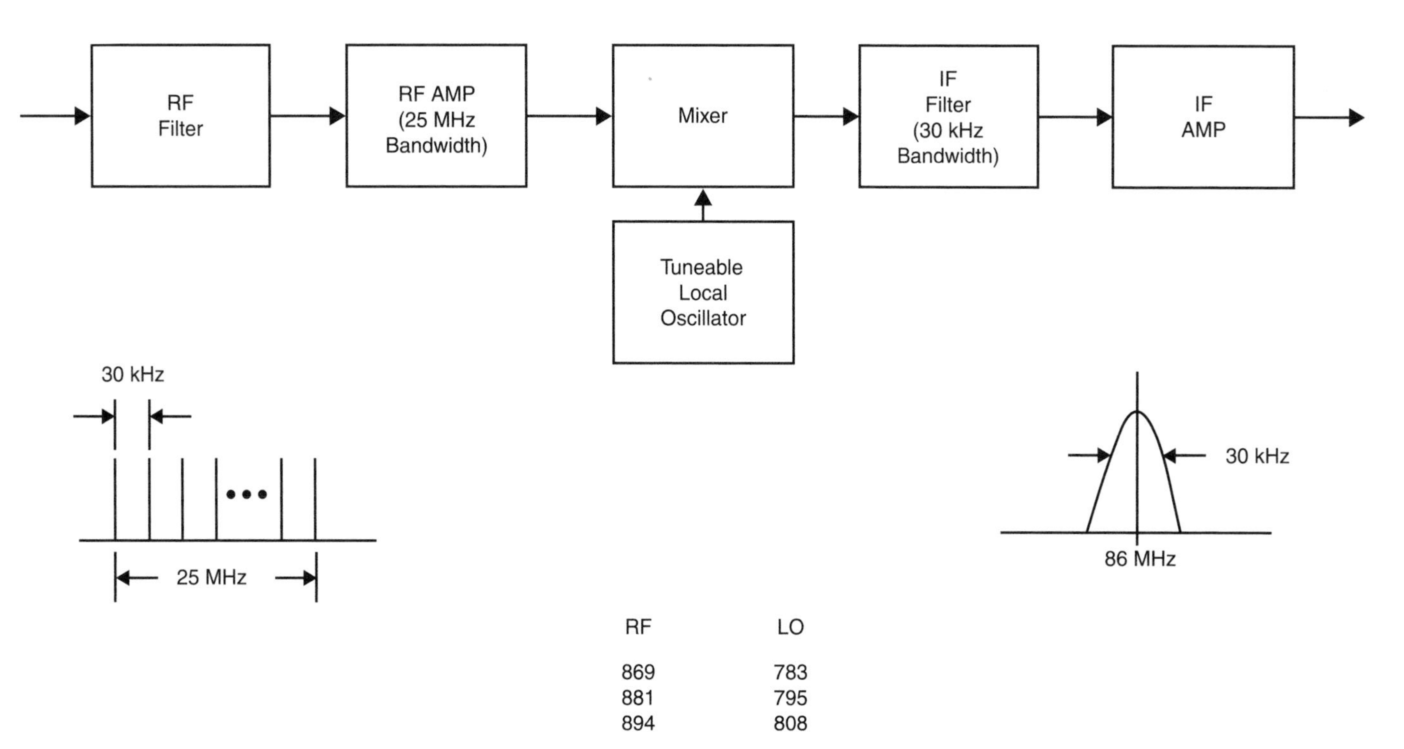

Figure 1.22 Selection of individual voice channels.

For many years, the RF signal has been shifted down to the IF frequency range in the mixer, as shown in the example above. Many systems still use this design. However, most RF communications systems now use a "zero IF" (ZIF) mixer design, where the local oscillator is set to nearly the same frequency as the RF frequency, so that the IF frequency coming out of the mixer is nearly 0. The filtering and amplification of the desired channel are done by digital processing. This ZIF design significantly reduces the cost of the receiver.

1.29 SUMMARY OF CHAPTER 25: NOISE FIGURE MEASUREMENT

A typical noise figure test setup was shown in Figure 1.13. The measured noise figure of the LNA whose gain and S-parameters are discussed in Chapter 23 is provided in Figure 1.23.

The measured noise figure of the other components in the overall receiver are as follows:

Noise Figure and Gain of Individual Components and Complete Receiver at 2.45 GHz

Component	Noise Figure (dB)	Gain (dB)
RF filter	0.7	−0.7
LNA	4.4	20.5
RF filter + LNA	5.4	19.5
Mixer	8.9	0.5
Complete receiver	6.5	16.9

1.30 SUMMARY OF CHAPTER 26: INTERMODULATION PRODUCT MEASUREMENT

The receiver in a wireless mobile unit must operate under a variety of conditions. When the mobile unit is at the edge of the coverage area, the receiver must have a low noise figure to receive the very low signal from the base station transmitter with a satisfactory S/N for achieving the required BER. When the mobile unit is close to the base station, the received signals intended for all the mobile units in the cell are very large, and two signals may mix in a mobile receiver to form a signal that jams the signal intended for the mobile. This jamming signal is called an intermodulation product. Figure 1.24 shows three measured spectra of the output of an LNA, each at different input power levels, and shows the development of the intermodulation products. Figure 1.25 shows a graph of the fundamental and third-order intermodulation products as a function of the input fundamental signals and illustrates how the output IP3 (OIP3) value is defined.

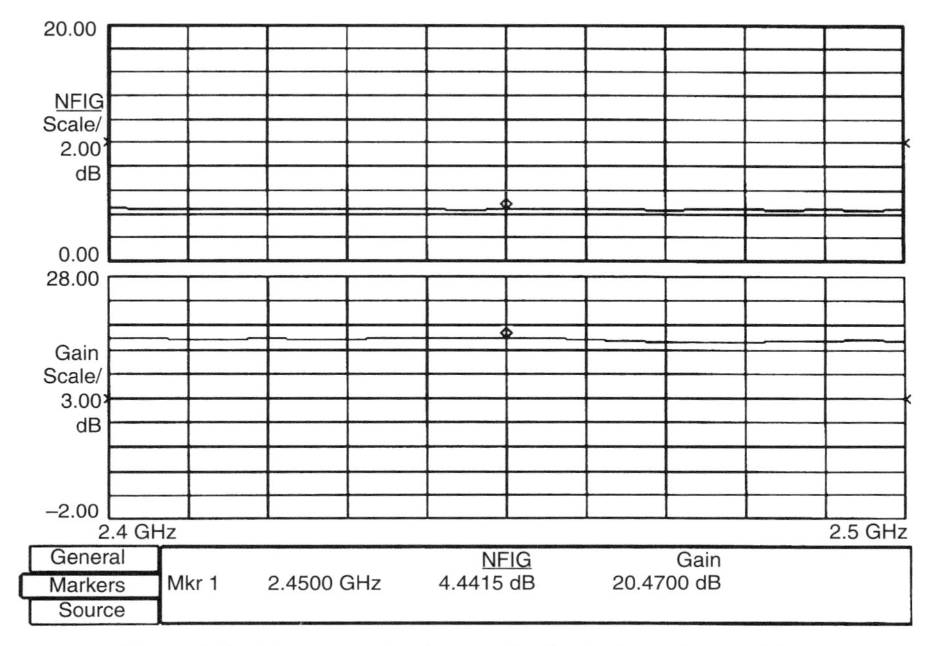

Figure 1.23 Measured noise figure and gain of a low noise amplifier.

1.31 SUMMARY OF CHAPTER 27: OVERALL RECEIVER

All of the individual receiver components (RF filter, LNA, and mixer) contribute to the overall gain, noise figure, and IP3 of the receiver in a complicated way, which are demonstrated by the following example:

Device	RF Filter	LNA	Mixer	Total
Noise figure (dB)	1.00	4.50	8.50	5.62
Gain (dB)	−1.00	19.00	2.00	20.00
IP3 (dB)	—	13.00	10.00	8.81

An available design program is described that allows the measured values of noise figure, gain, and intermodulation products to be combined to yield the overall system performance.

1.32 SUMMARY OF CHAPTER 28: RFICs and SOC

Many of the individual devices in an RF communications system are combined into RFIC chips. Figure 1.26 displays a photograph of the complete RF communications system shown in Figure 1.14 but integrated into two RFICs and two RF filters. The

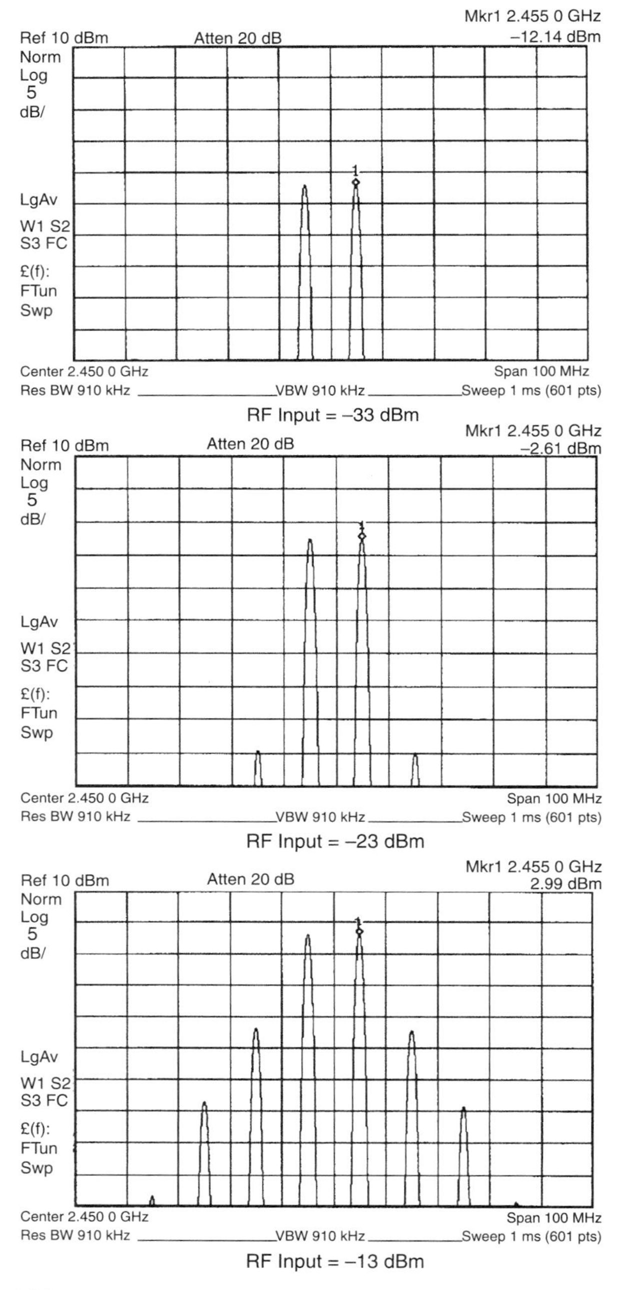

Figure 1.24 Two tone intermodulation products from a low noise amplifier.

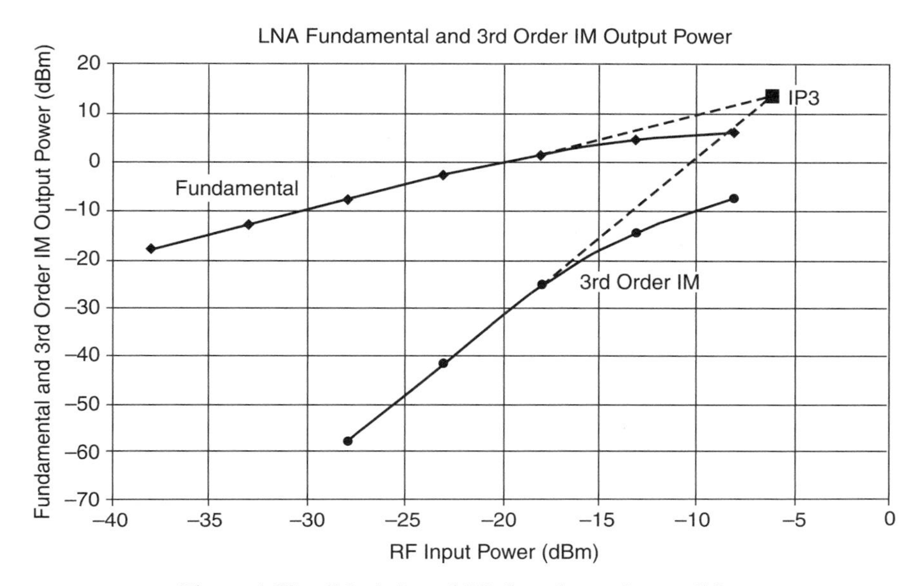

Figure 1.25 Calculation of IP3 for a low noise amplifier.

block diagram of this particular RFIC chip is also shown. In such cases, the RF measurements must be done on the complete RFIC.

1.33 PART IV: TESTING OF DEVICES AND SYSTEMS WITH DIGITALLY MODULATED RF SIGNALS

Part IV describes testing of RF devices with RF signals that are carrying digital modulation. RF devices may distort the digital signal. A particular RF device can distort different signals in different ways.

Chapter	Topic
29	Digital communications signals
30	Multiple access techniques—FDMA, TDMA, CDMA
31	Orthogonal Frequency Division Multiplexing (OFDM) and Orthogonal Frequency Division Multiple Access (OFDMA)
32	Adjacent channel power (ACP)
33	Constellation, vector, eye diagrams, and EVM
34	Complementary cumulative distribution function (CCDF)
35	BER
36	Measurement of GSM evolution signals

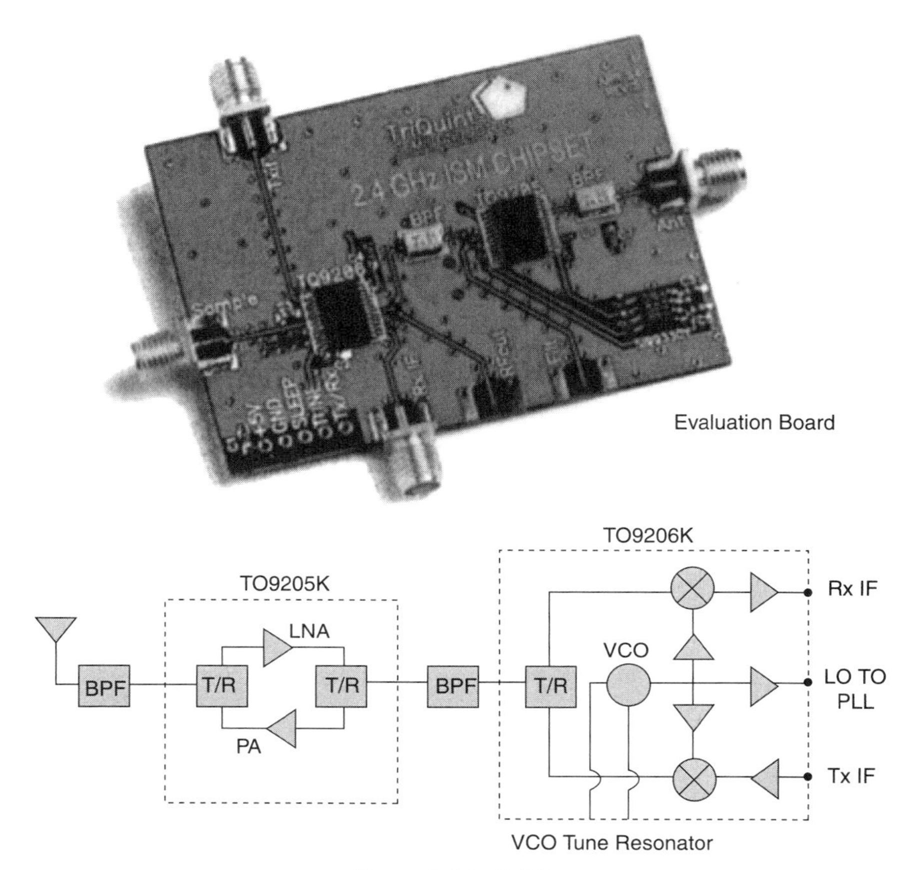

Figure 1.26 RFICs.

1.34 SUMMARY OF CHAPTER 29: DIGITAL COMMUNICATIONS SIGNALS

Digital communications systems carry voice, video, and digital data. At the beginning of 2006 about 90% of all cellular phone capacity was used for voice signals, but data transmission usage is growing rapidly. This chapter explains what the digital data rates are for voice, video, and data carrying signals.

The frequency and data rates for voice signals are as follows:

Analog sound waves from a human speaker: 30 Hz to 6 kHz (note 1)
Telephone signal: 0 Hz to 4 kHz (note 2)
Pulse code modulation (PCM) digitized telephone signal: 64 kbps (note 3)
Adaptive differential PCM (ADPCM) digitized telephone signal: 16 kbps (note 4)
Synthesized telephone signal: 2 kbps (note 5)

Cell phone compressed speech (note 6)
GSM: 13 kbps; CDMA: 9–13 kbps

Note 1: The frequency of the analog sound waves from a human speaker are determined by their vocal cords, diaphragm, tongue, and lips.

Note 2: The analog telephone voice signal is deliberately filtered, so that more voice channels can be transmitted over a given bandwidth.

Note 3: The PCM digitized telephone signal using 64 kbps sounds exactly like human voice on a telephone circuit.

Note 4: The ADPCM sends only the change in the signal from one sample to the next, but it requires a microprocessor on each end of the transmission channel.

Note 5: Synthesized speech is perfectly understandable but sounds like a machine.

Note 6: Compressed telephone speech using 13 kbps accurately reproduces accents of all world languages.

Forward error correction (FEC) bits are added to the digital speech signal to allow transmission errors to be corrected at the receiving end, increasing the bit rate by about 1.8 times. Additional signal control bits, which allow continuous signaling between the transmitter and the receiver, are also added, so that the total data transmission rate for telephone speech is about 2 times the digitized voice rate.

The frequencies and data rates for video signals are as follows:

Analog broadcast quality TV: 0–4 MHz
PCM digitized broadcast quality TV: 56 Mbps
MPEG-2 compressed video: 1.5–7 Mbps
MPEG-4 compressed video: 64 kbp for low definition cell phones
High-definition, digital broadcast TV: 18 Mbps

Live telephone calls and real-time TV signals must be transmitted as the signals are generated. These signals use "circuit switched" connections. Two-way telephone calls have a time delay of several seconds as the connection path is set up between the sender and the receiver. Two separate circuits are dedicated to the call, one in each direction of transmission, and each circuit is used an average of only 40% of the time.

When recorded voice, video, or data is being transmitted, it does not have to be received in real time, so packet switching is used. The digital transmission is broken up into groups of bits, called packets, each with a group of header bits specifying the receiving location and the serial number of the particular packet. All transmitting and receiving points are permanently connected to the network, so the connection delay time is eliminated and there is no idle time in the channels. Familiar wired examples of this type of data transmission are Ethernet and TCP/IP (an Internet protocol). With packet switching, error correction bits can be reduced to simple parity, which indicates that a group of bits has an error, but not which bit is incorrect, so the group of bits must be retransmitted. This correction technique is called automatic repeat request (ARQ).

Chapter 29 describes the details of all digital signals for cell phones and wireless LANs. These various signal protocols can be applied to the transmitted RF signal from the RF signal generator that is displayed in Figure 1.6.

1.35 SUMMARY OF CHAPTER 30: FDMA, TDMA, AND CDMA MULTIPLE ACCESS TECHNIQUES

This chapter describes multiple access techniques, which control how many users can use a licensed RF frequency band at the same time. The individual user channel bandwidth is much smaller than the total licensed RF bandwidth.

One basic multiple access technique is FDMA. It is used by all systems either as the only multiple access technique or in combination with other multiple access techniques. With FDMA, the total RF bandwidth is divided into smaller bands, each of which uses only the bandwidth required for a single user. A simple example is the original analog cell phone used in North America. The licensed bandwidth, divided between two competing service providers, was 25 MHz for each direction of transmission. Each user required 30 kHz for transmission in each direction, so the available number of user channels was $25\ \mathrm{MHz}/30\ \mathrm{kHz} = 832$ channels.

The RF electronics of the base station can be significantly simplified (and therefore reduced in cost) by dividing the transmission into time slots within a wider FDMA channel. This multiple access technique is called TDMA.

The multiple access technique providing the greatest user capacity in a cell phone environment for a given RF bandwidth is CDMA. With this multiple access technique, all users use the same RF frequency at the same time. However, each user is assigned a unique 128 bit code by the base station. Each user multiplies each bit of his digitized voice signal by his unique assigned code. The bandwidth of the transmission is spread in frequency by the voice data rate of 16 kbps $\times$ 128 bits, which is equal to approximately 2.5 Mbps. At the receiving end the signal is decoded to recover the original voice transmission. Every other user's coded RF signal is also received, but it is not decoded correctly and looks like noise that is $1/128 = 21$ dB below the correctly coded signal.

Figure 1.27 illustrates this coding and decoding process. The upper three signals are at the transmitter site. The top signal shows 2 bits of the information data stream. The middle signal shows the unique PN code assigned to the user. (Only 16 coding bits are shown in the drawing, instead of 128, to simplify the drawing.) The data signal is coded by an exclusive NOR process, which generates a digital 1 if the data bit and coding bit are the same and a digital 0 if they are different. This coded signal is shown on the third line. This bitstream is then modulated onto the RF carrier and transmitted.

The lower three lines in Figure 1.27 show the conditions at the receiver. The top line shows the demodulated coded bitstream, removed from the RF carrier. The middle line are PN coding bits, and they are the same as the coding bits used at the transmitter. The

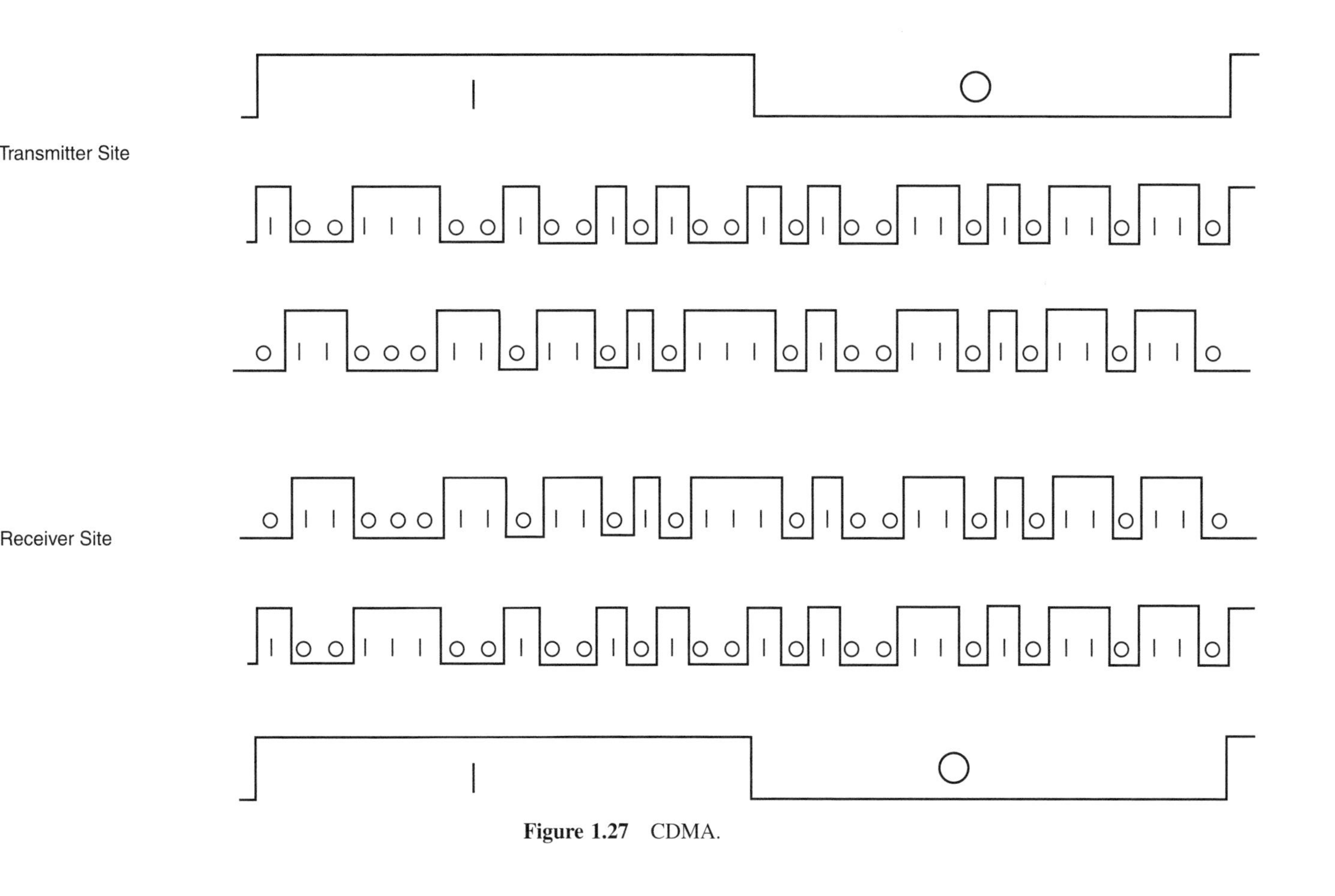

Figure 1.27 CDMA.

bottom line contains the decoded bits, which are the same as the original data bits at the transmitter. Other signals with different codes are not despread. Each of the other signals produces an interference of 1/128 with the 128 chip spreading.

The above description of the CDMA process was for digitized voice signals. For data transmission the CDMA system uses only a short PN code of about 16 chips, and uses packet data techniques to accommodate multiple users.

A complete description of these multiple access techniques is given a Chapter 30, and the necessary measurements that must be made for TDMA and CDMA systems are explained.

1.36 SUMMARY OF CHAPTER 31: OFDM AND OFDMA

OFDM is a multiple access technique that provides greater data rates in a given RF frequency band than TDMA or CDMA. Features of OFDM include the following:

- There are multiple subcarriers to carry the digital bitstream.
- Four pilot tones at different frequencies across the bandwidth continuously monitor transmission quality.
- Modulation can be adjusted between BPSK, QPSK, 16QAM, and 64QAM.
- Various levels of FEC can be used, depending on signal transmission quality.
- The modulation type and FEC can be changed every 4 ms, based on transmission quality.
- OFDM has less spreading of power into adjacent channels than CDMA.

Figure 1.28 shows the spectrum of an OFDM signal carrying a single set of data. The available bandwidth is filled with a set of subcarriers. Fifty-two subcarriers are commonly used, spaced 312.5 kHz apart. Four of the subcarriers are pilot carriers, which give channel condition information and serve as a phase reference for the modulation. The data, with error correction and control bits added, is divided into sets and each set is modulated onto 1 of the 48 data channels. These 48 channels are modulated with BPSK, QPSK, 16QAM, or 64QAM, depending on the S/N of the system. The noise in the channels is determined by the number of other users on the system. The type of modulation is the same for all 48 channels, and it is changed with each 4 ms transmission. OFDM is currently used for short-range high-speed wireless LANs.

OFDM allows only one user on the channel at any given time. Multiple users are accommodated by using packet switching techniques. To accommodate multiple users more efficiently, OFDMA is used. OFDMA is a technique that allows multiple access on the same channel (a channel being a group of evenly spaced subcarriers as shown in Figure 1.26 for OFDM). It distributes the subcarriers among all users, so all users can transmit and receive at the same time. The subcarriers can be matched to each user to provide the best performance, meaning the least problems with fading and interference based on the location and propagation characteristics of each user.

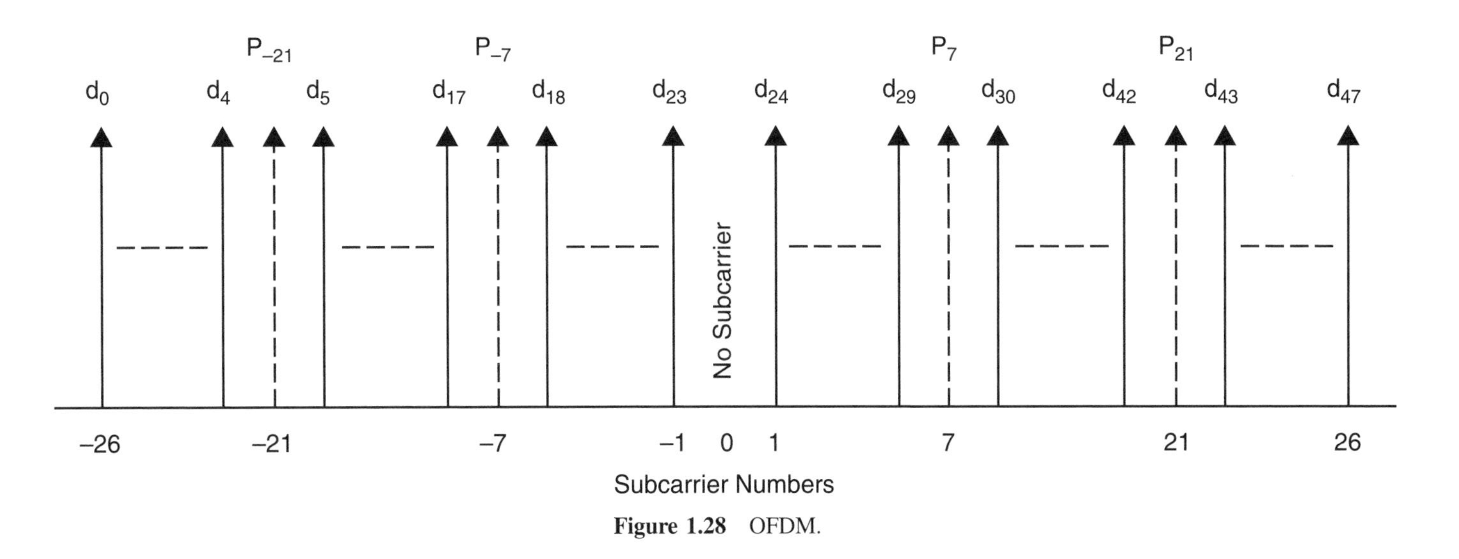

Figure 1.28 OFDM.

1.37 SUMMARY OF CHAPTER 32: ACP

Chapters 32–35 show various measurements defining the quality of modulated RF signals when they are processed by RF components. Chapter 32 describes ACP.

To accommodate the maximum number of users in a given licensed RF frequency band, it is desirable to leave a minimum amount of frequency band between each user's channel. When digital signals are modulated onto a RF carrier, some of their carrier power spreads out into adjacent user channels. The ratio of this interfering power to the power in the desired channel is called ACP.

Figure 1.29 shows the effect of power amplifier saturation on ACP. The upper left curve shows the test signal. By careful filtering, the power in the next lower adjacent frequency channel or in the next higher adjacent frequency channel is maintained 30 dB below the power in the desired channel.

The upper right-hand curve shows the signal coming out of the amplifier when it is operating in its linear range at 16 dB below saturation, where the output power is directly proportional to the input power. Note that there has been no growth in the ACPs when the linear amplifier is used.

The lower right-hand curve shows the ACP when the amplifier is operating within 3 dB of its maximum power capability. The ACP is now significantly greater and is only 25 dB below the power in the main channel. The lower right-hand curve shows the output power of the amplifier when it is operating at saturation. Now the ACP is only 18 dB below the power in the main channel. For most cellular phone systems, this performance is not satisfactory. The practical solution to this problem is to design the system so that the amplifier is never operated within 3 dB of its maximum output power.

1.38 SUMMARY OF CHAPTER 33: CONSTELLATION, VECTOR, AND EYE DIAGRAMS, AND EVM

RF devices can cause spectral regrowth and create interference between user channels. They can also distort the modulation information within the desired channel.

Figure 1.30 shows various ways of observing these distortions using a VSA. The upper figure shows the constellation diagram of a $\pi/4$DQPSK modulated RF signal. The pattern has eight phase positions, each 45° apart. The angle of each symbol point gives a phase of the RF signal, and the distance from the center to each symbol point gives the amplitude. The crosshairs show a perfectly demodulated signal; the data points show a set of actual signal values.

The lower left panel of Figure 1.30 is a vector diagram. It shows not only the constellation points but also the variation of phase and amplitude of the signal during the transition time between data points. The lower right panel of the figure shows an eye diagram, which is commonly used in analyzing a baseband data stream. The $\pi/4$DQPSK eye diagram is more complicated than the baseband case.

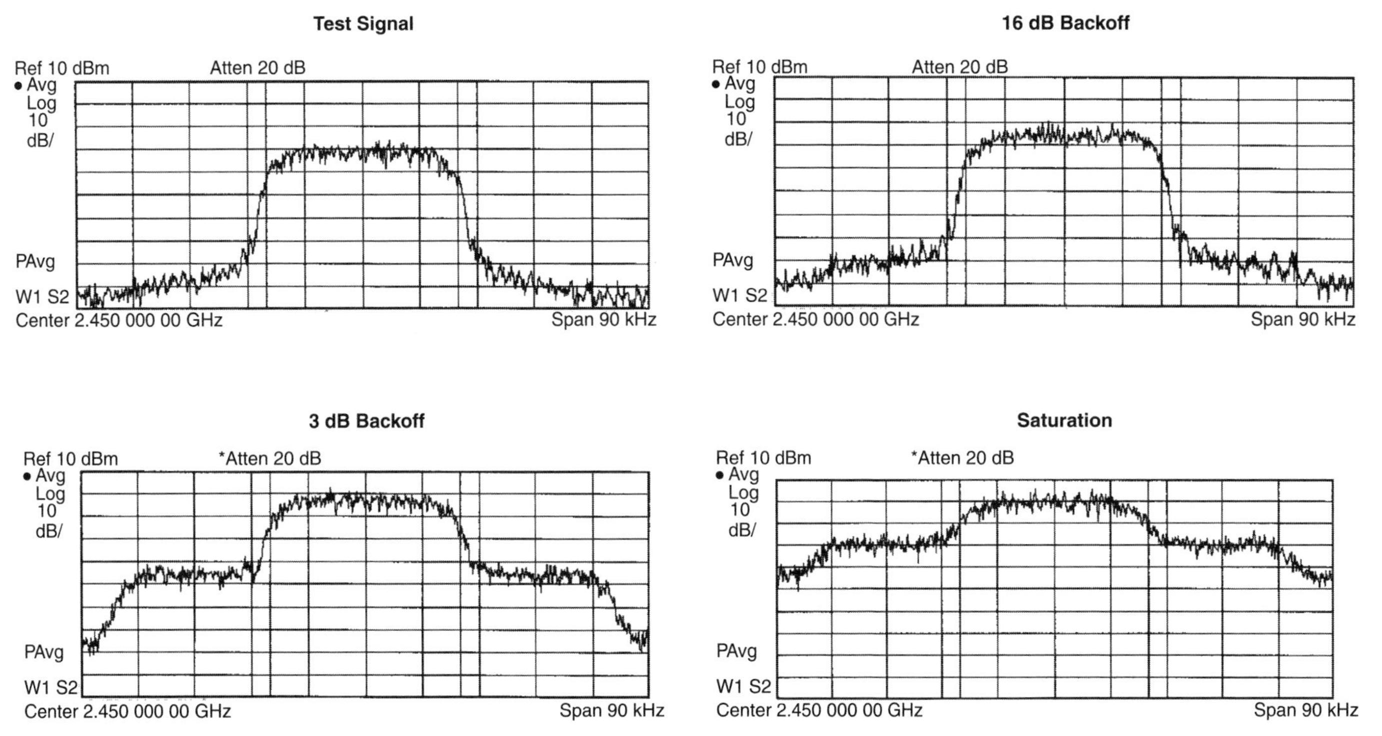

Figure 1.29 ACP of the $\pi/4$DQPSK signal for various amounts of backoff.

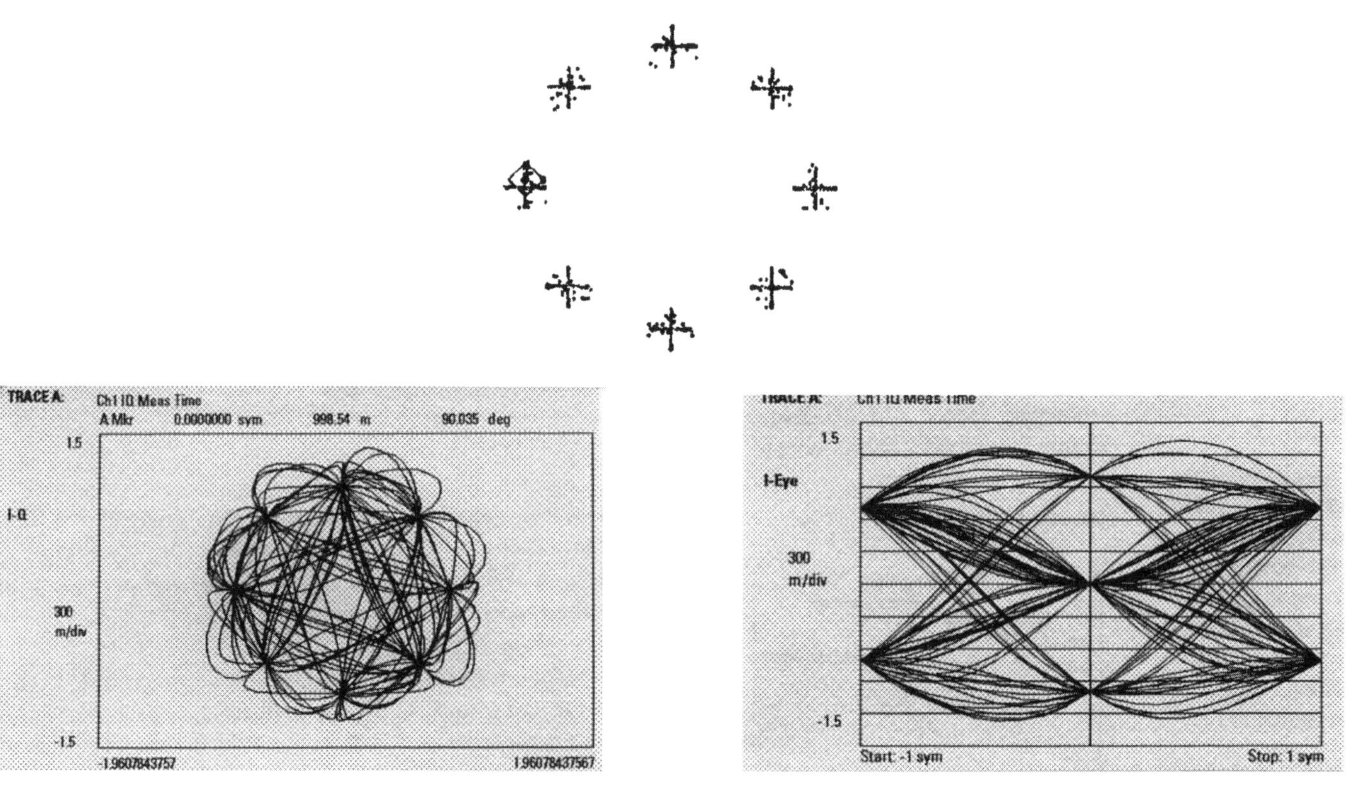

Figure 1.30 Constellation, vector, and eye diagrams.

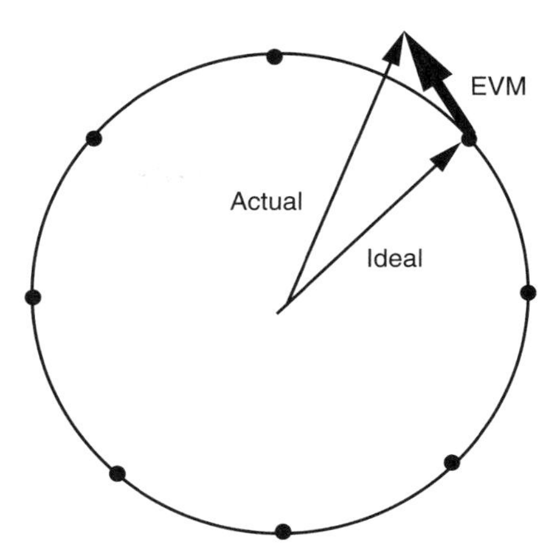

Figure 1.31 Explanation of EVM.

Any of the three displays provided in Figure 1.30 can be obtained by simply adjusting the VSA. With these diagrams a trained engineer can determine the cause of the signal distortion.

The analysis diagrams described above suggest a possible cause for the bit errors in a wireless data transmission system, but they give no quantitative information about how bad the distortion is.

The quantity EVM provides the necessary quantitative information. EVM is explained in Figure 1.31, which is a constellation diagram for $\pi/4$DQPSK modulation. The eight ideal points are shown. Each can be represented by a vector of normalized signal amplitude extending from the origin to one of the eight phase locations. The amplitude and phase of an actual signal point are also shown. Because of signal distortions, the amplitude of the vector and the phase are greater than they should be. Both of the phase and amplitude differences between the ideal and the actual signal could be measured, but an "error vector" is defined to simplify the analysis. The EVM can then be determined. Note that the amplitude of the EVM is affected by both phase and amplitude errors. The EVM is a single metric quantifying how bad the error is.

The allowed EVM in a particular system depends on how much information is to be obtained from the modulated signal. With QPSK modulation, where 2 bits/symbol are to be obtained, the allowed EVM is about 7 dB. With 64QAM modulation, where 6 bits/symbol are to be obtained, the allowed EVM is about 0.5 dB.

1.39 SUMMARY OF CHAPTER 34: CCDF

In certain RF wireless communication systems, such as CDMA, many signals intended for different users are transmitted on the same frequency at the same time. The individual signals are separated by using their CDMA code, as described

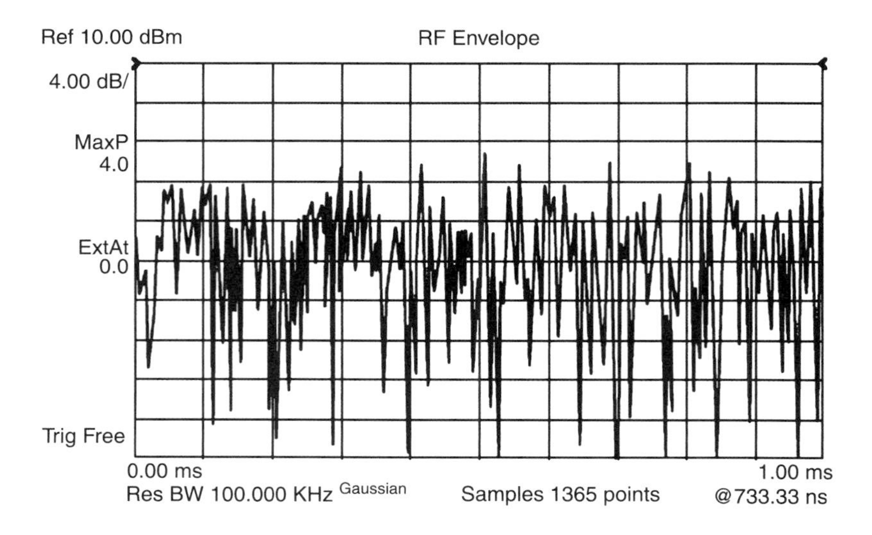

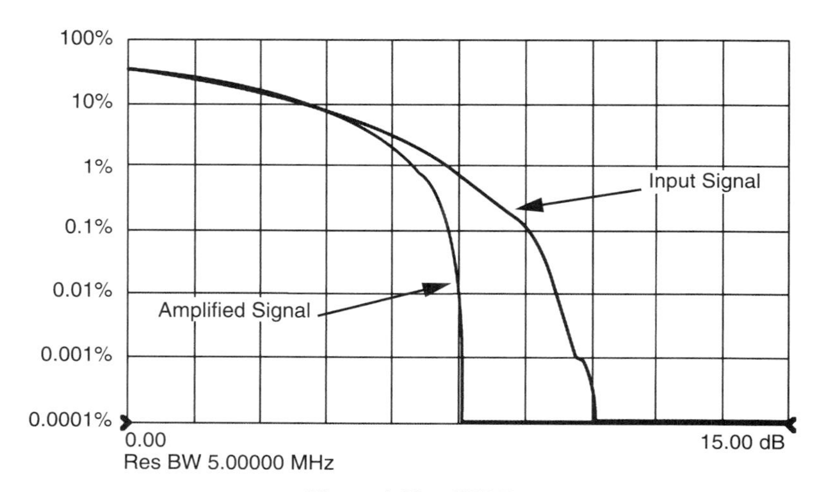

Figure 1.32 CCDF.

in Chapter 30. The top graph of Figure 1.32 shows the resulting signal as a function of time. An individual signal can be detected by despreading with its unique code. However, it is difficult to determine the degradation caused by RF components on this mixture of many signals using the analysis diagrams or the EVM measurement. Using a spectrum analyzer, a simple way of determining if an RF component is distorting the CDMA signal is to use the CCDF measurement shown in the bottom of Figure 1.32. CCDF simply shows the percentage of the time (on the vertical scale) that the signal is a specified dB above its mean value (on the horizontal scale). Two CCDF measurements are shown in the lower part of the top figure. One is the input RF signal to a power amplifier, and the other is the output RF signal from the amplifier. Notice that these two curves are quite different, indicating that the amplifier is distorting the signal, because of its saturation characteristic. A higher power amplifier with more "head room" would be required for the system to work properly.

1.40 SUMMARY OF CHAPTER 35: BER

The ultimate measurement that must be made on a digital communication system is the BER. In this measurement a known data stream is transmitted through the equipment. The received bitstream is compared to the transmitted bitstream and a BER is determined.

The unit under test (UUT) is usually a complete system. The digital test signal is a pseudorandom bit sequence (PRBS) that eliminates the need to time synchronize the transmitted and received bitstream, because the sequence has an easily detectable pattern at the start. PN9 or PN15 are common PRBS signals.

Figure 1.33 shows a typical BER measurement setup. An RF signal generator (see Fig. 1.6) generates an RF signal with the correct frequency, power level, and type of modulation. A PRBS signal is modulated onto the RF carrier. The signal is transmitted to a UUT as shown. The incoming RF signal passes through the receiver in the UUT, and the received bitstream is remodulated onto the UUT's transmitted RF signal. The transmitted signal is received, the PRBS digital bitstream is compared to the signal that was originally transmitted, and the BER is measured.

The BER test measures whether the UUT is performing to specifications, and it is basically a go–no go test. If the unit does not pass, the BER test gives no indication of the cause of the problem. To determine the cause of the problem, analysis diagrams and EVM measurements must be made.

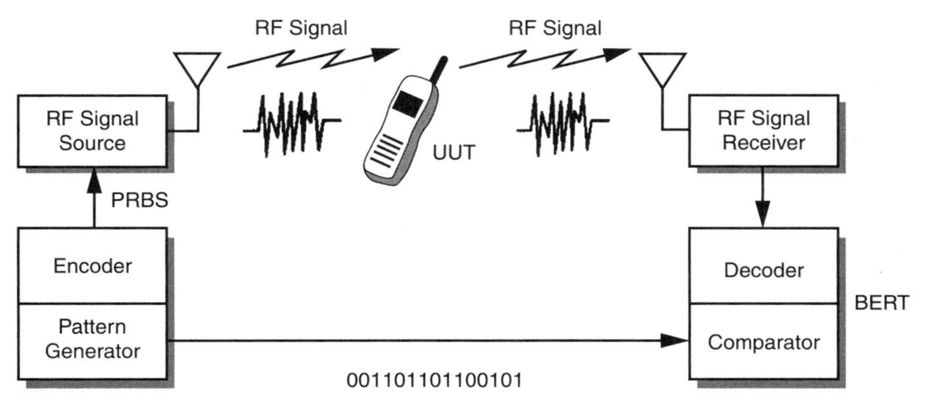

Figure 1.33 BER measurement.

1.41 SUMMARY OF CHAPTER 36: MEASUREMENT OF GSM EVOLUTION COMPONENTS

The VSA displays constellation, vector, and eye diagrams and measures EVM. However, it can also be supplied with software that will perform tests to specific system requirements, including making pass/fail measurements.

This type of measurement to system specs is illustrated in Chapter 36. Measurements are shown for the GSM evolution systems described in Chapter 30, including Enhanced Data Rates for GSM Evolution (EDGE), WCDMA, and HSDPA. Special Agilent software is used to control the VSA (Fig. 1.34). The test signals are provided by the RF signal generator described in Chapter 5, using special software to generate the EDGE, WCDMA, and HSDPA modulated RF signals. Measurements were made in the 1.9 GHz Personal Communication Services (PCS) frequency band.

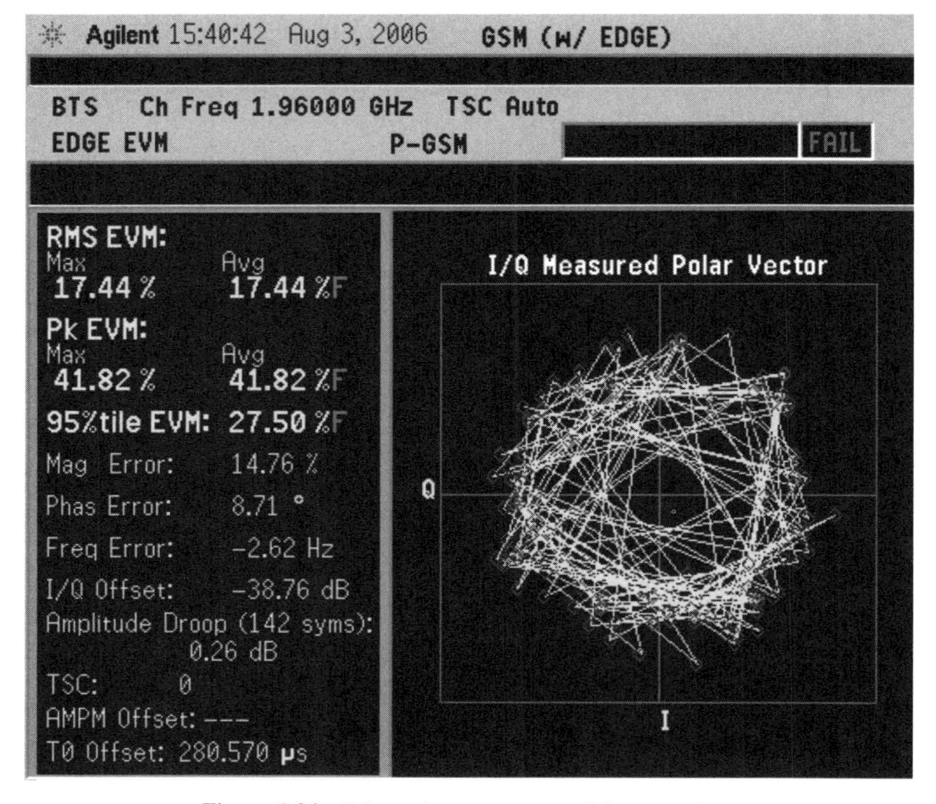

Figure 1.34 Edge polar vector at amplifier saturation.

1.42 ANNOTATED BIBLIOGRAPHY

1. Weisman, C. J., The Essential Guide to RF and Wireless, Prentice Hall, Upper Saddle River, NJ. 2000.
2. Micelli, A., Wireless Technicians Handbook, Artech House Norwood, MA. 2000.
3. Besser Associates, BesserNet.com, http://www.bessernet.com.
4. Besser, L. and Gilmore, R., Practical RF Circuit Design for Modern Wireless Systems, Vols. I and II, Artech House, Norwood, MA, 2003.

Reference 1 is an excellent reference on RF and wireless at an elementary level. It provides an overview of various cellular and wireless data systems.

Reference 2 is a good reference on cellular phone systems at the technician level.

Reference 3 is a free website with free interactive design programs, and it refers to many RF and wireless design articles.

Reference 4 provides an excellent background on passive and active RF circuit design principles.

PART I

RF AND WIRELESS PRINCIPLES

CHAPTER 2

CHARACTERISTICS OF RF SIGNALS

This chapter explains the characteristics of RF signals and how they differ from low frequency electrical signals. If the reader is unfamiliar with RF principles and terminology, this chapter should be studied carefully. If the reader is already familiar with RF, this chapter will be a useful reference.

Chapter 2 is organized into five sections. Section 2.1 describes the characteristics of electric and magnetic fields. Section 2.2 explains how electric and magnetic fields are interrelated by Maxwell's equations and how they generate electromagnetic waves. Section 2.3 describes the properties of electromagnetic waves, including

Frequency
Wavelength
Impedance
Power density
Phase

Section 2.4 describes the dB and dBm system of units used to characterize RF power. Finally, Section 2.5 shows an example of using RF wave properties to determine received power in an RF communication system.

RF Measurements for Cellular Phones and Wireless Data Systems. By A. W. Scott and R. Frobenius

2.1 ELECTRIC AND MAGNETIC FIELDS

Electric fields are defined in Figure 2.1. The left-hand sketch shows two electrons. They both have negative charges, so they repel each other. The force exerted on the upper electron by the lower electron is represented by an arrow in the direction of the force. In the right-hand sketch, the effect of the lower electron on any other electron at some distance from the lower electron is represented by arrows. The direction of the arrow shows the direction of the force, and the strength of that force is represented by the density (or number) of arrows in a unit area. The effect of the electron as a force field is more than just a visual aid. Its presence can be detected as a form of energy, and the electric field is as real as the electron itself.

Magnetic fields are defined in Figure 2.2. The magnetic field is the force on a moving charge (such as electron current flowing in a wire) that is due to other moving charges. The magnetic field has no effect on a stationary charge. Figure 2.2a shows the magnetic field around a wire, generated by the electrons flowing in the wire. Figure 2.2b and c show the magnetic field around a loop (or several loops) of wire that are wound into a coil. Winding the wire into a coil concentrates the magnetic field along the center of the coil.

Figure 2.3 shows combined electric and magnetic fields. In the left-hand sketch, a 10 V battery is connected to a 100 Ω resistor through two large wires. One wire is connected to the positive terminal of the battery and thus is at +10 V; the other wire connected to the negative (ground) terminal is at 0 V. Current flows through the wire from the positive terminal of the battery through the resistor and returns to the battery through the other wire. The electric and magnetic fields around these wires are shown in the right-hand drawing. Because one wire is positively charged and the other is negatively charged, a positive charge would be repelled from the positive wire and attracted to the negative wire, as shown by the solid

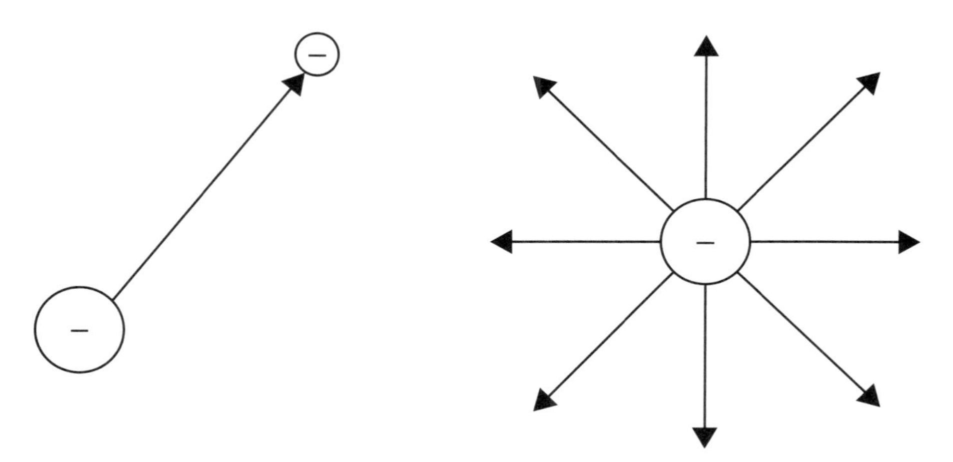

Figure 2.1 Electric fields.

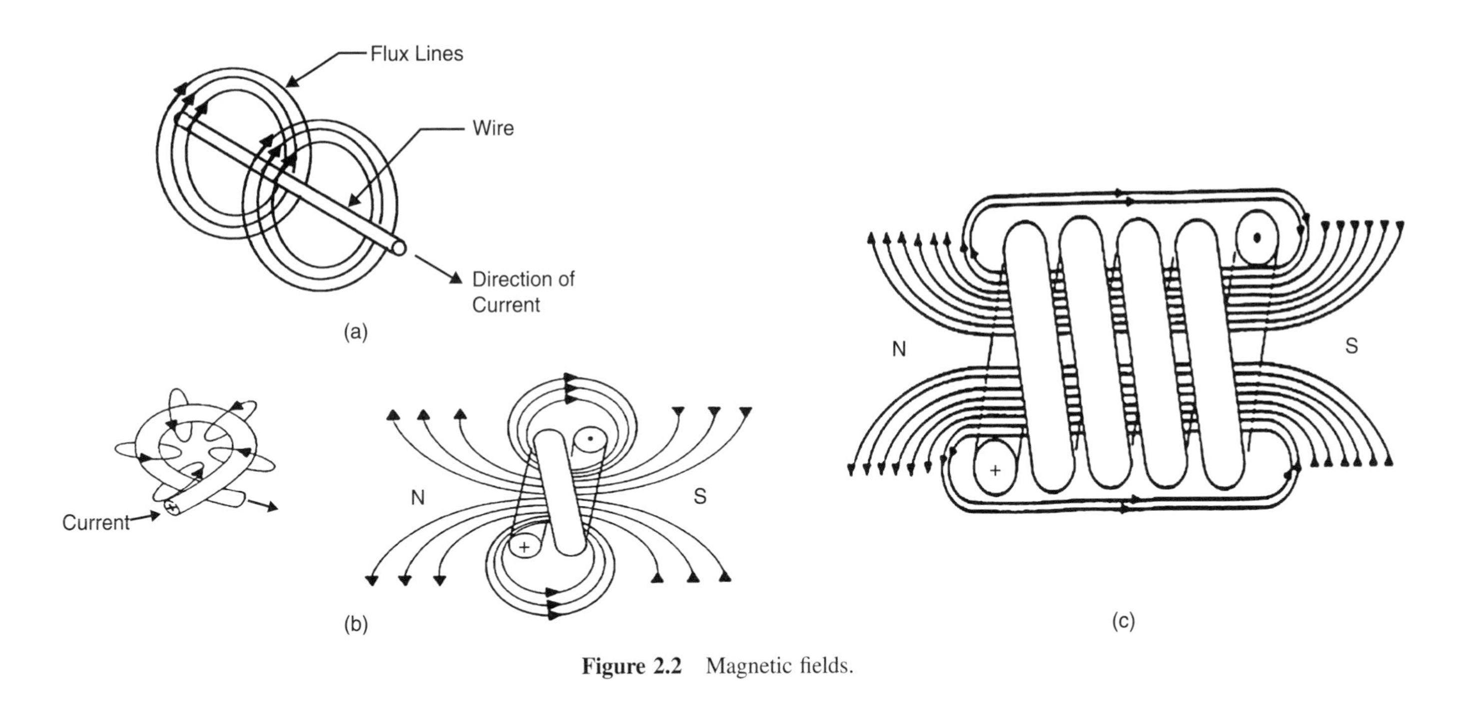

Figure 2.2 Magnetic fields.

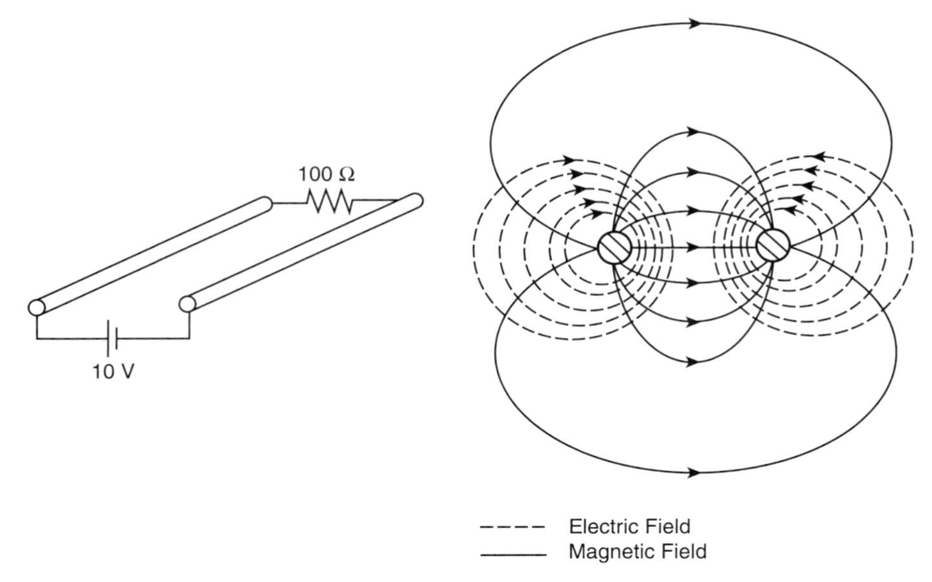

Figure 2.3 Combined electric and magnetic fields.

arrows. The electrons are moving in opposite directions in the wires, so the magnetic fields are oppositely directed around each wire, which is represented by dotted lines.

Low frequency electronics is concerned with the voltage and currents in the wire, and it is less concerned about the electric and magnetic fields. However, at RF frequencies, the voltage and current are difficult to define, and the electric and magnetic fields are dealt with directly.

2.2 ELECTROMAGNETIC WAVES

If a magnetic field is changed, an electric field is generated. This is the principle of operation of all motors and generators. A generator generates electricity by moving a coil of wire through a magnetic field, so that the magnetic field, as seen by the coil of wire, is continually changing. This changing magnetic field generates an electric field that will push current through an external load.

The four laws that completely describe electric and magnetic fields are called Maxwell's equations:

Law 1: The electric field depends on stationary charges.
Law 2: The magnetic field depends on moving charges (current).
Law 3: The electric field depends on a changing magnetic field.
Law 4: The magnetic field depends on a changing electric field.

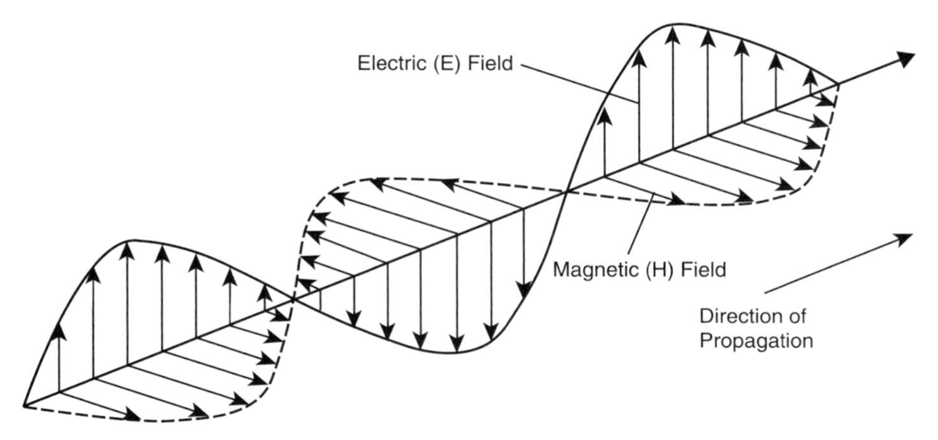

Figure 2.4 Electromagnetic waves.

The first three laws have already been discussed, and they can be demonstrated experimentally. The last law states that the magnetic field depends on a changing electric field; Maxwell added it to complete the symmetry of the equations.

The solutions of Maxwell's laws are electric and magnetic fields that travel together through space in a wavelike fashion at the speed of light. These waves are called electromagnetic waves. The field configurations in an electromagnetic wave traveling from left to right are illustrated in Figure 2.4. At the left-hand side the electric field points upward and the magnetic field points out of the plane of the paper. The electric and magnetic fields are always at right angles to each other. They decrease to zero and then increase in opposite directions, so in the middle of Figure 2.4 the electric field is pointing downward and the magnetic field is pointing into the plane of the paper. Proceeding along the wave, the electric and magnetic fields again decrease to zero and then increase in the direction that they originally had on the left-hand side, with the electric field pointing up and the magnetic field pointing out of the paper. This wave behavior of the combined electric and magnetic fields is a consequence of the simultaneous solution of Maxwell's equations, which describe electric and magnetic fields individually, based on stationary charges, moving charges, and the effect on one of the fields when changing the other field. Maxwell's laws were conceived approximately 150 years ago. Shortly thereafter electromagnetic waves were demonstrated experimentally by the generation, transmission, and reception of radio signals.

2.3 PROPERTIES OF RF WAVES

The properties of electromagnetic waves are frequency, wavelength, impedance, power, and phase.

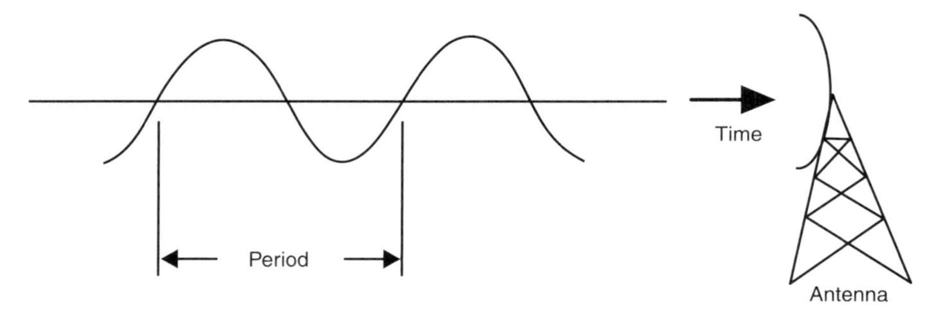

Figure 2.5 Frequency.

Frequency

Frequency is defined in Figure 2.5. In low frequency electronics, where electron currents flow through wires, frequency is defined as the number of oscillations that an electrical signal completes in 1 s. An alternative definition, which is more useful for electromagnetic waves, is that frequency is the number of electromagnetic waves that pass a given point in 1 s. The unit of frequency is the Hertz; it is the number of oscillations per second or the number of waves per second. Typical frequencies in electronics are 1000 Hz, which is 1 kHz; 1 million Hz, which is 1 MHz; and 1 billion Hz, which is 1 GHz.

The reciprocal of frequency is called the period, which is the time for an electrical signal to complete one oscillation, or the time between one electromagnetic wave and the next electromagnetic wave passing a given point. For example, if the frequency is 1 Hz, the period is 1 s; if the frequency is 1 kHz, which means 1000 waves come by in 1 s, then the period is 1 ms.

Wavelength

Wavelength is defined in Figure 2.6 as the distance in which the fields of an electromagnetic wave repeat themselves. Frequency and wavelength are related

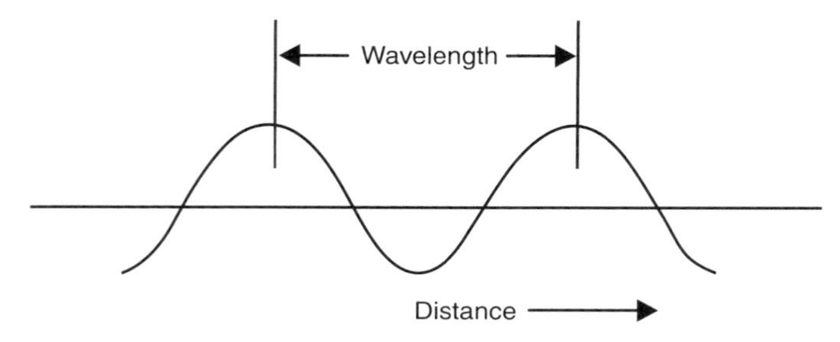

Figure 2.6 A wavelength is the distance between wave peaks at an instant of time: frequency $\times$ wavelength $=$ velocity of wave.

by the formula

$$\text{frequency} \times \text{wavelength} = \text{velocity}$$

In free space, electromagnetic waves travel at the velocity of light (3×10^8 m/s), a consequence of the physical constants in Maxwell's equations describing the electric and magnetic forces on charged particles.

A numerical formula relating wavelength to frequency when the wave is traveling in free space, or in air, is

$$\text{wavelength (m)} = c/f = 300 \times 10^6/\text{frequency (Hz)}$$

The above formula can be reduced to a more useful form for RF calculations:

$$\text{wavelength (cm)} = 30/f \text{ (GHz)}$$
$$\text{wavelength (in.)} = 1.18/f \text{ (GHz)}$$

At 1 GHz, the wavelength is 30 cm = 11.8 in. At 10 GHz, the wavelength is 3 cm = 1.18 in. Wavelengths are shorter when the RF wave travels through the plastic support materials of coaxial cable or microstrip. For coaxial cable, the wavelength is reduced by the square root of the dielectric constant of the support material. If the support material is Teflon, with a dielectric constant of about 2, the velocity of the RF wave is reduced to about 70% of the free space value. With microstrip, the calculation is more complicated than with coax, because the dielectric is only on one side of the trace.

Figure 2.7 shows the electromagnetic wave spectrum from 300 kHz to over 300 THz (300×10^{14} Hz). Electromagnetic waves exist below 300 kHz, but their wavelengths are so long that practical half-wavelength antennas are difficult to build and install.

The first consumer application of RF waves was AM radio, which was established in the 1920s. At that time there was no practical equipment to measure frequency, so wavelength, which could be measured, was used to classify regions of the electromagnetic spectrum. These wavelengths are shown along the bottom of Figure 2.7. Decade ranges of wavelengths are labeled as follows:

Label	Wavelength Range	Frequency Range
Medium frequency (MF)	1000–100 m	300 kHz to 3 MHz
High frequency (HF)	100–10 m	3–30 MHz
Very high frequency (VHF)	10–1 m	30–300 MHz
Ultrahigh frequency (UHF)	1 m to 10 cm	300 MHz to 3 GHz
Superhigh frequency (SHF)	10–1 cm	3–30 GHz
Extremely high frequency (EHF)	1 cm to 1 mm	30–300 GHz

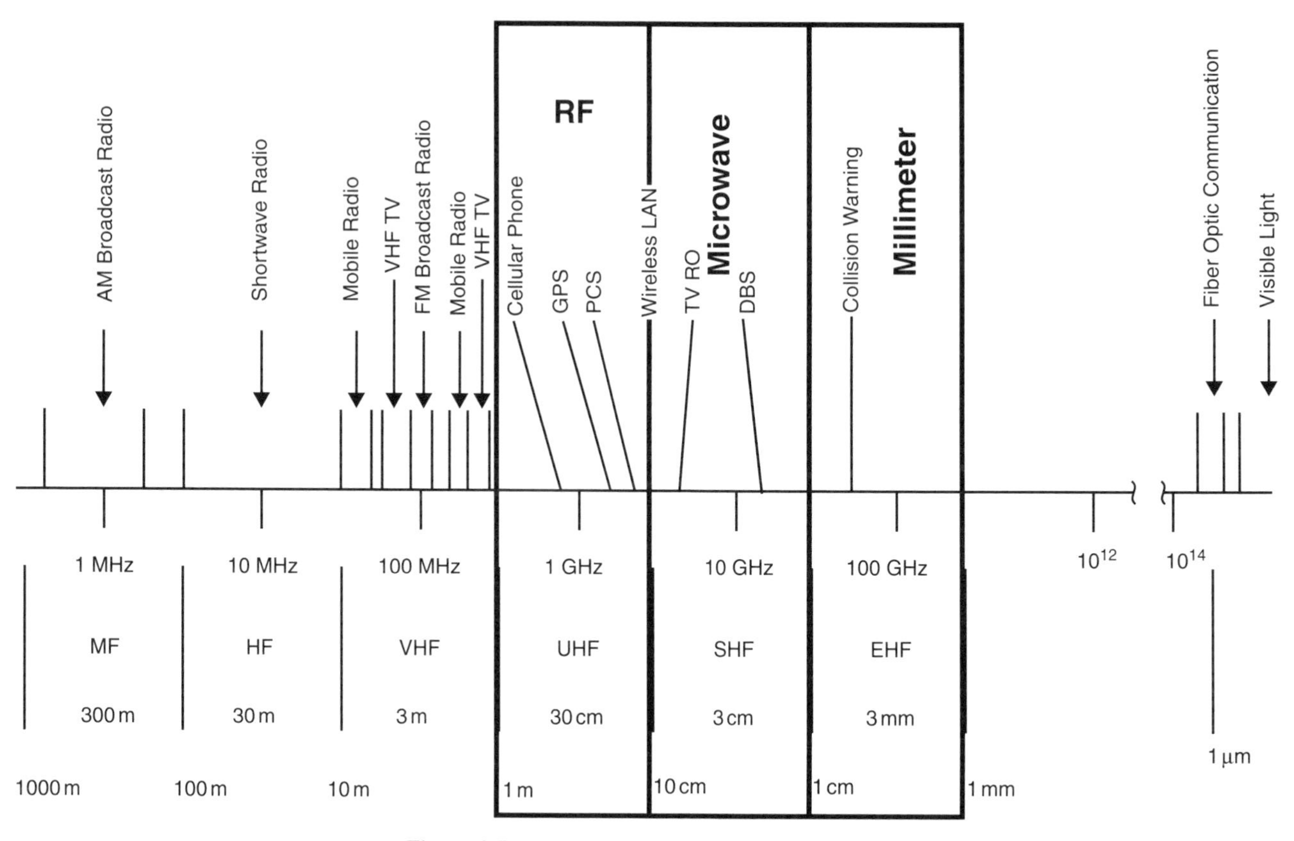

Figure 2.7 Electromagnetic spectrum.

These band designations are still used today. The test equipment and procedures described in this book relate to the UHF, SHF, and EHF frequency bands shown in the boxed regions of Figure 2.7. These bands are popularly called the RF, microwave, and millimeter (mm) wave bands.

The frequency/wavelength scale has a break between 1 mm and 1 μm wavelengths. This range is called the infrared band. It is the range of thermal energy, and it is not useful for communications on the surface of the Earth because the RF wave is highly attenuated by air.

The frequency range around a wavelength of 1 μm is the frequency range of fiber optics. The communication signals are carried inside glass fibers. The frequency range at this wavelength is 300 THz.

The wavelength range around 0.5 μm is visible light. At shorter wavelengths, ultraviolet waves, gamma waves, and cosmic waves exist.

The discussions in this book will concentrate on the RF and microwave bands from 300 MHz to 30 GHz.

Impedance

Impedance (Z) is defined as the ratio of the electric field to the magnetic field. The unit of the electric field is volts per meter (V/m), and the unit of the magnetic field is amps per meter (A/m). Therefore, the unit of their ratio is ohms (Ω). (Low frequency impedance is equal to the voltage in volts divided by the current in amps, so the unit of impedance is ohms, as in the RF case.)

In free space, the impedance of an electromagnetic wave is 377 Ω. The ratio of the electric and magnetic fields is determined by the fundamental nature of electromagnetic waves, and the value of 377 Ω is a physical constant. In contrast, in a transmission line the impedance depends on the dimensions and material of the line. Therefore, impedance can be controlled by the transmission line designer. More will be said about this subject in Chapter 3.

Power Density

Power density is defined as the power carried by an electromagnetic wave. It is equal to the electric field (E) multiplied by the magnetic field (H):

$$\text{power density} = E \times H$$

The unit of power density is thus

$$(\mathrm{V/m})\ (\mathrm{A/m}) = \mathrm{V} \times \mathrm{A/m^2} = \mathrm{W/m^2}$$

The low frequency definition of power by Ohm's law is voltage times current.

$$\text{power} = \text{volts} \times \text{amps} = \text{watts}$$

Power density can be appreciated by considering an electromagnetic field broadcast from a base station to a cell phone. As the RF wave travels through space, it spreads so that by the time it reaches the cell phone it covers an area of hundreds of square meters. The power received by the cell phone is equal to the power density of the RF wave times the area of its antenna. By definition, power density $= E \times H$, but $H = E/Z$, so

$$\text{power density} = E^2/Z = H^2 \times Z$$

At low frequencies, the strength of the electrical signal is specified by voltage and current. These two quantities are easily measured. Voltage and current are related by the impedance of the circuit (voltage = impedance × current), and power is the product of voltage and current.

At RF frequencies, the equivalents to voltage and current are electric field and magnetic field, respectively. Unlike their low frequency equivalents, electric and magnetic fields cannot be measured directly. Only power can be measured at RF frequencies, and the electric and magnetic fields are calculated from the measured power, using the value of impedance.

Phase

Figure 2.8 provides and illustration of phase. Phase is the time difference between two electrical signals at the same frequency. A single electromagnetic wave is characterized by frequency, wavelength, impedance, and power density. However, for phase to be specified, there must be two waves. The phase of one wave is relative to the other wave or to the same electromagnetic wave at another instant in time. Phase is expressed in degrees, with 360° equal to a time difference of one period. In the figure, signal A leads signal B by 90°, which means that when the two

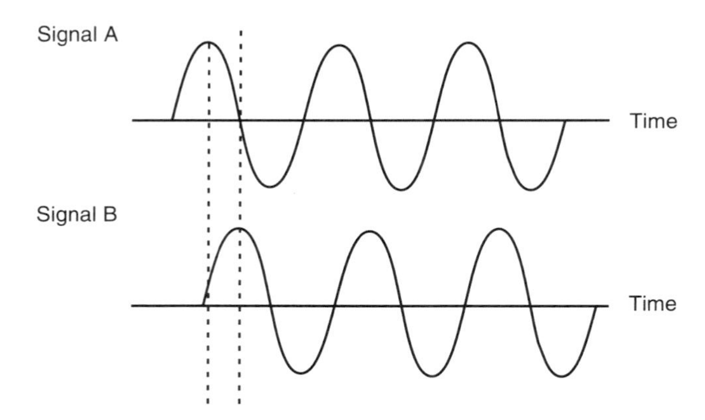

Figure 2.8 Phase is the time difference between two electrical signals at the same frequency. It is expressed in degrees, with 360° equal to a one period time difference. In this example: signal A leads signal B by 90°.

electromagnetic waves are compared on the same time scale, signal A reaches its maximum one-quarter of a cycle, or 90°, before signal B reaches its maximum.

2.4 RF POWER EXPRESSED IN dB AND dBm

When the RF wave is carried by a transmission line, the total power is the power density of the RF wave integrated over the cross-section area of the transmission line. The units of power traveling down the transmission line are specified in watts.

Figure 2.9 shows the range of power encountered in RF equipment. At RF frequencies, the reference level of power is 1 mW, which is 1/1000 W. The reason is that 1 mW of power is enough to operate a telephone, video display, or computer,

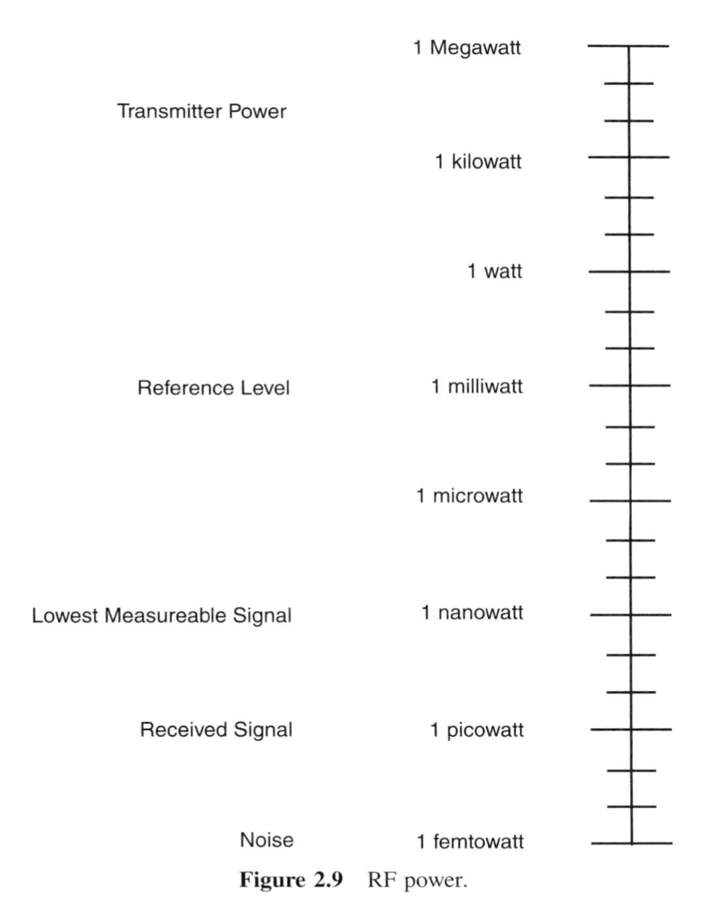

Figure 2.9 RF power.

so the standard power reference in the telecommunication industry is 1 mW. RF test equipment is usually designed to operate at the milliwatt power level and is calibrated in milliwatts.

However, RF systems operate at powers much greater than or much less than 1 mW. Each line on the scale of Figure 2.9 represents a 10-fold change in power. Starting at the reference level of 1 mW and going up in steps of 10, the scale lines represent 10, 100, and 1000 mW (which is 1 W); then 10, 100, and 1000 W (which is 1 kW); and finally 10, 100, and 1000 kW (which is 1 MW).

Below 1 mW, the scale lines represent one-tenth of a milliwatt, one-hundredth of a milliwatt, and one-thousandth of a milliwatt, which is a microwatt. Further down from a microwatt is a nanowatt, the lowest RF signal that can be measured. Still lower is a picowatt, which is approximately the level of the RF signal in communication receivers. Finally, below a picowatt is a femtowatt, which is approximately the level of noise power in RF receivers; it sets the limit on the lowest possible power that a signal can have and still be detected. If a signal is less than a femtowatt, it will be lost in the noise; when the signal is amplified, the noise is amplified with it, and the amplified signal will always stay lost in the amplified noise.

dB Terminology

All of the zeros before the decimal point for high powers or after the decimal point for low powers make calculations cumbersome. To solve this problem, the dB and dBm system of units is used. All RF equipment performance is specified in dB and dBm, and all RF measurement equipment have their scales calibrated in dB and dBm.

dB notation compresses the wide range of power values that occur in RF equipment into a practical range of numbers. It allows femtowatts and megawatts to be dealt with in the same calculation. dB notation also allows addition to be used instead of multiplication when tracing a signal through an RF system or test setup.

Three number systems are compared in Figure 2.10. The decimal system is shown in the left column. Scientific notation is shown in the middle column, and the dB number system is shown in the right column. These are three different ways of representing a number. The decimal number system is well understood. Its problem is that very large numbers have many zeros before the decimal point, and very small numbers have many zeros after the decimal point. It is very difficult to do arithmetic with large and small numbers and keep all of the zeros straight.

The problem with keeping all the zeros correct is usually solved by scientific notation, which is shown in the middle column. The dB representation shown in the left-hand column is just 10 times the exponent of the scientific notation. To convert a decimal number to its dB representation, represent the number as 10 raised to its logarithm, and multiply the logarithm by 10 to get the dB value.

To multiply numbers, add their dB values. For example, to multiply 10 times 10, add their dB equivalents: 10 dB + 10 dB = 20 dB. To multiply 10 times 0.1,

Decimal	Scientific Notation	dB
1	10^0	0 dB
2	$10^{0.3}$	3 dB
10	10^1	10 dB
20	$10^{1.3}$	13 dB
100	10^2	20 dB
1000	10^3	30 dB
1/10	10^{-1}	−10 dB
1/1000	10^{-3}	−30 dB

10^4	40 dB
1000	30 dB
100	20 dB
10	10 dB
1	0 dB
.1	−10 dB
.01	−20 dB
.001	−30 dB

Figure 2.10 dB notation.

add their dB equivalents 10 dB − 10 dB = 0 dB, which is the dB equivalent of 1.

To divide numbers, subtract their dB values. Note that 100 divided by 10 is 20 dB − 10 dB = 10 dB, which is 10.

The previous examples have been for numbers that were exact powers of 10. What about other numbers? Any number can be expressed in dB by finding the logarithm of the number (which can be done on a hand calculator by entering the number and pressing the log key) and then multiplying the logarithm by 10.

The dB value of any number can also be obtained by using Figure 2.11, which shows the dB value of any number between 1 and 10. Any other number can be represented by a number from 1 to 10 multiplied by 10 raised to some power. For example,

$$525 = 5.25 \times 100 = 5.25 \times 10^2$$

$$8000 = 8 \times 1000 = 8 \times 10^3$$

$$0.03 = 3/100 = 3 \times 10^{-3}$$

The dB value of 10 to some integer exponent can easily be determined from Figure 2.10. The dB value of the multiplying number (5.25, 8, 3 in the examples) can be determined from the graph of Figure 2.11. Because multiplying numbers is

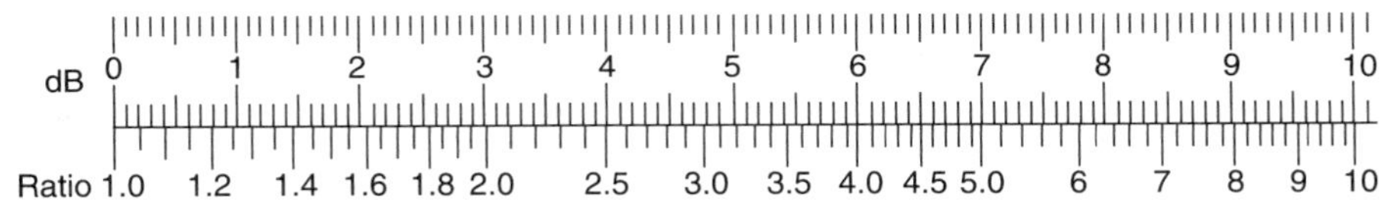

Figure 2.11 Number to dB conversion.

equivalent to adding their dB values, the dB values of the above three examples are the following (from Fig. 2.11):

Number	dB Value of Multiplier	dB
$5.25 = 5.25 \times 10^2$	7.2	$7.2 + 20 = 27.2$
$8000 = 8 \times 10^3$	9.1	$9.1 + 30 = 39.1$
$0.03 = 3/100$	4.8	$4.8 - 20 = 15.2$

The dB equivalents of numbers from 1 to 10 can be estimated without using a calculator or Figure 2.11 by just remembering that the dB equivalent of the number $2 = 3$ dB. Then,

dB equivalent of $4 = 2 \times 2$: $3\ \text{dB} + 3\ \text{dB} = 6\ \text{dB}$
dB equivalent of $8 = 2 \times 2 \times 2$: $3\ \text{dB} + 3\ \text{dB} + 3\ \text{dB} = 9\ \text{dB}$
dB equivalent of $5 = 10/2$: $10\ \text{dB} - 3\ \text{dB} = 7\ \text{dB}$
dB equivalent of $3 = (2 \times 2 \times 2 \times 2 \times 2)/10 = 32/10 = 3.2 = \sim 3$:
$3\ \text{dB} + 3\ \text{dB} + 3\ \text{dB} + 3\ \text{dB} + 3\ \text{dB} + 3\ \text{dB} - 10\ \text{dB} = 5\ \text{dB}$
dB equivalent of $6 = 2 \times 3$: $\sim 3\ \text{dB} + 5\ \text{dB} = 8\ \text{dB}$
dB equivalent of $7 = 8.5$ dB by extrapolation
dB equivalent of $9 = 9.5$ dB by extrapolation

The reverse conversion of a dB equivalent to a number can be done easily on a calculator. Enter the dB value and divide by 10, which converts the dB value to a logarithm. Then use the inverse logarithm function to get the number.

The conversion of dB equivalents to numbers can also be done by remembering that 3 dB is 2. For example, to convert 6 dB to a number, note that 6 dB is 3 dB + 3 dB. Because 3 dB is 2, $2 \times 2 = 4$, so 6 dB is equal to 4.

dBm Terminology

dB terminology is used for the ratio between two power levels in RF equipment, for example, to compare the RF power coming out of a component to the RF power going into it.

The dB number system can also be used to express absolute values of RF power as dB relative to 1 mW. As discussed previously, the reference level for RF power measurements is 1 mW. Power measured in dB relative to 1 mW is the power value that is being measured referenced to 1 mW. To see how dBm is used, refer to Figure 2.12. On the right-hand scale the RF power is shown in watts, kilowatts, megawatts, and so on, as was shown in Figure 2.9.

On the right-hand scale in Figure 2.12, the same power levels are shown in dBm. The reference level of power is 1 mW, which is 0 dBm. For example, a

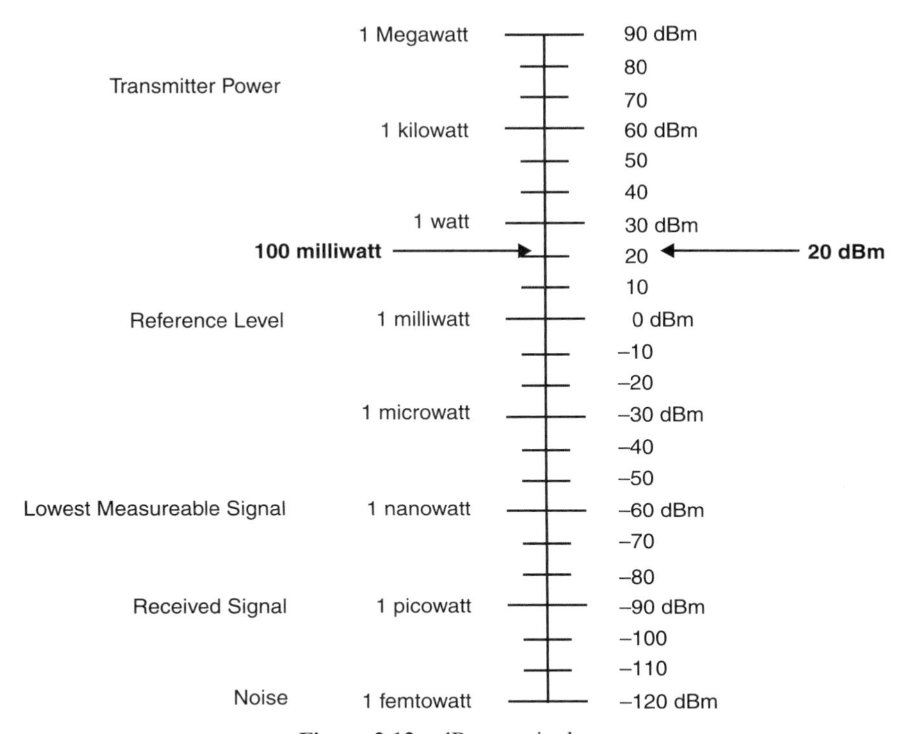

Figure 2.12 dBm terminology.

power of 10 dBm is 10 × 1 mW = 10 mW. A power of 20 dBm is 20 dB above 1 mW and is 100 mW. The entire range of RF power from 1 fW to 1 MW is now represented by a range of dBm units from −120 dBm through 0 dBm up to 90 dBm.

2.5 USING dB AND dBm TO DETERMINE AN RF LINK BUDGET

Figures 2.13–2.17 show the use of dB and dBm relative to 1 mW to determine the power link budget of a RF communication system. Figure 2.13 shows a typical cellular communication system. A base station is trying to communicate with a mobile unit. Unfortunately, a hill is blocking the direct RF signal. However, the RF signal transmitted from the base station can reflect off a nearby hill and still reach the mobile unit as shown. The reflection off the hill sends the RF power in all directions, so the received power at the mobile will be less than if there was a direct path.

Figure 2.14 shows the mean path loss of the RF signal under various propagation conditions for the situation demonstrated in Figure 2.13. The curves

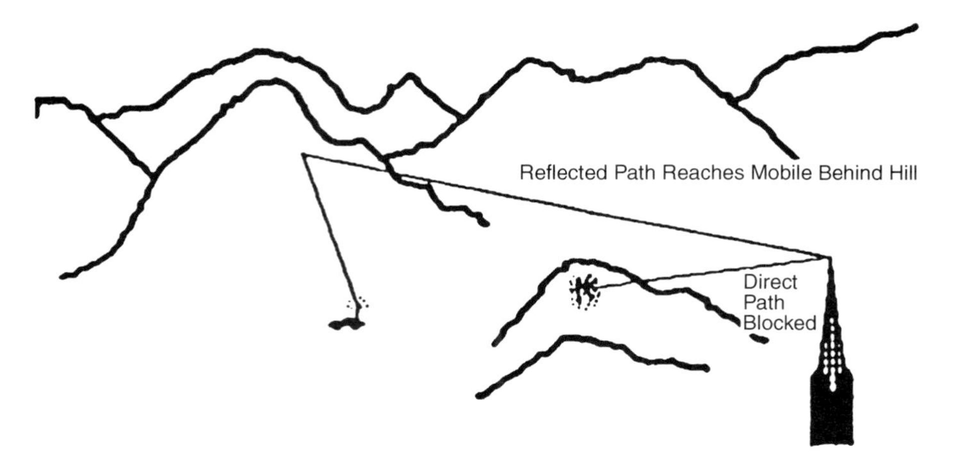

Figure 2.13 RF propagation. The direct propagation path is blocked, but the signal reaches the receiver by reflection.

are normalized to a unity gain antenna on the base station and a 1 square wavelength antenna on the mobile receiver. The first solid curve with no data points is the free space case with no obstructions. The free space loss can be calculated directly by comparing the area of an idealized 1 square wavelength

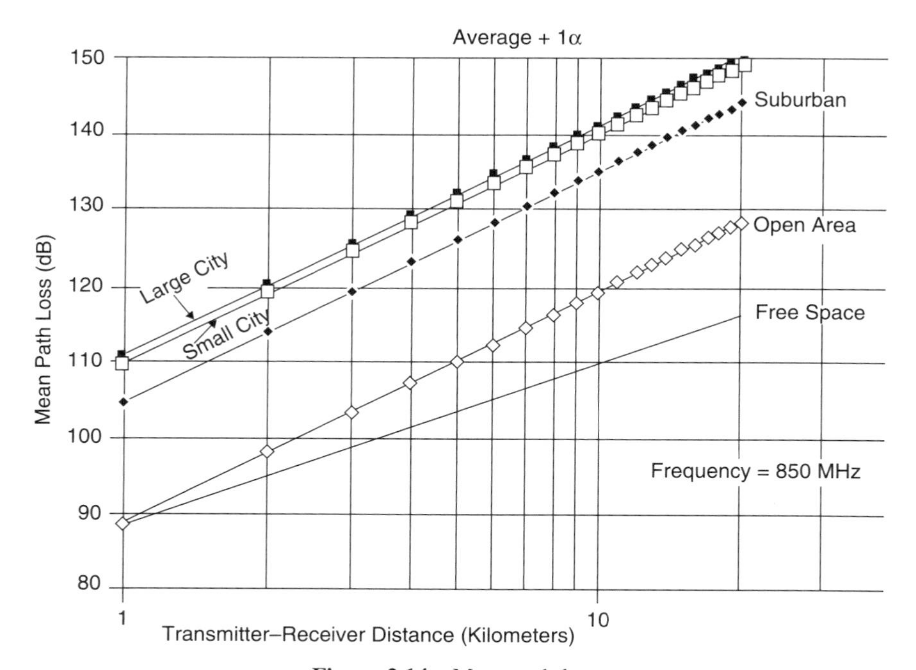

Figure 2.14 Mean path loss.

antenna on the mobile unit to the area of the sphere with a radius equal to the transmitter–receiver distance. At a 10 km transmitter–receiver separation, which is approximately 6 miles, the mean path loss is 10^{-11}, which is equal to a loss of 110 dB.

The second curve on the graph shows measurements in an open area. The obstacles blocking the propagation path are hills, trees, barns, vehicles, and animals. The measurements vary considerably about the average value depending on the obstacle locations at the exact measurements time, so a pessimistic statistical value of the average plus 1 standard deviation (σ) is plotted. The obstacles in an open area add an additional 10 dB to the mean path loss.

The third curve shows measurements in an urban area, which has an increase of 25 dB over the free space case. Here, the obstacles are houses with spaces between them. The RF wave can propagate between and over the houses. The fourth and fifth curves show mean path loss in a city. The obstacles are now large buildings and add 30 dB to the free space case. The mean path loss is now 140 dB.

Multipath fading causes additional path loss as illustrated in Figure 2.15. There may be multiple paths for the signal to reach the mobile unit, including the original reflection off of a hill, a reflection from a building, a reflection from an airplane, and even a direct path. These multipath signals may add or subtract, depending on their phase relationship.

Figure 2.16 charts the multipath fading signals at a receiver. The conditions are a receiver in a vehicle moving at 12 mph in a downtown location, as it travels a distance of 5 m. The RF frequency is 850 MHz. The 0 dB level on the received RF signal axis is the average value. Half of the time the combination of the signals is greater than the average value, and it may be 6 dB (4 times) greater when the four signals are adding constructively. The other half of the time the four signals are

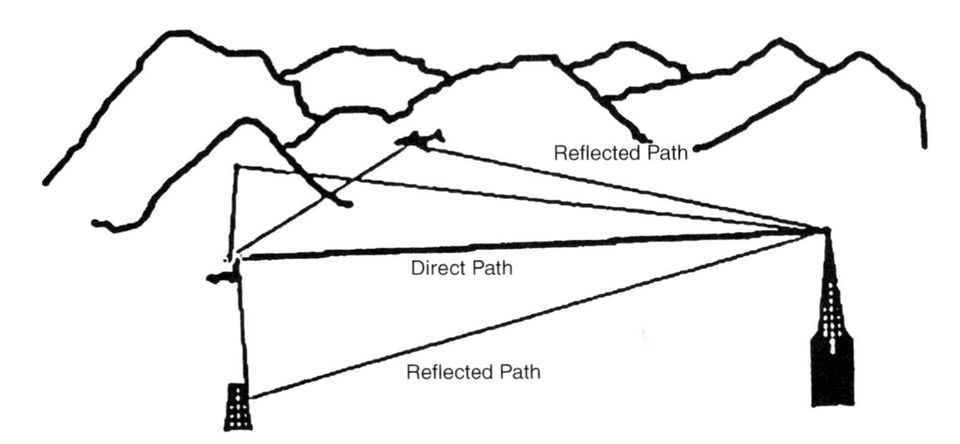

Figure 2.15 Multipath fading. Signals may reach the receiver by multiple paths and the multiple signals may add or subtract, depending on their phase.

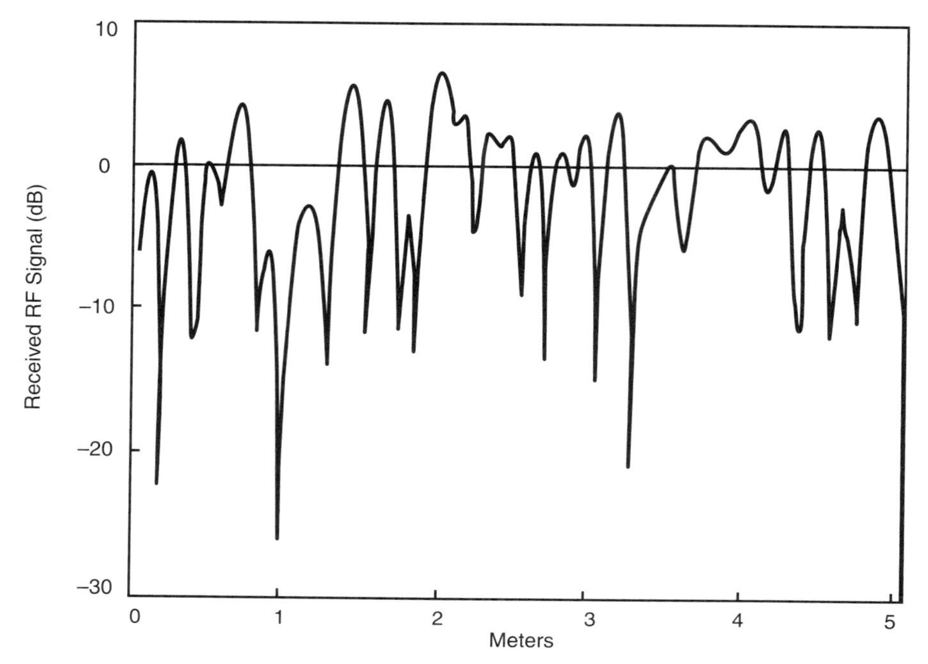

Figure 2.16 Fading rate. The speed is 12 mph, the RF frequency is 850 MHz, and the location is downtown.

adding destructively, with occasional fades greater than 20 dB below the mean value of the signals.

Figure 2.17 provides the statistics resulting from the multipath signals, when several levels of diversity receivers are used. For all conditions the combined signal is greater than the mean signal for 50% of the time. The lower curve is the result when just one receiver is used. In this case the signal is almost 20 dB below the mean for 1% of the time, which is about 15 min/day.

When two space diversity receivers are used, with their antenna's 0.5 wavelength or more apart, the fading is reduced from 20 to 10 dB for 1% of the time. This two antenna diversity is used in all base stations, but it is not physically practical in mobile receivers. When four diversity receivers are used, multipath fading is practically eliminated. This is achieved in CDMA receivers, not by using multiple hardware, but by signal processing (as discussed in Chap. 30).

The performance of the link from the mobile unit to the base station in a cellular phone system can be calculated. The received power at the base station (C) is given by the formula

$$C(\text{dBm}) = P_{\text{T}} + G_{\text{T}} + G_{\text{R}} + P_{\text{L}} = -110\,\text{dBm}$$

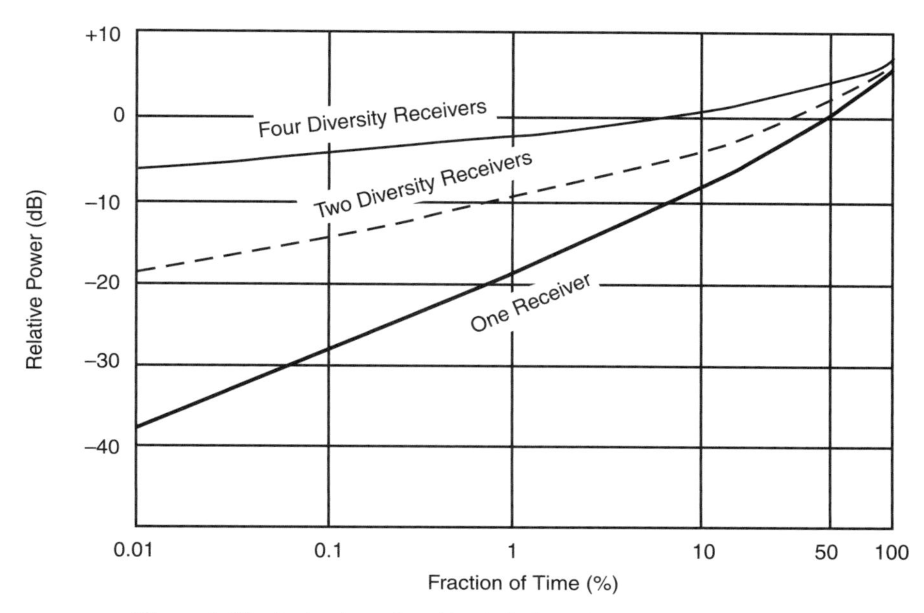

Figure 2.17 Reduction of multipath fading with diversity receivers.

where

$$\text{transmitter power } (P_T) = 27 \text{ dBm}$$
$$\text{transmitter antenna gain } (G_T) = 0 \text{ dB}$$
$$\text{receiver antenna gain } (G_R) = 13 \text{ dB}$$
$$\text{path loss } (P_L) = -150 \text{ dB}$$

For this example the transmitted power from the mobile unit is 27 dBm or 0.5 W. The transmitter antenna gain of the handset is 0 dB. This antenna cannot have gain without being directional, and a directional antenna cannot be used on the mobile unit because the mobile unit has no idea where the base station is. In contrast, the base station antenna can be directional by covering just a sector of the cell and with limited vertical coverage, so it can have 13 dB gain. The transmitter power, transmitter antenna gain, and receiver antenna gain are known exactly because they are hardware items. Conversely, the path loss can only be estimated. Assuming that the base station receiver uses two antennas and two receivers, the path loss at 10 km is −150 dB when the 10 dB of multipath loss is included. Adding these values together shows that the received power is −110 dBm.

Whether this is enough power depends on the noise in the receiver and the S/N required to achieve the system BER. Anticipating the results of the overall base station receiver given in Chapter 27, the overall receiver noise level is −122 dBm and the S/N will be 12 dB. Considering the results of Chapter 4 showing that for QPSK modulation the S/N must be 7 dB for a BER of 10^{-3}, the system will have a 5 dB performance margin.

2.6 ALTERNATE NAMES FOR dB AND dBm

The dB and dBm number system was originally developed by audio engineers. Audio engineering also has to deal with sound pressure levels that vary over many orders of magnitude from that of a jet aircraft engine to that of a human whisper. The audio engineers originally used the unit "decibel" as the unit for comparing sound levels. The "bel" was used to honor the inventor of the telephone, Alexander Graham Bell. The prefix "desi" stood for the multiplying factor of 10. The term "decibel" is still used by sound engineers today, but it has mostly been replaced by the abbreviation "dB".

Because RF and microwave engineers also face the need to deal with a large range of RF powers, they quickly adopted the "decibel" system. Early RF literature is filled with the term decibel, but since the beginning of the cell phone era in 1970, it has almost exclusively used the dB terminology to represent power ratios and the dBm terminology to represent absolute power. One reason for this is that all of the test equipment used for RF measurements and described in this book, such as power meters, vector network analyzers, spectrum analyzers, and vector signal analyzers measure and display RF power in some way. The terms "decibels" and "decibels above 1 mW", instead of dB and dBm would not fit well on the instrument displays.

Therefore the terms dB and dBm will be used almost exclusively in this book.

2.7 ANNOTATED BIBLIOGRAPHY

1. Scott, A. W., Understanding Microwaves, Wiley–Interscience, New York, 1993.
2. Frobenius, R., dB and dBm Calculator, Besser Associates, http://www.bessernet.com, 2006.

Chapters 2 and 3 of reference 1 cover the characteristics of RF signals in greater detail than this book.

Reference 2 is a free Web-based interactive program for converting between dB/dBm and power ratios/milliwatts.

CHAPTER 3

MISMATCHES

Chapter 3 is organized into 10 sections. The mismatch problem is explained in Section 3.1. Ways of expressing mismatches are defined in Section 3.2. Conversions between mismatch units are presented in Section 3.3. Section 3.4 defines S-parameters. Matching designs to reduce mismatches, using the Smith Chart, are given in Sections 3.5–3.10.

3.1 THE MISMATCH PROBLEM

The mismatch problem is explained in Figure 3.1. Mismatches are unique to RF equipment and do not occur in low frequency electronic equipment operating below 300 MHz. As shown in Figure 3.1a, RF power travels down a transmission line from the left to enter an RF component. However, some power enters, but some is reflected because the fields between the transmission lines and the component do not match. Mismatch is a measure of the RF power that gets reflected when it reaches an RF component.

A mismatch occurs every time a transmission line is connected to a microwave part. In the simplest RF subassemblies there may be 10 connections where mismatches occur. In an RF subsystem there are probably 100 connections, and a mismatch occurs at every one.

To better understand how mismatches occur, refer to Figure 3.1b, which shows a cross section of a coaxial cable and a microstrip with the microwave field

---

RF Measurements for Cellular Phones and Wireless Data Systems. By A. W. Scott and R. Frobenius

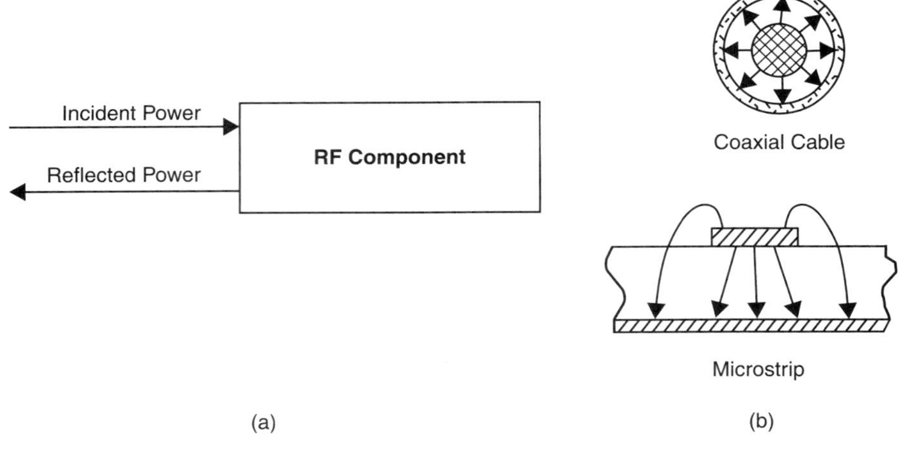

Figure 3.1 Mismatches.

configurations sketched in. When the RF power traveling down the coaxial cable reaches the microstrip, all of the power is supposed to go from the coaxial cable into the microstrip, but the fields in these two transmission lines do not match. In the coaxial cable, the RF fields extend radially in all directions from the center conductor. In the microstrip, most of the fields extend from the conductor into the insulator. Hence, at the junction between the two transmission lines, the fields in the bottom part of the coaxial cable approximately match the fields in the microstrip, so that part of the power will go directly from the coaxial cable into the microstrip. The fields that point upward in the coaxial cable find no matching fields in the microstrip, so they do not leave the coaxial transmission line because there is no matching field configuration in the microstrip line. Consequently, they are reflected back into the coaxial transmission line.

The incident power (Fig. 3.1) is the RF power trying to enter the component. The reflected power is the RF power reflected from the component. The mismatch is the ratio of reflected power to incident power. For example, assume the incident power is 1 mW and half gets reflected, so 0.5 mW enters the component. The mismatch is then $0.5\ \text{mW}/1.0\ \text{mW} = 0.5 = 50\%$. Thus, 50% of the power is reflected.

3.2 WAYS OF SPECIFYING MISMATCHES

The various ways of specifying reflected power are given in Table 3.1 for four different mismatches. The most common way of specifying mismatch is return loss expressed in dB. It is simply 10 times the logarithm of the percentage of reflected power, which is shown in column 3.

Column 4 in Table 3.1 provides the SWR. SWR can be understood from Figure 3.2, which shows the electric field of the RF waves along the input

TABLE 3.1 A Comparison of Different Ways of Expressing RF Mismatches

Power Reflected (%)	Power Transmitted into Component (%)	Return Loss (dB)	SWR	Reflection Coefficient
1	99	20	1.25	0.10
5	95	13	1.58	0.22
10	90	10	1.95	0.32
50	50	3	5.80	0.71

The simplest way to express the reflected power is as a percentage of the incident power, as shown in column 1. The RF power transmitted into the component is shown in column 2.

transmission line. The total electric field is the sum of the incident electric field, which is traveling down the transmission line toward the mismatch point, and the reflected electric field, but these two fields add up in phase and out of phase. If the reflected electric field is in phase with the incident electric field, then the two fields add, resulting in a maximum electric field in the transmission line. However, if the two fields are 180° out of phase, they subtract, resulting in a minimum total electric field. Between these points where the fields exactly add or exactly subtract are intermediate values, depending on the phase relationship.

The ratio of the maximum electric field to the minimum electric field is the SWR, and its value is related to the percentage of the power reflected at the mismatch. As

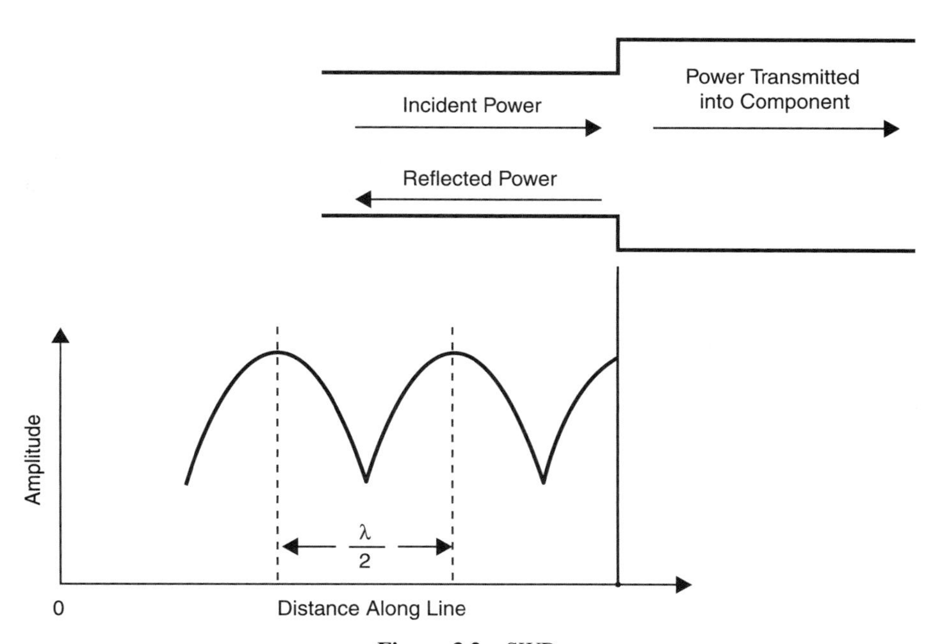

Figure 3.2 SWR.

column 4 in Table 3.1 shows, if 1% of the power is reflected, the SWR is 1.22, which means that the maximum of the total electric field is 1.22 times greater than the minimum value.

The SWR can be measured as follows. A small slot is cut in the input transmission line and the field is measured by inserting a small sensing probe in the slot. Note that the pattern of field maxima and minima repeats itself every 0.5 wavelength. The pattern might have been expected to repeat every full wavelength, but, in moving a 0.5 wavelength down the transmission line, the incident electric field has traveled 0.5 wavelength and the reflected electric field has traveled 0.5 wavelength in the other direction. Therefore, the total phase difference between the two fields changes by a full wavelength.

The last way to specify reflected microwave power is by the reflection coefficient (ρ). The reflected electric field or the incident electric field cannot actually be measured separately, but if they could, their ratio would be the reflection coefficient. The reflection coefficient can only be calculated from other measured quantities. As shown in column 5 of Table 3.1, if 1% of the power is reflected, then the reflection coefficient is 0.10, which means that the reflected electric field is 10% of the incident electric field.

3.3 CONVERSION BETWEEN DIFFERENT WAYS OF EXPRESSING MISMATCH

The relationships between reflection coefficient and the other ways of specifying the mismatches are shown in Figure 3.3. The ratio of the reflected electric field (E_r) and the incident electric field (E_i), E_r/E_i, is the reflection coefficient ρ, which can take any value between 0 (which means none of the incident RF field is reflected) and 1 (which means all of the incident RF field is reflected). Because RF power is proportional to the electric field squared (recall that in low frequency AC circuits power is proportional to voltage squared), the expression for reflected power over incident power is proportional to the reflected electric field squared divided by the incident electric field squared, so the percentage of reflected power is just equal to ρ^2. The return loss is 10 times the log of the percentage of reflected power. The absolute value is taken because the reflected power is less than the incident power, and taking the log gives a minus sign, which has already been taken into account by using the word loss. The reflected power divided by the incident power equals ρ^2, so the return loss in dB is 10 log ρ^2 or 20 log ρ.

At any point in the transmission line, the total electric field is equal to the sum of the incident electric field and the reflected electric field, but how they add up depends on their phase relationship. Therefore, the total electric field has maxima and minima along the transmission as the incident and reflected fields go in and out of phase. This field variation is called a standing wave pattern. The ratio of the maximum field to the minimum field is called the SWR, and it is equal to $1 + \rho$ (which is the maximum electric field when the fields are in phase) divided by $1 - \rho$ (which is the minimum electric field when the fields are out of phase).

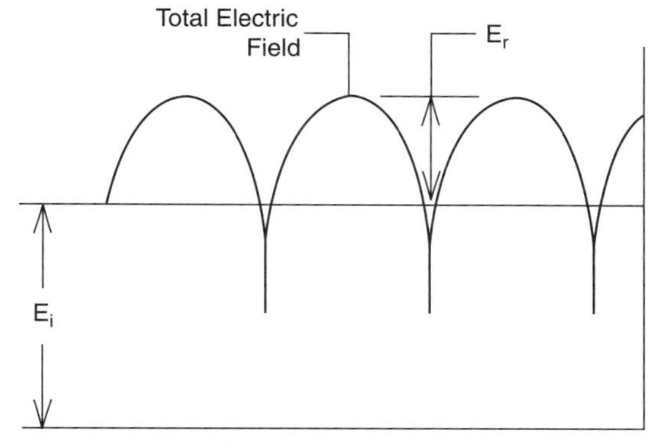

Distance Along Line

$$\text{Reflection Coefficient} = \rho = \frac{E_r}{E_i}$$

$$\frac{\text{Reflected Power}}{\text{Incident Power}} = \left(\frac{E_r}{E_i}\right)^2 = \rho^2$$

$$\text{Return Loss} = \left[10 \log \frac{\text{Reflected power}}{\text{Incident Power}}\right]$$

$$= |10 \log \rho^2| = |20 \log \rho|$$

$$\text{SWR} = \frac{1+\rho}{1-\rho} \quad \text{and} \quad \rho = \frac{\text{SWR}-1}{\text{SWR}+1}$$

Figure 3.3 Conversion between ways of specifying mismatches.

Figure 3.4 shows a useful nomograph of the mismatch conversion formulas of Figure 3.3. The left-hand side of the figure is a perfect match. The percentage of reflected power is 0, the reflection coefficient is 0, the SWR is 1, and the return loss is infinity. The right-hand side of the figure is a short. The percentage of reflected power is 100, the reflection coefficient is 1, the SWR is infinity, and the return loss is 0.

Three significant mismatch values are shown by the three vertical lines (Fig. 3.4). The line at 1% reflection is generally considered to be an acceptable match. Of course, 0% reflection is better, but it is expensive to achieve in actual equipment. The line at 4% reflection is often used as a specification for an overall RF system mismatch. The third line at 10% is the maximum allowable mismatch. Greater values of mismatch must be reduced by using the matching techniques described in Sections 3.5–3.10.

Return loss, SWR, and reflection coefficient are merely different ways of specifying the ratio of the reflected power to the incident power at a mismatch. To characterize a mismatch completely, however, the amplitude of the mismatch (as expressed by return loss, SWR, or reflection coefficient) and the phase of the reflected RF field relative to the incident field must be specified. Thus far, the discussion has covered only the amplitude of the reflected and incident signals.

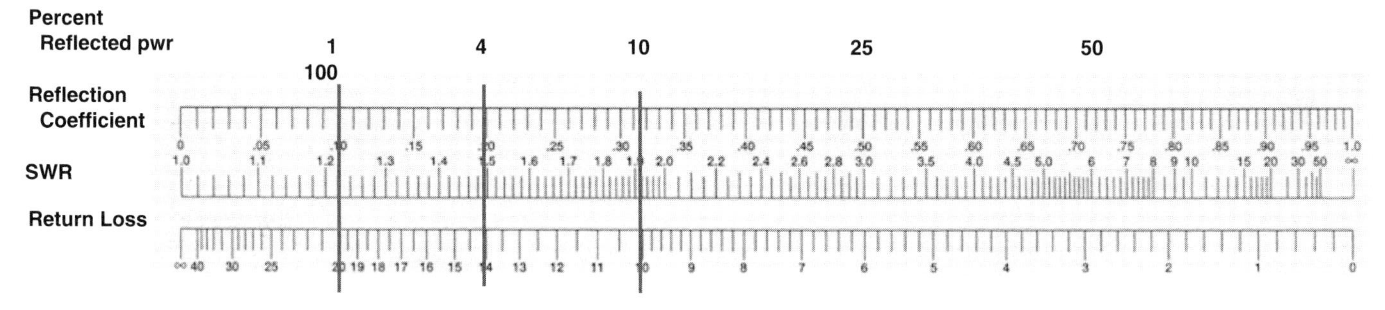

Figure 3.4 Conversion of mismatches.

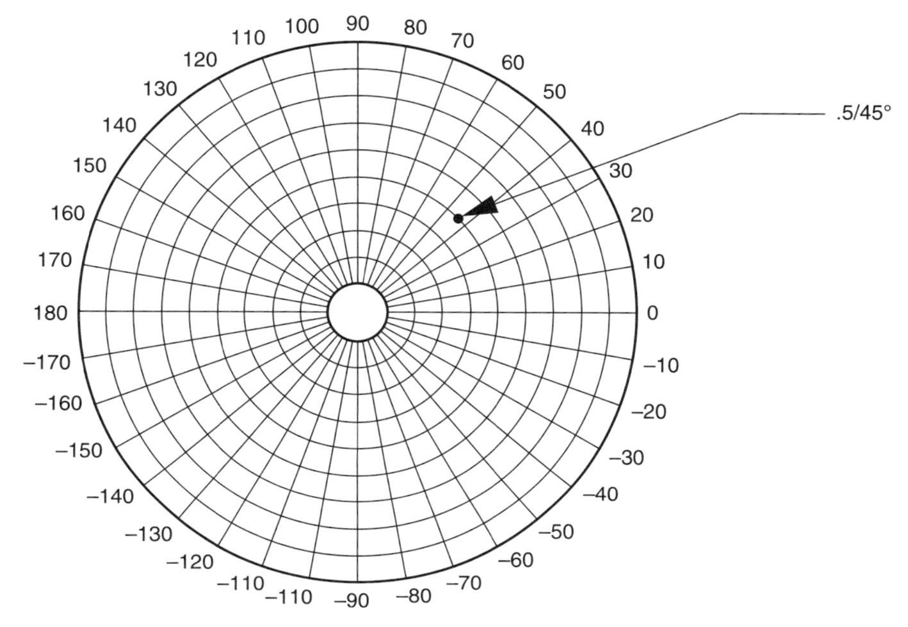

Figure 3.5 Reflection coefficient shown on a polar plot.

When the amplitude and phase of the mismatch are to be specified, it is customary to use reflection coefficient terminology. Consider a mismatch with a reflection coefficient of $0.5/45^\circ$. The first number (0.5) is the ratio of the amplitudes of the reflected and incident electric fields. It states that half of the electric field is reflected. If half of the electric field is reflected, then half of the magnetic field is also reflected, because the electric and magnetic fields are related by the impedance of the circuit. Because power is the electric field times the magnetic field, and half of each is reflected, one-quarter of the power is reflected, which is a return loss of 6 dB. The second number (45°) states that the phase of the reflected electric field leads the incident electric field by 45°.

The reflection coefficient is often displayed on a circular graph as shown in Figure 3.5. The distance out from the center of the graph is the amplitude of the reflection coefficient; the angle around the chart is the phase of the reflection coefficient.

3.4 S-PARAMETERS

S-parameters are another way of specifying return loss and insertion loss. Figure 3.6 defines S-parameters, which shows RF signals entering and exiting an RF component in both directions. If an RF signal is incident on the input side of the component, some of the signal is reflected and some is transmitted through the component. The ratio of the reflected electric field to the incident field is the reflection

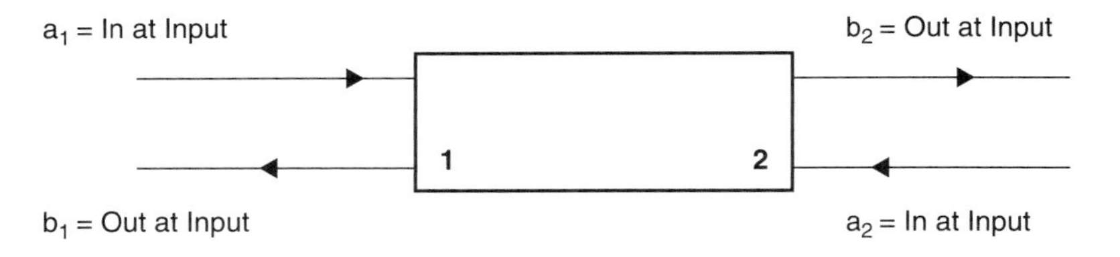

$S_{11} = \frac{b_1}{a_1}$ Input Reflection Coefficient

$S_{21} = \frac{b_2}{a_1}$ Gain/Loss

$S_{12} = \frac{b_1}{a_2}$ Isolation

$S_{22} = \frac{b_2}{a_2}$ Output Reflection Coefficeint

S-Parameters Have Both Amplitude and Phase

Figure 3.6 S-parameters. The performance specification of an RF component in terms of incident, reflected, and transmitted electric field.

coefficient. The ratio of the transmitted electric field to the incident field is the transmission coefficient.

The signals entering the input and leaving the input and output are easily understood in terms of reflection and transmission coefficients. However, how can there be an input signal at the output? The answer is that as the microwave signal travels down the output transmission line some of it is reflected at some point in the line because of a mismatch further down the output line by another component and it comes back into the original component's output.

To characterize the component completely, the reflection and transmission coefficients must be specified in both directions. In other words, such expressions as "the input going into the output," "the input going into the input," or "the output coming out of the input" must be used. S-parameter terminology is designed to avoid these cumbersome descriptions.

The S-parameters are now defined (Fig. 3.6). RF signals going into or coming out of the input port are labeled by a subscript 1. Signals going into or coming out of the output port are labeled by a subscript 2. The electric field of the RF signal going into the component ports is designated a; that leaving the ports is designated b. Therefore, S_{11} is the electric field leaving the input divided by the electric field entering the input, under the condition that no signal enters the output. Because b_1 and a_1 are electric fields, their ratio is a reflection coefficient.

Similarly, S_{21} is the electric field leaving the output divided by the electric field entering the input, when no signal enters the output. Therefore, S_{21} is a transmission coefficient and is related to the insertion loss or the gain of the component.

In like manner, S_{12} is a transmission coefficient related to the isolation of the component. It specifies how much power leaks back through the component in the wrong direction. Coefficient S_{22} is similar to S_{11}, but it is the mismatch looking in the other direction into the component.

It is important to realize that the most important of the S-parameters is S_{21}. It specifies how well the component is doing its intended job. The component was not purchased to reflect RF signals at its input or output ends or to have leakage in the reverse direction. Parameters S_{11}, S_{22}, and S_{12} are just parasitic performance features that tend to degrade performance.

3.5 MATCHING WITH THE SMITH CHART

Matching at the transmission line–component interface to reduce or eliminate reflected RF power involves measuring the reflected power and calculating where to add a "matching element" that compensates for the field mismatch and forces the power from the transmission line into the component. The Smith Chart allows this calculation to be made.

The Smith Chart is provided in Figure 3.7b. It is derived from the reflection coefficient chart of Figure 3.7a. The reflection coefficient chart shows all of the information about the mismatch. The distance out from the center of the chart is the amplitude of the reflection coefficient ρ, and the angle around the chart is the phase. In the figure $\rho = 0.5/45^\circ$. Recall that this means the reflected electric field is half of the incident electric field, and the reflected field is shifted in phase by 45° from the incident field at the mismatch.

One way to design the element that would force all of the RF waves into the component would be to solve Maxwell's equations for the electric and magnetic fields in the transmission line, in the component, and in the transition region, and then get consistent solutions where the field configurations match from one region to the other. This task is quite difficult.

An easier way is to use the Smith Chart. The measured RF mismatch is plotted in the same location, but the Smith Chart scales allow this mismatch to be represented as a low frequency equivalent of the RF problem. In other words, the Smith Chart represents the mismatch as an "equivalent circuit" of a resistor and inductor or a resistor and capacitor.

Figure 3.8a shows a mismatch represented by an equivalent series circuit of a $50\ \Omega$ resistor and an inductor with a reactance of $50\ \Omega$. The voltage across the resistor is in phase with the current, but the voltage across the inductor is 90° out of phase. The well-known vector triangle is used to solve for the total impedance of $71\ \Omega$.

There would be no mismatch (i.e., no reflected power) if the equivalent circuit of the transition between the transmission line and the component was just a $50\ \Omega$ resistor, because the component would match the $50\ \Omega$ characteristic impedance of the transmission line. The equivalent circuit of the mismatch could be changed into a $50\ \Omega$ resistor (Fig. 3.8b) by adding a capacitor to cancel out the inductance.

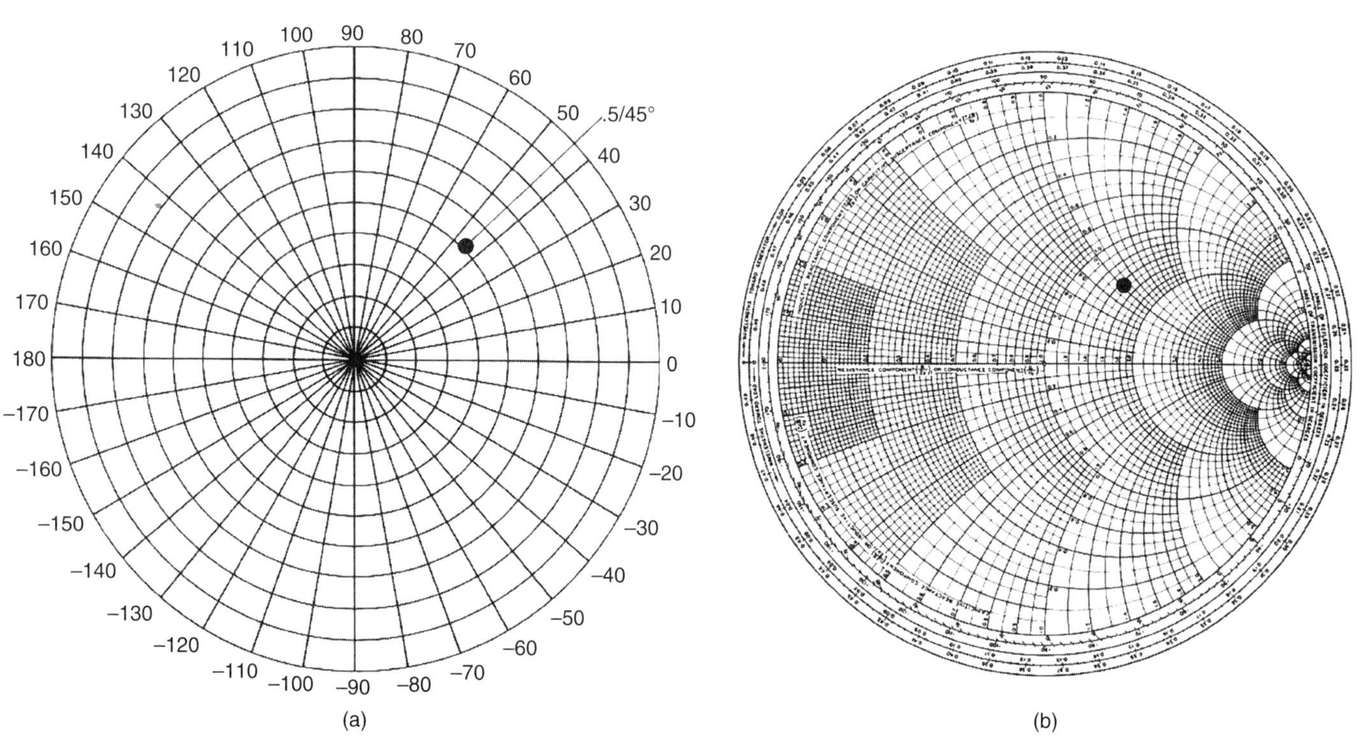

Figure 3.7 Comparison of reflection coefficient and Smith Charts.

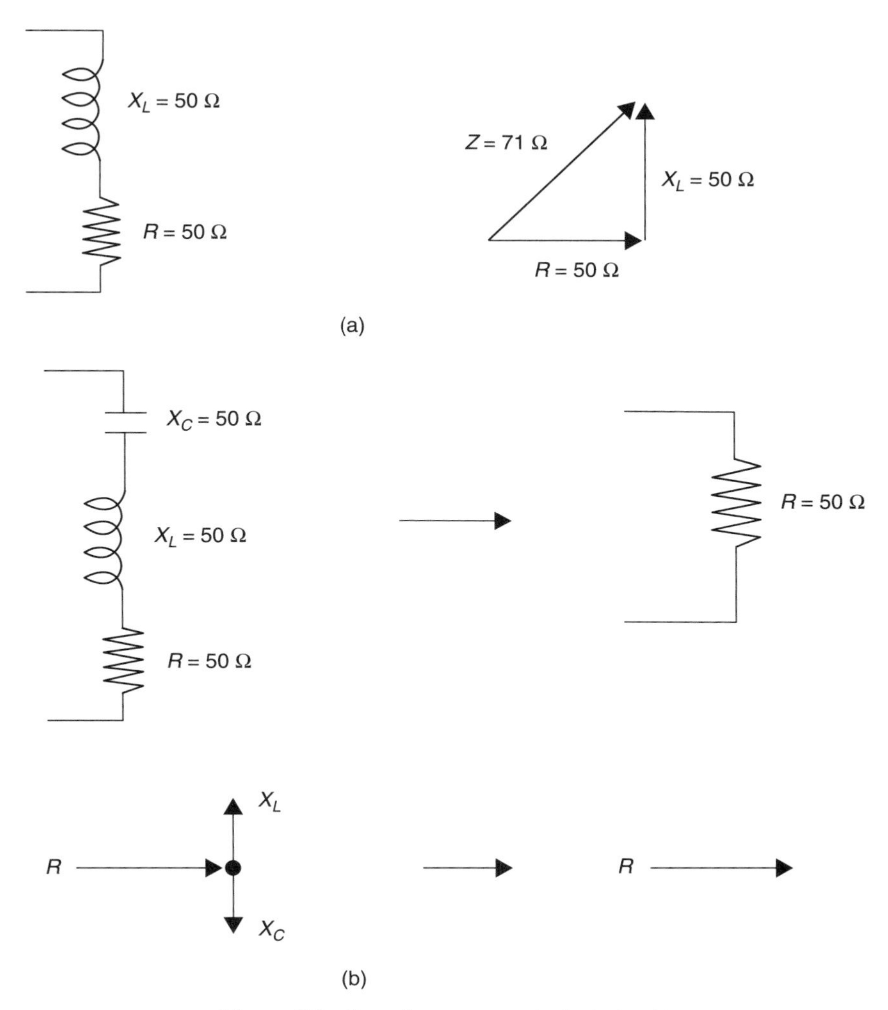

Figure 3.8 Low frequency equivalent circuit.

The capacitance must have a reactance of 50 Ω at the frequency of operation; because the voltage is 180° out of phase with the inductor reactance, the reactances cancel.

Thus, the Smith Chart transforms a difficult RF problem into a simple low frequency AC circuit problem. The next sections show how the Smith Chart is derived, how to plot mismatches on the chart, and how to use it to design various matching elements.

3.6 DERIVATION OF THE SMITH CHART

The measured mismatch, as shown in the reflection coefficient graph of Figure 3.7a, is the amplitude and phase of the reflected electric field relative to the incident electric field. The Smith Chart provides an equivalent impedance for the mismatch.

This impedance can be derived mathematically from the reflection coefficient. In this section, this impedance–reflection coefficient relationship is described in physical terms.

The impedance of a transmission line is the ratio of the electric and magnetic fields in the transmission line. The units of electric field are volts/meter, and the units of magnetic field are amps/meter, so their ratio is volts/amps or ohms. For an infinitely long transmission line, or one with a perfectly matched termination, there is no reflected signal, so the electric field and the magnetic field are both constant all along the line. Therefore, their ratio, and hence impedance, is constant all along the line. This constant impedance is called the characteristic impedance of the transmission line (Z_0). The characteristic impedance depends on the dimensions of the transmission line and on the dielectric material that fills it, so it is completely under the designer's control. In free space, the characteristic impedance of an electromagnetic wave (the ratio of the electric field to the magnetic field) is 377 Ω.

Coaxial cable and microstrip are usually designed to have a characteristic impedance of 50 Ω. In waveguides the characteristic impedance varies with frequency, but in the center of the band of a standard waveguide the impedance is about 150 Ω.

If the transmission line is not matched and part of the RF signal is reflected, then the electric field in the transmission line is the sum of the incident and reflected electric fields. At different locations along the transmission line, these add up differently, depending on their phase relationship, and a standing wave pattern is obtained.

The incident and reflected magnetic fields also add, but their phase relationship is different from that between the electric fields. Thus, the maximum and minimum of the magnetic fields occur at different positions in the transmission line, which is illustrated in Figure 3.9. Because the total electric fields and the total magnetic fields vary differently along the transmission line, their ratio, which is impedance, also varies.

Figure 3.10 shows a polar plot of the reflection coefficient, which is the amplitude and the phase of the reflected electric field relative to the incident electric field.

The reflected magnetic field can also be shown on a reflection coefficent chart (dotted line, Fig. 3.10). Its amplitude is equal to the electric field reflection coefficient, because the same percentage of magnetic field as electric field is reflected, but the phase of the magnetic field is 180° different.

Figure 3.11a provides the vector addition of the electric fields and magnetic fields at the mismatch. Remember that the total electric field is the vector sum of the incident and reflected fields, and the amplitude and phase of the reflected field relative to the incident field is given by the reflection coefficient. In the example, the reflected electric field is half of the incident electric field and leads it by 45°.

The total magnetic field is calculated in exactly the same way. The reflected magnetic field is also equal to half of the incident magnetic field, but its phase angle lags the incident field by 135°.

As stated earlier, the impedance is the ratio of the total electric field to the total magnetic field. The total electric and magnetic fields are vector quantities, having amplitude and phase. In Figure 3.11b the total electric field is divided into two components: one in phase with the total magnetic field and the other 90° out of phase with it. The impedance then consists of two components: one component in which the fields are in phase and one in which the fields are 90° out of phase. In Figure 3.11b, the

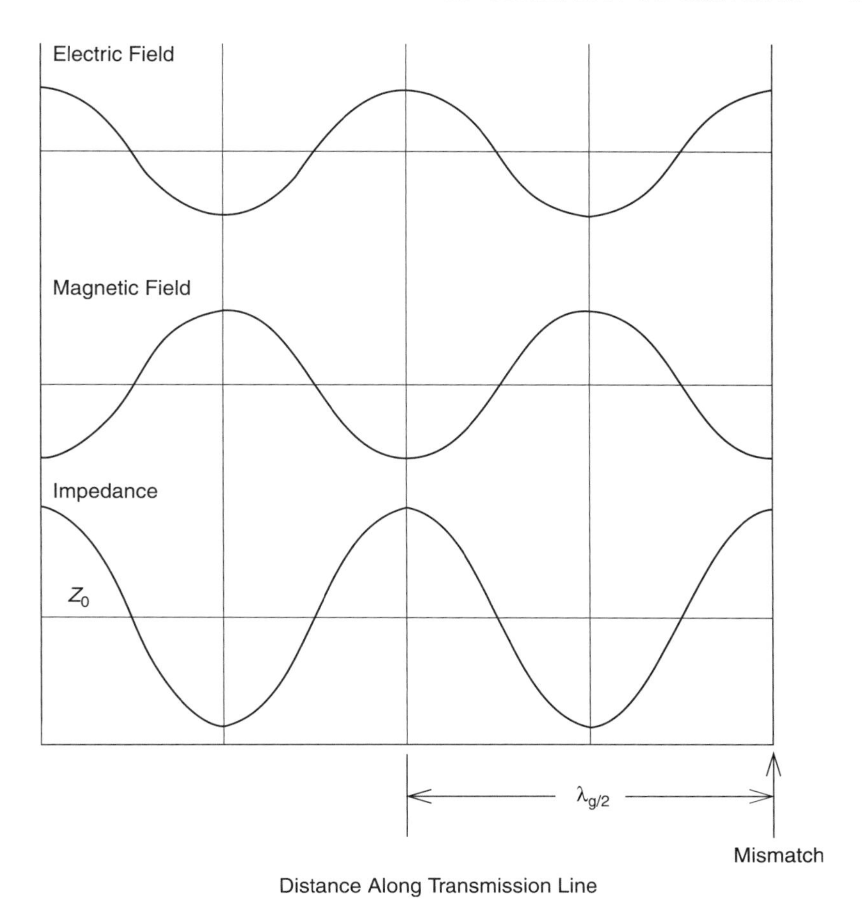

Figure 3.9 Impedance versus distance along a matched transmission line.

in-phase component has a value of about 1.5 (i.e., the strength of the electric field is about 1.5 times the strength of the magnetic field) and the out-of-phase component also has a value of about 1.5. Compare these impedances to the impedances of an AC circuit.

The impedance of an AC circuit consists of a resistive part (where the voltage and current are in phase) and a reactive part (where the voltage and current are 90° out of phase). It is helpful to remember, when contrasting RF problems to low frequency AC circuit problems, that the electric field of an RF signal and the voltage of an AC circuit are related, and the magnetic field of an RF signal and the current of an AC signal are related.

An electric field can be generated by applying a voltage across a pair of capacitor plates, and a magnetic field can be generated by passing a current through an inductor coil. Often the terms voltage and current are used to describe the RF fields along the transmission line, but actually electric field and magnetic field are the correct terms.

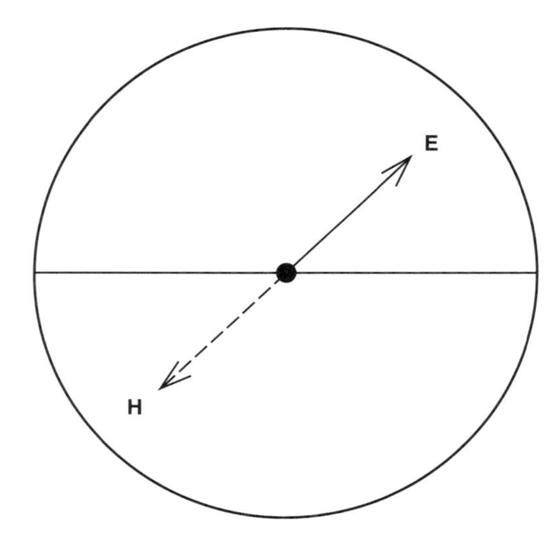

Figure 3.10 Polar plot of reflection coefficient.

For every value of the reflection coefficient (i.e., every different amplitude and phase combination), there is a specific value of impedance. From calculations like those of Figure 3.11, the resistive and reactive parts of the impedance can be calculated. This is exactly how the Smith Chart is made. The resistive and reactive parts of the impedance were calculated for hundreds of reflection coefficient points. Then all

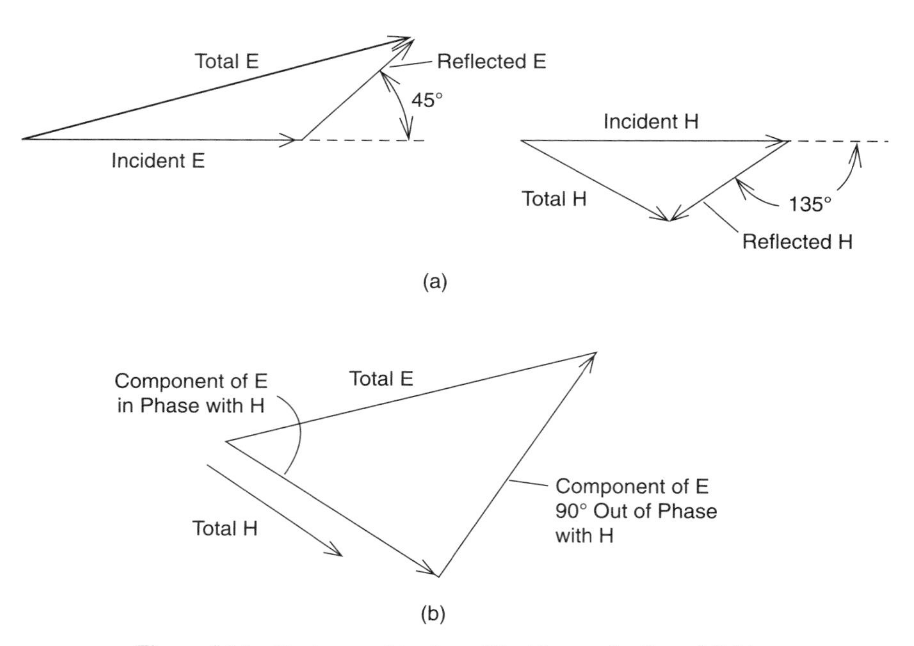

Figure 3.11 Vector combination of incident and reflected fields.

points of equal resistance and all points of equal reactance were connected, and the result is the Smith Chart shown in Figure 3.7b.

The distance out from the center of the Smith Chart is the amplitude of the reflection coefficient (i.e., the amplitude of the reflected electric field relative to the incident electric field), and the angular location around the chart is the phase of the reflected electric field relative to the incident electric field. However, instead of amplitude and phase coordinates, the Smith Chart uses coordinates of resistance and reactance. Then, if the reflection coefficient of a mismatch is known, it can be plotted on the Smith Chart to determine the resistive and reactive parts of its impedance.

All lines of equal resistance and equal reactance are "normalized," and the meaning of this normalization is shown in Figure 3.12. The actual resistances and reactances are not plotted, but they are divided by the characteristic impedance of the transmission line. For example, if $R = 75\ \Omega$ and $X = 25\ \Omega$ in a transmission line where $Z_0 = 50\ \Omega$, the normalized impedance (z) is

$$\begin{aligned} z &= \frac{75\Omega}{50\Omega} + j\frac{25\Omega}{50\Omega} \\ &= 1.5 + j0.5 \end{aligned}$$

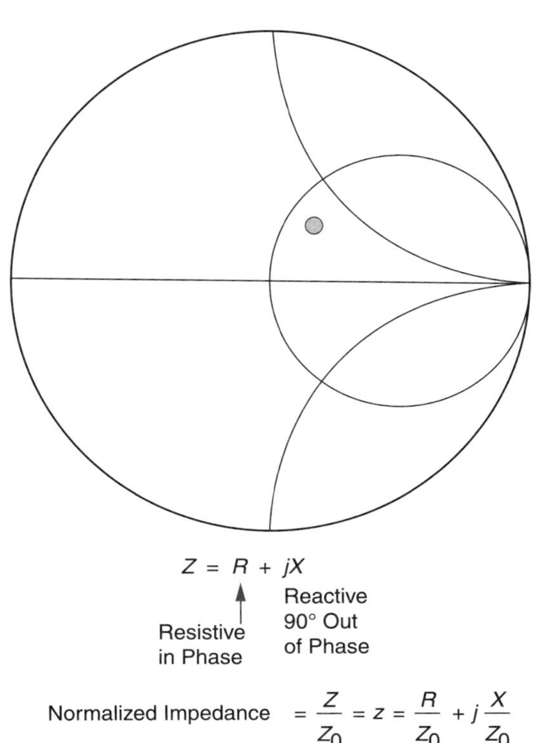

Figure 3.12 Normalization.

The reactive part of the expression is preceded by the letter j to indicate that the voltage and current are 90° out of phase.

3.7 PLOTTING MISMATCHES ON THE SMITH CHART

Figure 3.13 explains the Smith Chart scales. Some resistance and reactance lines have been darkened and numbered. Circle 1 is called the unity-resistance circle; every point lying on that circle has the resistive part of its impedance equal to 1. Because the Smith Chart is normalized, this means 1 times the characteristic impedance of the transmission line. With coax or microstrip, which have characteristic impedances of 50 Ω, every point on circle 1 has 1 times 50 Ω or 50 Ω resistance. The points all have different values of inductive or capacitive reactance, but they all have 50 Ω resistance. Correspondingly, all of the points along circle 2 have 2

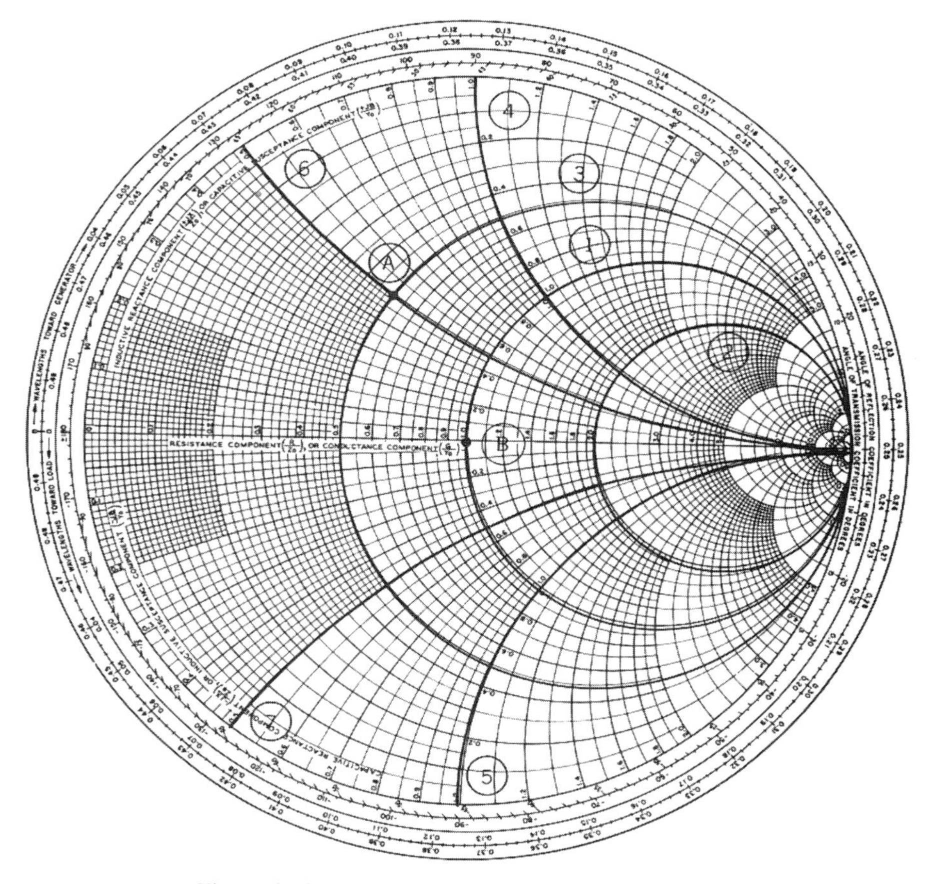

Figure 3.13 Constant resistance and reactance curves.

times the characteristic impedance of the transmission line for the resistive component of their impedance. Thus, for a 50 Ω line, all points along circle 2 have 100 Ω resistive impedance. Similarly, on circle 3, which has a normalized resistance of 0.5, all points have a 25 Ω resistance in a 50 Ω line. All of the points along curve 4, where the electric and magnetic fields are 90° out of phase, have a normalized reactive component of 1, which in a 50 Ω system means that they have a reactance of 50 Ω. All points in the upper half of the chart have a positive reactance, which means they look like inductors (electric field leads the magnetic field). In curve 5 in the lower part of the chart, the reactance is also 1, but it is capacitive because the electric field lags the magnetic field. Curve 6 is a normalized inductive reactance of 0.5, and curve 7 is a normalized capacitive reactance of 0.5.

Point A, which is a mismatch point lying at the intersections of curves 3 and 6, has a resistive component of 0.5 and an inductive reactive component of 0.5 (Fig. 3.13). Thus, in a 50 Ω system a mismatch represented by this point has 25 Ω resistance and 25 Ω inductive reactance. The resistance and reactance of any other mismatch point can be determined in the same way.

Consider point B at the center of the chart (Fig. 3.13). Because the distance out from the center represents the magnitude of the reflected wave, which is 0 at the center, this point represents a perfect match: the reactance is 0, and the resistive component is Z_0.

Figure 3.14 shows some additional points of interest on the Smith Chart. Figure 3.14a shows the pure resistance line. Every point on this line has resistance only, the reactance is 0, and the electric and magnetic fields are in phase. However, $R = Z_o$ only at the center; thus, although there are no inductive or capacitive reactive components, points along this line are still mismatched. Mismatches along the pure resistance line are at SWR maximums to the right of center, because the incident and reflected electric fields are in phase. Mismatches to the left of the center are SWR minimums, because the reflected field is 180° out of phase with the incident electric field.

Figure 3.14b shows that the outer boundary of the Smith Chart is a pure reactance; that is, the resistance is 0. This outer boundary of the chart represents a reflection coefficient of 1, which means that all the power is reflected. The reflected wave is equal to the incident wave, and the electric field is 180° out of phase with the magnetic field. Figure 3.14c again shows that the upper half of the Smith Chart has inductive reactance values and the lower part has capacitive reactance values. Figure 3.14d provides the unity-resistance circle; although the impedance has an inductive or capactive component, the resistance component is equal to the characteristic impedance of the transmission line. The unity-resistance circle is an important part of the chart for matching.

Figure 3.15 shows an RF transistor mounted in a microstrip test fixture. The mismatch at the junction of the test fixture line and the transistor, measured as a reflection coefficient, is $0.44/117^\circ$. This mismatch is plotted on the Smith Chart of Figure 3.16. Note that the distance from the chart center to a mismatch point is the amplitude of the reflection coefficient, and the direction or angle around the Smith Chart is the phase angle. The phase angle is shown around the outer circumference, labeled "angle of

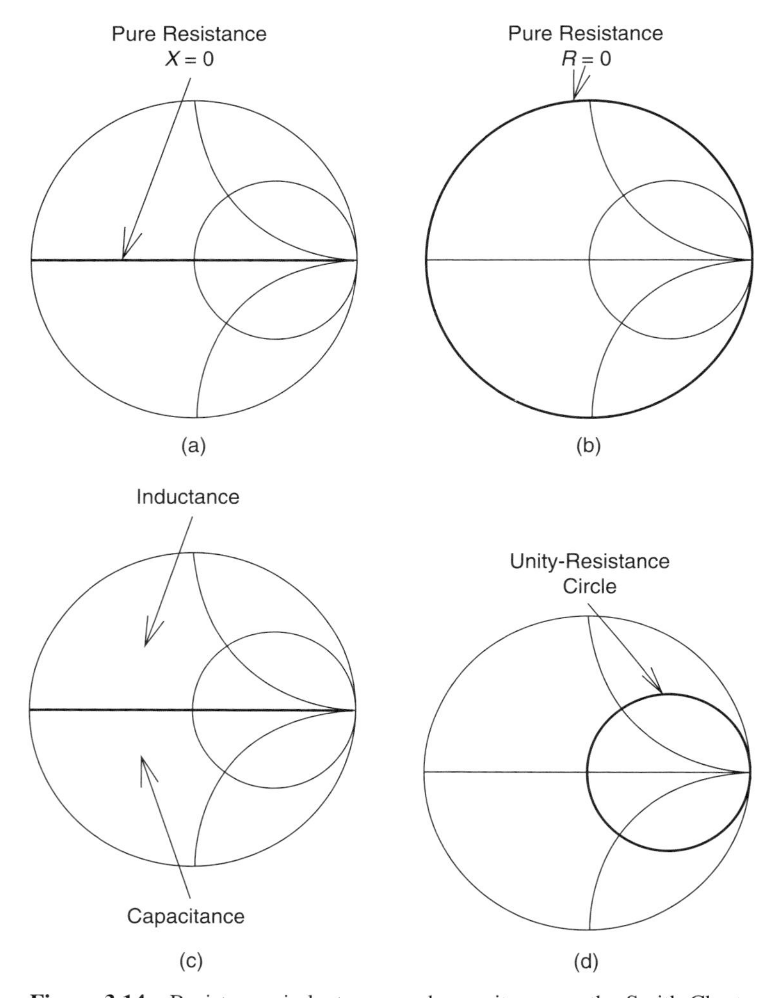

Figure 3.14 Resistance, inductance, and capacitance on the Smith Chart.

reflection coefficient in degrees." The angle is positive around the upper half of the chart, indicating that the reflected wave leads the incident wave, and negative around the lower part of the chart, indicating that the reflected wave lags the incident wave. This angle scale is the same as that on the polar plot of the reflection coefficient in Figure 3.7. In this example the phase angle is 117°. The exact amplitude of the reflection coefficient is measured by using the reflection coefficient scale along the bottom of the Smith Chart; it is the third scale down on the left-hand side and is marked "reflection coefficient." Note that it is a linear scale from the center to the outside of the chart. With a compass or divider, the length of the reflection coefficient arrow from the center out to the reflection coefficient point can be measured along

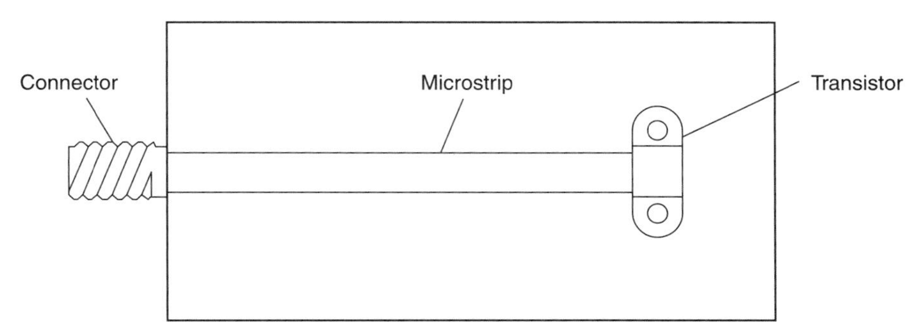

Figure 3.15 An example of an RF mismatch. At the microstrip–transistor junction, the reflection coefficient is $0.44/117^{\circ}$. The impedance, determined from the Smith Chart, is $0.5 + j0.5$, $R = 25\ \Omega$, and $X = 25\ \Omega$.

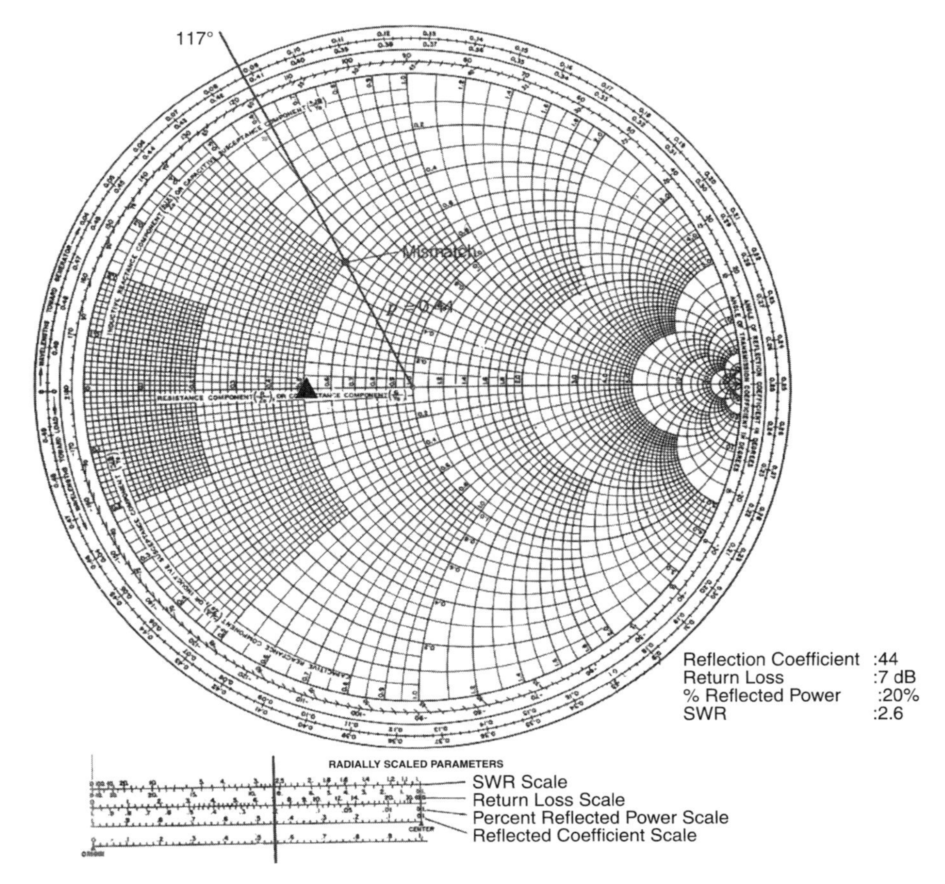

Figure 3.16 An example mismatch plotted on the Smith Chart.

the reflection coefficient scale. The mismatch point therefore lies along the 117° phase angle line, at a distance of 0.44 from the center, as shown in Figure 3.16.

The equivalent values of resistance and reactance of the mismatch can then be read directly from the Smith Chart's scale. The mismatch point lies on the 0.5 normalized resistance circle and the 0.5 normalized inductive reactance circle. The resistance in a 50 Ω transmission line is therefore 25 Ω and the inductive reactance is 25 Ω. The mismatch can be represented as a 25 Ω resistor in series with an inductance that, at the frequency of operation, has a reactance of 25 Ω.

Figure 3.17 relates movement along the transmission line with rotation around the Smith Chart. Rotating once around the Smith Chart is equivalent to moving 0.5 wavelength down the transmission line. In moving 0.5 wavelength down the line, the incident wave travels 0.5 wavelength and the reflected wave travels 0.5 wavelength in the other direction, so the reflected wave moves 1 full wavelength relative to the incident wave.

Rotating halfway around the chart is equivalent to moving 0.25 wavelength down the transmission line. Moving away from the mismatch toward the generator corresponds to a clockwise rotation around the Smith Chart, and moving from the mismatch toward the load corresponds to a counterclockwise rotation. Note the outermost set of numbers around the circumference of the Smith Chart labeled at the left-hand side of the chart as "wavelengths towards generator." Starting at the left side, these numbers are 0.04, 0.05, 0.06, and so forth. One-quarter of the way around the chart the wavelength is 0.125; halfway around the chart the wavelength is 0.25, which is 0.25 wavelength. Three-quarters of the way around the chart in a clockwise direction the wavelength is 0.375, and finally back to the left-hand side again the wavelength is 0.5 wavelength.

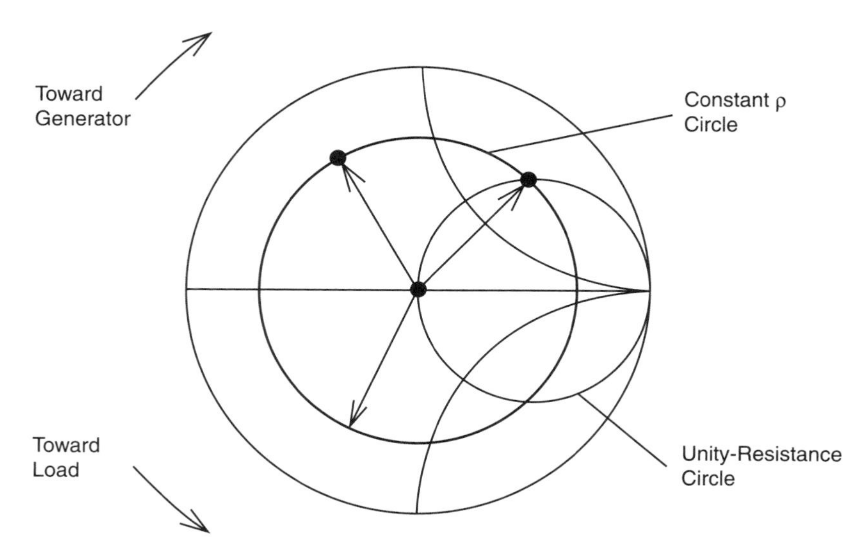

Figure 3.17 Rotation around the Smith Chart.

3.8 MATCHING CALCULATIONS WITH THE SMITH CHART

Matching means reducing or eliminating the reflected power, which means moving the mismatch point to the center of the Smith Chart where the reflection coefficient is 0. To illustrate the use of the Smith Chart for matching design calculations, the mismatch plotted in Figure 3.16 will be matched by adding a capacitor in series with the transmission line.

The reflection coefficient amplitude of the mismatch of Figure 3.16 is 0.44, which is equivalent to an SWR of 2.6 or a return loss of 7 dB, which means that 20% of the power is reflected at the mismatch. This is really a terrible match. This mismatch could also be expressed as an impedance consisting of a 25 Ω resistance and a 25 Ω inductive reactance. It is still a terrible match; converting a mismatch from a reflection coefficient to an impedance does not make it better.

A capacitor could be added to cancel the effect of the mismatch inductance. In this example the capacitor must have a reactance of 25 Ω at the operating RF frequency. The mismatch, now only the 25 Ω resistance, would then appear at the point marked by the triangle in Figure 3.16. The new mismatch is slightly better, but it is still bad because the resistive component is not 50 Ω.

A perfect match can be obtained by moving back down the transmission line, away from the junction where the transmission line joins the component, to a location at which the resistance is 50 Ω and then adding a capacitor to cancel the inductance. To find this point, consider what happens to the impedance along the transmission line. Along the transmission line, the amplitude of the reflection coefficient remains the same, because the amount of reflected power remains the same, but the phase relationship between the reflected and incident fields changes constantly. Thus, the total electric field and the total magnetic field change, so the resistive and reactive components of the impedance change. This is illustrated in Figure 3.18. The mismatch does not improve along the transmission line, but a point exists where the resistance is exactly equal to Z_0. At that point a capacitance can be added to cancel the inductance. Moving along the transmission line, the amplitude of the reflection coefficient remains constant, because the reflected power stays constant, so the move is represented as traveling around the constant ρ circle, that is, the constant reflection coefficient circle. This traveling is equivalent to rotating around that circle, and the phase of the reflection coefficient, and therefore the impedance, changes. Matching is achieved by moving along the transmission line to the intersection of the ρ circle and the unity-resistance circle. At this intersection, the resistive part of the impedance is 50 Ω and the inductance can be canceled with a matching capacitance.

A summary of the matching procedure follows:

1. Locate mismatch on the Smith Chart.
2. Determine wavelength position of mismatch
3. Move down the transmission line toward the generator until the p circle intersects the unity-resistance circle.

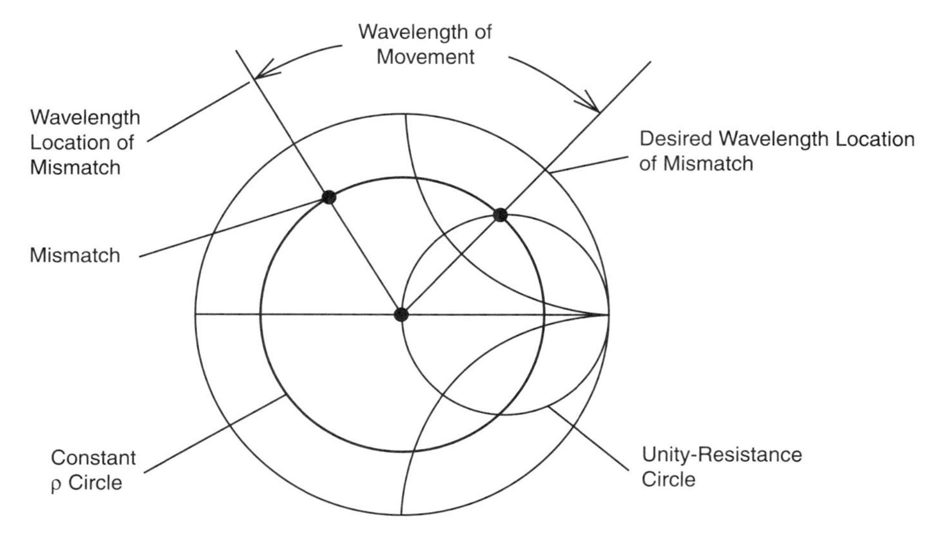

Figure 3.18 Matching with the Smith Chart.

4. Determine the number of wavelengths moved down the transmission line.
5. Add normalized capacitance to cancel inductance.
6. Determine the capacitance.

Consider a numerical example of this calculation. A transistor mounted in a 50 Ω microstrip line (Fig. 3.19) presents a mismatch at its input of $0.5 + j0.5$. This means that the resistive component is 25 Ω and the reactive component is 25 W. This is the same mismatch shown in Figure 3.16. The return loss of this mismatch is 7 dB. The first step in matching this transistor is to plot its mismatch on the Smith Chart, which is shown as point 1 of Figure 3.20. The next step is to move down the transmission line, away from the transistor, to the unity-resistance circle. At this point, the resistive component is 50 Ω. The inductive reactance can be cancelled with a capacitor, which is added into the transmission line in Figure 3.19.

To determine how far to move down the transmission line to reach the unity circle, the location of the mismatch in wavelengths is determined. As shown by point 2 on the Smith Chart in Figure 3.20, it is 0.088 wavelength toward the generator. This is the starting point. Moving down the transmission line means moving around the constant ρ circle to point 3 on the Smith Chart, where the ρ circle and the unity-resistance circle intersect. At this point, the resistive component of the impedance is 1 and is exactly equal to 50 Ω. The distance moved down the transmission line from 0.088 to 0.162, as shown by point 4, is 0.074 wavelength. The Smith Chart calculation shows how far to move in wavelengths. This normalized wavelength must then be converted to actual physical dimensions, so the guide wavelength of the transmission line must be known.

At point 3, where the constant p circle and the unity circle intersect, the correct value of capacitance must be added to cancel the inductance. At point 3 in

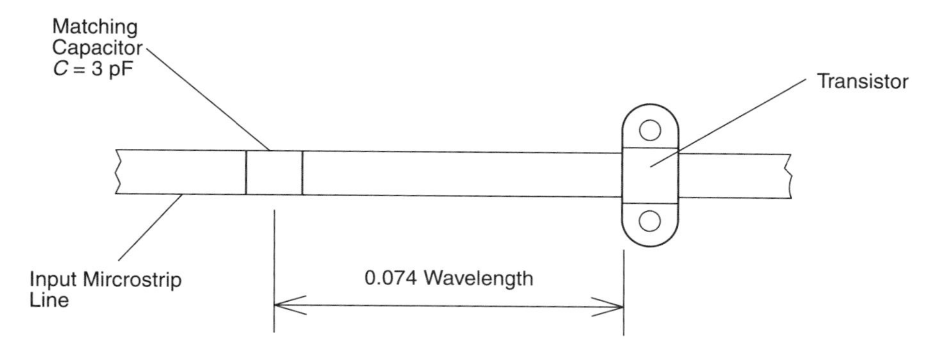

Figure 3.19 Matching of example mismatch.

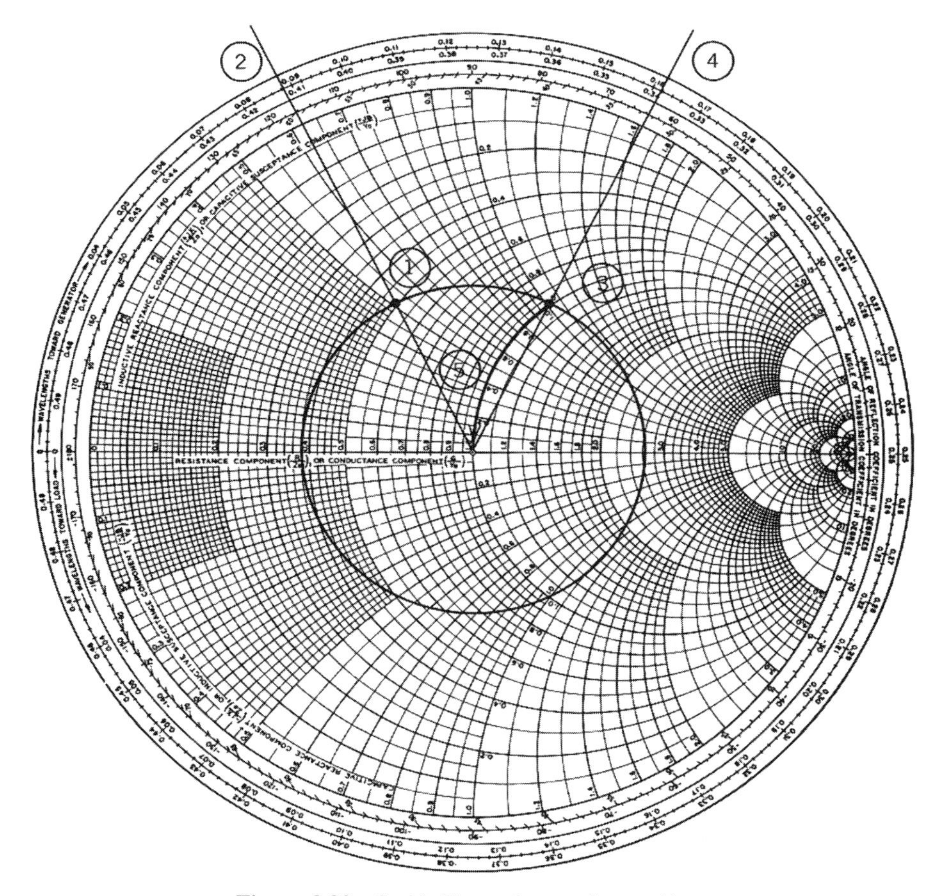

Figure 3.20 Smith Chart of example matching.

Figure 3.20, the normalized inductive reactance is 1, so a capacitance must be added whose normalized capacitive reactance is 1. The actual capacitive reactance is $1 \times 50\,\Omega$, so the value of the capacitance is

$$\begin{aligned} C &= \frac{1}{2\pi f X_c} \\ &= \frac{1}{2\pi(10^9)(50)} \\ &= 3.2\,\text{pF} \end{aligned}$$

at an operating frequency of 1 GHz. Adding this capacitance is equivalent to moving (arrow 5, Fig. 3.20) along the unity-resistance circle from the point where the reactive component is 1 down to the point where the reactive component is 0, which is the center of the chart, a perfect match.

Figure 3.21 shows a photograph of matching elements for an RF transistor. Both the transistor input and output must be matched. The input RF signal enters the transistor from the microstrip line at the upper right of the photograph. The RF input

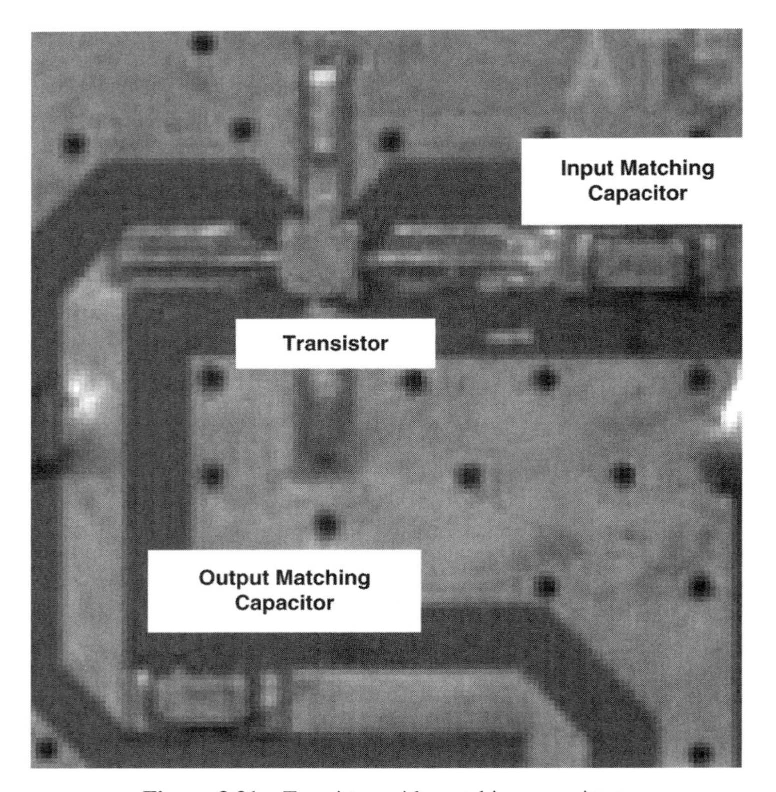

Figure 3.21 Transistor with matching capacitors.

signal passes through the input matching capacitor to the transistor. The capacitor is located at the correct distance and has the correct capacitance to match the transistor to the 50 Ω input microstrip line.

In a similar way the amplified RF signal from the transistor passes through the matching capacitor at the lower left-hand side of the figure. Note that the output capacitor is located in a different phase position and has a different capacitance value, because the input and output impedances of the transistor are not the same.

3.9 USING PARALLEL MATCHING ELEMENTS

In the matching techniques considered thus far, matching elements have been added in series with the transmission line. A better match, and one that is easier to fabricate, is often obtained by adding a matching element in parallel.

Designing must now be done in the "admittance" plane of the Smith Chart, which is shown in Figure 3.22. Admittance (Y) is the reciprocal of impedance (Z) and therefore is equal to the magnetic field divided by the electric field:

$$Y = \frac{1}{Z} = \frac{\text{magnetic field}}{\text{electric field}}$$

Admittance is used when RF circuit elements are combined in parallel. To see the significance of using admittance when elements are combined in parallel, recall the formula for combining resistors in parallel:

$$\frac{1}{R_T} = \frac{1}{R_1} + \frac{1}{R_2} = G_T + G_1 + G_2$$

Notice that the reciprocals of the resistances are combined to get the reciprocal of the total resistance. Alternatively, conductances could have simply been added to attain the total conductance.

The conversion of impedance to admittance is easily accomplished on the Smith Chart. The admittance coordinates are obtained by a rotation of the impedance coordinates by 180°, as shown in Figure 3.22. This transformation intuitively appears reasonable considering that the phase of the reflected magnetic field (relative to the incident magnetic field) is 180° different from the phase of the reflected electric field (relative to the incident electric field), which was shown in Figure 3.10.

Admittance (Y) has two components: conductance G (where the magnetic field is in phase with the electric field) and susceptance B (where the magnetic field is 90° out of phase with the electric field). The unit of admittance is Siemens (S), which is equal to $1/\Omega$. The characteristic admittance (Y_0) of a transmission line is equal to $1/Z_0$. Thus, if $Z_0 = 50\ \Omega$, then $Y_0 = 1/50\ \Omega = 0.02$ S.

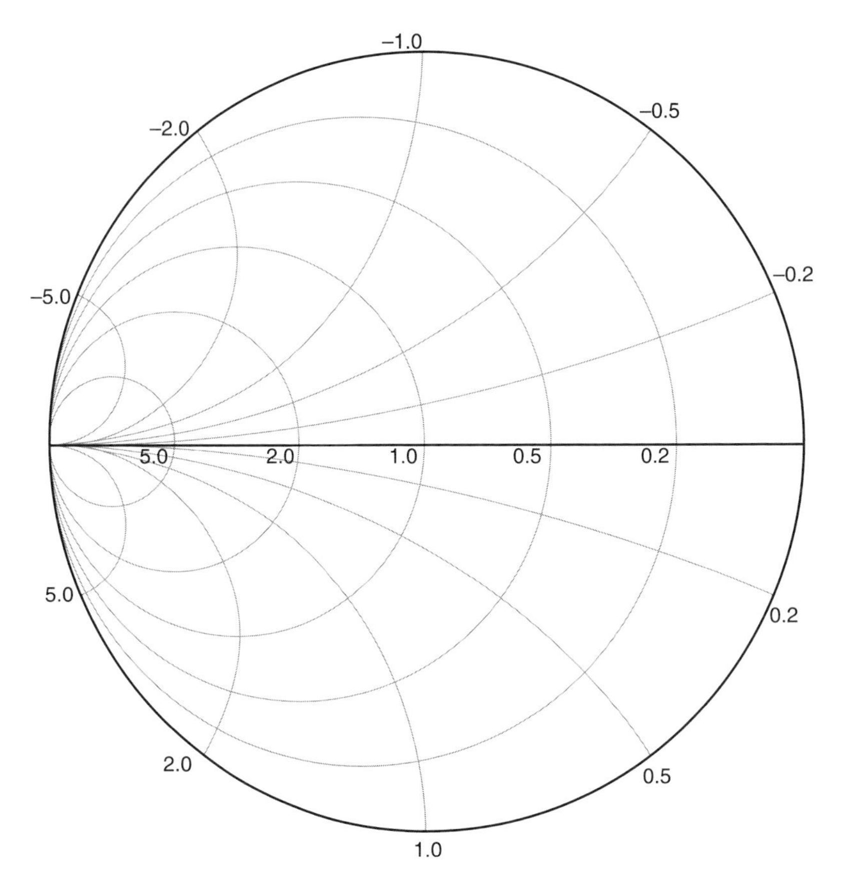

Yo = 0.02 Siemens = 1/Zo
G/Yo = Normalized Conductance
B/Yo = Normalized Susceptance

Figure 3.22 Admittance plane for parallel matching.

3.10 LUMPED ELEMENTS IN COMBINATION

In all previously discussed matching techniques using inductors or capacitors, there was movement down the transmission line until the location of the unity-resistance or conductance circle was reached, where the resistive part of the mismatch exactly equaled the characteristic impedance or admittance of the transmission line. Then an inductor or capacitor was added to cancel the unwanted capacitance or inductance. At RF frequencies from 300 MHz to 3 GHz, this is often impractical because the distance of movement is often greater than the equipment package size.

Another matching technique uses an inductor or capacitor to move the mismatch point to the unity-resistance or conductance circle rather than moving down the transmission line. For this matching technique, the "immitance" Smith Chart is used, which is shown in Figure 3.23. The immitance Smith Chart has both impedance

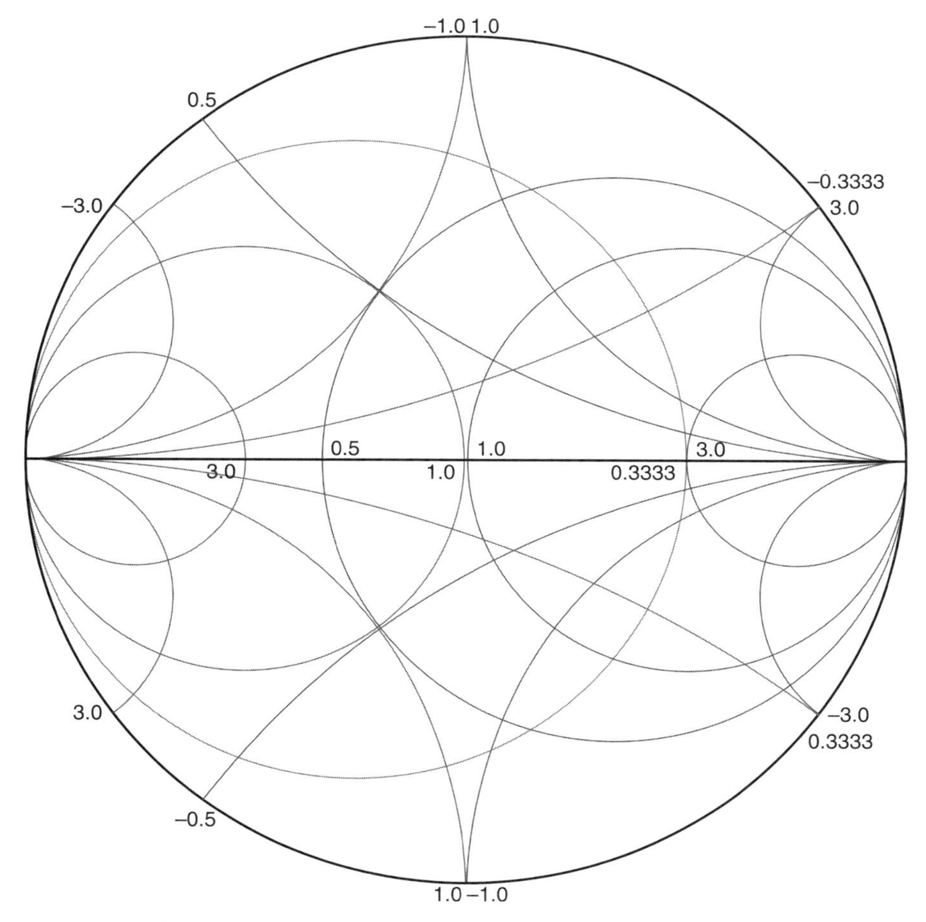

Figure 3.23 Immitance Smith Chart for using series and parallel components.

and admittance circles on the same chart. Usually the impedance circles are printed in red and the admittance circles are printed in green.

Figures 3.24 and 3.25 show the use of the immitance Smith Chart. The mismatch at the transistor is plotted as a reflection coefficient, shown as point 1 in Figure 3.24. A capacitance is added in series to the transistor to move the mismatch point to point 2 on the unity conductance circle. As shown in Figure 3.25, a capacitance is then added at point 2 to move the mismatch to the center of the chart, which is a perfect match.

3.11 SMITH CHART SOFTWARE

For many years, Smith Chart matching was done with paper Smith Charts with detailed scales like those in Figure 3.16 and with compasses and rulers to locate and draw the necessary construction lines. Today software programs are available for Smith Chart matching, with varying degrees of complexity. Figures 3.26 and 3.27 show one such program (see ref. 2). The calculation shown is for the same mismatch calculated with the paper Smith Chart in Section 3.7.

The mismatch point can be located simply by dragging it with the mouse pointer to the reflection coefficient point or by typing it into the data table on the right-hand side of Figure 3.26. The boxes along the top of the page show matching elements that can be added. These are, from left to right,

Series inductance
Series capacitance
Series resistance
Length of transmission line (moving back from the mismatch toward the source)
Shunt inductance
Shunt capacitance
Shunt resistance
Shorted length of parallel transmission line
Open length of parallel transmission line

A schematic drawing of the matching circuit is shown in the lower right-hand side of Figures 3.26 and 3.27. The left-hand arrows show the reflection coefficient at the original mismatch, which is a reflection coefficient of $0.44/117^\circ$. Next, 21.78 mm of movement down the transmission line is shown. This is the fraction of a wavelength in a free space transmission line, and it must be corrected by a separate calculation to the actual physical length of the microstrip line, based on its construction. The movement down the transmission line converts the mismatch to the unity-resistance circle, where the impedance is 50 Ω resistive and 50 Ω inductive reactance. A capacitance of 3.23 pF, which has a capacitive reactance of 50 Ω, is now added in series, which cancels out the inductive reactance and gives a perfect 50 Ω resistive match.

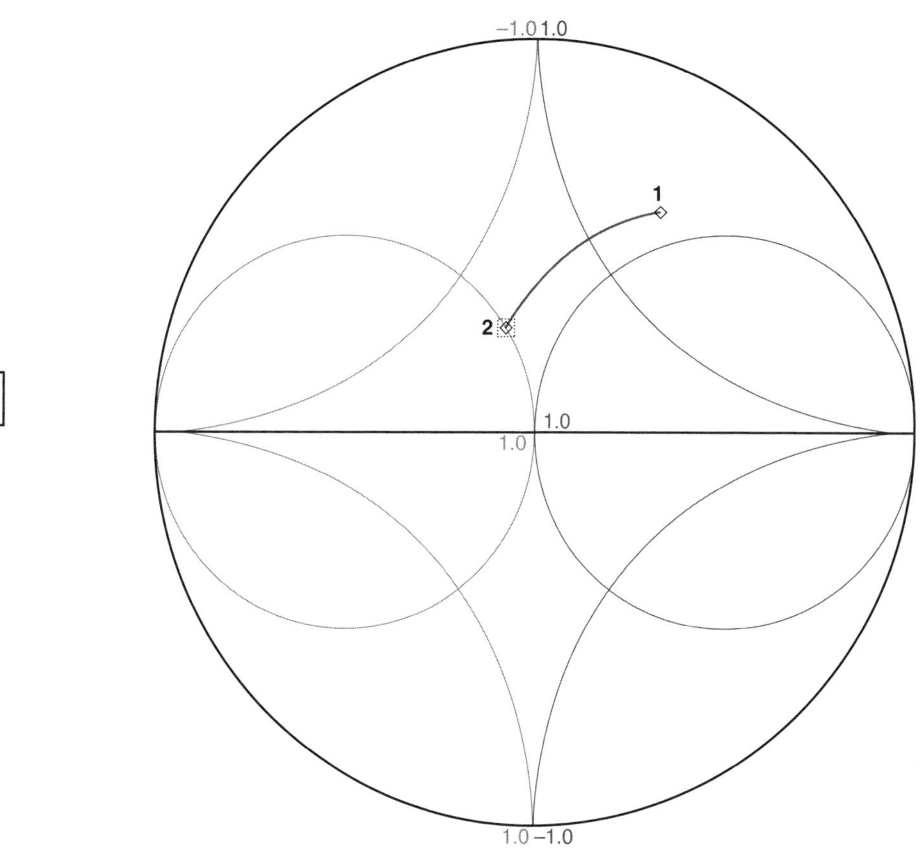

Figure 3.24 Using series capacitance to move to the unity conductance circle.

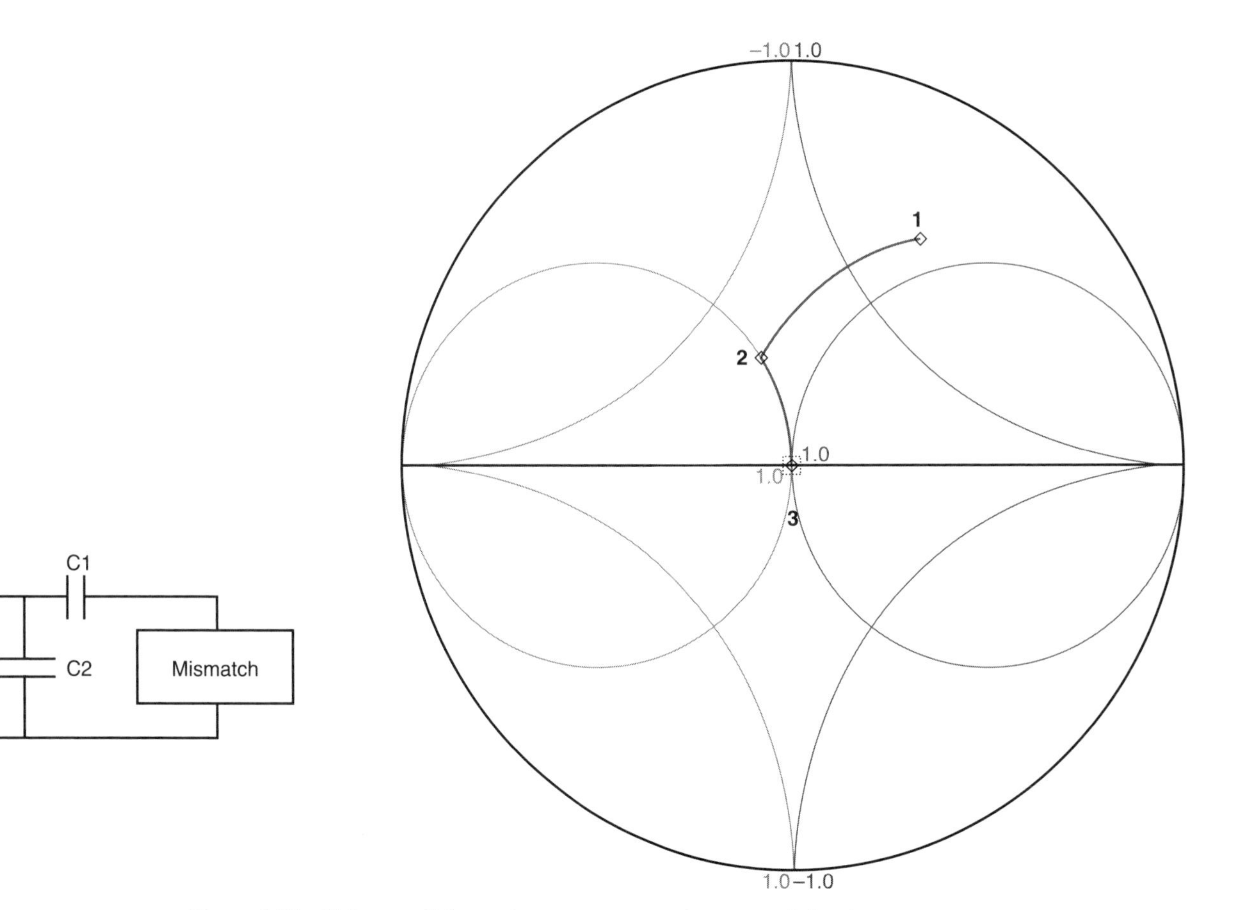

Figure 3.25 Using parallel capacitance to move to the center of the chart.

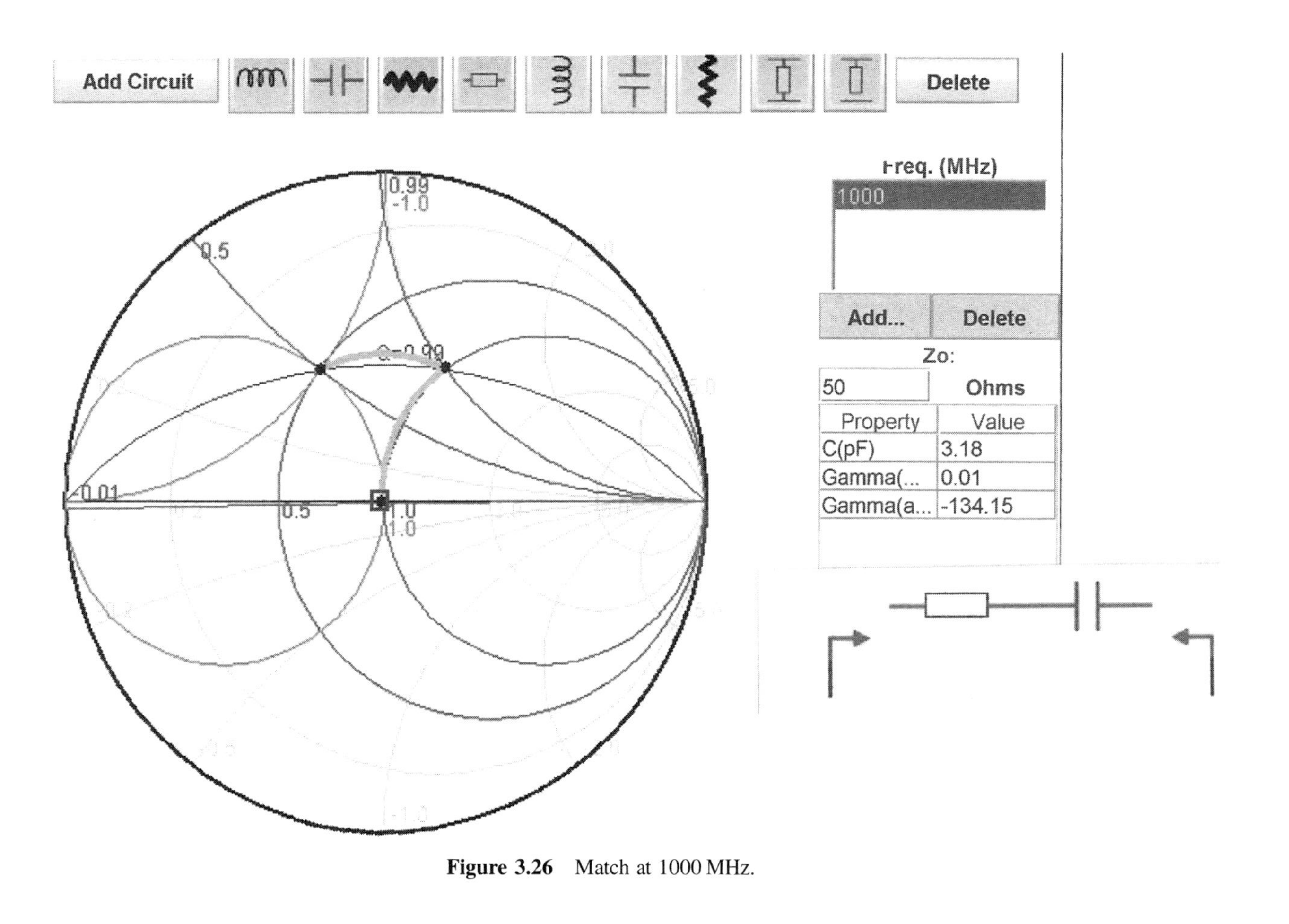

Figure 3.26 Match at 1000 MHz.

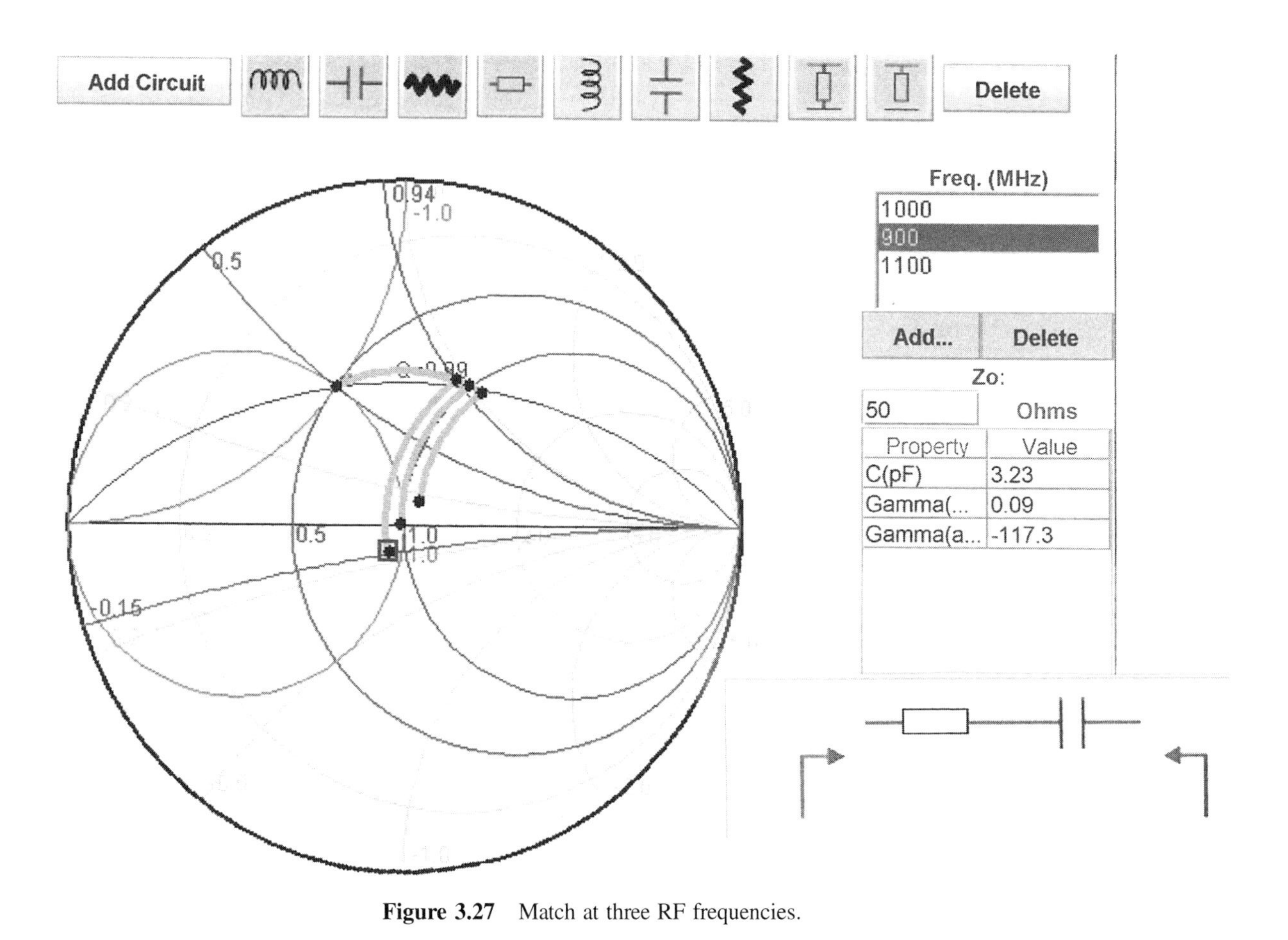

Figure 3.27 Match at three RF frequencies.

Figure 3.27 shows the final match at three frequencies (900, 1000, and 1100 MHz), which is a bandwidth of $\pm 10\%$. Any mismatch can be reduced to a perfect match at a single frequency. However, because the length of transmission line and the matching capacitor change their values with frequency, the match cannot be perfect over a range of frequencies.

The calculation of 900 MHz is highlighted in Figure 3.27. The resulting mismatch with the length and capacitance that gave a perfect match at 1000 MHz gives a reflection coefficient of 0.09 at 900 MHz. The goal of matching is to achieve a reflection coefficient of 0.10, which is a 20 dB return loss, and the illustrated match achieves this across the full 900–1100 MHz range.

3.12 ANNOTATED BIBLIOGRAPHY

1. Scott, A. W., Understanding Microwaves, Wiley–Interscience, New York, 1993.
2. Besser, L. and Gilmore, R., The Smith Chart and S-parameters, Practical Circuit Design for Modern Wireless Systems, Artech House, Norwood, MA, 2003.
3. Frobenius, R., Smith Chart Calculator, Besser Associates, http://www.bessernet.com, 2007.

Chapters 4 and 5 of reference 1 cover mismatches and matching with the Smith Chart in more detail than this book.

Reference 2 provides in-depth coverage of the development of the Smith Chart and S-parameters.

Reference 3 is a free Web-based Smith Chart matching program written by author Rex Frobenius.

CHAPTER 4

DIGITAL MODULATION

4.1 MODULATION PRINCIPLES

The purpose of an RF communications system is to transmit voice, video, or data signals wirelessly from one location to another. To achieve this, the transmitted RF signal must be modulated by the analog or digital data streams that represent the information. The original cellular telephone systems used analog FM, because digital modulation IC chips were not available. Currently, all wireless communications systems use digital modulation because of the improved performance it offers.

Figure 4.1 shows types of digital modulations on an RF wave. The RF wave is called the carrier, because it is carrying digital information by its modulation. The upper curve shows a bipolar digital data stream that is to be transmitted. In a wired communications system, this digital signal is simply transmitted as a voltage through wire, coaxial cable, or optical fiber. The data bitstream in this example is 110100.

The second curve in Figure 4.1 shows amplitude modulation of a wireless carrier. There are various types of amplitude modulation. The simplest type shown here is OOK. When the RF signal is on, the data is a 1; when the RF signal is off, the data is a 0. OOK is used in synchronous optical network (SONET) fiber optic networks. A timing reference is provided on each network node by an atomic clock. In contrast, all wireless communications systems are "asynchronous." It is not feasible in a wireless network to provide an atomic clock at each wireless receiver. In wireless communication, the timing reference is obtained from the digital bitstream

RF Measurements for Cellular Phones and Wireless Data Systems. By A. W. Scott and R. Frobenius

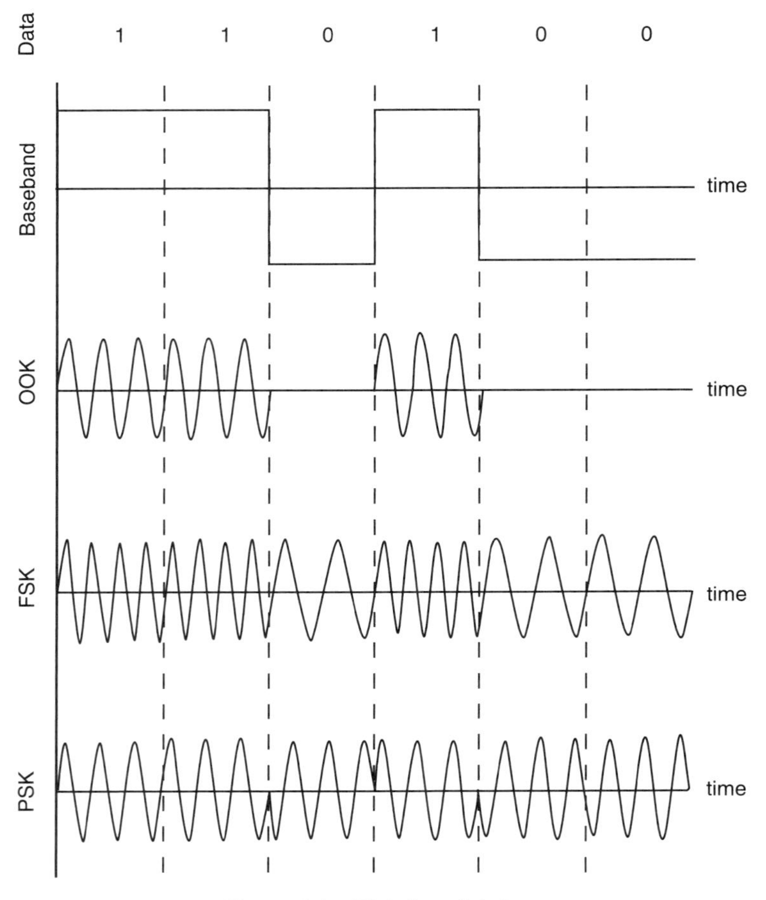

Figure 4.1 Digital modulation.

itself. It is not possible to use simple on–off amplitude keying, because if the transmitted signal contains a long string of zeros, the wireless system would lose its timing reference. A multilevel amplitude modulated technique would avoid this problem. However, in normal use, multilevel amplitude modulation is combined with multiple level phase modulation. The resulting modulation in which both amplitude and phase is varied simultaneously is called QAM, which is discussed below.

The third curve in Figure 4.1 shows digital frequency modulation, called FSK. The amplitude (or power level) of the FSK modulated wave is constant, but the frequency is changed to represent the digital information. When the frequency is low, the data is a 0. When the frequency is high, the data is a 1.

The fourth curve in Figure 4.1 shows phase modulation. The amplitude and the frequency of the wave are constant, but the phase is changed to represent information. When the phase is 0°, the data is a 0. When the phase is 180°, the data is a 1. Actually there is no way of telling from the RF wave whether the phase is 0° or 180°. The change can be detected, but not the absolute value. Therefore, a second phase reference signal must be transmitted along with the phase modulated wave.

Note that the amplitude of the RF wavelets is constant when phase modulation is used, but there is a variation of RF amplitude during the phase transition between data pulses. This amplitude change creates difficult design problems for the power amplifier, which amplifies the digitally modulated RF signal before transmission.

4.2 MULTILEVEL MODULATION

Each time the amplitude, frequency, or phase of the RF carrier is changed, approximately 1 Hz of bandwidth is used. Therefore, if the data rate is 1 Mbps, the required RF bandwidth to transmit the information is about 1 MHz.

To reduce the RF bandwidth requirements for transmission of a given data rate signal, multiple levels of amplitude, frequency, or phase modulation are used. Sometimes two of these parameters are changed simultaneously (i.e., amplitude and phase). The number of bits that can then be transmitted in a 1 Hz bandwidth is increased, and this increase is called the "spectral efficiency" of the modulation system. The information rate (bits per second) is the "bit rate." The rate at which the amplitude, frequency, or phase is changed is called the "symbol rate." The spectral efficiency is the bit rate divided by the symbol rate.

Constellation diagrams of various multiple level modulation systems are provided in Figure 4.2. The constellation diagrams show the phase of the RF signal in the angular direction and the amplitude of the signal in the radial direction.

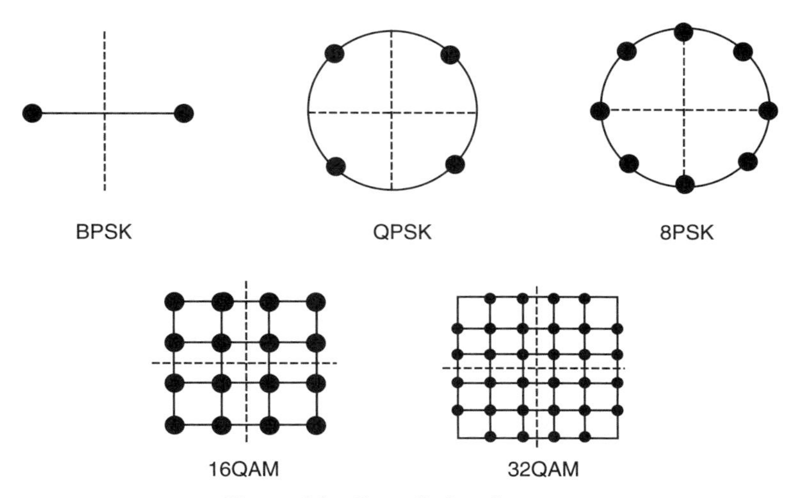

Figure 4.2 Constellation diagrams.

The upper left-hand drawing in Figure 4.2 shows the simplest phase modulation scheme, which is BPSK. One bit is transmitted for every 180° phase change of the carrier.

The upper center drawing in Figure 4.2 shows QPSK modulation with four phase positions: 45°, 135°, 225°, and 315°, which represent bits 00, 01, 10, and 11, respectively. The upper right-hand drawing shows 8PSK modulation, where the constellation points are 45° apart. With 8PSK, 3 bits are transmitted every time 1 Hz of bandwidth is used.

The lower two drawings in Figure 4.2 show constellation diagrams where the amplitude and phase are changed simultaneously. These modulation schemes are called QAM. In the lower left-hand drawing there are 16 possible phase and amplitude positions, so that 4 bits are transmitted every time 1 Hz of bandwidth is used. In the lower right-hand drawing there are 32 possible phase and amplitude positions, so that 5 bits are transmitted every time 1 Hz of bandwidth is used.

The more complex the constellation diagram is, the greater the spectral efficiency of the modulation system. However, the more complex the constellation diagram is, the greater the S/N of the received signal needs to be to achieve a given BER. For example, it is relatively easy to distinguish between the four different phase states of QPSK, where the phase states are 90° apart, but much harder to distinguish between the 64 different phase states of 64PSK, where the phase states are 5.625° apart.

Figure 4.3 shows the obtainable BER as a function of received S/N. Different wireless communications systems require different BERs as shown in the following table:

Information Being Transmitted	Required BER
Telephone voice	10^{-3}
Web browsing	$10^{-4}-10^{-6}$
Engineering data	
Entertainment video	
Corporate business data	$10^{-9}-10^{-12}$
Bank financial data	

Wireless data systems cannot easily provide BERs better than 10^{-6}. For transmission of better bit rates, fiber-optic systems are required.

The moderate BER requirements for cellular phones arises from two factors. First, voice phone conversations are self-correcting. If a listener does not understand a phone message, he asks for it to be repeated. Second, cellular phone uses FEC. Extra bits, about as many as are required for the voice signal itself, are transmitted to provide error correction at the receiver end. This error correction improves the BER from the 10^{-3} value of the received signal to the corrected value of 10^{-8}.

Table 4.1 compares the bits per Hertz and the required S/N for multilevel modulation at a 10^{-3} BER. The S/N required to achieve a given BER shown in Figure 4.3 represents the minimum S/N required for any given BER, and it is valid only under the following conditions:

1. The channel filter bandwidth is exactly equal to the RF signal bandwidth.
2. No multipath signal is present.

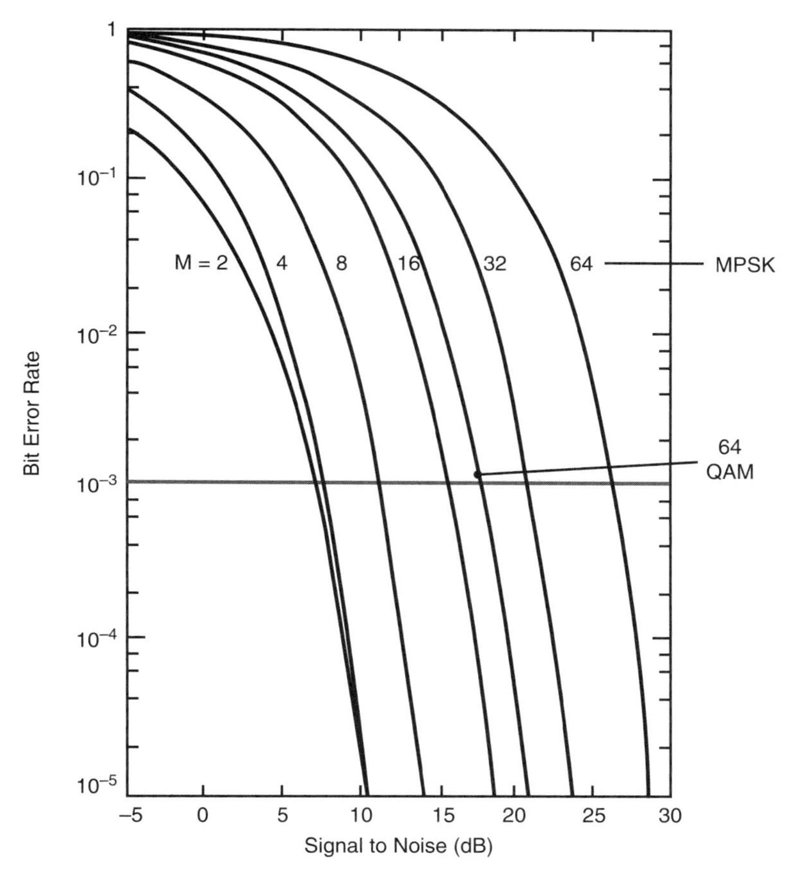

Figure 4.3 BER versus S/N for various digital modulations.

TABLE 4.1 Bits/Hertz and Required S/N for Multilevel Modulation for a 10^{-3} BER

Modulation	Bits/Hz	S/N
BPSK	1	7
QPSK	2	7
8PSK	3	11
16PSK	4	16
32PSK	5	21
64PSK	6	27
64QAM	6	17

Because of practical hardware considerations, the filter bandwidth cannot match the bandwidth of the signal exactly. Usually an increase of 2 dB is required to account for this fact.

As was discussed in the path loss calculation in Chapter 2, the effect of multipath can be reduced by using multiple receiving antennas. It can also be reduced in CDMA systems by using Rake receiver signal processing as discussed in Chapter 29. Either technique will bring the required S/N closer to the minimum values of Figure 4.3.

4.3 SPECIAL PHASE MODULATION TECHNIQUES

DPSK

A special form of phase modulation called DPSK eliminates the requirement of sending a separate unmodulated signal to provide a phase reference. This is illustrated using BPSK as an example in Figure 4.4.

The upper curve in Figure 4.4 shows standard BPSK that requires a phase reference. The data sequence being transmitted is 11001. The phase change occurs between every 1 to 0 or 0 to 1 transition, but without a separate phase reference signal there is no way of knowing what the phase change means.

The lower curve shows differential BPSK (DBPSK). In this case the phase shifts by 180° if the new symbol is a 1, but remains the same if the new symbol is a 0. The receiver tracks relative phase not absolute phase, so there is no need to transmit an additional phase reference signal.

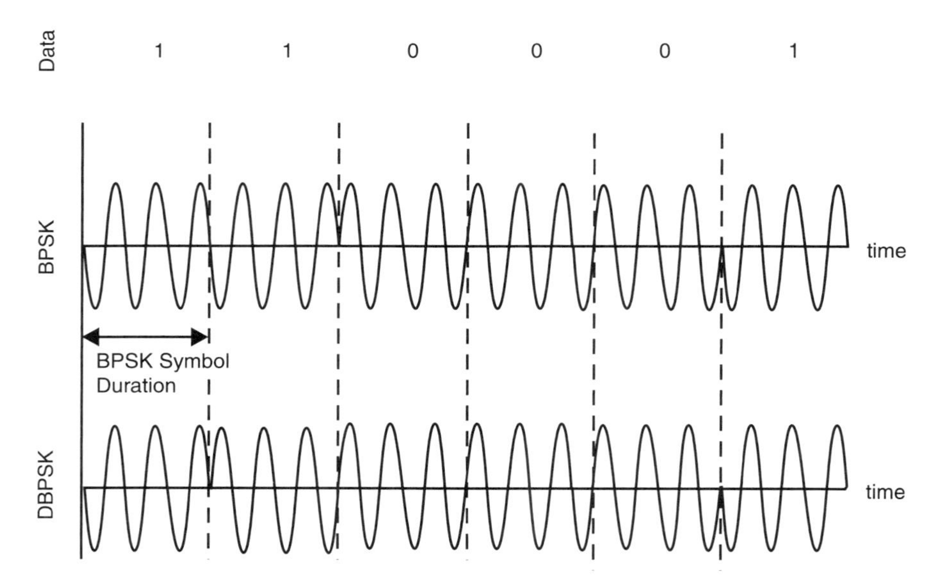

Figure 4.4 DPSK.

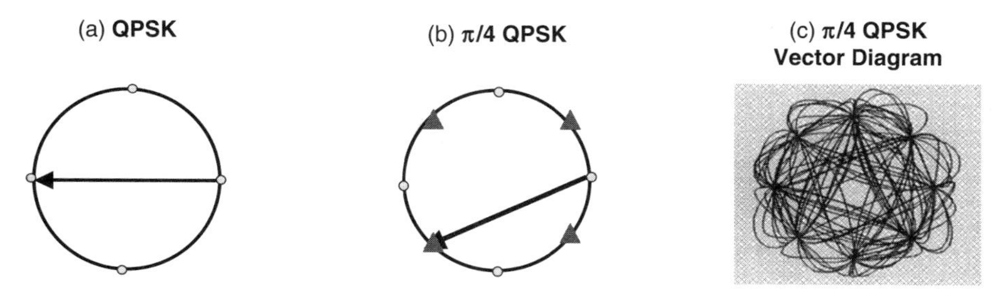

Figure 4.5 Modulation techniques to minimize amplifier backoff.

Figure 4.4 illustrates DBPSK because the waveforms are easy to draw. The same principle also applies to higher level DPSK, for example, DQPSK.

$\pi/4$QPSK

During the transition in phase from points that are directly across the constellation diagram from each other (e.g., $0^\circ - 180^\circ$, $90^\circ - 270^\circ$ for QPSK), the amplitude of the RF wave must pass through 0. Such a transition is shown in Figure 4.5a. If the system's power amplifier is being operated at saturation to obtain optimum battery efficiency, it is difficult to reproduce the amplitude changes that accompany the phase transitions without distortion. The only solution is to back the amplifier below its optimum power operating range into its linear range, which results in decreased efficiency.

To reduce this degradation of power amplifier performance a modified modulation scheme $\pi/4$QPSK is used, which is shown in Figure 4.5b. The constellation now consists of two QPSK constellations rotated by 90° ($\pi/4$) from each other. One constellation is shown by the circle points, the other by the triangle points. The QPSK data is divided into two bitstreams. Two bits are transmitted using the circular point constellation, and the next 2 bits are transmitted by the triangular point constellation. In this way, the amplitude of the RF wave, represented by the distance from the center of the constellation diagram, never passes through 0.

Figure 4.5c provides a "vector diagram" of a $\pi/4$QPSK modulation, showing the amplitude and phase variation of the modulated RF signal during all possible transitions. The complicated shape of the transition trajectories is caused by the filtering of the signal to reduce its bandwidth requirements. Note that the signal amplitude never passes through 0 amplitude during the transition from one phase symbol to the next.

The original digital cellular phone systems in both the United States and Japan used DQPSK to eliminate the requirement for a phase reference and used the $\pi/4$ shifted constellation to minimize power amplifier requirements.

$3/8\pi$ 8PSK Modulation for EDGE

Figure 4.6 shows the constellation diagram for the $3/8\pi$ 8PSK modulation used with GSM-EDGE cellular phones. Eighty percent of all cellular phone systems in the

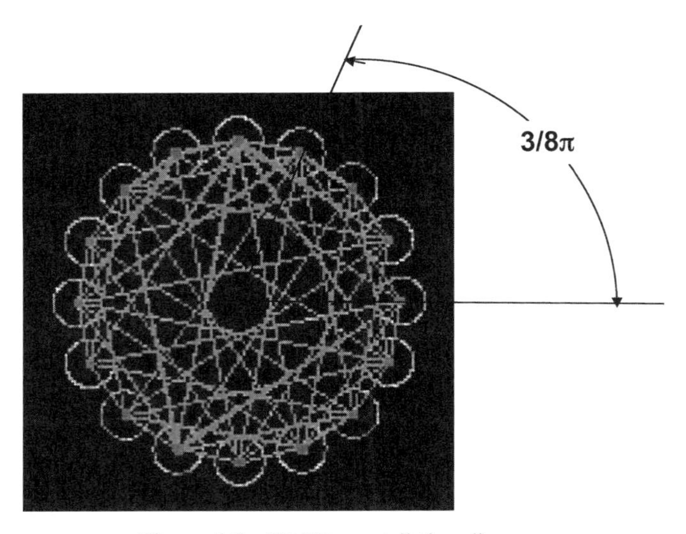

Figure 4.6 EDGE constellation diagram.

world are GSM. Originally, all of these systems used digital frequency modulation, which is described in Section 4.4. This modulation system required no amplifier backoff from saturation, so the power amplifiers were half as big and used half as much battery power as cellular phones with phase modulated signals. However, digital frequency modulation is much less spectrally efficient than the various types of QPSK modulation described earlier. The bandwidth capability of the GSM frequency modulated phones was adequate for voice phones, but not for combined voice and data phones. The original GSM phones using digital frequency modulation are currently being retrofitted with EDGE $3/8\pi$ 8PSK modulation.

The basic EDGE constellation is 8PSK, which was shown in Figure 4.2. It has a spectral efficiency of 3, meaning that every time the phase is changed, 3 data bits are sent. To prevent the problem of the power amplifier passing through 0, the entire 8PSK constellation is rotated at $3/8\pi$ (67.5°) from one symbol period to the next.

The $3/8\pi$ 8PSK modulation technique was retrofitted into the wide 34 kHz channels of the original GSM digital frequency modulated channels, so there was sufficient bandwidth to achieve sharp transitions between states, unlike the $\pi/4$QPSK transition of Figure 4.5c.

By using $3/8\pi$ 8PSK modulation, the EDGE cellular phone achieves three times higher bit rate in their existing 34 kHz wide frequency channels band than GSM does. However, these EDGE phones require approximately twice as much battery power to provide sufficient RF transmitter power when the power amplifiers are operated in the required linear range.

4.4 DIGITAL FREQUENCY MODULATION

Digital frequency modulation (FSK) has both advantages and disadvantages compared to digital phase modulated systems. The advantage of FSK is that it allows its power amplifiers to operate at saturation where their amplitude is distorted, because the amplitude carries no modulation information. The disadvantages of FSK are the following:

1. Because frequency modulation is used, it spreads the signal over a much wider bandwidth than PSK modulation does.
2. The use of multiple frequency levels is not effective in increasing the data rate.

The achievable data rate for different multiple frequency modulation levels with FSK is as follows:

Frequency Levels	Spectral Efficiency
2	0.80
4	1.14
8	1.10
16	0.84

Figure 4.7 shows the waveform and frequency levels for FSK. The upper curve shows the waveform. Notice that there is no change in the RF amplitude even during the transition times from one frequency level to another. The lower curve shows the frequency levels. Normally the high frequency represents a digital 1 and the low frequency represents a digital 0. The amount of frequency shift from the low frequency level to the high frequency level depends on the data rate of the modulating signal. The minimum allowable value that allows the carrying of modulation is one-half of the difference between the upper and lower modulating RF frequencies. This case is called "minimum shift keying" (MSK). GSM uses MSK. The data rate is 34 kHz, so that total swing between the lowest frequency and the highest frequency is 17 kHz. In other words, the modulated signal swings 8.5 kHz above and below the carrier.

If the modulation makes an abrupt transition from maximum to minimum frequency, the bandwidth of the modulating signal would be very large. Therefore, the digital bitstream is filtered so that the transitions from low to high modulating

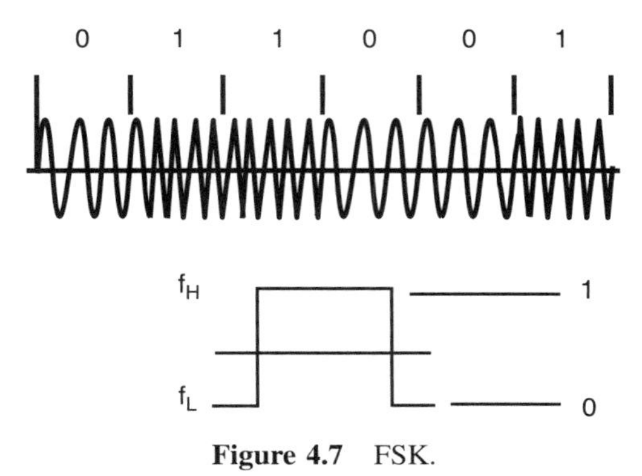

Figure 4.7 FSK.

frequencies are not as abrupt as the ones shown in Figure 4.7. For GSM, a Gaussian filter is used, so the name of the GSM modulation is Gaussian MSK (GMSK).

4.5 UPCONVERSION REQUIREMENTS

The complicated modulations that are used in cellular phones and wireless data transmission systems are difficult to generate at RF frequencies. Thus, in most RF communication systems, the modulated signals are generated at low frequency with digital processing chips that do not work at RF and then upshifted to the desired RF frequency in an upconverter. RF upconverters are discussed in Chapter 18.

4.6 ANNOTATED BIBLIOGRAPHY

1. Burns, L., Digital modulation and demodulation, In Larson, L., Ed., RF and Microwave Circuit Design for Wireless Communications, Chap. 4, Artech House, Boston, 1996.
2. Rappaport, T. S., Modulation techniques for mobile radio, In Wireless Communications Principles and Practice, Chap. 5, Prentice Hall, Upper Saddle River, NJ, 1996.

Reference 1 explains modulation types, performance, and circuits in a thorough and easy to understand way.

Reference 2 provides a detailed theoretical explanation of modulation types.

PART II

RF MEASUREMENT EQUIPMENT

CHAPTER 5

RF SIGNAL GENERATORS

5.1 WHAT AN RF SIGNAL GENERATOR DOES

To test any RF device or system, an RF signal is required, which is provided by an RF signal generator. Figure 5.1 shows a typical RF signal generator for testing cellular phones and wireless data transmission equipment. The characteristics of the RF signal can be controlled by the user. The RF signal generator provides a single RF signal whose characteristics are selected by the user and remain constant until the user changes them. Typical characteristics that can be adjusted are as follows:

Frequency
Power
Type of AM or FM analog modulation
Type of digital modulation

Digital modulation types can be specified for particular cellular phone or data systems. New modulation techniques can be programmed into the signal generator.

The RF signal generator can also be adjusted to supply a fixed set of multiple signals. BER testing on systems can also be done by the RF signal generator. Because the RF signal generator can apply phase modulation to the RF signal, it is often called a "vector signal generator."

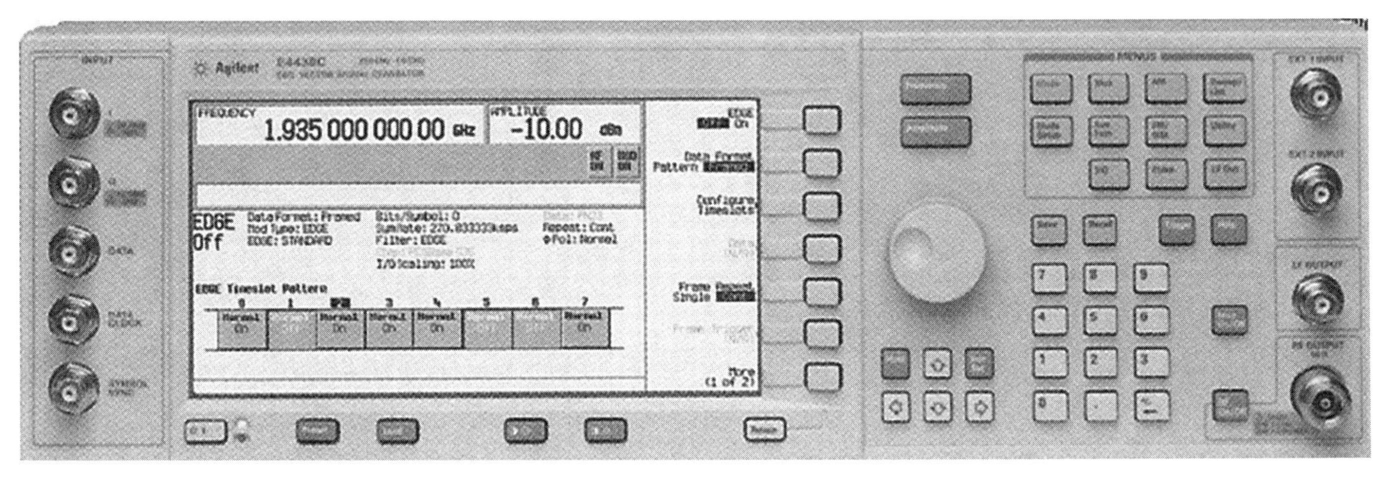

Figure 5.1 RF signal generator. Agilent Technologies © 2008. Used with permission.

5.2 SUPPORTED WIRELESS COMMUNICATION FORMATS

The RF signal generator shown in Figure 5.1 provides modulated RF signals for the following digital wireless communication system formats:

GSM, GPRS, EDGE, WCDMA, HSDPA
cdmaOne, cdma2000, EV-DO
NADC
Bluetooth
802.11a,b,g
WiMAX

The generator can be supplied with all or only some of these formats as required. New modulation formats can be installed as they are developed. Special software is also available for generating multitones for intermodulation testing (as discussed in Chap. 26) and for BER testing (as described in Chap. 35).

5.3 RF SIGNAL GENERATOR DISPLAYS

The characteristics of the signal generated by the RF signal generator are displayed on the LCD screen on the left-hand side of the front panel, as shown in Figure 5.1. The frequency and amplitude (power level) of the generated RF signal are shown at the top of the display. The frequency is shown by 12 figures, and all of these are significant. The RF signal is synthesized by adding together the harmonics of quartz crystal oscillators in the kilohertz, megahertz, and gigahertz ranges.

The power reading is less accurate, and it is obtained from a built-in power meter. It is accurate to only a few tenths of a decibel (as described in Chap. 13). The rest of the LCD displays the modulation characteristics of the generated signal. In the illustrated case the modulation is 8PSK, which is used in the EDGE cellular phone system.

5.4 RF SIGNAL GENERATOR CONTROLS

Two important controls are located on the left-hand side of the RF signal generator, just under the display. The main on–off power key is at the far right. The "green key" next to it, marked Pre-set, adjusts the signal generator to a standard set of operating conditions. Almost all RF test equipment has such a preset key. It is important to always preset the instrument before beginning a new set of measurements. Otherwise, various unwanted measurement settings from the previous measurements will remain active.

Most of the rest of the controls are located on the right-hand side of the front panel, which is shown in Figure 5.2. These are divided into hard keys and soft keys. There

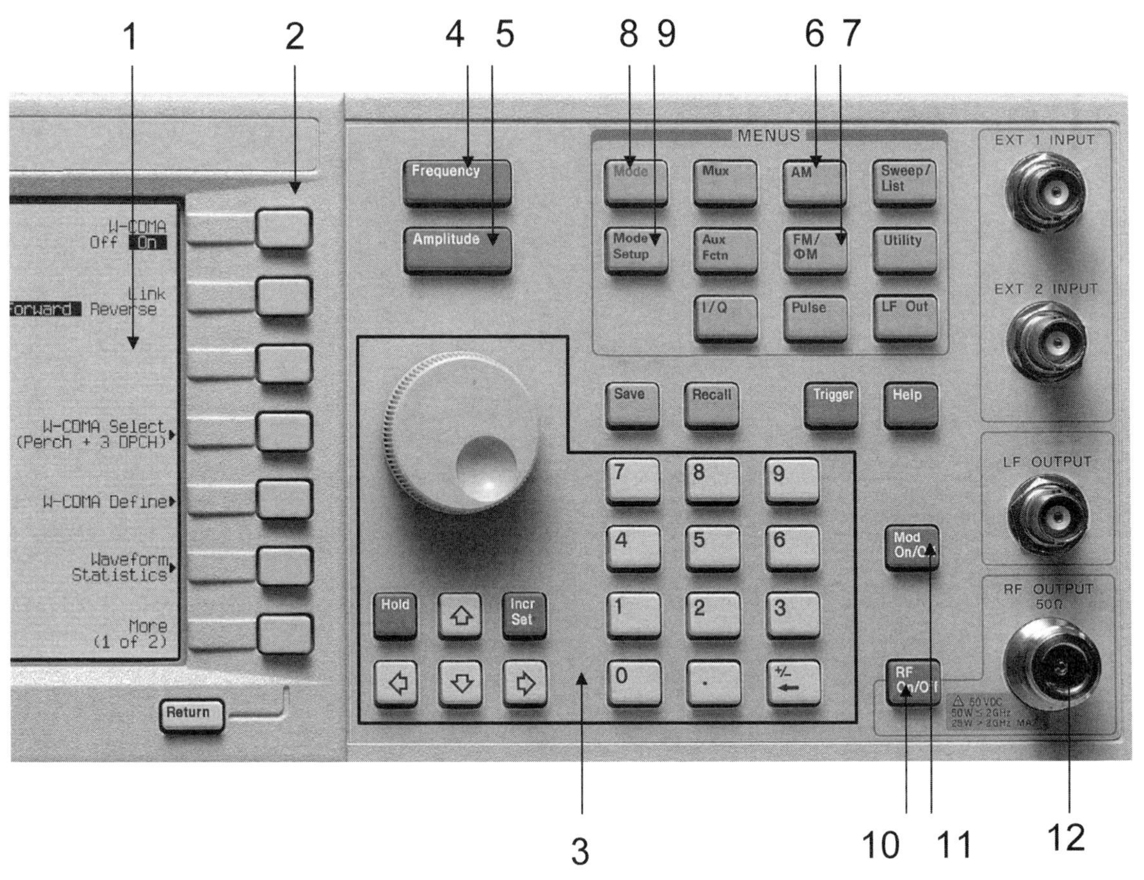

Figure 5.2 RF signal generator controls.

are several hundred keystrokes available to adjust the characteristics of the test signal. There would be no practical way of providing hard keys for each of these adjustments and no easy way of finding a particular key. For this reason, each of the hard keys (except those in the numeric keypad) generates functions that are displayed on the right-hand side of the LCD, each next to a unmarked soft key (2). The soft keys control different functions, depending on the setting of the hard keys. For example, to set the RF frequency of the generated signal, the frequency key (4) is pressed. The numerical value of the frequency is entered using the numeric keypad (3). The soft key display (1) changes to gigahertz, megahertz, or kilohertz next to the top three soft keys, so the correct units of frequency can be selected.

In a similar way, when the amplitude soft key (5) is pressed the soft key label displays a selection of the desired power units such as milliwatts or decibels relative to 1 mW.

The AM hard key (6) sets up a soft key menu for setting the modulation characteristics of an analog amplitude modulated RF signal. The FM/ϕM hard key (7) sets up a soft key menu for analog FM or phase modulated signals. The soft key menu allows the FM rate and the FM deviation as well as other FM modulation parameters to be selected.

The Mode key (8) and Mode Setup key (9) set up various types of digital modulations based on particular digital communication systems, such as GSM, EDGE, cdma2000, EV-DO, WCDMA, 802.11, and WiMAX.

The RF test signal is supplied for testing through the RF output port (12). The RF On/Off (10) hard key allows the RF signal to be disconnected from the RF output port to permit adjustments to be made on the DUT without the need to reset the RF signal generator for further testing. The Mod On/Off (11) hard key allows the modulation to be toggled on and off to observe the effect of modulation on the RF component being tested.

5.5 MODULATION ARCHITECTURES

Two architectures are typically available for producing digital modulation formats inside the signal generator. One path is to use the same type of "IQ modulator" hardware as would be seen in an actual radio transmitter to modulate the bits onto the carrier. The advantage of this approach is that the signal generator can accept external data as an input and modulate it onto the signal in real time for testing purposes. The disadvantage of this approach is that the signal formats that can be output are limited by the hardware that is installed in the machine.

Another approach is to use an "arbitrary waveform generator" to calculate the shape of the waveform for a given series of input data bits, then synthesize that waveform and output it at the desired frequency. The benefit of this approach is that virtually any waveform can be created to correspond to any modulation format. In addition, signal impairments can be recreated and included in the signal to see how devices perform when the input signal is distorted. The disadvantage of this approach is that the ability to input digital data to the signal in real time is lost.

5.6 PHASE NOISE OF THE RF SIGNAL GENERATOR

The important characteristic of the phase noise of the RF signal generator is discussed in detail in Chapter 17.

5.7 ANNOTATED BIBLIOGRAPHY

1. Agilent Vector Signal Generator Brochure, Product Note 5988-3935EN, Agilent Technologies, Santa Clara, CA.
2. Generating Digital Modulation with the Agilent ESG-D Series Dual Arbitrary Waveform Generator, Agilent Product Note 5966-4097E, Agilent Technologies, Santa Clara, CA.

Reference 1 provides detailed performance specifications of a typical RF signal generator.

Reference 2 provides additional background information about the arbitrary waveform generator and IQ modulator functions of the signal generator.

CHAPTER 6

RF POWER METERS

6.1 RF POWER METER BASICS

Power meters provide the most accurate measurement of RF power of any of the types of RF measurement equipment. Power meters can be stand-alone instruments, such as the power meter shown in Figure 6.1, or they can be built into other instruments like signal generators, frequency counters, and spectrum analyzers.

RF power meters provide absolutely *no* information about the frequency distribution of the RF power. The indicated RF power is the total power incident at the power meter. If the signal is a single frequency, the power meter displays its power. However, if multiple signals are present at different frequencies, the power meter displays the total RF power of all the signals together.

RF power meters consist of two parts:

1. sensors, which convert the RF power to an electrical voltage that is proportional to the RF power, and
2. a power meter unit, which analyzes the electrical voltage, calculates the RF power it represents, and displays the calculated power.

Different sensors can be used with a given power meter, depending on the type of power measurement that needs to be made.

RF Measurements for Cellular Phones and Wireless Data Systems. By A. W. Scott and R. Frobenius

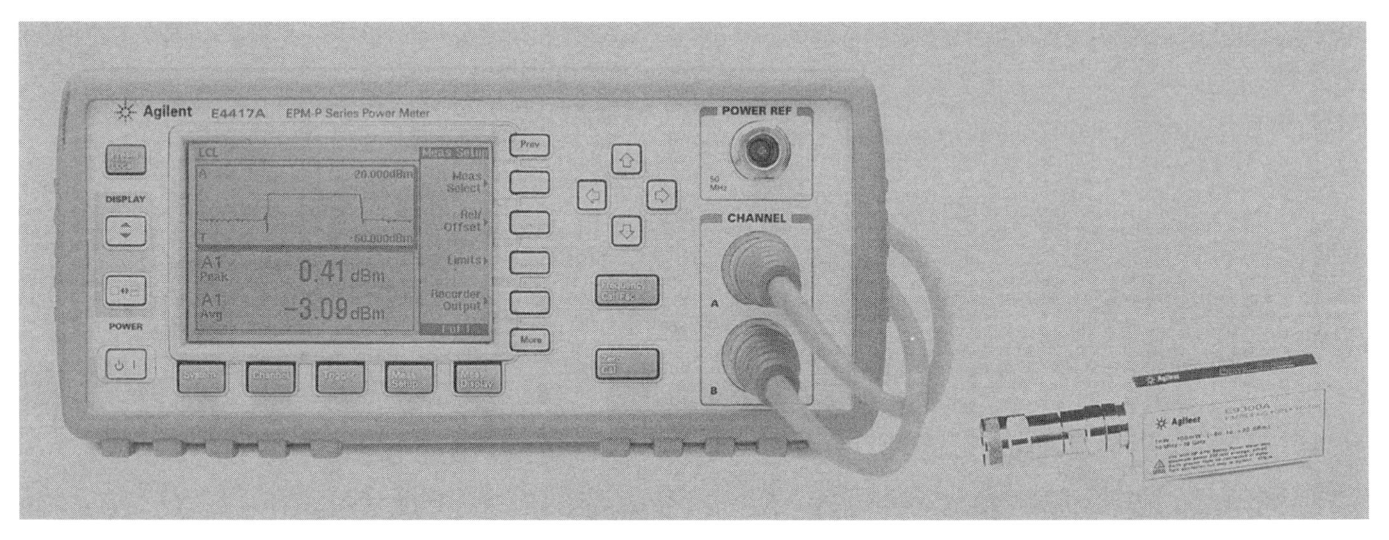

Figure 6.1 Power meter. Agilent Technologies © 2008. Used with permission.

6.2 POWER METER SENSORS

The most commonly used sensor for power measurement of wireless telephone and high-speed data communications equipment is the Schottky diode. Figure 6.2 shows what a Schottky diode is and what its electrical and RF characteristics are. Figure 6.2a shows that a Schottky diode is similar to the PN junction of semiconductor technology, except that the P region is replaced by a conductor. Figure 6.2b shows current as a function of applied voltage. Like a PN junction, the Schottky junction is a rectifier, but with proper doping of the N region, conduction begins as soon as the applied voltage becomes positive. This is in contrast to a PN junction where the voltage must be about 0.6 V for conduction to start. The rectifying action converts the RF signal into a series of positive halves of the RF wave, which when applied across a capacitor gives a DC voltage that is proportional to the applied RF power.

Figure 6.2c shows a typical Schottky diode performance curve. The input RF power is shown horizontally starting at the noise level of −70 dBm and increasing to +20 dBm. The detector output voltage is shown vertically. Initially, with small input RF power, the detector output voltage is linearly proportional to the RF input power. This is called the "square law" region of operation. At about −20 dBm of input RF power, the diode's transfer characteristic changes from a linear function.

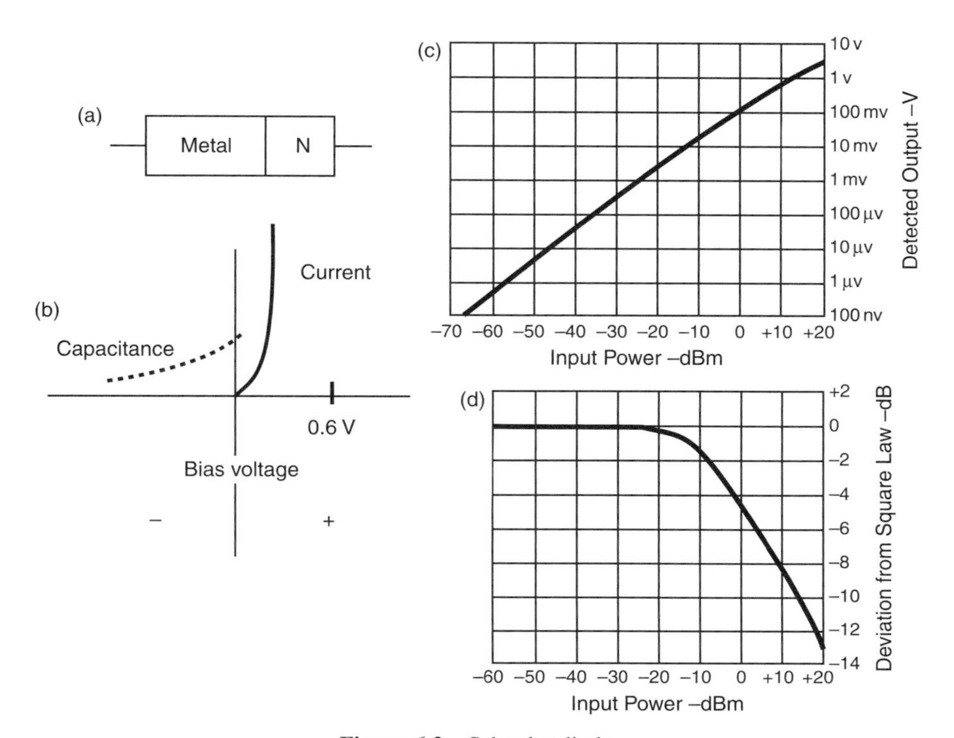

Figure 6.2 Schottky diode.

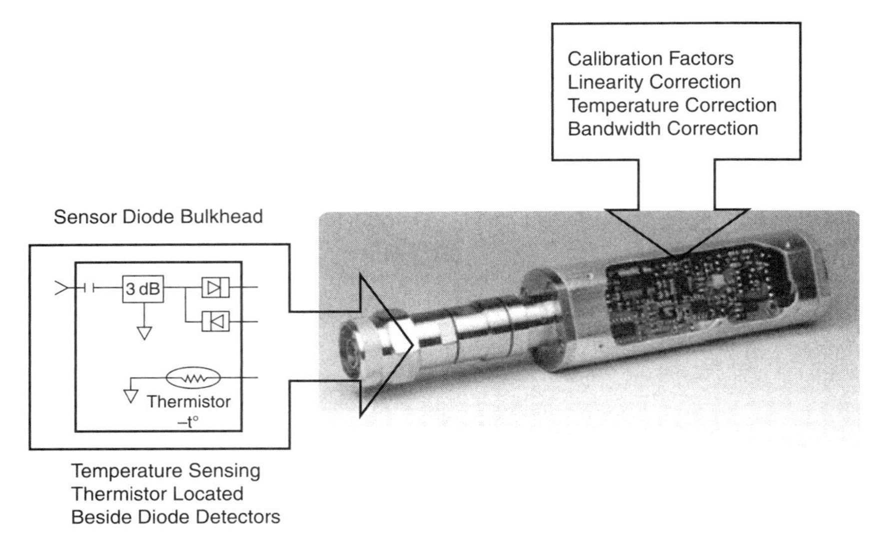

Figure 6.3 Internal details of Schottky diode sensor. Agilent Technologies © 2008. Used with permission.

Figure 6.2d shows the deviation from square law performance of the Schottky diode as a function of RF input power.

At an input RF power above +20 dBm, the diode will be damaged, and at −70 dBm the rectified RF signal will be lost in the noise of the Schottky diode. Therefore, the range of power measurements is 90 dB from −70 to +20 dBm. Most RF power meters using Schottky diode sensors switch a precision fixed resistor into the measurement path at high input power levels to attenuate the applied signal level to less than −20 dBm to keep the diode in its linear range. With this approach the Schottky diode sensor can measure power levels up to +40 dBm. For higher power levels, precision external attenuators can be added before the sensor.

Each individual Schottky diode has a slightly different sensitivity because of manufacturing tolerances. The Schottky diode sensitivity also varies with power level; that is, it is not quite linear. Its sensitivity also varies with temperature and with the bandwidth of the applied RF signal. All of these variations are measured during fabrication of each individual sensor and are recorded in an EEPROM that is mounted in the sensor assembly, as shown in Figure 6.3. Note the thermistor mounted near the Schottky diode to measure its temperature. Correction signals are all sent to the power meter to make the necessary corrections to the displayed signal.

6.3 A SCHOTTKY DIODE FOR POWER MEASUREMENTS IN CELLULAR PHONE SYSTEMS

The particular RF power meter shown in Figure 6.1 was designed to measure modulated RF signals for TDMA and CDMA cellular phones (which are described in

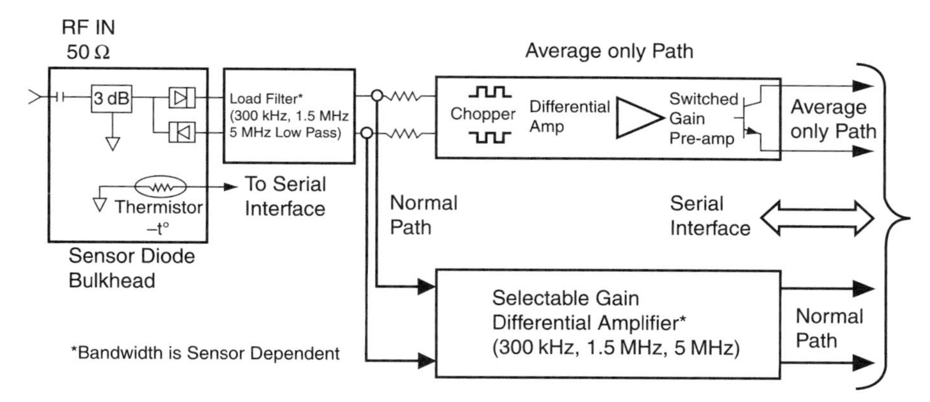

Figure 6.4 Block diagram of dual sensors in a single package.

Chap. 30). A block diagram of its sensor is shown in Figure 6.4. It contains two sensors in one package. The RF signal that enters the sensor diode bulkhead is converted to an electrical signal in the diode and then passes through the load filter that controls the low pass bandwidth of the signal. The 300 kHz setting is for TDMA signals such as GSM. The 1.5 MHz setting is for narrowband CDMA signals. The 5 MHz setting is for wideband CDMA signals.

Two measurement paths are provided in the sensor head. The "normal path" is for high-speed sampled measurements that can be time gated to show amplitude variation as a function of time. This is the default path.

The "average only" path measures the total RF power. This is the most accurate measurement of total power because none of the power is filtered out by the time gating. Power can be measured from -65 to $+20$ dBm in this mode.

6.4 THE POWER METER UNIT

The power meter unit shown in Figure 6.1 has two input ports. Models are also available with only one input port. The multiple signals from the sensor(s) are brought to the power meter unit by special shielded cables.

The advantage of having two ports is that two independent signals can be measured simultaneously. This is of particular value, for example, in amplifier testing. One port can measure the input power and the other port can measure the output power. The power meter can compute and display the ratio of the output power to the input power, which is the gain of the amplifier. Alternatively, as shown in the display of the power meter in Figure 6.1, the peak output power can be measured and displayed from a one port and the average power can be measured and displayed from the other port.

The power meter contains a 50 MHz, 1.00 mW reference source that is traceable to NIST standards. This reference is used to calibrate the power meter and its sensors. Prior to taking a measurement, the sensor (or sensors in sequence) is connected to the power reference output, and the instrument automatically calibrates itself. Another step that is taken prior to taking a measurement with the power meter is to "zero" the sensor. In this step, all signal power is shut off and any offsets in the sensor from temperature variations are corrected. This step can be important when trying to measure a very weak signal immediately after measuring a high power signal. The energy from the high power signal may heat up the sensor, introducing slight DC offsets to the final reading.

Power meters are not limited to individual power measurements; they can also show the variation of power with time. Figure 6.5 provides measurements of a pulsed signal showing the power level and time at the leading edge and at the trailing edge of a pulse. Another measurement that can be made is the peak power of the signal. It should be noted that when trying to measure peak power (as opposed to the average power), the response time of the sensor must be taken into consideration. This is often referred to as the "video bandwidth" of the sensor. This parameter indicates the widest bandwidth over which peak power can be measured accurately. For example, if two signals are separated by 10 MHz, then the amplitude envelope of the sum of the two signals will vary sinusoidally at a rate of 10 MHz. If you try to measure the peak power of this setup with a sensor that has a 5 MHz video bandwidth, the peak value will be wrong because the sensor can only accommodate variations in the amplitude that occur at a rate up to 5 MHz. The peak value will be too low and the minimum value will be too high, because the sensor is not tracking the amplitude envelope as quickly as it is changing. If you measure the average power of this amplitude envelope, the value will still be correct, because both the peak value error and minimum value errors will cancel each other out. In general, only peak or

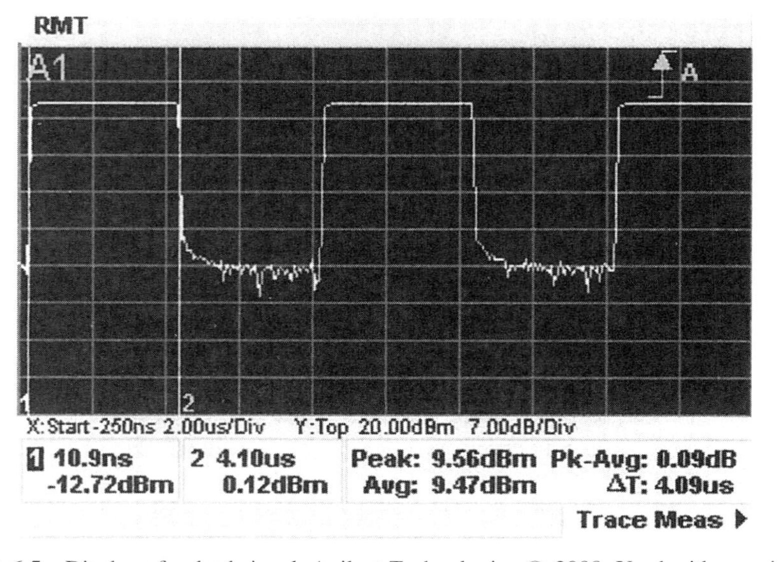

Figure 6.5 Display of pulsed signal. Agilent Technologies © 2008. Used with permission.

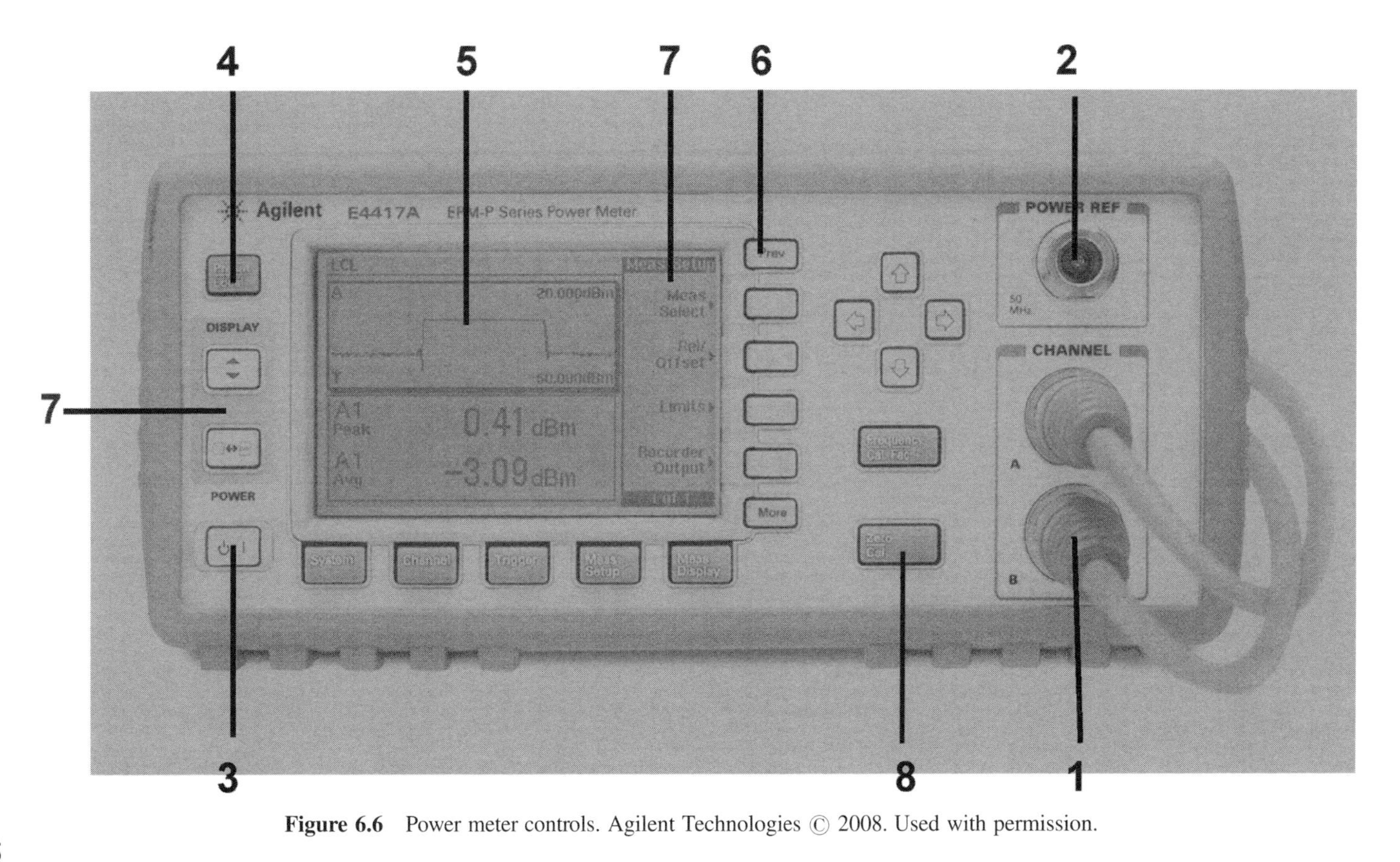

Figure 6.6 Power meter controls. Agilent Technologies © 2008. Used with permission.

minimum value power measurements are affected by the video bandwidth limitation of the sensor.

6.5 POWER METER CONTROLS

Figure 6.6 shows the most important power meter connections and controls:

1. Ports for connecting the power sensors to the power meter
2. The power reference port for calibrating the sensors
3. On–off prime power switch
4. Preset key
5. LCD display
6. Soft keys for menu selection
7. Display setting keys
8. Zero/cal activation keys

6.6 ANNOTATED BIBLIOGRAPHY

1. Product Note: Choosing the Right Power Meter and Sensor, Document 5968-7150E, Agilent Technologies, Santa Clara, CA.
2. Application Note 64-1: Fundamentals of RF and Microwave Power Measurements, Document 5965-6630E, Agilent Technologies, Santa Clara, CA.
3. Application Note 64-4: 4 Steps for Better Power Measurements, Document 5965-8167E, Agilent Technologies, Santa Clara, CA.
4. Power Measurement Basics, Agilent Technologies eSeminar Presentation, Agilent Technologies, Santa Clara, CA.

References 1–3 cover various topics on selection and use of RF power meters.

Reference 4 provides background on measurement uncertainties and sensor bandwidth limitations.

CHAPTER 7

FREQUENCY COUNTERS

7.1 FREQUENCY COUNTER OPERATION

RF frequency counters measure the frequency of a single RF signal. At RF frequencies up to about 500 MHz, frequency meters simply count the cycles of the single frequency RF wave with a digital counter. Accuracy can be as good as 1 part in 1 million. Digital counting circuits do not work above about 500 MHz, so for counting higher RF frequencies, some type of downconversion is used.

One type of downconversion is "prescaling." Prescaling involves simple division of the input frequency by an integer N to reduce the frequency to a value that can be counted by a digital counter. Typically, N ranges from 2 to 16. The counted prescaled value is then multiplied in a signal processing circuit by the integer N and displayed. This technique allows counting to about 1.5 GHz.

For counting to higher RF frequencies, a heterodyne converter is used. The counter contains a signal generator, a mixer, and a lower frequency digital counter. The RF signal to be counted is mixed down to a lower frequency that can be counted, and the displayed signal is the sum of the frequency of the lower frequency signal generator and the difference frequency of the mixer. Accuracy is determined by the frequency accuracy of the internal RF signal generator.

Figure 7.1 shows a frequency counter that can count up to 12 GHz, using the mixing process described in the previous paragraph. The counter has two input channels. Channel 1 connects directly to a digital counter, and it can measure

---

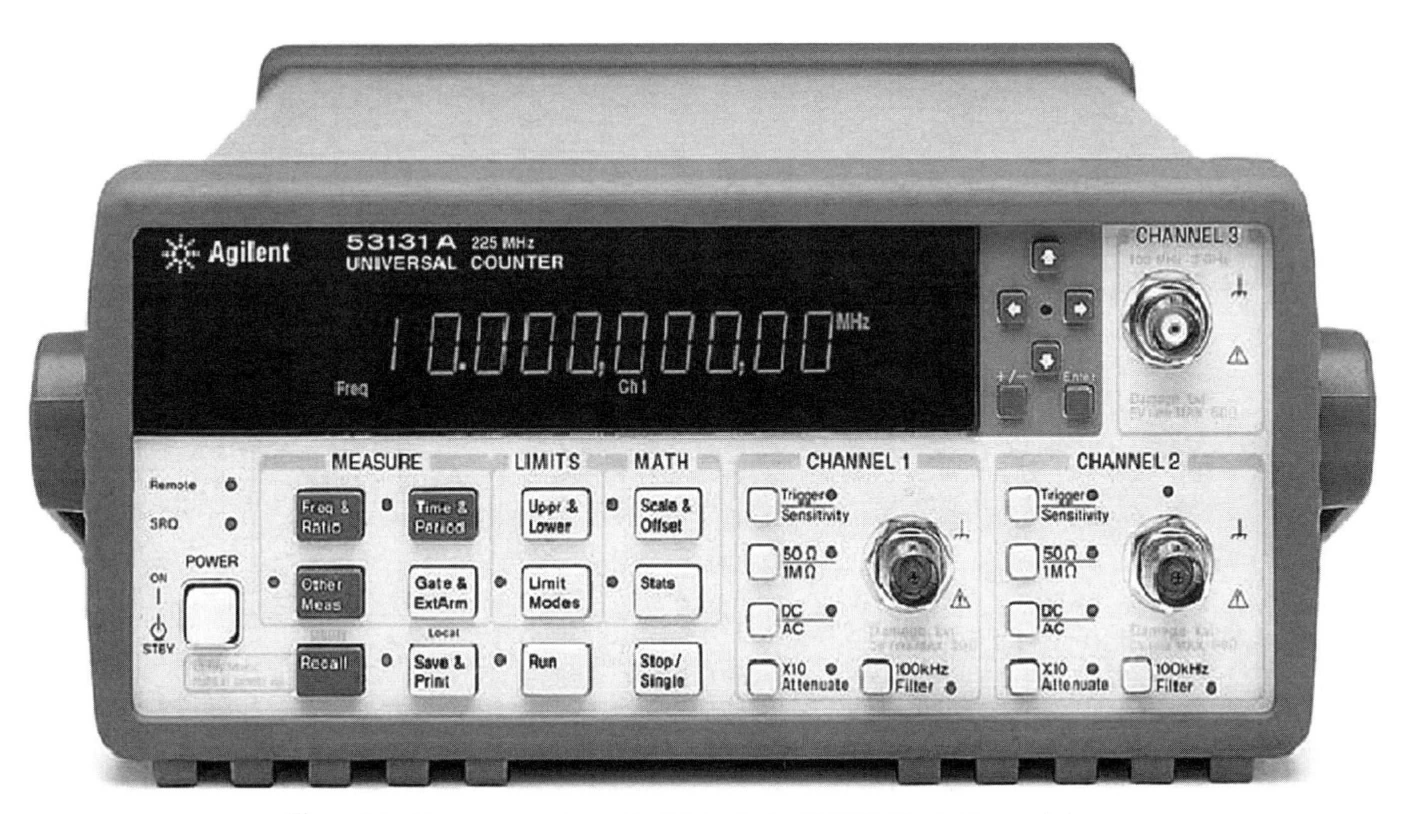

Figure 7.1 Frequency counter. Agilent Technologies © 2008. Used with permission.

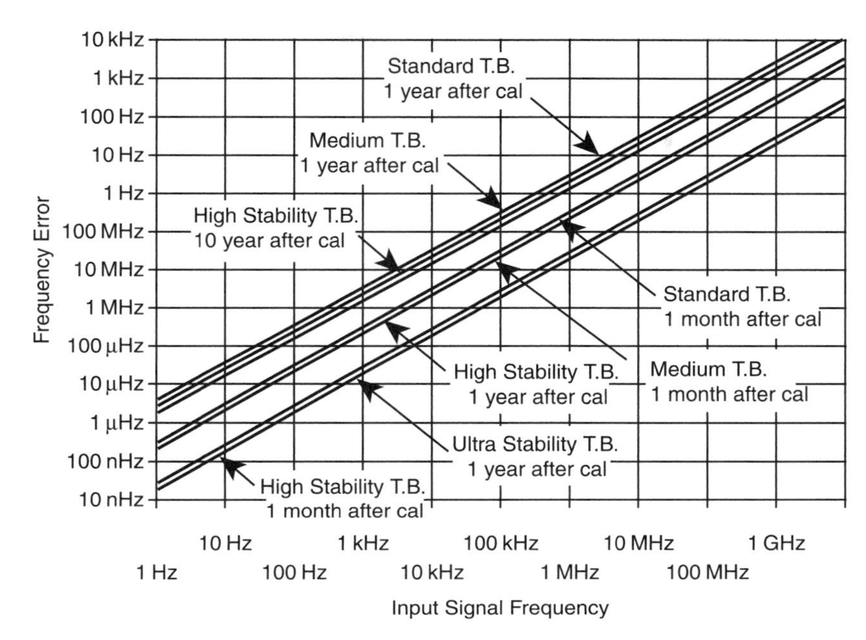

Figure 7.2 Frequency counter error.

frequency up to 225 MHz. Channel 2 connects to a heterodyne converter that can count up to 12 GHz.

The accuracy of an RF counter is limited by the stability of its time base. The frequency counter shown in Figure 7.1 can be procured with one of several different time bases. Figure 7.2 shows the frequency accuracy that can be obtained with various time bases and time after calibration. If multiple signals are present, the counter will be able to count the frequency of the highest power signal if it is 10 dB greater than any other signals.

It should be noted that the particular frequency counter illustrated cannot accurately measure the frequency of the signal generator shown in Figure 5.1, because the frequency accuracy of the signal generator is better than that of the internal frequency source in the frequency counter.

7.2 ANNOTATED BIBLIOGRAPHY

1. Application Note 200-1: Fundamentals of Microwave Frequency Counters, Agilent Technologies, Santa Clara, CA.

A free copy of Reference 1 can be obtained from http://www.agilent.com.

CHAPTER 8

VNAs

8.1 WHAT A VNA DOES

An RF VNA measures the response of RF devices or networks (which is a group of devices) as a function of the frequency of an applied continuous, nonmodulated, RF signal. The VNA measures the response of the network one frequency at a time, but it varies the measurement frequency over a user adjusted RF bandwidth very rapidly, making hundreds of measurements in 1 s. The term vector designates the fact that the VNA measures both the amplitude and *phase* of the RF signal. A VNA is shown in Figure 8.1.

8.2 WHAT A VNA CAN MEASURE

The VNA measures the incident test signal, the reflected signal, and the transmitted signal from the RF device and then automatically reverses the connections to measure the same quantities looking into the device from the opposite direction. The VNA can display these measure quantities as a function of frequency. However, it usually processes the information first to display derived quantities such as return loss, insertion loss, S-parameters in amplitude and phase, Smith charts, group delay, and other performance characteristics.

The frequency range over which the VNA sweeps can be adjusted by the user. Alternately, the frequency can be fixed at a constant value and the power level can

RF Measurements for Cellular Phones and Wireless Data Systems. By A. W. Scott and R. Frobenius

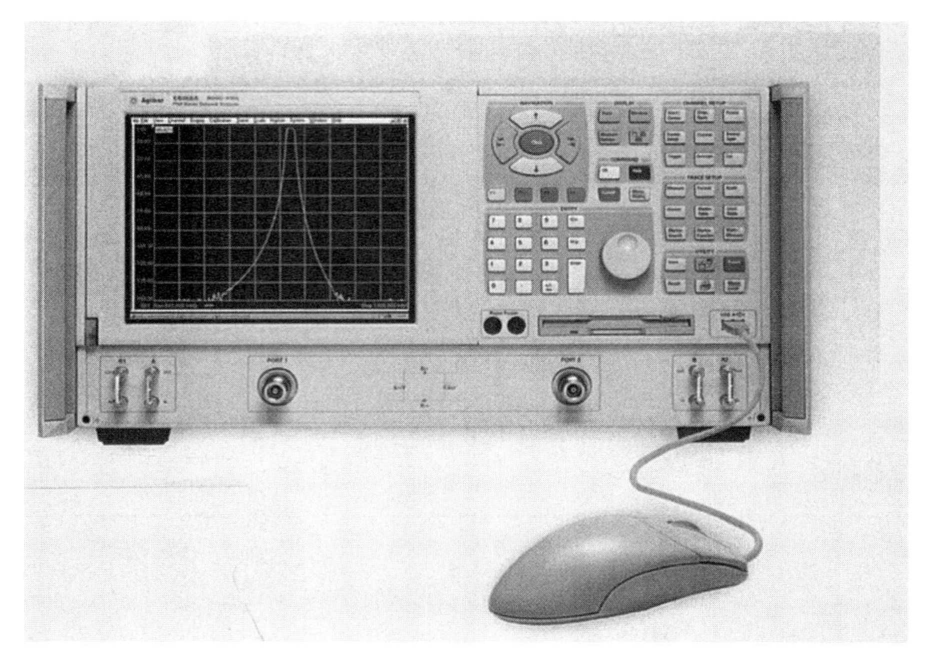

Figure 8.1 VNA. Agilent Technologies © 2008. Used with permission.

be swept so that the display shows the desired performance as a function of power at a fixed frequency.

8.3 VNA CONTROLS

Figure 8.2 shows a larger photograph of the controls of the VNA shown in Figure 8.1. The controls are divided into five groups:

Display
Channel setup
Trace setup
Entry
Utility

Control can be achieved by the use of the instrument keys or a standard computer mouse using a Windows interface.

Display Control

The performance of a DUT can be analyzed in many different ways. Each analysis window can be displayed one at a time on the screen or several at a time.

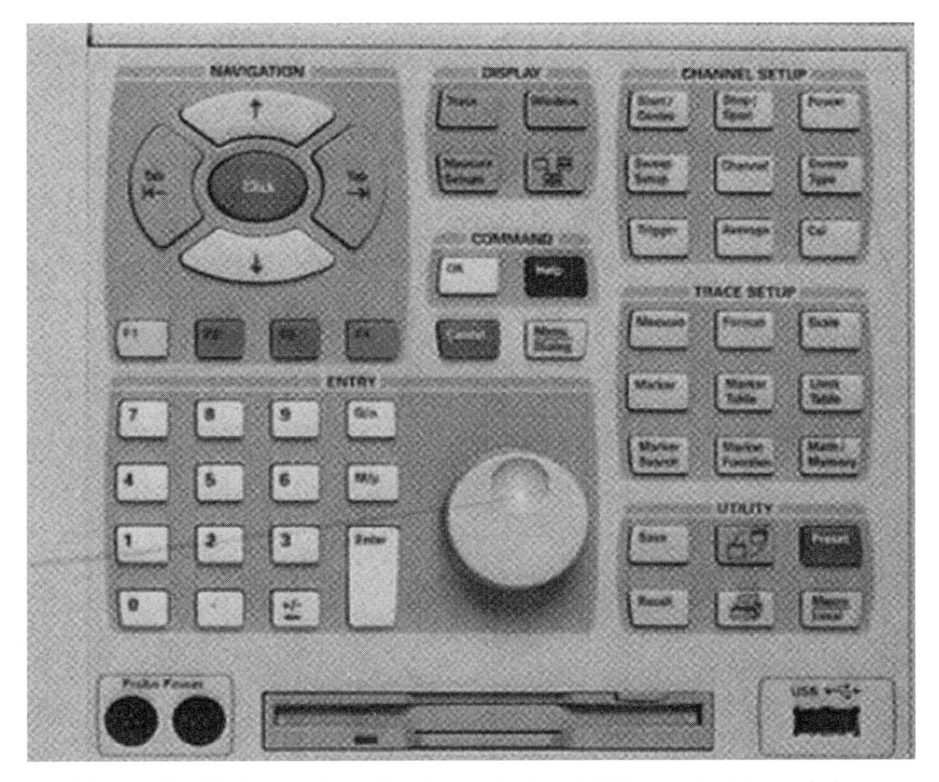

Figure 8.2 VNA controls. Agilent Technologies © 2008. Used with permission.

However, the display of the device performance can only be adjusted one trace at a time. This trace is called the "active trace," and it is selected by the using the mouse.

Channel Setup

The channel setup changes the characteristics of the RF signal that is being applied to the DUT and that is displayed along the horizontal axis. A menu of discrete stimulus characteristics is provided in Figure 8.3. The Windows interface allows any signal characteristics to be obtained with a standard mouse control.

Trace Setup

The trace setup sets the quantity to be measured, which is displayed along the vertical axis, and in what units it is to be displayed. Two response menus are shown in Figure 8.4. The menu on the left shows the "format" in which the measurement will be displayed: decibels, SWR, phase, group delay, polar, or Smith Chart. The

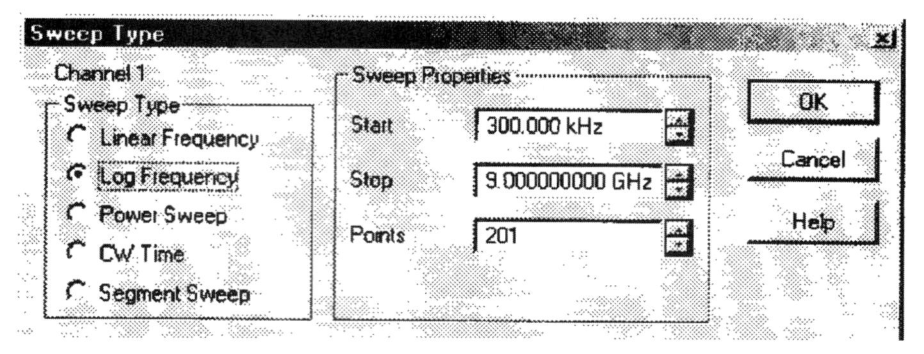

Figure 8.3 Computer screens for sweep control.

menu on the right shows the "power" scale to be used. Other response menus include

"measure" control, which determines which S-parameter or channel power will be measured;

"display" control, which sets the display to show one or more channels, overlaid or on a split screen, and other display parameters;

"average" control, which averages the measurement over several scans before displaying it; and

"marker" control, which turns on the marker function, and the "marker function" control, which sets what the marker key is marking.

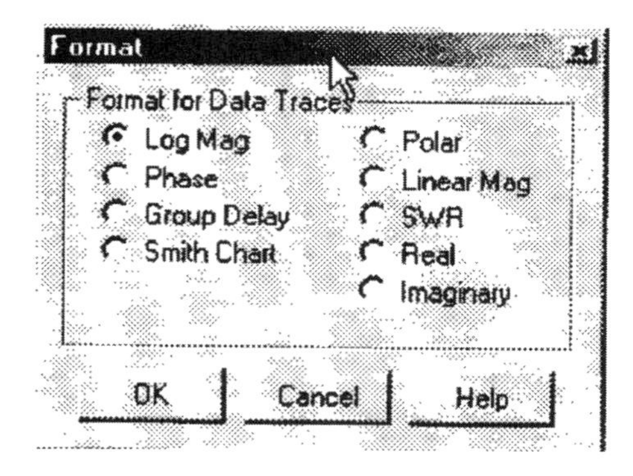

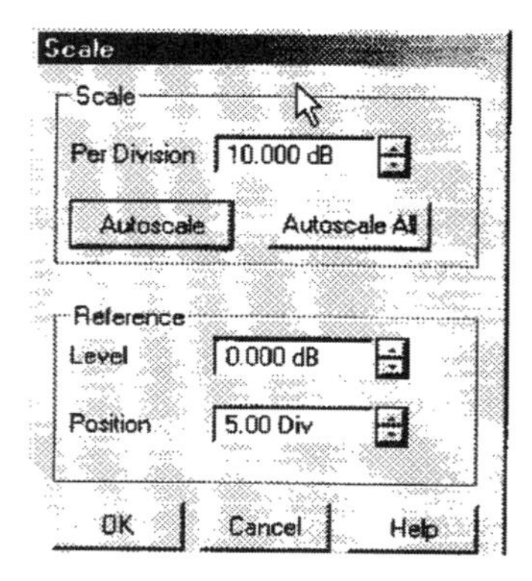

Figure 8.4 Computer screens for format and scale control.

Entry Control

The entry keypad allows numerical values to be assigned to each stimulus or response entry. Numerical values can be entered by the keypad, the rotary analog knob, or by the up-and-down step controls.

8.4 VNA DISPLAY NOTATIONS

The value set for each stimulus and response function is displayed on the screen. Figure 8.5 shows these quantities for a particular bandpass filter measurement. The start and stop frequency sweep are shown along the bottom of the display. For this particular measurement, the power is swept from 2.0 to 3.0 GHz. The active channel (which is what is being measured), the format, the scale, and the reference level and type of correction are shown at the top left of the scale. Note that the active channel is channel 1. The quantity being measured is S_{21}, and the measurement is expressed in log mag (dB). The calibration is response. The scale is shown vertically along the left-hand edge of the screen, and it is 1 dB/division. The reference line of 0 dB is one line down from the top of the scale, which is shown on the right-hand edge of the screen. Two markers are shown on the trace. The values of these markers are shown at the top of the screen.

8.5 ERROR CORRECTION

The uncertainty of power measurements made with a VNA is about ± 1 dBm, which is obviously not very good. The reasons for this uncertainty are explained in Figures 8.6 and 8.7. Techniques for reducing measurement uncertainty using a calibration procedure are briefly explained here, and they are explained in detail in Chapter 13.

Figure 8.6 shows a block diagram of a VNA. The signal used for testing is generated by a synthesized RF sweeper shown in the lower left. The signal is swept in frequency across the desired measurement range in a time interval of the order of milliseconds. The signal is formed by combining the harmonics of quartz crystal oscillators so that the signal does not actually sweep, but it steps from one frequency to the next in small intervals. At a particular instant in time, the frequency will be at a certain value within the desired sweep range. The signal is sent through a directional coupler, where a sample of the signal is coupled to a power detector, and its power is measured and most of the signal is sent on through a second directional coupler to the DUT. As the arrows in Figure 8.6 show, this second directional coupler is reversed in direction and should not couple any of the incident power into its detector. The second directional coupler is used to measure the power that is reflected from the DUT. The ratio of the forward transmitted power (R) to the reverse reflected power (A) is calculated from the power in the detectors, and it is the measured mismatch at the input to the DUT. The remaining incident power passes through the DUT and is sampled by the directional coupler (B). The ratio of the output power (B) to the incident power (R) is the loss or gain of the device. The network analyzer

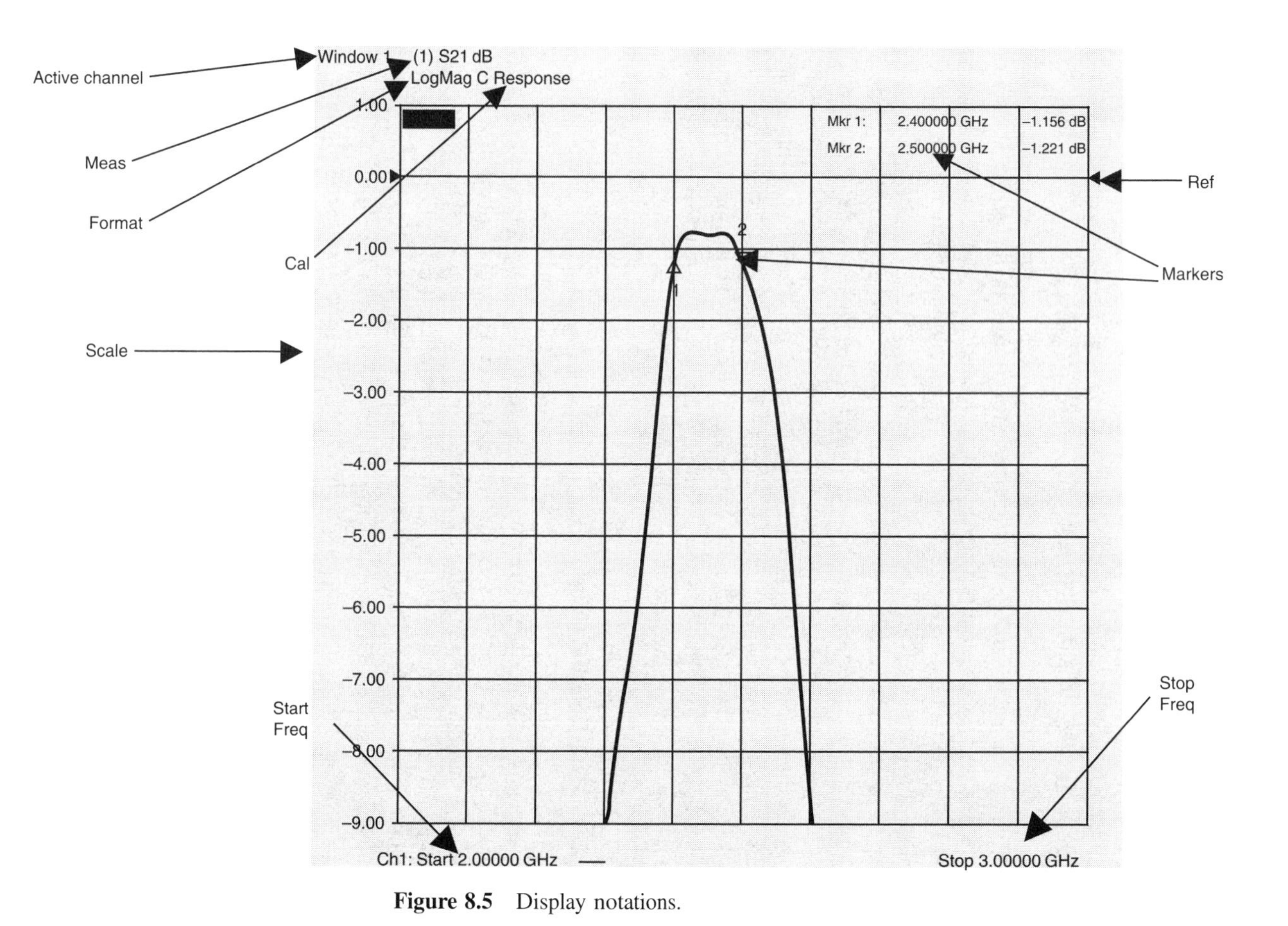

Figure 8.5 Display notations.

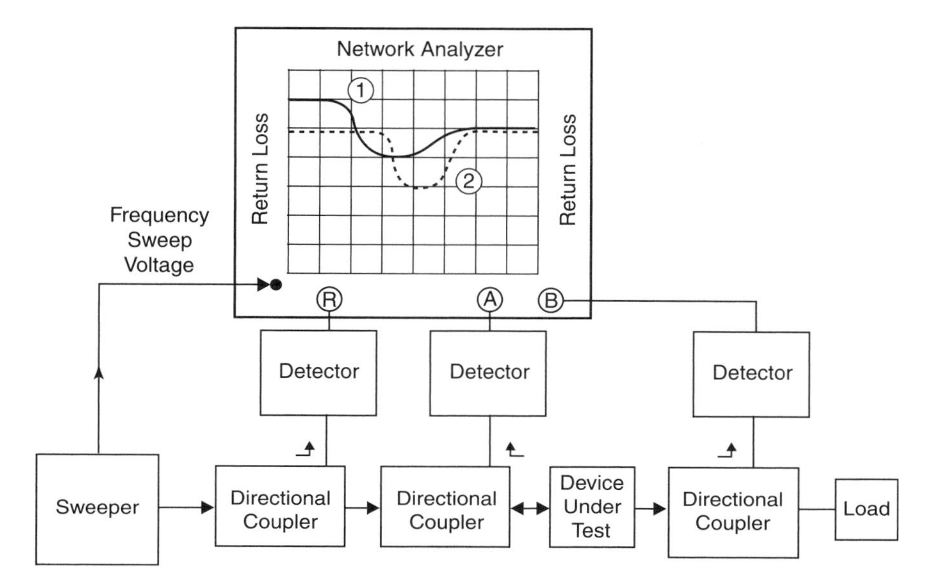

Figure 8.6 Block diagram of VNA.

itself is simply a computer, which processes the incident, reflected, and transmitted signals and presents their ratios on the display.

The uncertainty problem in the measurement occurs because the reversed directional coupler that is sampling the reflected power has some leakage in the wrong direction. Figure 8.7 is a visualization of the RF signal traveling through the reversed RF coupler that is supposed to be measuring the reflected RF power. The signal, represented by the solid arrow, is shown entering the directional coupler from the right. It is supposed to pass through the coupler and be reflected by the mismatch, and the reflected signal (shown as Ex) passes through the reversed directional coupler in the correct direction. Unfortunately, some of the incident signal leaks into the coupler in the wrong direction as shown by the dotted arrow.

The real reflected signal and the leakage of the incident signal both appear in the common arm that is supposed to be measuring reflected power only. These signals will add or subtract, depending on their phase. For small mismatches, the leakage power in the wrong direction may be comparable to the small reflected signal, and large measurement errors will occur.

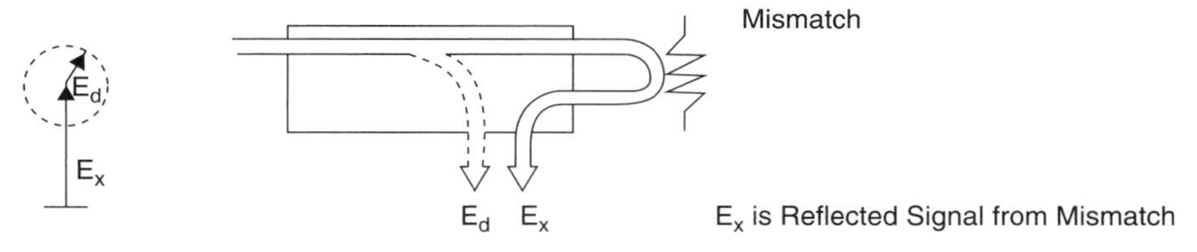

Figure 8.7 Return loss measurement error due to coupler directivity.

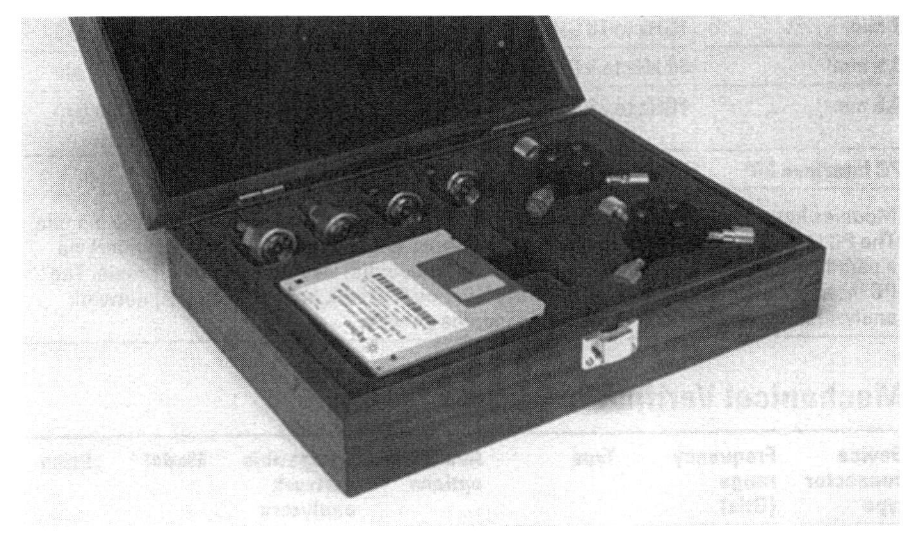

Figure 8.8 Manual calibration standards. Agilent Technologies © 2008. Used with permission.

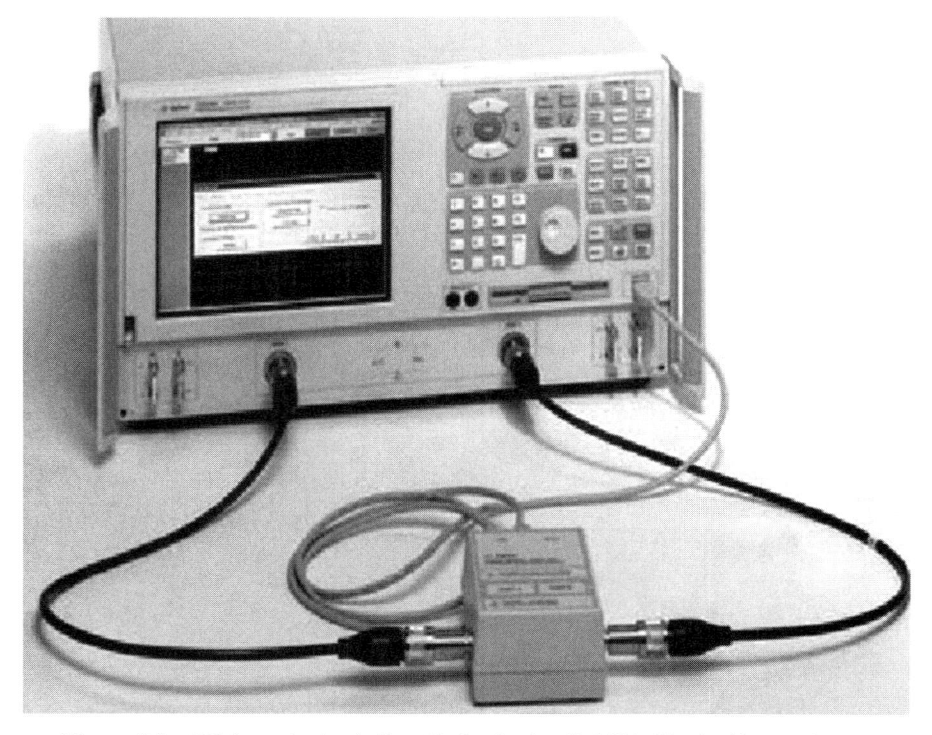

Figure 8.9 ECal standards. Agilent Technologies © 2008. Used with permission.

The practical solution to this coupler leakage problem, and other measurement errors of reflected power, is to calibrate the VNA with a set of known mismatches, determine what the error is, and use this information to correct the measurements of the actual DUT.

The traditional way of performing this error correction has been to use a set of precision SOLT standards that are sequentially attached to the VNA ports. A set of such standards are shown in their felt lined hardwood box in Figure 8.8. Measurements of phase and amplitude are made on each standard. The difference between the measurement and the standard is computed, stored in memory, and applied to correct each measurement of the DUT. Modern network analyzers, like that shown in Figure 8.1, display step-by-step instructions for attaching the standards, storing the measurement, and removing the calibration standards from the VNA ports. The correction factors are only calculated for the points along the current sweep setup. Thus, if the sweep frequencies for the measurement are changed to include frequencies outside the range that is currently being used, another calibration would need to be performed. Therefore, the normal procedure to perform a network analyzer measurement is to set up the measurement parameters such as sweep frequencies and power levels first, then perform a calibration and take the measurement. Calibration also needs to be repeated any time the measurement conditions, such as ambient temperature, drift from the original conditions at the time that the calibration was performed.

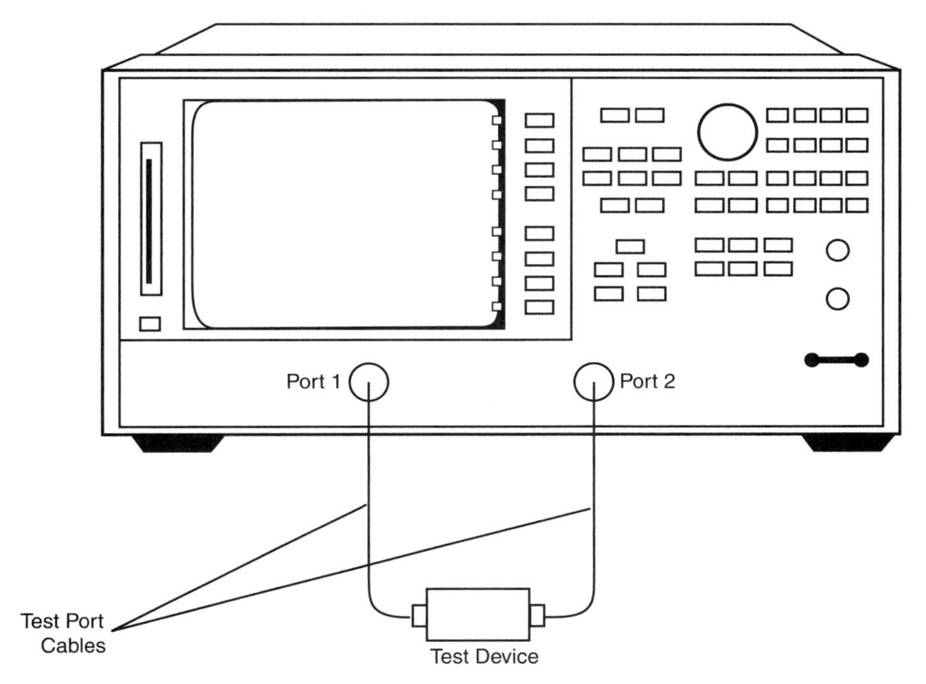

Figure 8.10 Measurement setup for VNA.

An improved calibration procedure is shown by the ECal system of Figure 8.9. The same SOLT standards are used, but they are mounted in a sealed box and sequentially connected and disconnected to the VNA ports by electronic switches. The ECal system eliminates human error due to dropping or other damage to the standards and due to connecting them in the wrong sequence.

The measurement uncertainty of using ECal is about 0.02 dB greater than that of the manual system due to wear of the internal switches, which is negligible in most testing compared to possible human error. Additional discussion of VNA measurement errors is presented in Chapter 13.

8.6 EXAMPLE OF VNA MEASUREMENTS ON AN RF PART

Figure 8.10 shows the measurement setup for using the VNA to measure the characteristics of a RF filter. The RF filter, which is the DUT, is simply connected

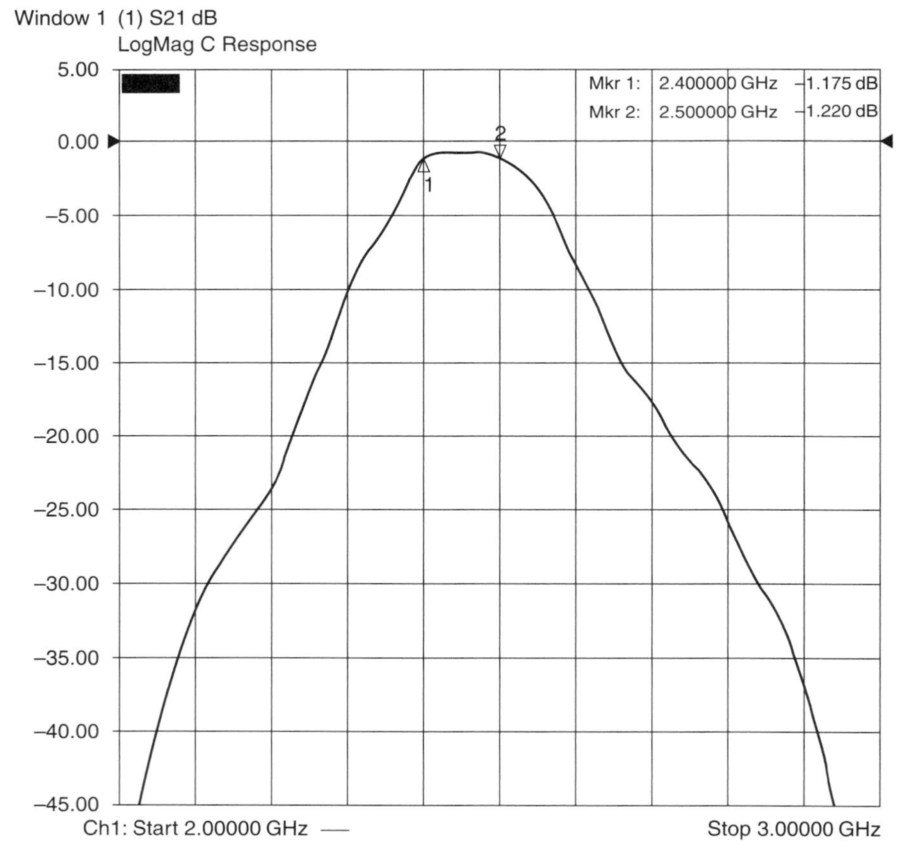

Figure 8.11 Insertion loss versus frequency.

to the test ports of the VNA. If port 1 of the VNA is connected to the input port of the device and port 2 is connected to the output port of the device, the VNA measures S_{11} and S_{21} of the device. If the port connections are reversed, the VNA measures S_{12} and S_{22} of the device. To obtain all four S-parameters without reconnecting the DUT, the VNA simply switches the connections internally, using electromechanical switches.

Figures 8.11–8.18 show measurements made with the VNA on the same RF filter whose basic characteristics were provided in Figure 8.5. All measurements are shown as a function of frequency from 2.0 to 3.0 GHz. Figure 8.11 shows insertion loss over a power range from 0 to 45 dB, instead of the limited range of 0–8 dB shown previously in Figure 8.5. Figure 8.12 shows the mismatch of the input expressed as return loss. Figure 8.13 shows the mismatch of the input expressed as SWR. Figure 8.14 shows the mismatch of the input as a function of reflection coefficient. Figure 8.15 shows S_{11} and S_{22} on a common display. Figure 8.16 shows the

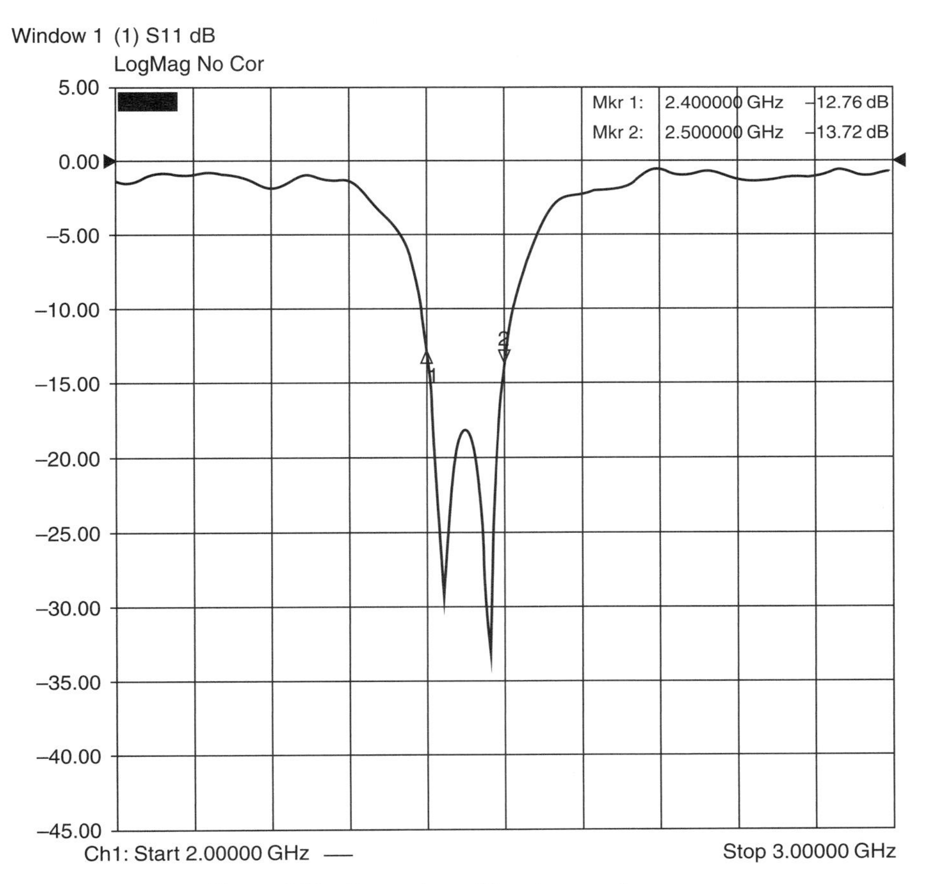

Figure 8.12 Return loss of input versus frequency.

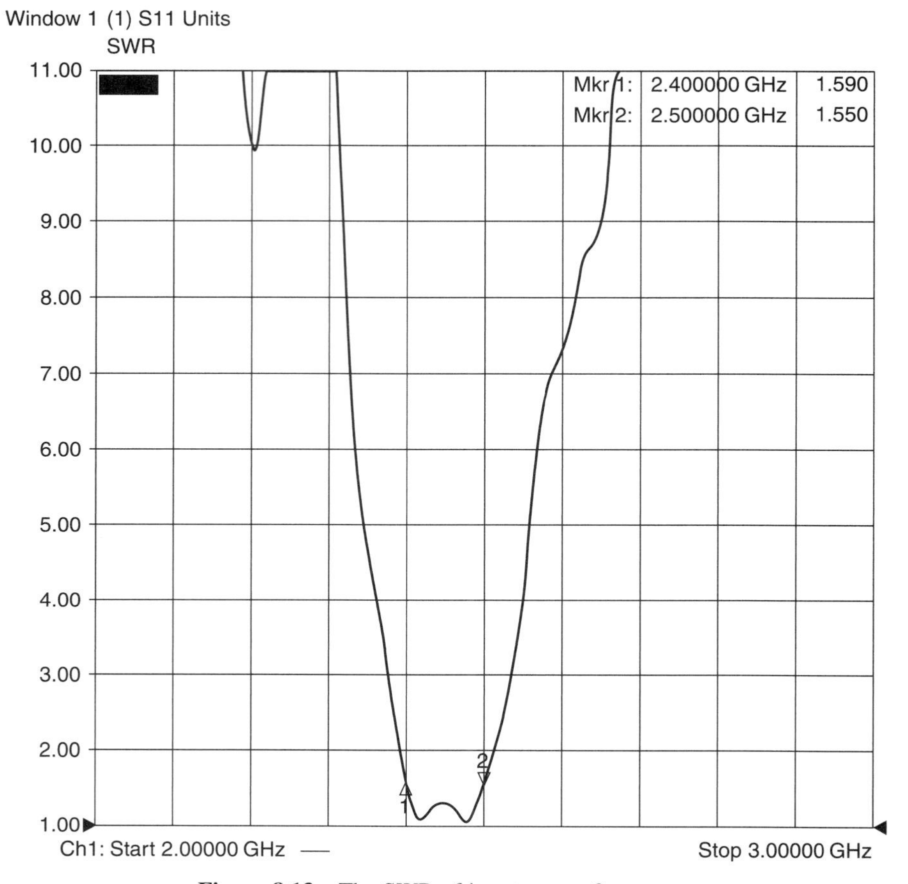

Figure 8.13 The SWR of input versus frequency.

amplitude and the phase of S11 in polar format, with frequency as the independent variable of the curve. Figure 8.17 shows the same curve as Figure 8.16, but with Smith Chart grid coordinates. Figure 8.18 shows the phase of S21 as a function of frequency over the frequency range from 2.3 to 2.6 GHz.

8.7 SWEPT MEASUREMENTS ON THE VNA AS A FUNCTION OF POWER

All of the measurements shown in Figures 8.11–8.18 were made on passive filters whose performance was independent of the RF power level. Such devices are called "linear." Many RF devices, like amplifiers, are nonlinear and their performance depends on their power level of operation. By changing the stimulus from a

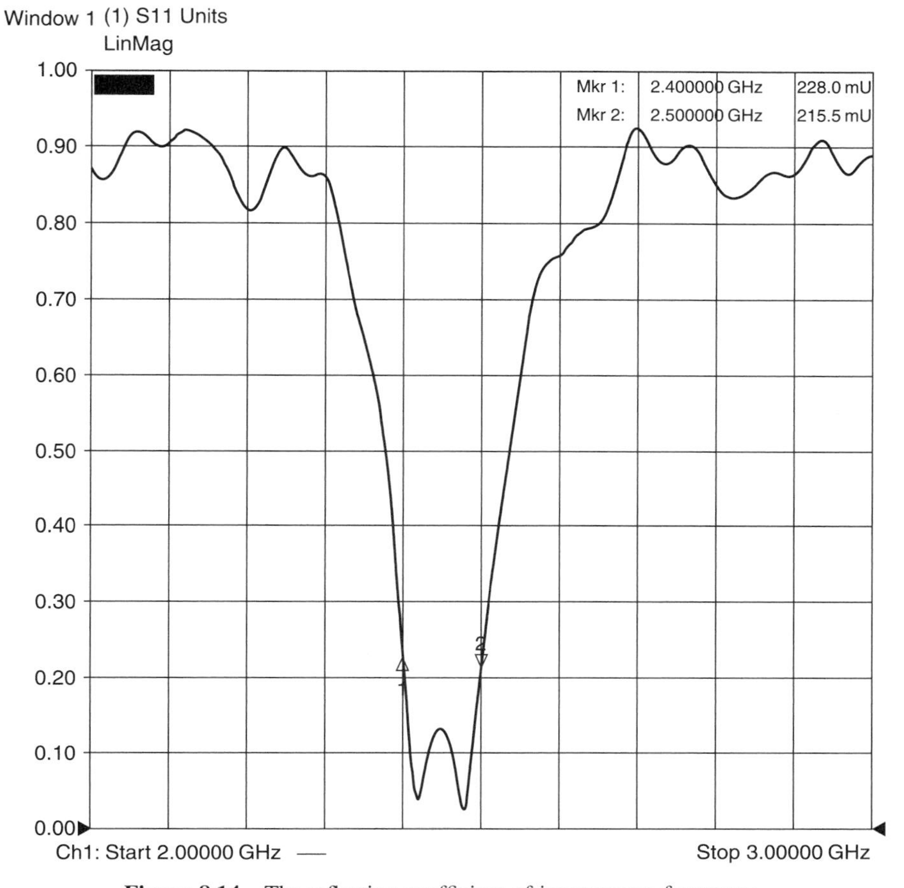

Figure 8.14 The reflection coefficient of input versus frequency.

frequency sweep to a power sweep, the VNA can measure the performance of nonlinear devices as a function of their input power level at a fixed frequency.

Figure 8.19 shows a VNA measurement of swept gain (left-hand graph) and swept power (right-hand graph) as a function of frequency, with a fixed input power level of 0 dBm. If the input power level were changed, the graphs would be different.

Figure 8.20 illustrates a VNA measurement of swept gain (left-hand graph) and swept power (right-hand graph), but as a function of the input power level with a fixed frequency of 2.45 GHz. From the left-hand graph note the decrease in gain as a power level is increased, and the marker shows the point where the gain has dropped by 1 dB from its small signal value. Note from the right-hand graph that the output power is proportional to the input power over most of the input power range and that the output power is approaching saturation at the 10 dBm input power level.

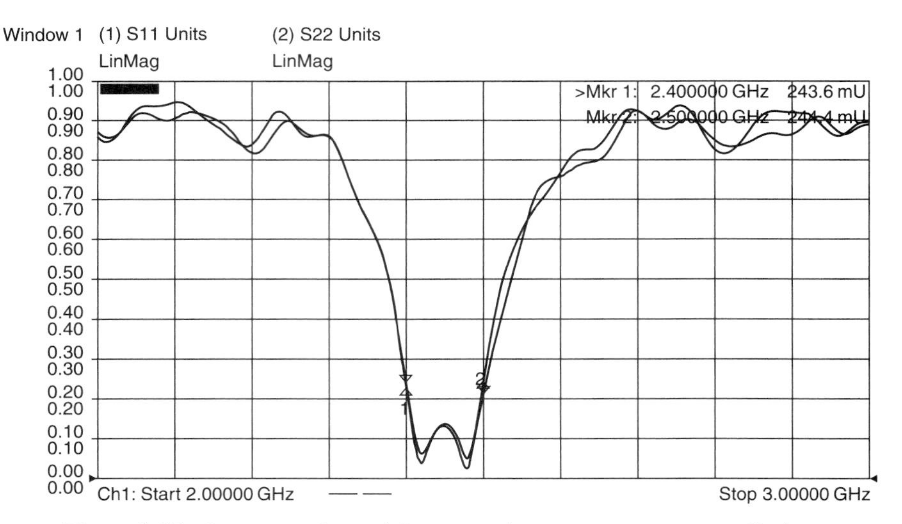

Figure 8.15 Parameters S_{11} and S_{22} versus frequency on a common display.

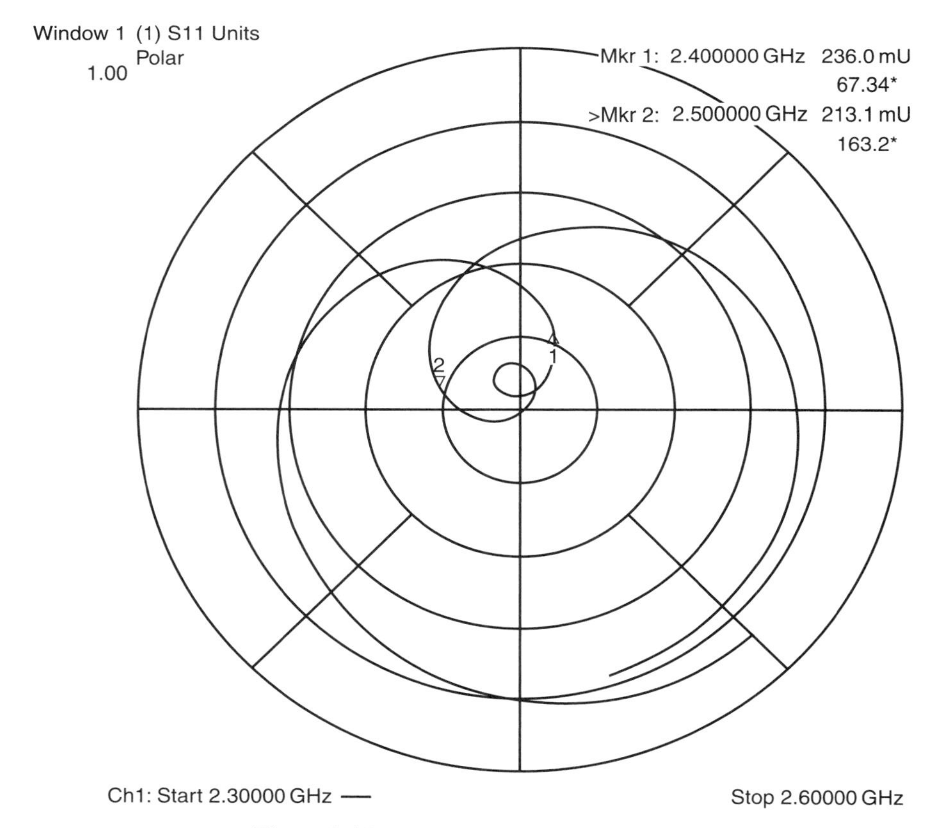

Figure 8.16 Parameter S_{11} in polar format.

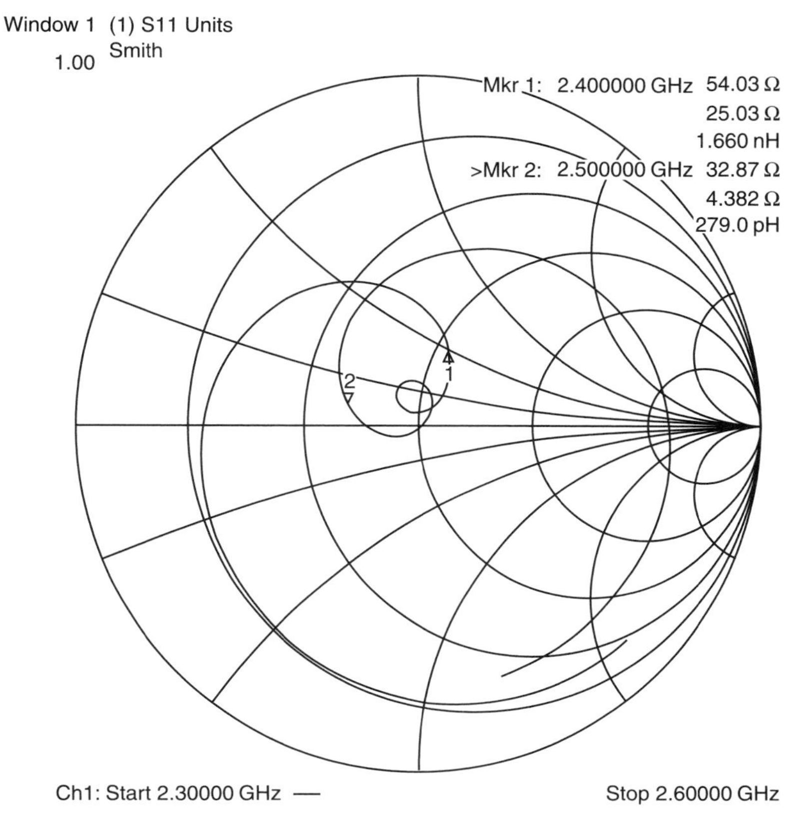

Figure 8.17 Parameter S_{11} in Smith Chart format.

8.8 EXAMPLE MEASUREMENT PROCEDURE USING THE VNA

Instructions for making the measurements shown in Section 8.6 are given in this section. The instructions are generic and apply to any modern VNA.

In this experiment the in-band and out of band loss and the input and output match of the RF filter are measured with the VNA.

Objective

Our objective is to demonstrate basic operation of the VNA by measuring the S-parameters of a passive RF filter. All connections in this course are using subminiature-A (SMA) connectors, and all DUTs have female connections. Therefore, the VNA is set up with male SMA connectors at the ends of both cables.

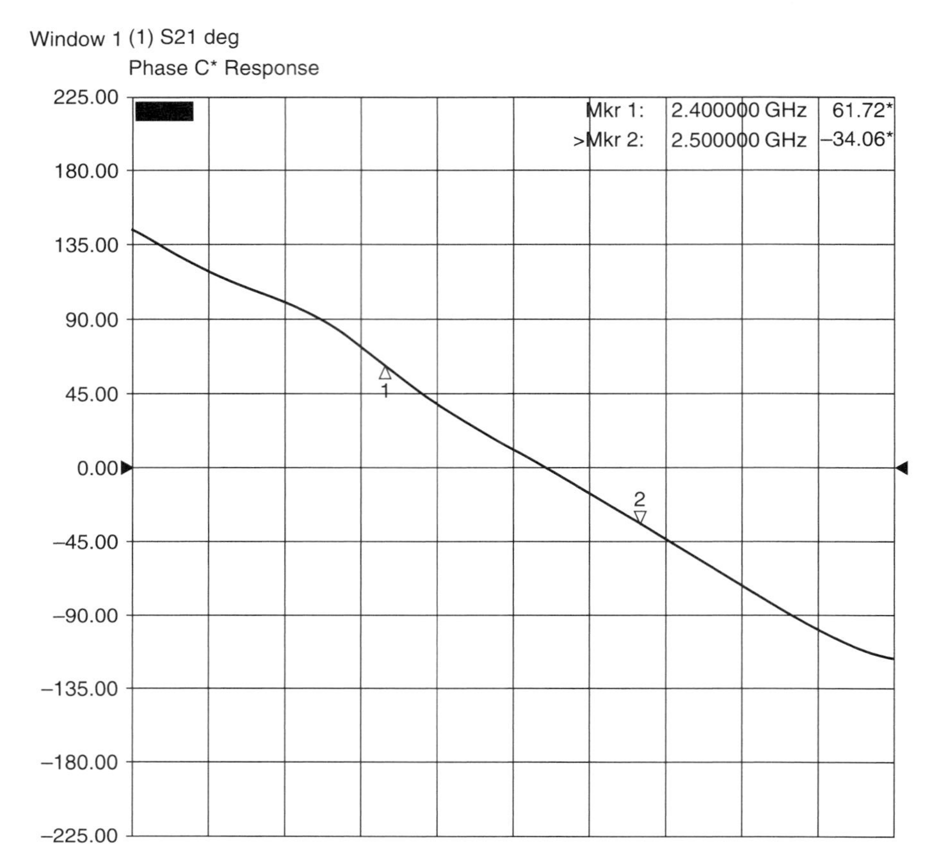

Figure 8.18 Phase of S_{21} versus frequency.

Measurements Being Demonstrated

Insertion loss versus frequency
Return loss versus frequency
Input SWR versus frequency
Input reflection coefficient versus frequency
S_{11} and S_{22} versus frequency on a single display
S_{11} polar format
S_{11} Smith Chart format
S_{21} phase versus frequency

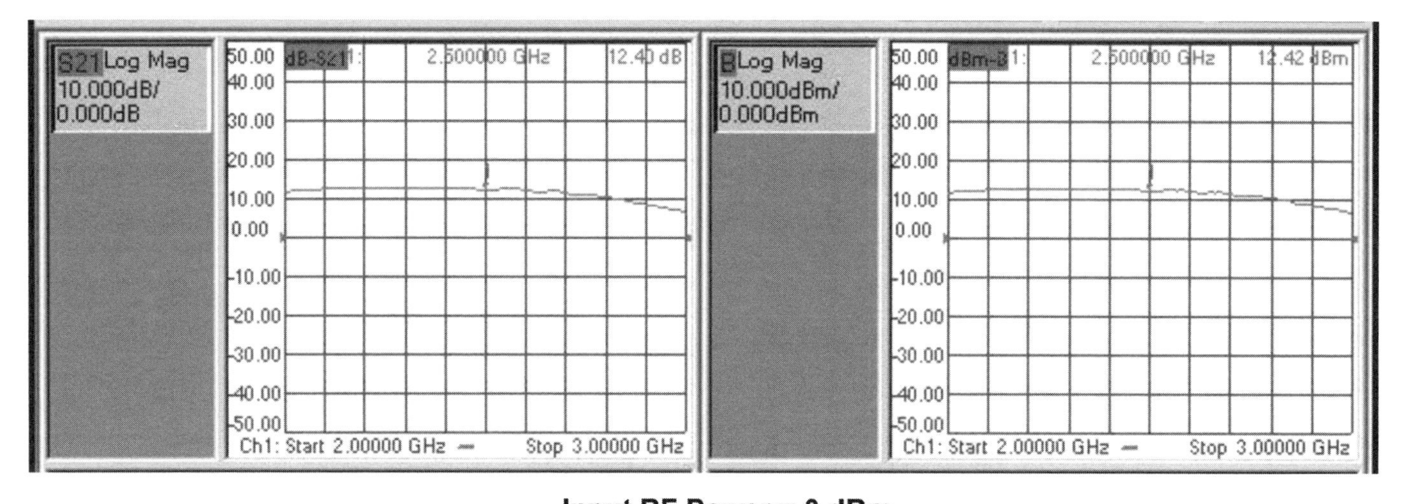

Input RF Power = 0 dBm

Figure 8.19 Power amplifier performance versus frequency.

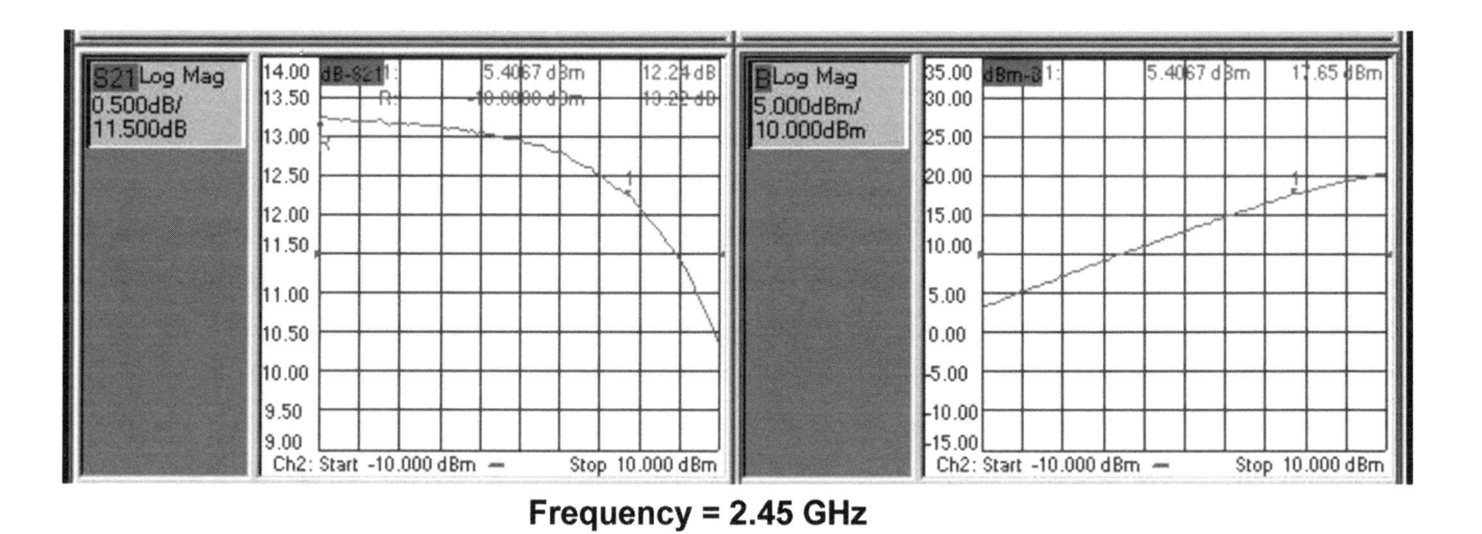

Figure 8.20 Power amplifier performance versus RF input power.

DUT Specifications

The Specifications for the RF Filter are summarized in the following table:

Manufacturer	Toyo
Model	TDF2A-2450T-10
Center frequency	2.45 GHz
Passband width	±50 MHz
Insertion loss in band	1.0 dB
Attenuation at ±280 MHz	20.0 dB

Significance to Wireless System Performance

The performance of the filter is determined by how well it passes the desired signal without unwanted attenuation (passband insertion loss), as well as by how well the filter stops unwanted signals (out of band rejection).

Generic Procedure

Steps to perform the measurements are listed in the following table. The left column lists the actions to be performed, and the right column lists the motivation for taking each step and provides additional background information in some cases. The instructions are intended to be generic across most network analyzer models.

Procedure	Notes
Preset the instrument by pressing the green Preset key.	Presetting the instrument returns all settings to their factory default. This is useful because you can set up your measurement without worrying about an obscure option that may have been set by another user affecting the measurement.
Select S_{21} as the parameter to be measured. Use the display scale functions to set the scale to 5 dB/division, the reference value to 0 dB, and the reference position to 9.	Because this is a passive device, we know that S_{21} will be nonpositive. Therefore, we set the scale of the display appropriately so that we make best use of the screen space.
Set the start and stop frequencies for the frequency sweep to 2 and 3 GHz, respectively. Most VNAs have a Start and Stop key to accomplish this.	On the network analyzer, you set all of the parameters for the measurement, then perform the calibration.
Check the power level for the built-in source signal. The Power key is often grouped with other buttons under a Stimulus or Channel Setup heading. The default is usually 0 dBm, which will be good for our measurement.	Older models access the power setting by pressing the Menu key and then selecting the Power soft key. The default power level for the built-in source varies from model to model, so it is important to check it prior to measurements.

(*Continued*)

Procedure	Notes
Start a full two-port SOLT calibration. Most analyzers have a Cal key to access this function.	
Select the type of calibration kit being used. On older models this is done via the Select Cal Kit soft key. On newer models, check the box for Select Cal Kit in the calibration dialog box.	For SOLT calibrations, detailed information about the physical dimensions/parasitics of the standards are recorded and taken into account when the correction factors are calculated. Typical values for a given model of Cal Kit are stored in the memory of the VNA and can be recalled by the menu. For even greater accuracy, the exact parameters are included with the Cal Kit on a floppy disk. These can be loaded and used with the calibration process.
Select a full two-port SOLT calibration from the Cal menu.	
Connect the standards for SOL to the end of the measurement cable and press the corresponding key on the VNA. The VNA takes a quick measurement and calculates error terms.	This process moves the "measurement plane," or the point from which phase is measured, to the mating surface of the connector for the standard.
Connect the two cables. Ideally, the cables would be of opposite gender so that the two mating surfaces would be touching. Because we are using male connectors on both cables, a thru adapter is needed. Perform the thru calibration.	The use of a thru adapter of finite length introduces an error into the calibration setup. This error can be compensated in five ways: 1. Use a very short thru (relative to wavelength). This is the technique that we use in this example. 2. Use a set of "swap-equal" adapters. This is a set of four adapters with all possible combinations of gender at each end. The adapters are all designed to be electrically equivalent and one of the connectors is always connected to the end of a cable, depending on which gender is needed. 3. Modify the cal kit data to include information about the thru adapter that you are using. 4. Use the "adapter removal" technique to remove the effect of the adapter. Basically, two calibrations are performed with the adapter on a different cable each time. The network analyzer has enough information at the end of the process to remove the adapter's effects. 5. Use an ECal module.

(*Continued*)

Procedure	Notes
Omit isolation.	The isolation calibration is used only when you need to test a device with >90 dB of isolation. A load standard is connected to each cable and the leakage internal to the VNA is measured.
Finish the calibration.	When the calibration is complete you can save the calibration data, but that is not necessary for this example. You could save the calibration info if you wanted to view two frequency ranges, then recall the calibration data that are most appropriate as you switch between frequency ranges. Newer models allow you to set multiple measurements up at once; older models are limited to two different sweeps at once.
Verify the calibration. With an open or short standard connected to the end of the cable at port 1, measure S_{11} magnitude (return loss). The result should be 0 dB for all frequencies. Repeat the measurement on port 2 and measure S_{22}. The result should also be 0 dB.	The network analyzer will trust that you have connected each standard properly during the calibration sequence. If the wrong standard was used, or a loose connection was made during the calibration process, erroneous correction factors will be calculated and applied to all measurements. This step is a quick check that can be performed to make sure that the measurement system is working predictably after performing a calibration. Another quick check that can be performed in the event that your measurement seems "irregular" is to simply turn the calibration off (using the calibration menu). If the measurement seems more in line with expectations after turning off the calibration, then repeating the calibration for the measurement would be justified to make sure it was done properly.
Connect the filter and observe S_{21} (insertion gain/loss). Set a marker for 2.45 GHz using the marker key. You can use the markers to determine whether the filter passes the data sheet specification.	You can use marker search functions to find bandwidths and specific attenuation values. Marker functionality varies from model to model.
Measure return loss (S_{11}).	Note that in the regions where the filter is rejecting the signal, the return loss is small (the signal is reflected). The converse is true in the passband.

(*Continued*)

Procedure	Notes
Measure the SWR. Adjust the scale as needed.	An SWR of 1 indicates a perfect match condition. Large SWR values indicate a mismatch.
Measure the reflection coefficient by setting the format to linear magnitude. Adjust the scale as needed.	
Set the format to log magnitude. Activate a new trace and plot it on the same display. Set the new trace to measure S_{22} in log magnitude format with the same scale as the first trace (overlay the first trace). On older models use the Channel buttons to manage two traces on the same display. Newer VNAs often have a key labeled Display that shows a menu where you can choose to have separate windows or overlay on a single window.	Because the device is symmetrical, S_{11} should be the same as S_{22}.
Turn off the second trace. View a single trace.	
Set the start and stop frequencies to 2.3 and 2.6 GHz, respectively.	Because we are choosing a narrower frequency range that is within the frequency range where we performed the calibration, the VNA simply interpolates the calibration data for the new, narrower sweep range. A more diligent and accurate procedure would be to perform a calibration for each frequency sweep setup. On older models, the calibration data is stored and then recalled as needed. On newer models, it is easier just to set up multiple channels and view all of the measurements at once.
Measure S_{11} in polar format.	The polar format displays both magnitude and phase information of the reflection coefficient.
Select the Smith Chart format.	Note that the Smith Chart format displays the same trace with "mapping" to the impedances that produce the measured reflection coefficient.
Measure S_{21}. Change the format to "phase."	Phase is displayed.

8.9 ANNOTATED BIBLIOGRAPHY

1. PNA-L Microwave Network Analyzer N5230A Brochure, Document 5989-0168EN, Agilent Technologies, Santa Clara, CA.
2. PNA-L Microwave Network Analyzer N5230A Data Sheet, Document 5989-0514EN, Agilent Technologies, Santa Clara, CA.
3. ECal Reference Guide, Document N4693-9001, Agilent Technologies, Santa Clara, CA.

Reference 1 describes the features of a popular VNA.

Reference 2 describes the performance of a two-port 300 kHz to 6 GHz VNA.

Reference 3 describes the use and accuracy of the ECal calibration.

CHAPTER 9

SPECTRUM ANALYZERS

9.1 SPECTRUM ANALYZER PRINCIPLES

Suppose we were given the task of designing an amplifier to faithfully reproduce the signal shown in Figure 9.1. One of the first questions we would need to ask is how often does the signal repeat itself? If the entire horizontal axis of the figure represents a 1 s interval, then would we say that because the first half of the trace is positive and the second half of the trace is negative, the signal therefore repeats itself once per second? Or would we say that perhaps the signal repeats itself five times because five separate excursions can be seen in the pattern. What about all of the smaller variations in the signal?

A more constructive way to analyze this signal is to view it as the summation of several perfect sinusoids at different frequencies. Figure 9.2 shows three sinusoidal signals that form the trace in Figure 9.1 when we add them together. There is a sinusoid with a frequency of 1 cycle/s, another at 5 cycles/s, and a third at 16 cycles/s (or Hertz). All signals can be analyzed as the sum of several sinusoidal signals at different frequencies. Even square waves are actually the sum of an infinite number of harmonically related sinusoids at different frequencies. Analyzing signals in terms of their frequencies is called spectrum analysis.

Analyzing the signal in Figure 9.1 with a spectrum analyzer tells the amplifier designer that they need to design an amp that can faithfully reproduce everything from 1 to 16 Hz.

RF Measurements for Cellular Phones and Wireless Data Systems. By A. W. Scott and R. Frobenius

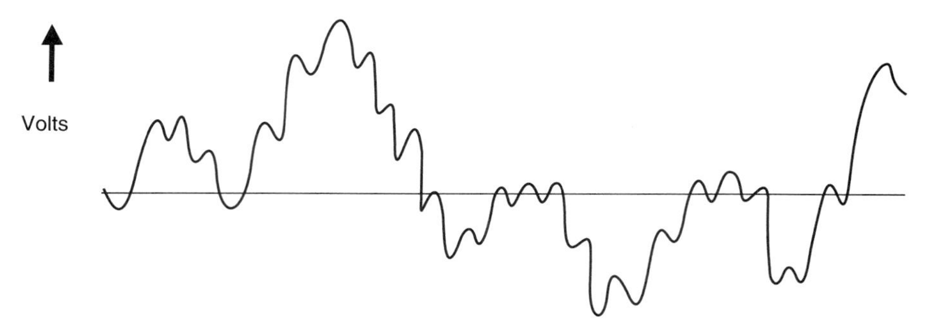

Figure 9.1 Frequency is how often an electrical signal repeats itself in 1 s.

9.2 WHAT A SPECTRUM ANALYZER CAN MEASURE

Figure 9.3 shows a spectrum analyzer, in this case a PSA series from Agilent Technologies. The spectrum analyzer can measure the frequencies present in a complex signal or the frequencies resulting from modulation on a carrier. Additional software loaded on the instrument can also integrate power over specified

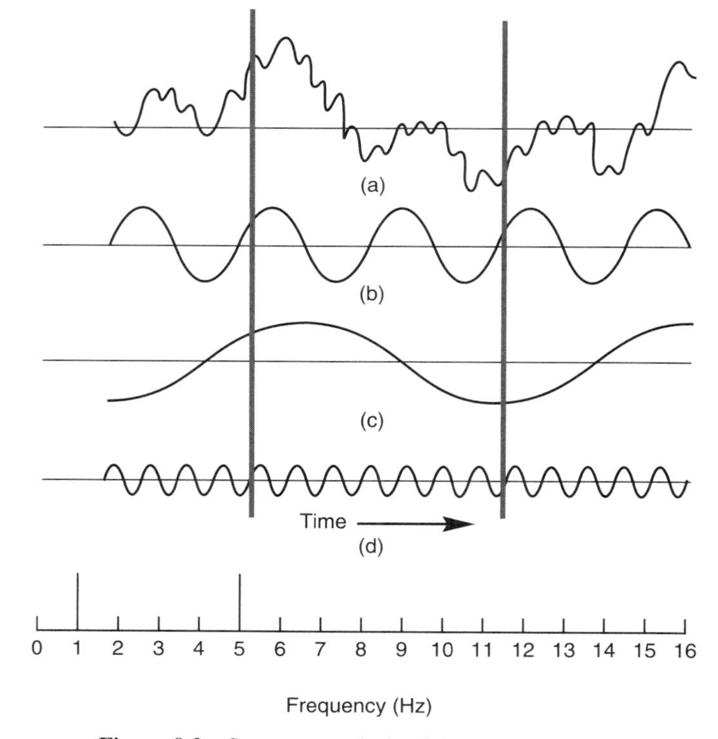

Figure 9.2 Spectrum analysis of the electrical signal.

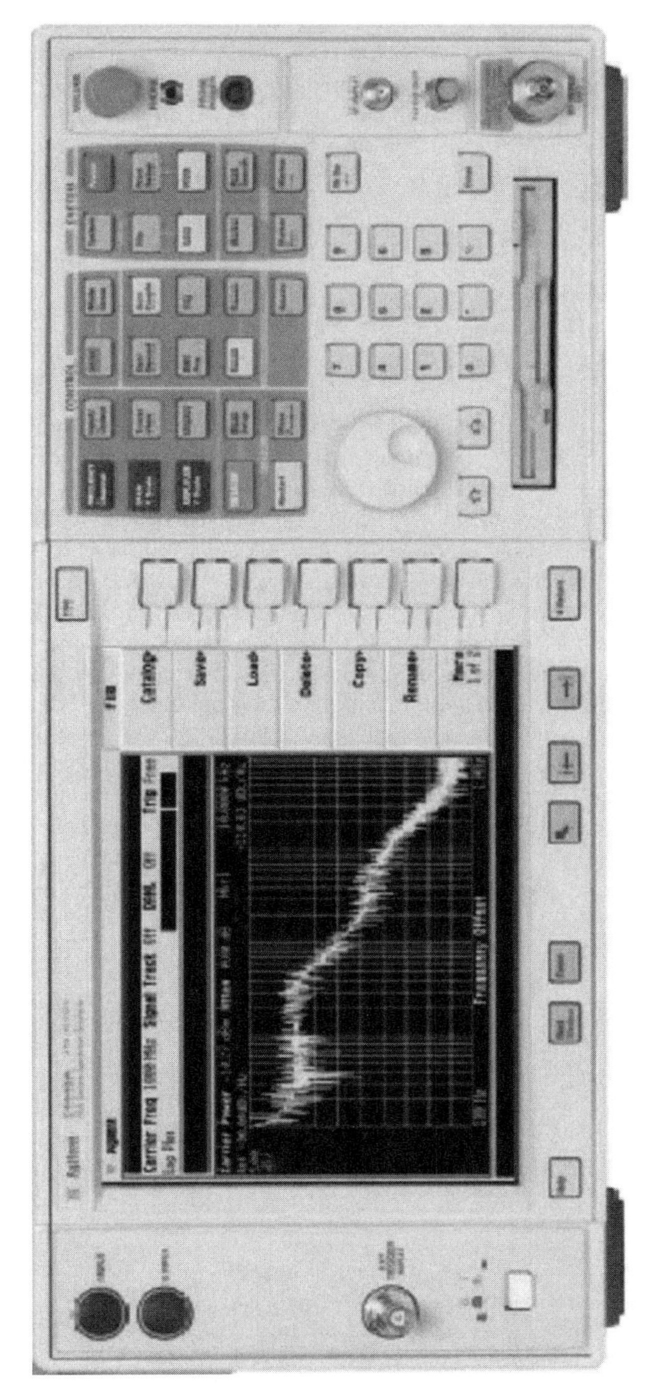

Figure 9.3 Spectrum analyzer. Agilent Technologies © 2008. Used with permission.

bandwidths to measure channel power or occupied bandwidth, for example. Device distortions that result in harmonic output can also be seen and measured. Very weak signals (−140 dBm) can be measured as well. Using special options, the instrument can perform specialized measurements such as noise figure and phase noise. Today's instruments can also be used simply as a data acquisition front end for other analysis, such as digital demodulation using vector signal analysis software.

9.3 SPECTRUM ANALYZER BLOCK DIAGRAM

One of the best ways to understand how to use a spectrum analyzer is to look at how to build one. We are only going to look at the basic functions, but we will see where most of the controls on today's sophisticated equipment come from. Figure 9.4 shows a simplified block diagram of a spectrum analyzer, which we use for this discussion.

At the heart of the spectrum analyzer is simply a power sensor, much like the variety that we would see with a power meter. A power sensor cannot tell us anything about the frequencies in a signal. However, if we put a narrow bandpass filter in front of the power sensor, then we would know that whatever power level we read on the power meter had to be within the passband of the filter. Thus, we could build a spectrum analyzer by putting a bandpass filter in front of a power meter, tuning the filter to different frequencies, and reading the power level at each frequency. Unfortunately, building a tunable filter of this type is not practically possible.

Instead, spectrum analyzers use a bandpass filter at a fixed frequency and then sweep the spectrum to be measured across this fixed "window" by shifting it up or down in frequency. The way that this is accomplished is by using a mixer and combining the measured spectrum with a controlled LO signal. A mixer is a device that takes two signals at its input and outputs the sum and difference frequencies between the two signals. In a spectrum analyzer, one of the signals is the

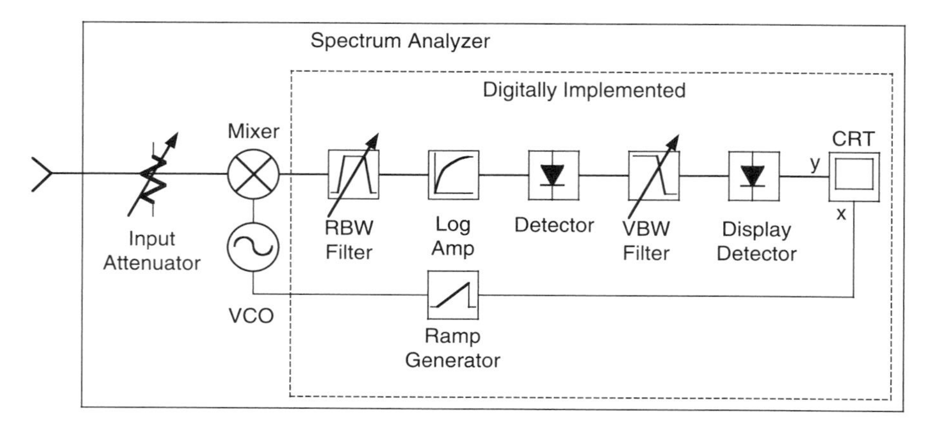

Figure 9.4 Spectrum analyzer block diagram.

spectrum that we want to measure, and the other signal is generated inside the analyzer: it is called an LO.

In the analyzer, we can control the LO. Let us say that we design our bandpass filter for a fixed center frequency of 100 MHz and we want to measure the spectrum from 2100 to 3100 MHz. Once again, a mixer outputs the difference frequency between two signals. Thus, if we set the LO to output a signal at 2000 MHz, then the difference between the LO signal and the lower end of the spectrum we want to measure would be 2100 – 2000 MHz = 100 MHz. Any signals that were at 2100 MHz would now be shifted to 100 MHz by the mixer and pass through the bandpass filter to be detected by the power sensor. The bandpass filter would block signals at all of the other frequencies, limited only by how narrow we make the passband of the filter.

Similarly, if we set the LO to 3000 MHz the difference between the LO and the upper end of the spectrum we wish to measure is 3100 – 3000 MHz = 100 MHz. Now the power sensor would detect any signals that were present at 3100 MHz. By sweeping the LO from 2000 to 3000 MHz, we achieve the effect of sweeping the spectrum that we wish to measure across the fixed bandpass window that is in front of the power sensor. This bandpass filter is called the "resolution bandwidth filter."

In the block diagram of Figure 9.4 a voltage from a ramp generator controls the LO output frequency. This same voltage controls the horizontal sweep of the gun on a CRT monitor. The output voltage from the power sensor controls the vertical motion of the gun on the screen. When we put the two together, the repeated ramping of the voltage causes the gun to scan horizontally on the CRT. As the gun is scanning, the oscillator frequency is sweeping and signals in the spectrum are allowed to pass through the bandpass filter. If a signal is detected at a given frequency, the output from the power sensor will cause the trace to rise vertically on the screen in proportion to the signal strength. We control the center frequency and span on the front panel of the spectrum analyzer, which in turn sets the appropriate voltage range on the ramp generator. This is the basic action of a spectrum analyzer. This simplified block diagram is helpful for understanding the controls that are available to "drive" the spectrum analyzer. On modern spectrum analyzers, the signal to be analyzed is digitally sampled after the first mixer and most of the elements of the block diagram are implemented in DSP. This improves the performance of the spectrum analyzer as well as offering more choices for settings such as resolution bandwidth and video bandwidth. Nonetheless, the controls on the instrument remain substantially the same.

9.4 SPECTRUM ANALYZER CONTROLS

Center Frequency and Span

When setting up a measurement on the spectrum analyzer, the horizontal axis typically represents a frequency span. The scale for the horizontal axis is determined by inputting a center frequency and a span over which the instrument will sweep. These controls are typically labeled frequency and span, respectively, and they are usually in a prominent location on the front of the instrument.

Reference Level and Attenuation

The vertical axis is adjusted by setting a reference power level. A well-established convention on spectrum analyzers is that the top line of the display represents the reference level. We can adjust this level by accessing what is typically labeled the Amplitude key on the front of the instrument and inputting whatever value we wish to set as the topmost value on the display scale. In addition, we can set the value of the scale per division, such as 10 dB/division. Figure 9.5 shows a signal with amplitude of −20 dBm displayed on the spectrum analyzer with a reference level of −20 dBm.

One of the main reasons that we are motivated to use a spectrum analyzer is to measure how well our components are performing with various input signals. We must always remember that the spectrum analyzer itself has many analog components "inside the box" and that these components can produce distortion if they are driven too hard by a large signal that we are trying to measure. For this reason, a built-in variable attenuator is located just inside the input of the analyzer, which allows us to reduce the level of the signal once it is inside. Because the analyzer "knows" the level of this attenuator, the display can compensate appropriately so the correct power level is represented to the user. As more attenuation is applied at the input of the spectrum

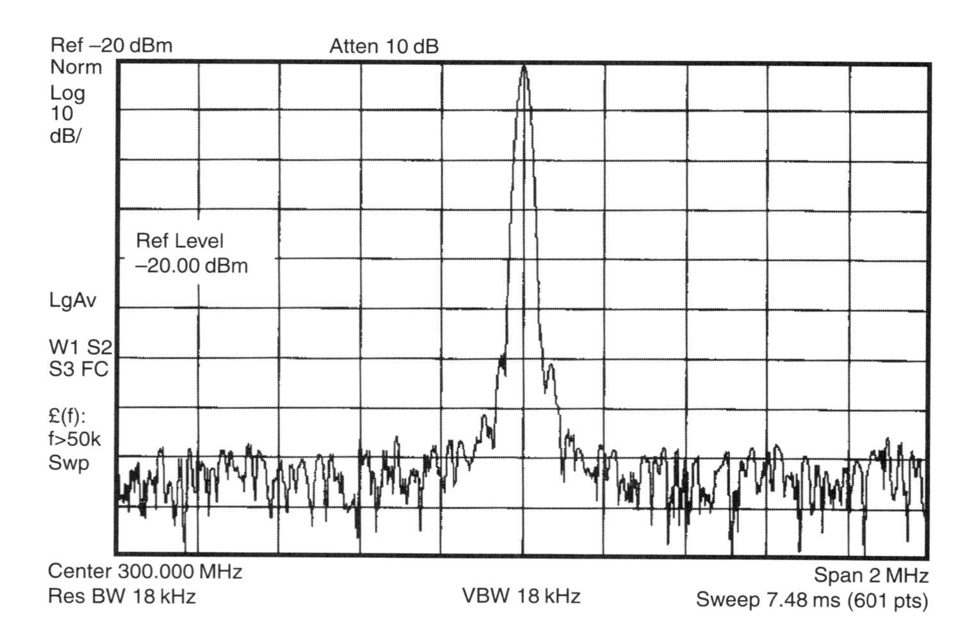

Figure 9.5 Display of a test signal.

analyzer, the S/N is reduced because the signal to be analyzed is weakened. Thus, even though the signal level remains constant on the display as the attenuation is adjusted, we will see the noise floor rise as more attenuation is applied to the input. Most spectrum analyzers will adjust the input attenuation based on the setting of the reference level power. More attenuation is applied for higher reference power levels, and less is applied when a lower reference power level is selected.

Resolution Bandwidth

Spectrum analyzers come with multiple settings for the width of the resolution bandwidth filter. The trade-off in setting the resolution bandwidth is between speed and detail. A narrow filter requires more time for the real-world filter elements to settle and obtain an accurate reading. If we are trying to sweep over a wide range of frequencies, then we will find that the sweep time will be very slow. In contrast, if we are searching for a signal that is very close to another signal, then we will need a narrow filter to be able to resolve the two signals on the display. Figure 9.6 presents charts of the resolution bandwidth.

Another trade-off when selecting the resolution bandwidth is that there is a fundamental relationship between the bandwidth and the noise floor. The wider the bandwidth of the filter is, the more noise energy is allowed to reach the detector, thus the higher the noise floor becomes. Therefore, in order to lower the noise floor on the display, we should choose a narrow resolution bandwidth filter, which in turn requires a slower sweep speed.

Video Bandwidth

The detector inside the analyzer outputs a voltage that corresponds to the amplitude of the signal envelope. At a single point in the frequency sweep, if a continuous tone that does not change amplitude is being measured, then the voltage from the detector will remain constant with respect to time. In contrast, if a noisy signal where the amplitude is rapidly changing is being measured, then the output voltage from the detector will also rapidly change. If a low pass filter to this signal is applied, the fast fluctuations would be removed. This is how averaging was accomplished in the early analog spectrum analyzers. Today, averaging can be done by storing successive sweep values in memory and calculating the average.

Nevertheless, the video bandwidth filter adjustment is still available to the user. Why? It turns out that using the video bandwidth filter can be useful for detecting low power signals that are close to the noise floor. In the block diagram of the spectrum analyzer, the video bandwidth filter appears after the signal has been adjusted to fit a logarithmic scale. This means that the averaging operation occurs on log values. Mathematically, if we take a series of numbers, convert them to a log scale, and then take the average of the log values, we will get a different result than if we average the numbers first and then take the log of that average. In the case of measuring Gaussian noise, the difference between the average of the log versus the log of the average is 2.5 dB. This turns out to be a useful trick, because using video bandwidth filtering

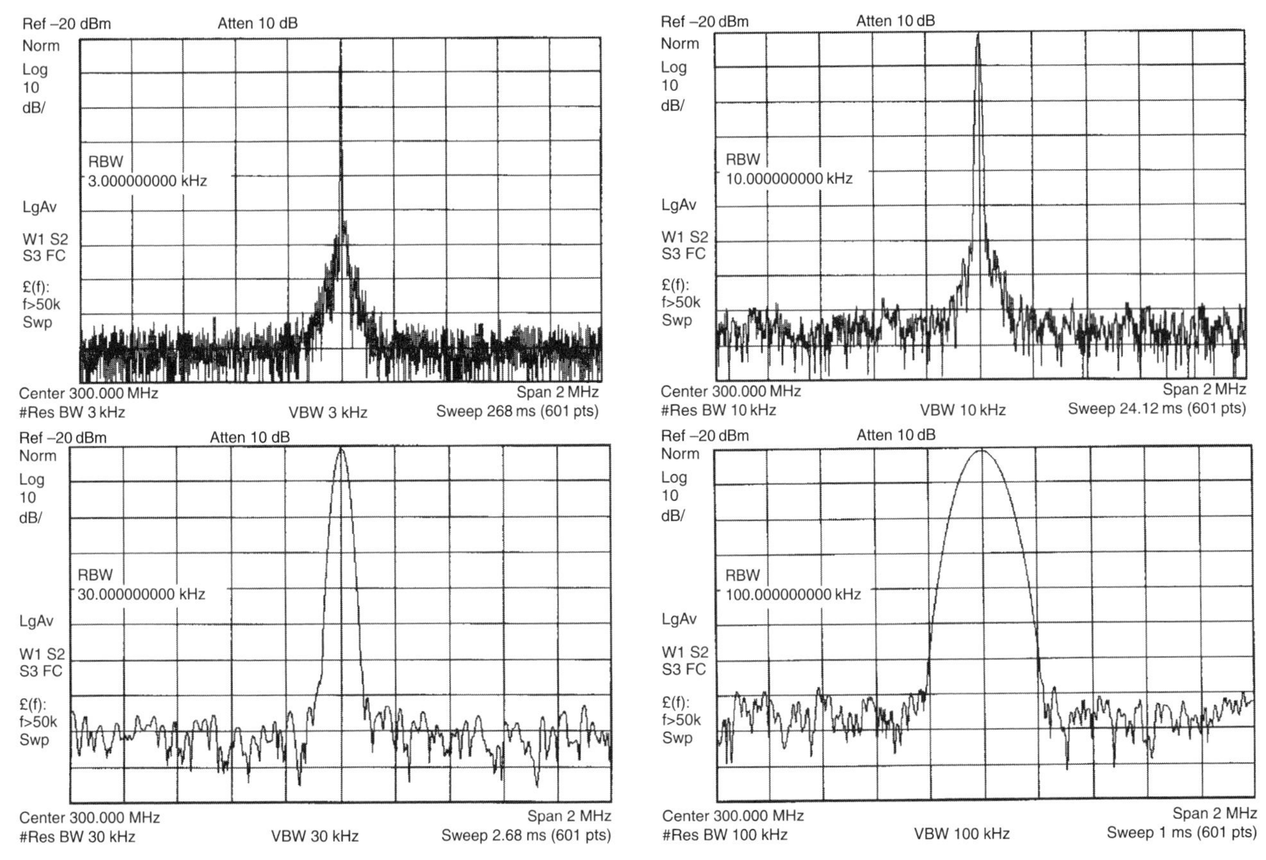

Figure 9.6 Resolution bandwidth.

will drop the noise floor of the measurement by 2.5 dB. In the case of measuring a continuous amplitude signal, there is no difference between the log of the average versus the average of the log. Thus, if we are trying to measure a spur or other small signal that is near the noise floor, using video bandwidth filtering will allow us to see 2.5 dB more of that signal than what we would be able to see otherwise.

Note that this 2.5 dB factor only applies to noise signals, with a Gaussian distribution of the signal energy. Today's digital modulation formats often appear noiselike when displayed on a spectrum analyzer. However, the statistics of the signal energy are not typically Gaussian, so the 2.5 dB correction factor cannot be used; it is some other unknown amount. Therefore, video bandwidth (log power) averaging should not be applied to measurements of digitally modulated signals in order to prevent underreporting of the power level. To ensure that video bandwidth filtering is not applied to the signal, the rule of thumb is that the video bandwidth should be set to 10 times larger than the resolution bandwidth for the measurement in question. Some modern high performance spectrum analyzers allow us to choose RMS power averaging, which applies the averaging before log scaling, thus avoiding the underreporting problem for digitally modulated signals.

Markers

Similar to other test instrumentation, the spectrum analyzer has a marker capability that can read out the power level at a specific frequency. Several markers can be set at different frequencies, and their values can be shown in a marker table. If we are trying to read a power level for a signal that is near the noise floor, the value may be difficult to read because the value fluctuates too much. The contribution of the noise energy passing through the resolution bandwidth filter is significant relative to a very weak signal, so the total energy hitting the sensor fluctuates. The solution is to use the narrowest resolution bandwidth filter that is reasonable, in order to lower the noise floor. The use of video bandwidth filtering can help in this case as well by lowering the noise floor by a further 2.5 dB.

9.5 POWER SUITE MEASUREMENTS

In addition to measuring power levels at individual frequencies using a marker, most spectrum analyzers include a suite of measurements that integrate the power over a set bandwidth so that we can measure the total power in a communications channel, for example. Another measurement that can be made in this way is to measure the occupied bandwidth for a given signal.

9.6 BASIC MODULATION FORMATS

Although measuring continuous wave [CW (unchanging)] signals is a valuable part of performing RF measurements, in practical applications RF signals need to carry information from one point to another in order to be useful. Figure 9.7 shows two

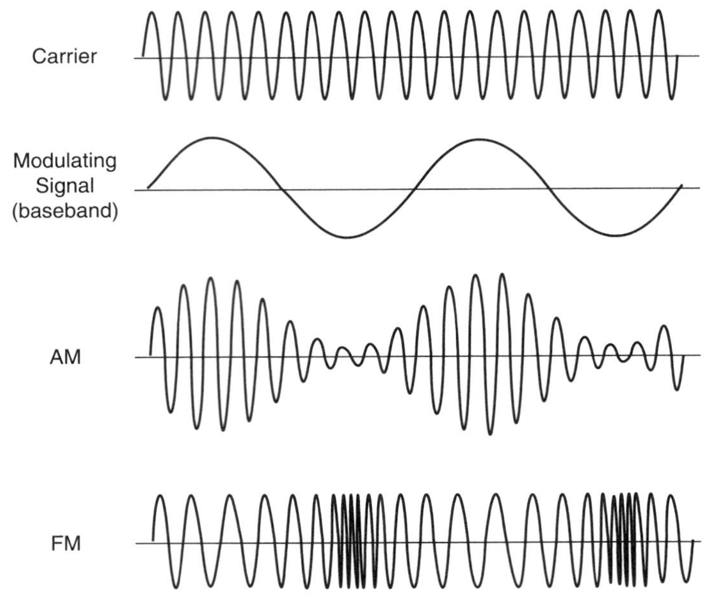

Figure 9.7 Modulated waveforms.

analog forms of modulation that have traditionally been used to "impress" information at a low frequency onto an RF carrier that is operating at a much higher frequency. Amplitude modulation is easy to visualize because as the information signal amplitude changes, the amplitude of the RF carrier is proportionally changed as well. Unfortunately, this format is susceptible to noise because most noise affects the amplitude portion of the signal.

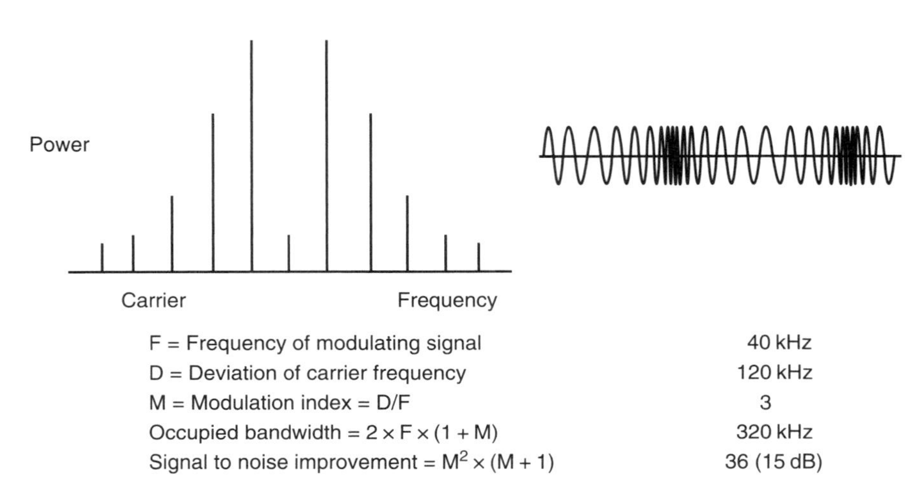

Figure 9.8 FM spectrum.

Frequency:	2.45 GHz
Power:	0 dBm
Internal modulation:	FM
Modulation frequency:	40 kHz
Carrier deviation:	40 kHz
RF:	On
Modulation:	Off/On

Figure 9.9 Settings for the signal generator.

Another traditional format is FM. In this case, as the information signal becomes positive in amplitude, the RF carrier is decreased in frequency. When the signal is negative, the frequency is increased. This process results in the generation of an infinite number of side bands, which can be predicted by Bessel functions. Fortunately, the side bands become weaker as we move away from the carrier frequency, so we only need to consider those side bands that contain 99% of the signal power. A Bell Labs engineer named Carson came up with the calculation to predict the bandwidth required for a given FM format. The calculation is based on the frequency of the modulating signal and the deviation of the carrier frequency (how much the frequency of the RF carrier is changed). Carson's rule is shown in Figure 9.8.

Using Carson's rule, we can set up an FM modulated signal and predict how much bandwidth the signal will occupy. We can then use the occupied bandwidth application on the spectrum analyzer to measure the signal bandwidth. Figure 9.9 shows the signal parameters that are set up on the signal generator. Using Carson's rule, the predicted occupied bandwidth for 99% of the signal power is 160 kHz. Using the power suite measurement for occupied bandwidth, the result is 160.1 kHz as shown in Figure 9.10.

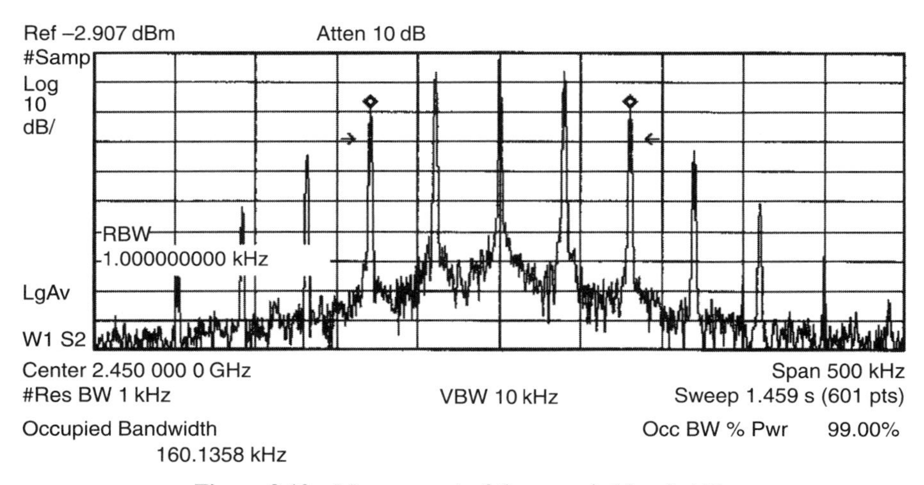

Figure 9.10 Measurement of the occupied bandwidth.

With the availability of built-in applications to integrate power over specific bandwidths, the spectrum analyzer is a useful and versatile instrument.

9.7 EXAMPLE SPECTRUM ANALYZER OPERATION AND FM SPECTRUM MEASUREMENT

Objective

The objective is to demonstrate the basic operation of the spectrum analyzer and signal generator by observing both a CW (unmodulated) signal and an FM modulated signal. The signal from the signal generator is input directly to the spectrum analyzer.

FM Theory

The formula for occupied bandwidth (OccupiedBW) of an FM signal is

$$\text{OccupiedBW} = 2 \times \text{FMrate} \times (1 + M),$$

where M is the FM deviation divided by the FM rate.

Measurements Being Demonstrated

Measurement of CW signal at 300 MHz

Observe the effect of spectrum analyzer controls: reference level, scale, resolution bandwidth, video bandwidth, attenuation

Demonstrate the spectrum analyzer display features: markers, peak search, tables

Demonstrate channel power measurement capability: occupied bandwidth for FM signal

Generic Procedure

Steps to perform the measurements are listed in the upcoming table. The left column lists the actions to be performed, and the right column lists the motivation for taking each step and provides additional background information in some cases. The instructions are intended to be generic across most spectrum analyzer models.

Procedure	Notes
Preset both the signal generator and the spectrum analyzer by pressing the green Preset key.	Presetting the instrument returns all settings to their factory default. This is useful because we can set up our measurement without worrying about an obscure option that may have been set by another user and affecting the measurement.

(Continued)

Procedure	Notes
Set the signal generator frequency to 300 MHz and the amplitude to −20 dBm.	
Set the spectrum analyzer center frequency to 300 MHz and the span to 2 MHz.	We set the center frequency to be able to view the signal. The span is chosen to be able to demonstrate the effect of different resolution bandwidths at a later step.
Using the amplitude key, set the reference level to −20 dBm.	This moves the peak of the signal to the top of the display. The reference level is fixed as the top line of the display.
Increase the attenuation by overriding the automatic setting.	The spectrum analyzer normally sets the input attenuator automatically based on the value that we choose for the reference level. If we choose a large value for the reference level, then the spectrum analyzer will increase the attenuation at the input to make sure that the analog components inside the spectrum analyzer do not become overloaded and distort the measurement. The spectrum analyzer keeps track of the attenuation and corrects for it so that the displayed values are unchanged. If we increase the input attenuation, the signal inside the spectrum analyzer is weaker, so we will see the noise floor increase on the display.
Set the attenuation back to the automatic setting.	
Activate a marker. Use the peak search button.	These features let us quickly read a value for a signal's frequency and amplitude.
Adjust the resolution bandwidth. This is usually found under Bandwidth, Averaging, or a similar heading. Set the resolution bandwidth to 3, 10, 30, and 100 kHz.	Note that in order to resolve closely spaced signals, a narrow resolution bandwidth setting must be used. The trade-off for using a narrow resolution bandwidth is that the sweep time is dramatically increased. The availability of narrow resolution bandwidth settings generally varies based on the cost of the test instrument (higher cost – narrower resolution bandwidth available).
Adjust the video bandwidth to a value that is one-third or smaller than the current resolution bandwidth setting.	The video bandwidth filter is a low pass filter that appears after the detector. By removing rapid changes in the amplitude of the detected envelope, this has the effect of averaging the signal. Because we are taking the average of a log-based value, random noise is reduced by about 2 dB.

(*Continued*)

Procedure	Notes
	Mathematically, this is because the log of a very weak signal can approach $-\infty$ dBm, whereas a very large signal is a finite number. This form of averaging is very useful for CW signals, but it should be avoided with random noise-like digitally modulated signals.
Set the resolution bandwidth and video bandwidth back to the automatic setting.	
Observation of FM signal:	
Set the signal generator frequency to 2.45 GHz and amplitude to 0 dBm.	
Turn on FM modulation with an FM rate of 40 kHz and FM deviation of 40 kHz.	
Set the spectrum analyzer to a center frequency of 2.45 GHz, a span of 500 kHz, and a resolution bandwidth of 1 kHz.	Observe the peaks of the FM modulated signal.
Set the scale to linear. This option is often found under the Amplitude key.	Observe that on the linear scale the side band amplitudes appear relatively smaller as we get away from the center frequency.
Set the scale back to log.	
Activate the built-in measurement for occupied bandwidth. Often, this can be found under the Measure key or a similar heading.	The spectrum analyzer can sum the power over a bandwidth and report either how much power is in a given bandwidth or how much bandwidth is used by a given signal power. Other channel power measurements are also available.
The occupied bandwidth should be 160 kHz.	This is because currently on the signal generator the FM rate = FM deviation, so $M = 1$. $2 \times (40) \times (1 + 1) = 160$
Set the FM deviation on the signal generator to 120 kHz. The occupied bandwidth should be 320 kHz.	$M = 3$. $2 \times (40) \times (1 + 3) = 320$

9.8 ANNOTATED BIBLIOGRAPHY

1. PSA Series Spectrum Analyzers Brochure, Document 5980-1283E, Agilent Technologies Santa Clara, CA.
2. PSA Series, Document 5980-1284E, Agilent Technologies Santa Clara, CA.
3. Self Guided Demonstration for Spectrum Analysis, Document 5988-3698EN, Agilent Technologies Santa Clara, CA.

4. Rappaport, T. S., Modulation techniques for mobile radio, In Wireless Communications Principles and Practice, Chap. 5, Prentice Hall, Upper Saddle River, NJ, 1996.

Reference 1 describes the family of Agilent PSA spectrum analyzers.

Reference 2 describes the detailed performance of a specific PSA model.

Reference 3 is a self-guided demonstration brochure for the PSA.

Reference 4 provides a detailed description of FM theory and implementation.

CHAPTER 10

VSAs

10.1 WHAT A VSA DOES

The VSA is one of the newest instruments that has been added to the RF measurement lab (Fig. 10.1). One of the key elements of the VSA that differentiates it from the spectrum analyzer is that the VSA captures measurement data in the time domain, whereas the spectrum analyzer works in the frequency domain. The VSA can record the amplitude envelope variations of the signal, as well as phase variations. This information can be used to demodulate any signal that can be generated, limited mainly by the size of the signal bandwidth. In order to successfully demodulate and analyze a signal, the instrument needs to have all of the modulation parameters entered, such as baseband filtering type and symbol rate. For each sweep of the analyzer display, the analyzer demodulates the signal and determines which symbols were sent in the measured signal. The analyzer then recreates a "perfect" reference signal based on those symbols. The difference in amplitude and phase between the ideal reference signal and the measured signal yields the error vector.

10.2 VSA EQUIPMENT

VSAs are available as stand-alone pieces of equipment, with the analysis software loaded into the instrument's firmware or hard drive. Currently, analysis software is also available as a separate product that can be run on a PC, with another instrument

RF Measurements for Cellular Phones and Wireless Data Systems. By A. W. Scott and R. Frobenius

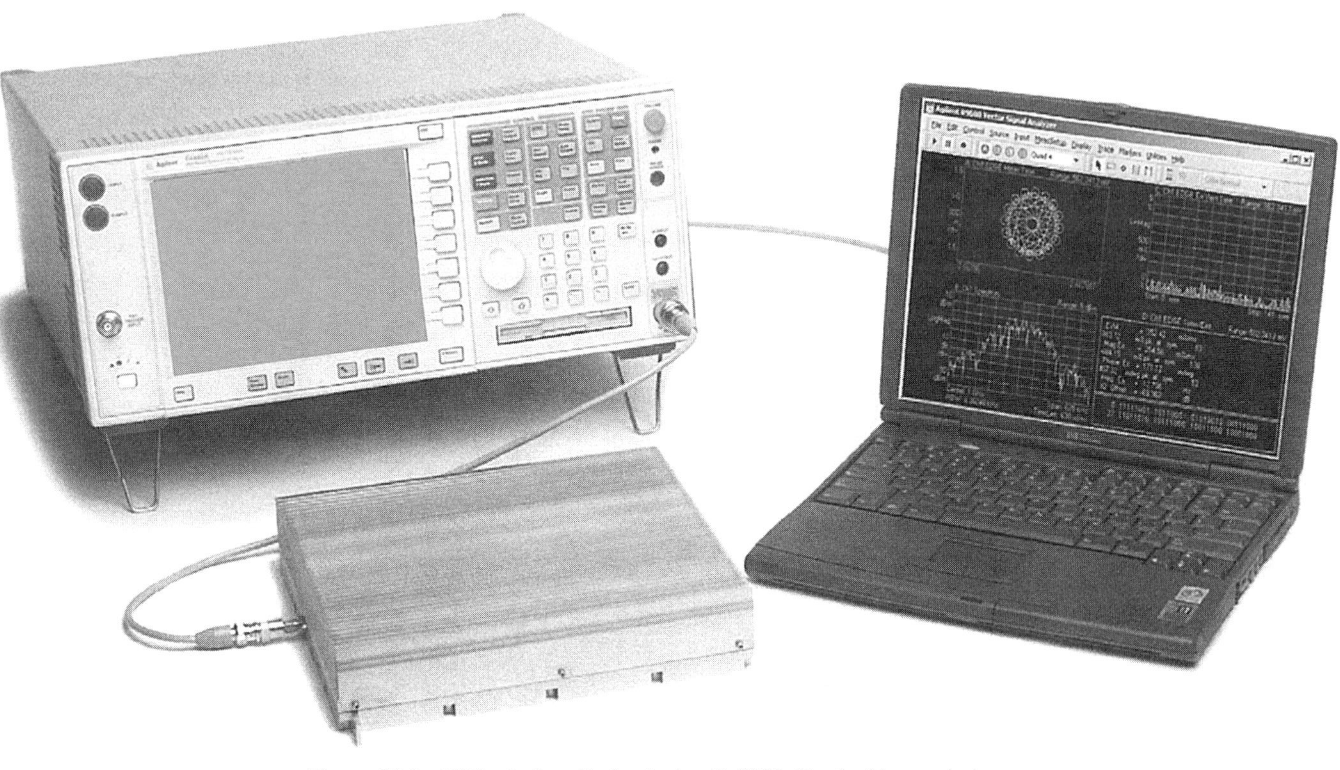

Figure 10.1 VSA. Agilent Technologies © 2008. Used with permission.

acting as a data acquisition hardware front end. For example, you can use the Agilent PSA series spectrum analyzer, with appropriate hardware options installed, as the front-end instrument to capture the signal from the output of your DUT. A properly configured PC can then run the analysis software, which connects to the PSA and assumes all control of the instrument. The front panel of the PSA is disabled when the software is connected to the machine. Other instruments that can act as a data acquisition front end for the analysis software include advanced digital oscilloscopes and specially designed hardware that works with the VXI chassis specification. The performance of these different configurations varies because of limitations of the hardware. The main parameter of interest in the performance is the maximum bandwidth of the signal that can be analyzed. In earlier commercial wireless systems, such as 2G cell phones, the bandwidth of the signal was limited to 30 kHz for IS-136 or 200 KHz for GSM. These relatively narrow signal bandwidths could be easily accommodated by the analysis hardware. Newer CDMA signals vary from 1.25 to 6 MHz, and broadband data systems continue to use broader bandwidths to achieve higher data rates. Earlier models of the VSA had analysis bandwidths of 4–5 MHz, whereas current models have approximately 80 MHz. The upper limit on signal analysis bandwidth is constantly improving.

10.3 WHAT THE VSA CAN MEASURE

Figures 10.2–10.10 show VSA measurements of a $\pi/4$DQPSK modulated signal used in the original digital cell phone systems in the United States and Japan.

This modulation was described in Chapter 4.

Because the VSA can measure the amplitude envelope and the phase of the RF carrier, this information can be presented in a variety of ways that help analyze and troubleshoot the system. Figure 10.2 shows a constellation diagram. This

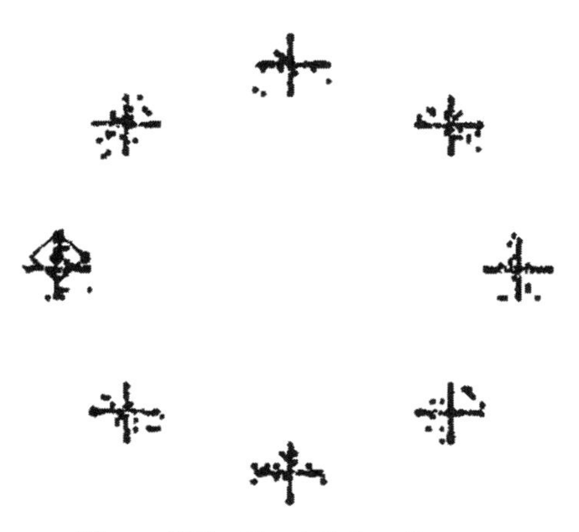

Figure 10.2 Constellation diagram.

diagram shows crosshairs for the ideal combinations of amplitude and phase that have been defined for that modulation format as representing symbols. In this case each represent 2 bits. The analysis software draws a dot at the sample times, which shows the amplitude as a radial distance from the center of the display, and the phase. Ideally, all of the dots will land exactly on the crosshairs. As the signal becomes distorted, the amplitude and/or phase will deviate from the ideal values and the dots will spread around the crosshairs. From this display you can visually see the effects of compression, where the dots will not reach out far enough to hit the cross, or phase distortion, where the dots will spread out in arcs through the crosses.

Figure 10.3 provides the I-Q diagram, which illustrates the same information as the constellation diagram along with the addition of traces showing the signal amplitude and phase between sample periods. Why does the signal seems so random and why does the amplitude overshoots the outermost sample targets on its way to the next symbol point? The answer lies in the fact that even though many modulation formats are phase-based modulation schemes, meaning they only rely on changes in phase to represent different data points, the amplitude of the signal varies substantially. The reason for the amplitude variation is that if we simply changed the phase of the signal abruptly, the signal would occupy an infinite bandwidth. In order to minimize the occupied bandwidth of the signal, filtering is applied to the baseband signal to make the phase transitions more gradual. The effects of this baseband filtering are shown in Figure 10.4. One of the effects of this filtering is to alter the amplitude of the signal, so that instead of sharp corners appearing in the constellation as in the left of the figure, the signal becomes smoother and curved, as in the right end of the figure. In general, the more abrupt a transition occurs in the time domain, the more spectrum is required in the frequency domain to represent that signal. To ensure that the signal

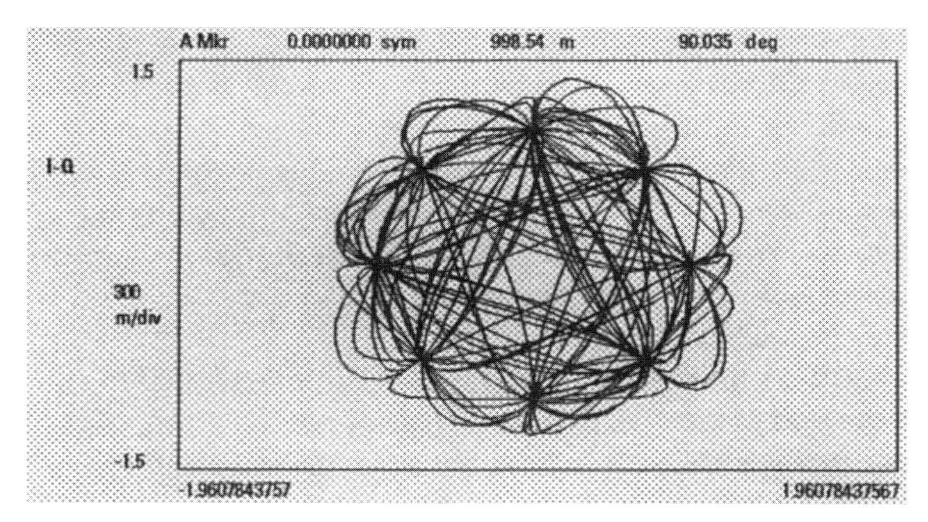

Figure 10.3 Vector diagram. Agilent Technologies © 2008. Used with permission.

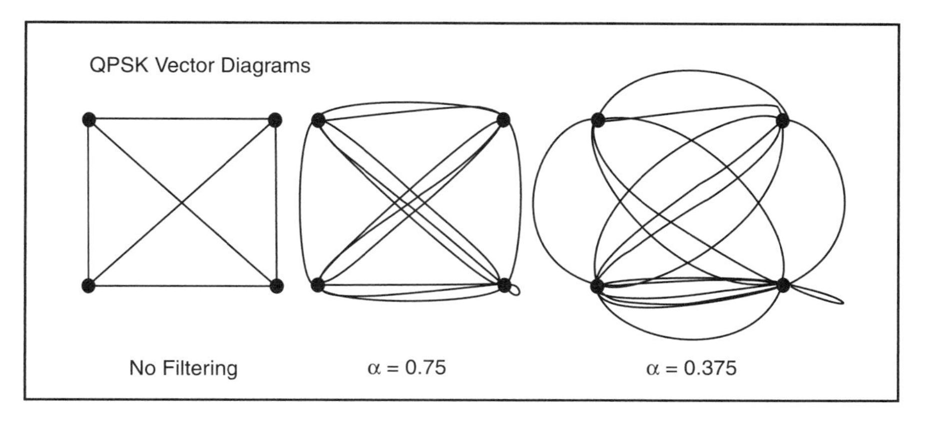

Figure 10.4 Effect of filtering on vector diagram.

occupies the predicted or specified amount of spectrum, the amplitude and phase of the signal need to track the ideal trajectory through the constellation at all times, not just at the sample times.

Figure 10.5 is an eye diagram. In the world of digital circuits, eye diagrams are widely used to analyze the quality of digital or baseband signals. Because baseband signals do not have a phase component to them, there is only one eye diagram that describes the signal amplitude versus time. RF signals have both amplitude and phase, so how can we represent the signal and account for the phase information? The answer lies in splitting the signal into two eye diagrams. One eye diagram represents the amplitude with respect to the horizontal, or the "in-phase" axis. Another eye diagram represents the amplitude with respect to the vertical, or the "quadrature" axis. Thus, at RF frequencies, an eye diagram can look at how the

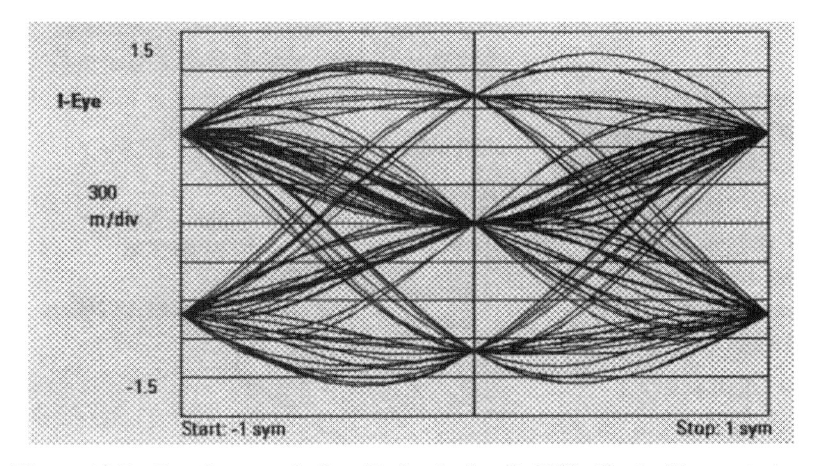

Figure 10.5 Eye diagram. Agilent Technologies © 2008. Used with permission.

signal is moving along either the vertical or horizontal axis of the constellation versus time. Because the horizontal axis of an eye diagram represents time, you can use eye diagrams to check on timing issues in the signal. We can tell whether the signal reaches a specific level too early or too late. Shifts in the timing of the signal can be caused by issues such as group delay from filters or other sources.

The quantity EVM provides the necessary quantitative information. EVM is explained in Figure 10.6, which is the constellation diagram for $\pi/4$DQPSK modulation. The eight ideal points are shown. Each can be represented by a vector of normalized signal amplitude extending from the origin to one of the eight phase locations. The amplitude and phase of an actual signal point is also shown. Because of signal distortions the amplitude of the vector is greater than it should be and the phase is also greater. Both of the phase and amplitude differences between the ideal and the actual signal could be measured; but to simplify the analysis, an error vector is defined. The EVM can then be determined. Note that the amplitude of the EVM is affected by both phase and amplitude errors. The EVM is a single metric quantifying how bad the error is. It is usually represented as a percentage of the distance from the origin of the constellation to its outermost point.

The allowed EVM in a particular system depends on how much information is to be obtained from the modulated signal. With $\pi/4$DQPSK modulation, where 2 bits/symbol are to be obtained, the allowed EVM is about 12%. With 64QAM modulation, where 6 bits/symbol are to be obtained, the allowed EVM is about 1%.

Figure 10.7 contrasts the EVM versus time. This display can be useful to troubleshoot pulsed systems, because if the errors occur mainly at the beginning or the end of the pulse, then we may have problems with the timing of amplifiers turning on too late or off too soon.

The tabulated data in Figure 10.8 tell us a great deal about the source of errors. The table shows the contributions of several factors to the overall error of the signal. If the error is primarily phase based, then we know to look for phase related problems, such

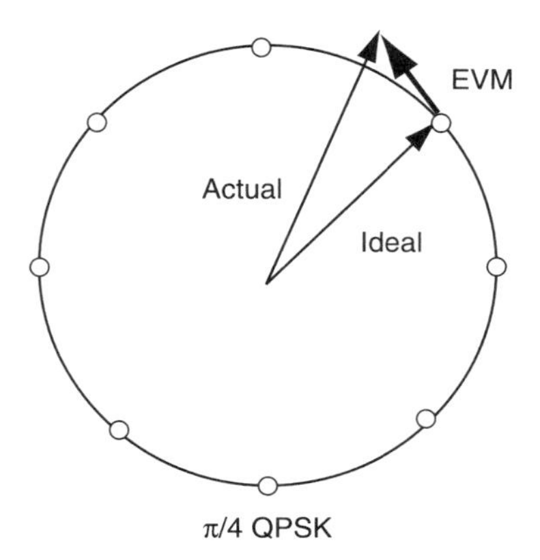

Figure 10.6 EVM is the ratio of the amplitude of the error vector relative to the actual signal, expressed as a percentage.

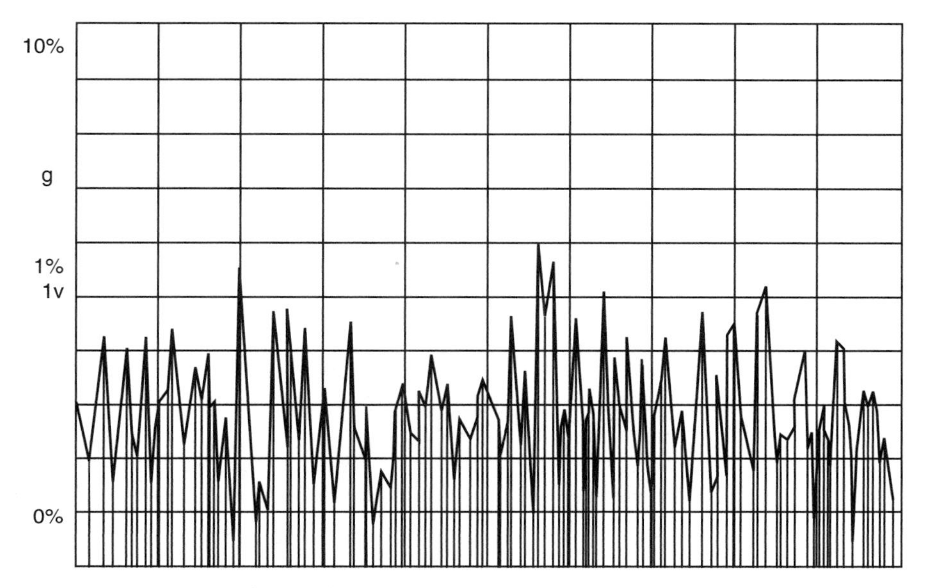

Figure 10.7 EVM for demodulated symbols.

as AM to PM conversion and group delay. If the error is magnitude based, then we can look elsewhere. The table also provides information about modulation quality. For example, it can tell is if our I and Q channels are out of balance, or if they are not exactly 90° apart in phase. The table also includes the demodulated bits. Depending on the type of signal that we are using, we may be able to recognize whether the correct bits were decoded or determine if a specific portion of the signal is causing a problem.

EVM	=	3.2620	%rms
5.3580	%	pk at sym	36
Mag Err	=	2.2522	%rms
–5.2352	%	pk at sym	36
Phase Err	=	1.3472	deg
2.8051	deg	pk at sym	35
Freq Err	=	5.0862	Hz
IQ Offset	=	–48.287	dB
Amp Droop	=	–130.5	udB/sym

0	01001000	01100111
16	00001011	11011011
32	00110100	00111011
48	11000011	11111110
64	00001111	01111100

Figure 10.8 Symbol table and error summary.

This is a beneficial analysis because it tells us troubleshooting information based solely on the output signal from the system. We do not need a probe point anywhere in the internal circuitry, which nowadays can be impossible because of the high levels of integration that are used.

Figure 10.9 shows the measurements we described as they might be used on the analyzer. Depending on the hardware configuration, four, six, or even more analysis traces can be displayed at the same time.

Figure 10.10 provides a measurement of an RF power amplifier with different signal levels applied to its input. As the signal level is increased, the ability of the power amplifier to accurately reproduce the signal at its output is taxed. As the signal swing becomes greater and greater at the input, the output amplitude swing is limited by the DC power supply and the amplitude envelope does not quite reach the level that it is supposed to reach. The result is increasing EVM as shown in the trace of EVM versus time and by the spreading of the sample points on the constellation. Another factor that affects power amplifiers is the AM to PM conversion, where the large signal swings inside the amplifier actually change the size of the junctions on the semiconductor. These changes affect the internal capacitances, which in turn cause an undesired change in the output phase. This phase distortion would show up as an increased phase component in the table of EVM errors and as the sample points spreading out in arc fashion on the constellation diagram.

The VSA is a useful tool for measuring RF system performance. This book is focused mainly on traditional RF component measurements. Several of the

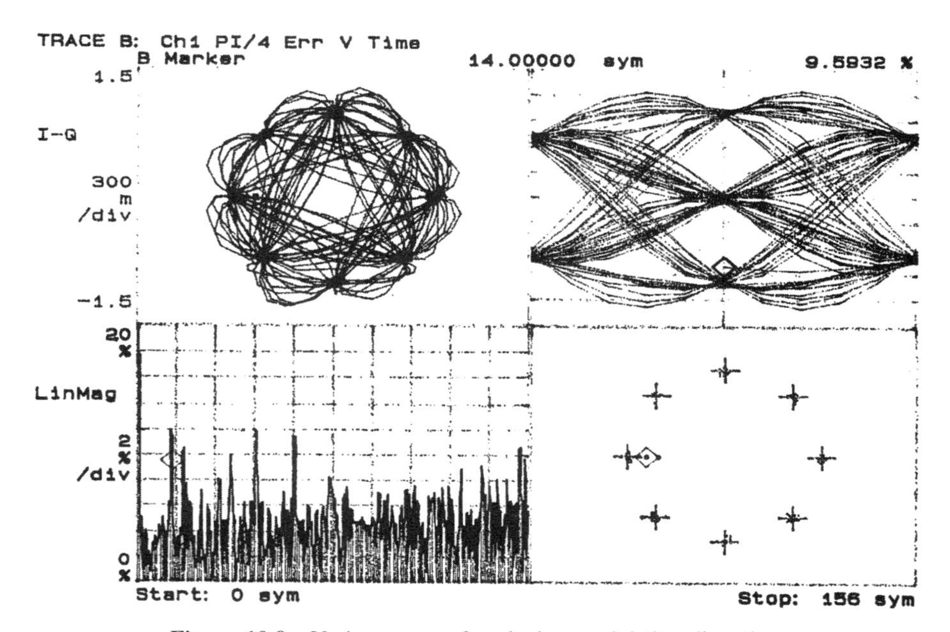

Figure 10.9 Various ways of analyzing modulation distortion.

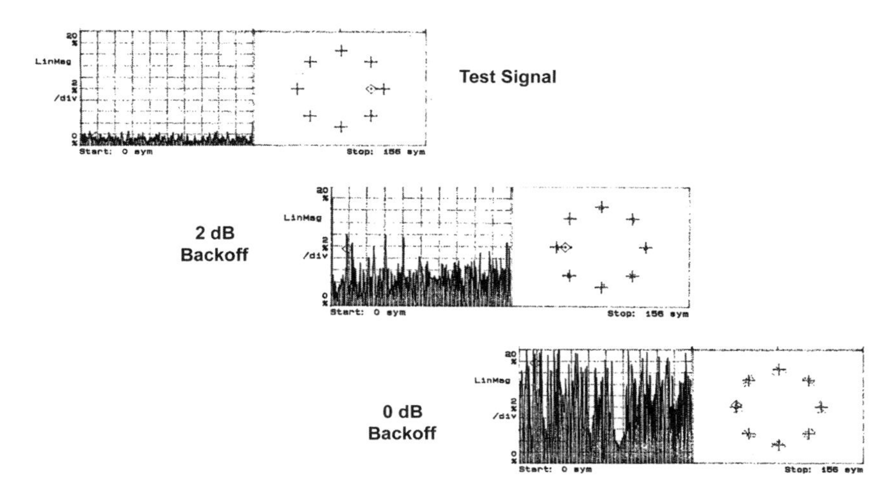

Figure 10.10 EVM of the $\pi/4$DQPSK signal for different backoffs from saturation.

component performance limitations affect system-level performance, and these limitations can be seen on the VSA.

10.4 ANNOTATED BIBLIOGRAPHY

1. Application Note 1314: Testing and Troubleshooting Digital RF Communications Receiver Designs, Agilent Technologies, Santa Clara, CA.
2. Product Note 8944-14: Using Error Vector Magnitude Measurements to Analyze and Troubleshoot Vector-Modulated Signals, Agilent Technologies, Santa Clara, CA.
3. Product Note 89400-14A: 10 Steps to a Perfect Demodulation Measurement, Agilent Technologies, Santa Clara, CA.

Reference 1 describes the application of EVM measurements to receiver tests, among other tests to be performed such as BER testing.

Reference 2 provides detailed information about setting up and troubleshooting with EVM measurements using a VSA.

Reference 3 can be considered a condensed version of the same information.

CHAPTER 11

NOISE FIGURE METERS

11.1 NOISE FIGURE METER SETUP

Figure 11.1 shows a typical noise figure test setup. The input to the DUT is a noise source, typically an avalanche diode. The special property of the avalanche diode is that when it is switched on the output is broadband noise at a level of about 14 dB above thermal noise. The noise diode has a control signal that turns it on and off. When the noise diode is turned off, the input to the DUT is thermal noise. When the noise diode is turned on, the input to the DUT is an approximately 14 dB larger signal with the same frequency characteristics as the thermal noise. The ratio between the two input signals is the input S/N to the DUT.

Note the input signal and the input noise are too small to be measured directly, but their ratio is set by the characteristics of the noise diode. The output S/N of the device can be measured, and the noise figure of the device can be determined. A photograph of the noise diode is provided in Figure 11.2.

11.2 NOISE FIGURE PRINCIPLES

The noise at the RF receiver determines the obtainable range of a communication system and the data rate that can be supported. Intuitively, the further a mobile receiver travels from a transmitter, the weaker the received signal becomes. Eventually the signal becomes so weak that it is buried in the noise energy. Even

RF Measurements for Cellular Phones and Wireless Data Systems. By A. W. Scott and R. Frobenius

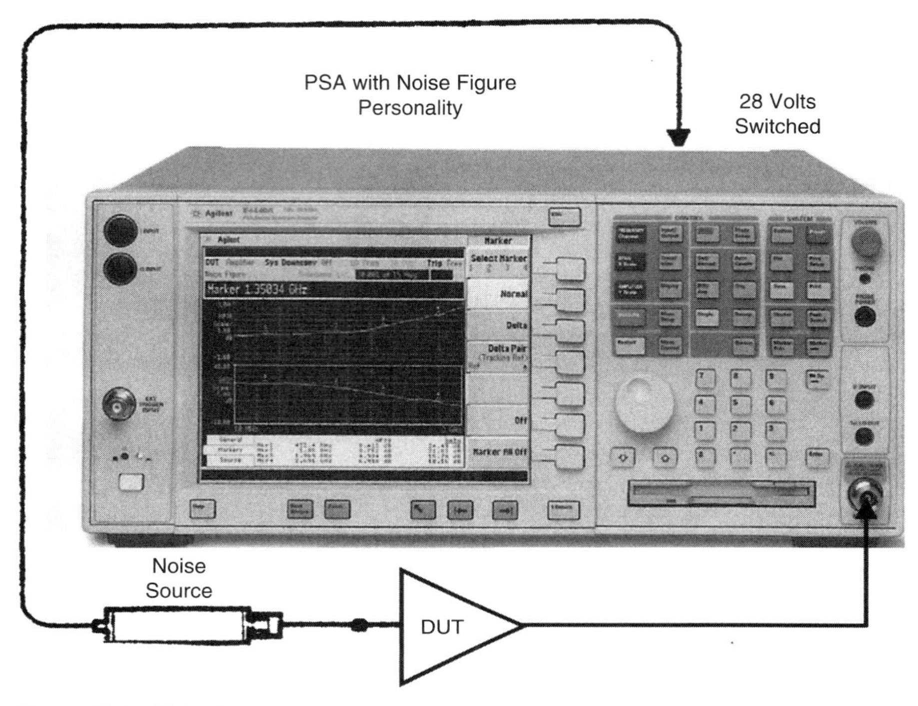

Figure 11.1 Noise figure test setup. Photo Agilent Technologies © 2008. Used with permission.

before the signal is buried, the noise energy distorts the signal, degrading both analog and digitally modulated information. The noise at the RF receiver determines both the maximum range of a wireless communications system and the data rate that can be supported. Noise comes from two sources: thermal noise radiating from all objects in the receiving antenna's field of view and noise generated by the first few RF components in the receiver.

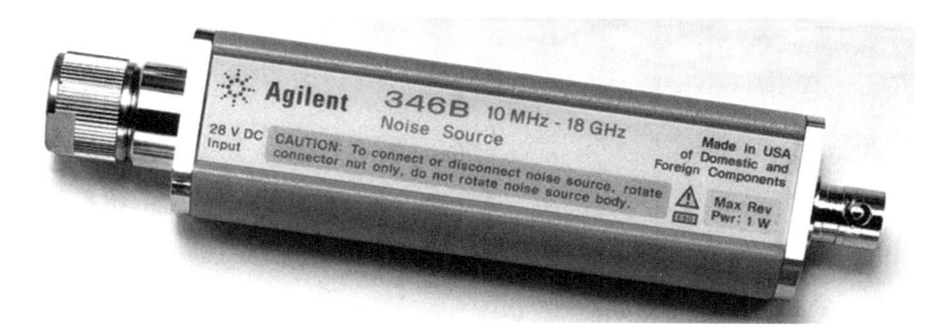

Figure 11.2 Noise diode. Agilent Technologies © 2008. Used with permission.

In the RF band, the thermal noise radiating from the surroundings is independent of frequency and is given by kTB, where k is the Boltzmann noise constant, which is a physical constant; T is the absolute temperature of the surroundings; and B is the RF bandwidth of the system (MHz).

Regardless of the actual temperature of the surroundings, the noise temperature is almost always assumed to be 293 K (20°C) for noise calculations. If we use the above formula to calculate the noise power in 1 Hz, the result is −174 dBm/Hz. If the system bandwidth is 1 MHz, then there would be 1 million times as much noise power. In the decibel system, multiplying by a factor of 1 million is the same as adding 60 dB, so at room temperature the thermal noise power in 1 MHz of bandwidth is −174 dBm/Hz + 60 dB = −114 dBm/MHz, in 2 MHz the noise is −114 dBm + 3 dB = −111 dBm, in 25 MHz the noise is −114 dBm + 14 dB = −100 dBm, and in 0.1 MHz (100 kHz) the noise is −114 dBm − 10 dB = −124 dBm.

Figure 11.3 shows the S/N as a signal passes through an amplifier with 30 dB gain. At the input in this example, the signal is at −110 dBm and the noise is at −124 dBm, so the S/N is 14 dB. At the output the signal has been amplified by 30 dB to −80 dBm. Similarly, we would expect the noise to be amplified to −94 dBm. Unfortunately, we instead observe that not only has the thermal noise at the input been amplified but also additional noise has been generated within the amplifier, resulting in an extra 4 dB of noise at the output. The end result is that the S/N has been degraded by 4 dB. By definition, this is the noise figure of the device. The noise figure shows how much a S/N is degraded after a signal passes through a device.

The noise figure of passive components like an RF filter or a length of transmission line between the antenna and the LNA also have a noise figure, which is equal to their attenuation. With a lossy component, the RF signal is attenuated in passing through the component. Unfortunately, we cannot attenuate thermal noise. Instead, output noise is regenerated in the component, which is at room temperature. The S/N degrades because the noise remains constant through the lossy component, but the signal is attenuated. According to the definition of noise figure, this means that the noise figure is the same as the attenuation.

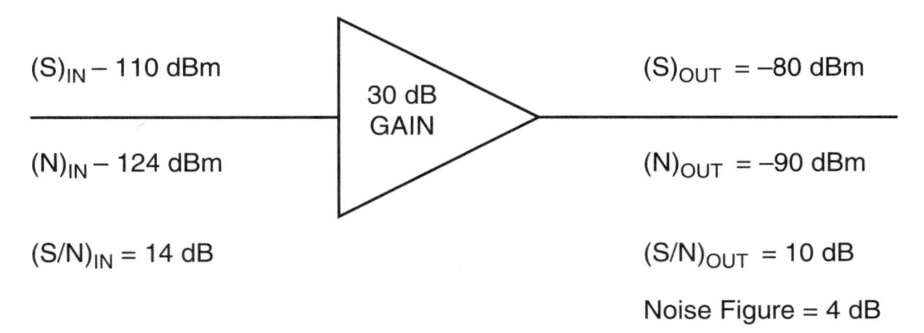

Figure 11.3 LNA. Noise figure is the ratio of the S/N power going into a device compared to the S/N coming out. Noise figure of a passive device, like a filter, is equal to its attenuation.

We might intuitively think that, because the noise figure is the comparison of S/Ns at the input and output of a device, the measurement of noise figure involves performing S/N measurements on our DUT. In practice, this is not the case. Instead, the "Y factor" method is a popular technique, which cleverly relies on two noise measurements to determine the noise figure, as follows: the fundamental formula for noise power is kTB. Note that k is a constant, and for a given system design, the bandwidth B is also a constant. Therefore, noise power simply increases linearly with temperature T, as shown in Figure 11.4. The slope of the line is kB, and if a device has gain then the slope versus temperature would be kBG, where G is the gain of the device. If we decrease the temperature to 0 K, then ideally the thermal noise power output of the device decreases until it goes to 0. For real devices, noise is also generated within the device; thus, instead of going to 0 the thermal noise power trace actually intersects the y axis at some value that is greater than 0. This y-intercept value represents the noise figure for the device. By calculating the slope of the line, we also determine the gain of the device, which is why noise figure meters usually report gain along with noise figure.

Realistically, we cannot attempt to reduce the temperature to 0 K and measure the noise figure. However, we can practically measure the noise power output from the device at room temperature and then raise the ambient temperature of the measurement setup by some known amount (i.e., 100°) and take another measurement. The thermal noise at the input would change, and we could measure the change in output noise power. Now there would be two points on the plot and we could extrapolate the line down to 0 K and determine the noise figure.

Needless to say, changing the ambient temperature of the test setup would be quite time consuming. Instead of actually changing the ambient temperature, an avalanche diode is used at the input of the device. When the diode is off, we can measure the noise output at room temperature. When the diode is biased with 28 V, it outputs a broadband noise signal that has the same properties as thermal noise at a much

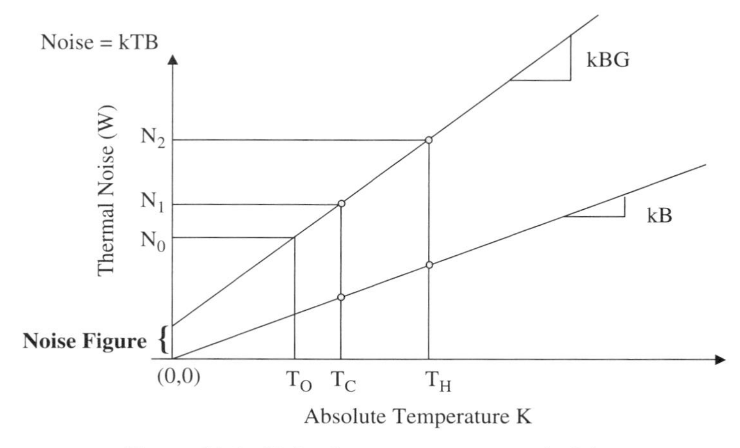

Figure 11.4 Noise figure measurement principles.

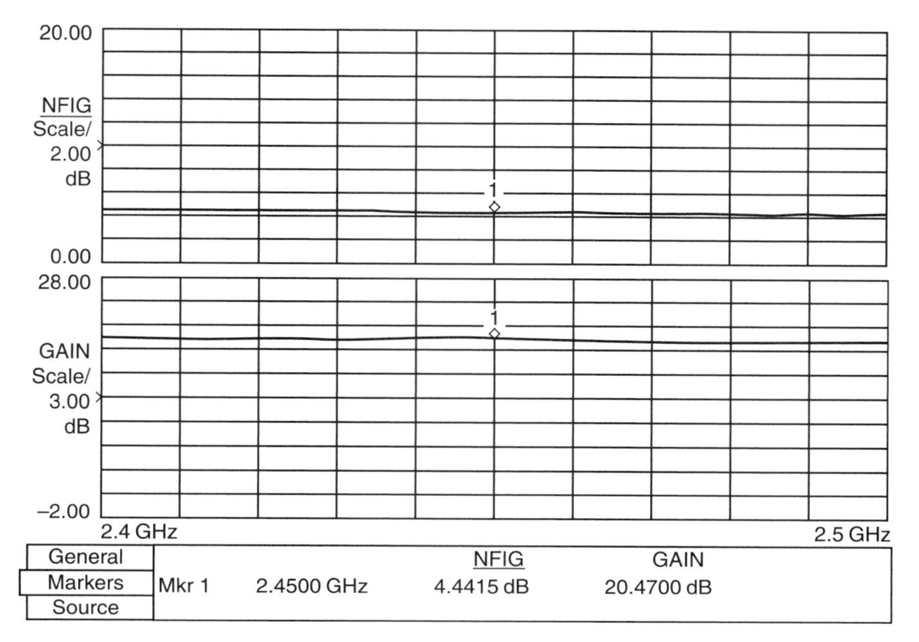

Figure 11.5 Measured noise figure and gain of LNA.

higher temperature. Using this diode we can rapidly take measurements at room temperature and at a predetermined higher temperature with well-defined accuracy.

The avalanche diode output is carefully measured by the manufacturer at several frequencies. The noise power when the diode is biased is reported in terms of ENR. The ENR value is a single number that tells about the equivalent hot and cold temperatures from the diode when it is biased and when it is off.

Figure 11.5 shows the spectrum analyzer display when it has been set to measure noise figure. The spectrum analyzer also measures gain while it is measuring noise figure, and it displays both as a function of RF frequency.

LNAs are discussed in detail in Chapter 23. Measurements of the noise figure of LNAs, RF filters, and mixers with the noise figure setup shown in Figure 11.1 are discussed in detail in Chapter 25.

11.3 ANNOTATED BIBLIOGRAPHY

1. Technical Overview with Self Guided Demonstration Option 212, Noise Figure Measurements Personality, Agilent Technologies, Santa Clara, CA.

Reference 1 is an Agilent brochure describing how to make noise figure measurements with the Agilent PSA E4440 spectrum analyzer, Option IDS built in preamplifier, and Option 219 noise figure measurement personality.

CHAPTER 12

COAXIAL CABLES AND CONNECTORS

12.1 COAXIAL CONNECTORS

Obtaining repeatable measurements hinges on being able to make connections between components and test instrumentation in a precise and reliable way. RF and high frequency signals also require maintaining a good impedance match each time a connection is made. High-quality cables and connectors will demonstrate low insertion loss and good matching properties across the frequency range of interest. As the frequency is increased, the wavelength gets smaller, and the physical tolerances become very demanding, especially for flexible test cables. Even with VNA measurements, where the effect of connectors and adapters is removed through two-port calibration, low-quality adapters will make calibration more sensitive to drift errors. Unfortunately, these demands are directly reflected in the price, with higher performing connectors and adapters costing proportionally more than those with poor matching and insertion loss specifications.

Connectors in particular are affected by proper care and cleaning. Because connectors by their very purpose are mated with other connectors in the lab, any dirt or damage on one connector will most likely propagate to other connectors in the lab, or even to the expensive front panel of one of the instruments. Therefore, it is important to keep connectors clean and to remove damaged connectors from the lab. As far as instrumentation is concerned, the connector on the front panel of your instrument can be protected by connecting a high-quality adapter to it and leaving it there

RF Measurements for Cellular Phones and Wireless Data Systems. By A. W. Scott and R. Frobenius

permanently. All connections are made to the adapter; if any damage occurs, simply replace the adapter instead of having to send the entire instrument for repair.

12.2 CABLES AND CONNECTORS BEST PRACTICES

The best practices for working with cables and connectors in the laboratory are summarized in the following table:

Do	Do Not
Keep connectors clean	Touch mating plane surfaces
Use plastic end caps during storage	Set connectors contact end down
Inspect all connectors carefully: look for metal particles, scratches, dents	Use a damaged connector (will damage other connectors)
Clean connectors with compressed air, isopropyl alcohol	Use abrasives
Clean connector threads	Allow liquid into plastic support beads
Use a gauge to verify connector before first use	Use an out of spec connector (will damage other connectors)
Align connectors carefully when making connections	Apply bending force to connection
Make preliminary connections lightly	Overtighten preliminary connection
Turn only the connector nut (keep center conductor stationary)	Twist or screw any connection
Use a torque wrench for final connect	Tighten past torque wrench "break" point
Turn off RF sources and active DUTs when making connections	Connect/disconnect loads from high power devices while signal is active

For maximum repeatability in measurements, many connector types specify a tightening torque value to be applied with each connection. Specialized torque wrenches are available to facilitate the connections. The general guideline is to apply torque until the wrench just begins to "give." Many experienced RF technicians often joke that their fingers have been "calibrated" to the correct torque for a given connection.

The frequency range of the connector is generally inversely proportional to the diameter of the conductors (smaller diameter = higher frequency capability). Note that in some cases the outer size of the connector may be several times larger than the mating surfaces that actually determine the frequency performance. Other factors such as the type of connection (threaded or snap-in) and conductor construction will limit performance.

12.3 POPULAR COAXIAL CABLE CONNECTORS

Popular coaxial cable connector types are presented in Figures 12.1–12.4.

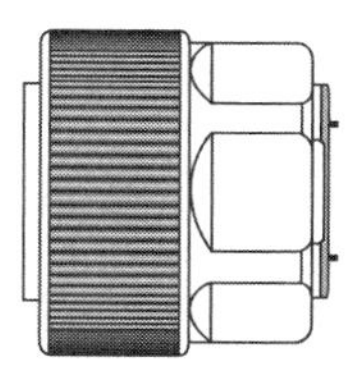

Figure 12.1 The Amphenol 7 mm precision connector (APC-7) provides the lowest reflection coefficient and highest repeatability for frequencies up to 18 GHz. Each connector is "sexless" because the threads can be retracted or extended to mate with the other connector. Note that when connected, one of the sleeves of each connected pair should spin freely and the other should be tight. If both sleeves are tightened, the mating surfaces can actually be pulled apart. Courtesy of Aeroflex-Inmet. Used with permission.

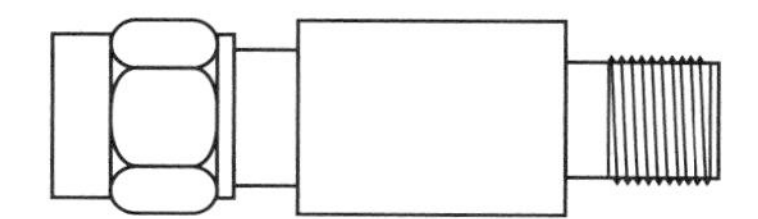

Figure 12.2 An SMA/3.5 mm coaxial connector is one of the most commonly used RF/microwave connectors. The frequency performance of SMA interconnects is useable to 18 GHz. Some high-quality SMA connectors are rated up to 26.5 GHz. The 3.5 mm connector was designed to provide performance to 34 GHz while still connecting compatibly with SMA connectors. This is achieved by adjusting the dimensions of the center and outer conductors, as well as using air dielectric instead of plastic. Note that the standard SMA connector must be within specifications; a damaged, bent, or improperly long center conductor will damage the 3.5 mm connector. Courtesy of Aeroflex-Inmet. Used with permission.

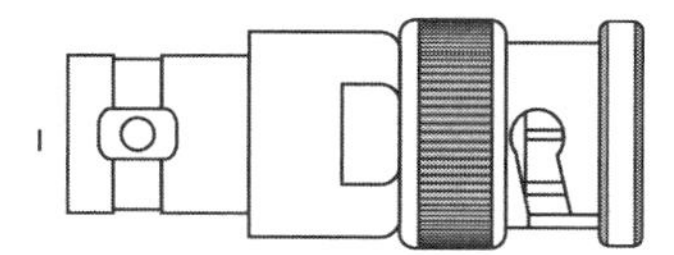

Figure 12.3 The bayonet Navy connector (BNC, also known as the bayonet Niel Concelman) is widely used for connections up to 2 GHz. Beyond 4 GHz, slots in the connector's conductor may radiate signals and the connector is generally not used above that frequency. A threaded version (TNC) helps resolve leakage for applications up to 12 GHz. Courtesy of Aeroflex-Inmet. Used with permission.

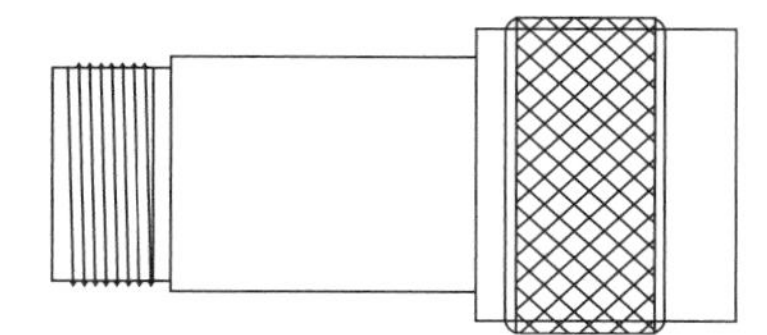

Figure 12.4 The Navy type (N) coaxial connector, which has been improved since its original military implementation in the 1940s, can now handle signals up to 18 GHz. Certain type N connectors that are designed for 75 Ω use are not compatible with the 50 Ω connectors because of different center conductor diameters. Courtesy of Aeroflex-Inmet. Used with permission.

12.4 COAXIAL CABLES

Similar to connector qualities, cables used in testing should also demonstrate good match and low insertion loss. In addition, high quality cables will provide stable performance when flexed in various ways while making connections to the DUT. Again, these qualities are more difficult to achieve for higher frequencies and result in a price premium for cables that perform consistently well. Even with high-quality cables and connectors, it is important to check all connections any time the test setup is moved or adjusted to make sure that none of the connections has become loosened.

12.5 ANNOTATED BIBLIOGRAPHY

1. Accessories Description, http://www.agilent.com/find/accessories, Agilent Technologies, Santa Clara, CA.
2. Williams, T., RF Connectors for Upper Frequencies, http://www.wa1mba.org/rfconn.htm.
3. Application Note 1287-3: Applying Error Correction to Network Analyzer Measurements, p. 5, Agilent Technologies, Santa Clara, CA.

References 1 and 2 provide additional information about connectors.

Reference 3 offers a brief discussion of the effects of connectors and adapters on network analyzer measurement uncertainty.

CHAPTER 13

RF MEASUREMENT UNCERTAINTIES

Care must be exercised to insure RF measurement accuracy. The correct technical term for the accuracy of RF measurements is "RF uncertainty." This chapter describes the uncertainty of RF measurements with the power meters, VNAs, spectrum analyzers, and noise figure meters described in previous chapters. Steps that can be taken to minimize measurement uncertainties are also presented.

With moderate care, RF power can be measured with an uncertainty of ± 0.5 dBm. Unless moderate care is taken, RF power measurements will be several dBm in error. With great care, RF power can be measured within an uncertainty of ± 0.2 dBm. By exercising extreme care, metrology labs can measure RF power to ± 0.1 dBm.

As discussed in Chapter 8, relative power (such as gain, loss, S-parameters) expressed in dB can be measured to within a few hundredths of a dB. To achieve this uncertainty level, relative power measurements must be made with a VNA using its calibration procedure to known standards. This calibration procedure was described in Chapter 8, and it is described in greater detail in Section 13.5.

This chapter is organized into six sections. Mismatch uncertainties, which affect all measurements of RF power, independent of the type of measurement instrument used, are discussed in Section 13.1 Measurement uncertainties of RF power meters, VNAs, RF spectrum analyzers, ratioed measurements (gain, loss, S-parameters), and noise figure measurements are described in Sections 13.2–13.6, respectively.

RF Measurements for Cellular Phones and Wireless Data Systems. By A. W. Scott and R. Frobenius

13.1 MISMATCH UNCERTAINTIES

The uncertainty of all RF power measurements, independent of the type of instruments used (power meters, network analyzers, spectrum analyzers), is affected by the RF mismatch of the source being measured and the RF mismatch of the instrument sensor.

This mismatch uncertainty is illustrated in Figure 13.1, which shows a generic setup for measuring power from a source with any type of instrument sensor. The source being measured has a mismatch reflection coefficient (ρ_1) looking into its output port, and the sensor has an input reflection coefficient (ρ_2). The bold arrow shows the RF signal coming from the source going to the sensor, which is the power to be measured. The other arrows show successive reflections back and forth between the instrument sensor and the source.

Consider first the bold arrow in Figure 13.1, which shows power coming from the source that is to be measured. As this power reaches the sensor, most of it enters the sensor, but some of it is reflected because of the sensor mismatch. The sensor mismatch can be independently measured to an accuracy of a few hundredths of a decibel and a correction made to the sensor reading. Although this is tedious and time consuming, this correction can be made.

However, there is more to the problem. As shown by the other arrows in Figure 13.1, this reflected power is reflected back to the sensor by the source mismatch and adds or subtracts from the original incident power, depending on the phase length of the connecting transmission line. The second reflection from the source repeats the process and so on.

As shown at the bottom of Figure 13.1, the measured source power is some value between $P(1 + \rho_1\rho_2)$ and $P(1 - \rho_1\rho_2)$, where P is the actual power of the source being measured, ρ_1 is the reflection coefficient of the source, and ρ_2 is the reflection coefficient of the sensor. If $\rho_2 = 0.2$, which a 14 dB return loss, and $\rho_2 = 0.1$, which is a 20 dB return loss, the uncertainty of the measurement will be ± 0.17 dB.

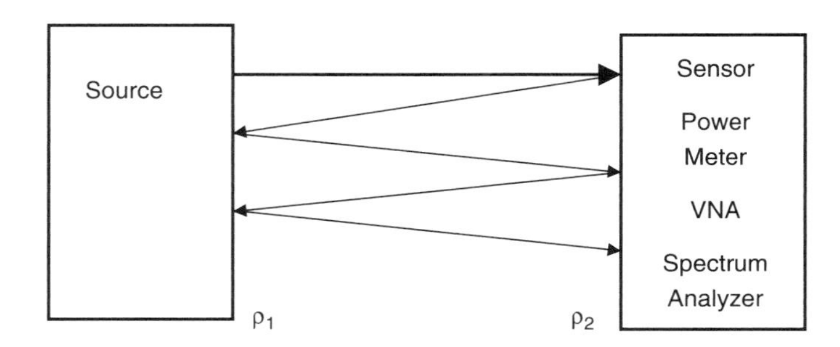

Figure 13.1 Mismatch uncertainty affects all power measurements. Measured power is some value between $P(1 + \rho_1\rho_2)$ and $P(1 - \rho_1\rho_2)$.

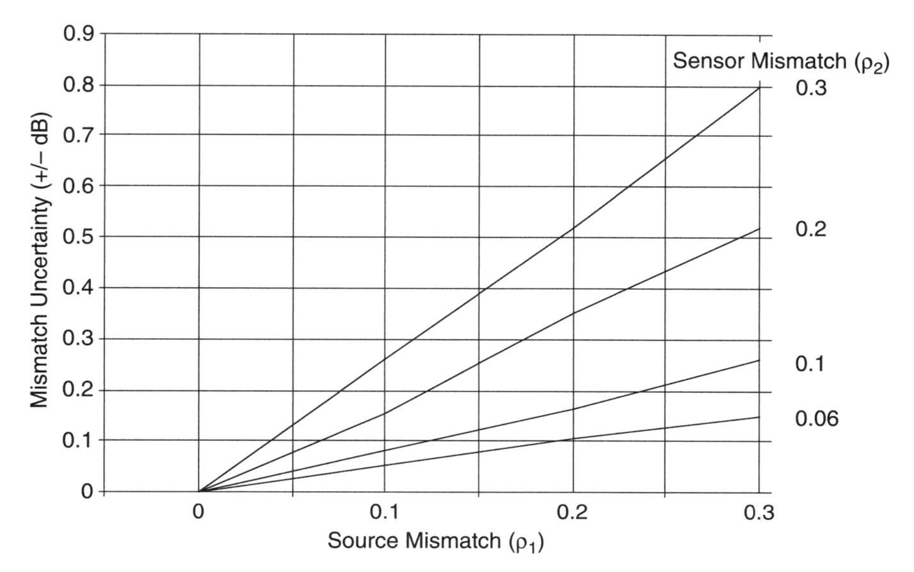

Figure 13.2 Mismatch uncertainty as a function of source mismatch.

Figure 13.2 shows measurement uncertainty as a function of the source mismatch for several values of sensor mismatch. The exact mismatch correction could be calculated by measuring the phase shift through the connecting cable at each measurement frequency. This would be a costly and time-consuming measurement procedure. Remember that the mismatch uncertainty is only part of the total measurement uncertainty, which must also include the inherent measurement uncertainties of the particular instruments used in the power measurement.

13.2 RF POWER METER MEASUREMENT UNCERTAINTIES

RF power meter measurement uncertainties include the following:

Mismatch uncertainties
Calibration factor
Magnification and offset

Mismatch Uncertainties

The mismatch uncertainties of RF power meters are those common to all RF measurement equipment, and they were discussed in the previous section. As discussed in Chapter 6, because power meters have relatively simple sensors, whose reflection coefficient can usually be maintained below 0.1, with reasonable source mismatches, the mismatch uncertainty can be held below ± 0.15 dB. For example, the power

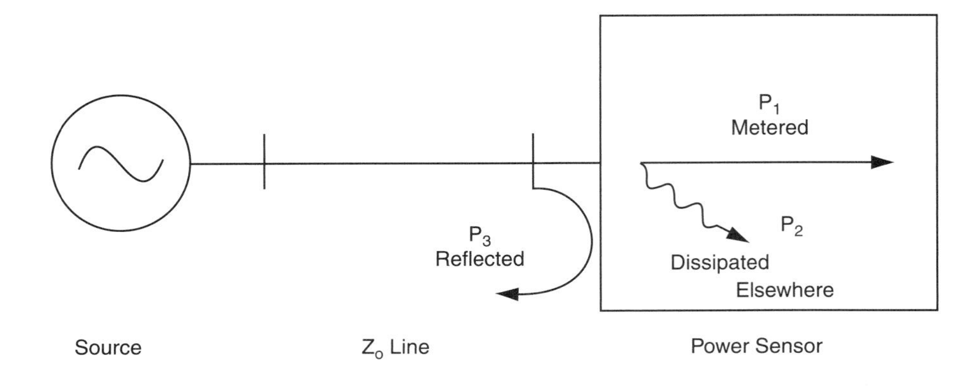

Figure 13.3 Power meter calibration factor.

meter sensor shown in Figures 6.3 and 6.4 has a reflection coefficient of 0.06 across the 50 MHz to 3 GHz frequency range. As shown in Figure 13.2, even with a source mismatch reflection coefficient of 0.2, the mismatch uncertainty is less than ± 0.07 dB.

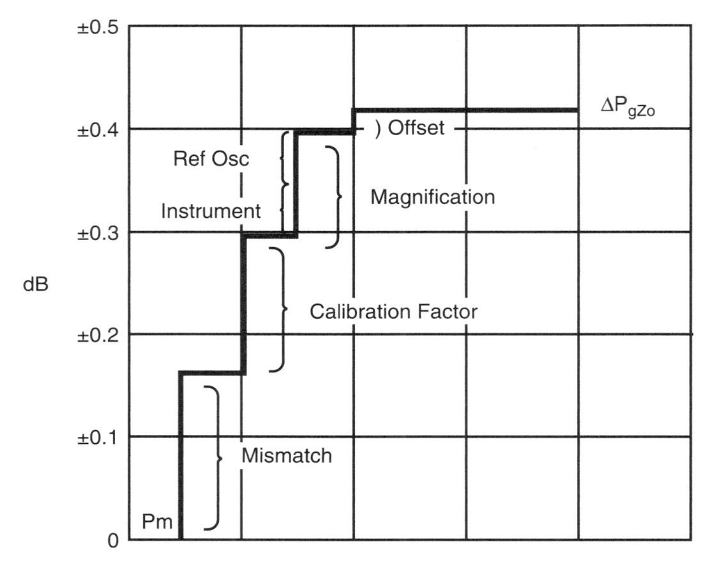

Figure 13.4 Individual power meter measurement uncertainties.

Calibration Factor Uncertainty

The calibration factor uncertainty is illustrated in Figure 13.3. The total power from the source is $P1 + P2 + P3$, where $P3$ is the RF reflected power that never gets into the power sensor, $P2$ is the RF power that gets absorbed as heat inside the power meter sensor, before entering the sensing diode, and $P1$ is the metered power.

As shown in the figure, the calibration factor is defined as

$$\frac{P1}{(P1 + P2 + P3)}$$

As described in Chapter 6, the power meter sensor manufacturer measures the uncertainty of all components of the power sensor and stores them in an EEPROM in the sensor as a function of frequency, temperature, and other operating conditions. However, these corrections can only be measured and stored at discrete power levels, frequencies, and temperatures. Under other conditions uncertainties still exist, which are called "sensor uncertainties."

Magnification and Offset

Finally, there are various uncertainties in the power meter electronics, including the accuracy of the 50 kHz reference source.

The effects of typical values of these three uncertainties that affect power meter measurements are shown in Figure 13.4. Note that this is just a representative example. Actual uncertainties can be greater or less than those shown. If all uncertainties added in the same direction, that is, all were plus or all were minus, the total uncertainty would be ± 0.4 dB $= \pm 10\%$. Typically these three uncertainties would be combined statistically, which can be estimated by a root sum of squares (RSS) combining that leads to an uncertainty for this example of ± 0.18 dB $= \pm 4.3\%$.

13.3 UNCERTAINTY OF VNA MEASUREMENT OF ABSOLUTE POWER

A VNA can measure relative power (gain, loss, S-parameters) with an uncertainty of a few hundredths of a decibel. This is accomplished using a manual or electronic calibration, referencing the measurements to known standards such as SOLT. This was discussed in Chapter 6 and is discussed in further detail in Section 13.5.

The VNA can measure absolute power on either of its measurement channels by setting the instrument to measure and display absolute power. The measurements are displayed as absolute power in dBm, but without calibration these measurements are only accurate to about ± 1 dBm. The manual or electronic calibration using standards only works for ratioed measurements like gain, loss, or S-parameters.

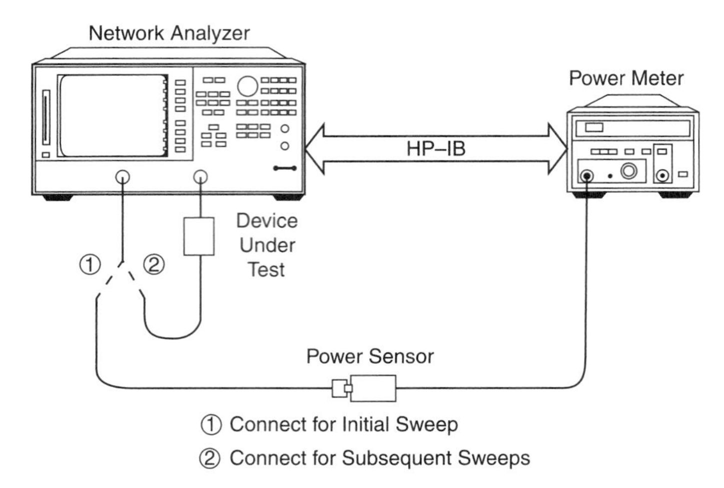

Figure 13.5 Use of a power meter to correct VNA power measurement.

However, the VNA can be calibrated with a power meter. This setup is provided in Figure 13.5. The power meter and the VNA are connected with a suitable data cable. Port 1 of the VNA is adjusted to measure *R*, which is the input signal. As shown in the figure, port 1 is first connected to the power sensor of a power meter, which has already been calibrated, and the VNA readings are then corrected to have power meter accuracy. The calibrated VNA source of port 1 is then used to calibrate port 2, which is set up to measure *B*, the output power of the DUT.

Figure 13.6 shows a swept measurement of an RF amplifier as a function of input power at a fixed RF frequency. The lower curve shows RF output power as a function of RF input power. The 1 dB compression point is shown by the marker, and the saturation characteristics of the amplifier are clearly displayed. The upper curve shows the phase shift of the RF output signal as a function of RF input power. The change of phase as the amplifier is driven into saturation, which is AM/PM conversion (°/dB), is clearly evident.

The calibration procedure described above achieves VNA power measurement uncertainty as good as the uncertainty of the power meter itself. With moderate care it can be as good as ± 0.2 dBm, instead of the ± 1 dBm power measurement uncertainty of the VNA alone.

The VNA calibration procedure takes about 15 min or longer, depending on the power range to be covered. However, once completed, it allows the signal processing capability of the VNA to be applied to the measurement. No built-in capability exists for postprocessing of the power meter measurement. In contrast, the VNA makes measurements over a range of frequency, can process the results to display them in many ways, and can transfer the information to a computer for postprocessing.

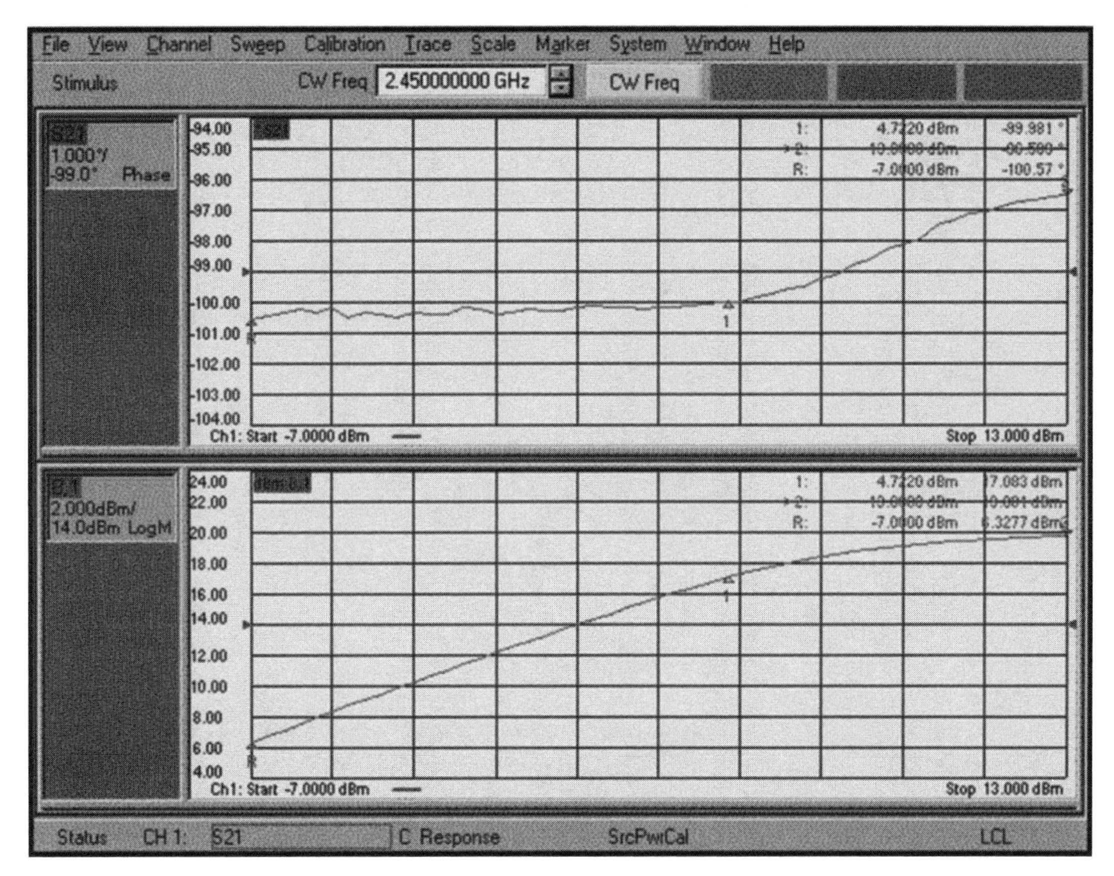

Figure 13.6 Power amplifier swept phase and output power.

13.4 UNCERTAINTY OF SPECTRUM ANALYZER MEASUREMENTS

As discussed in Chapter 9, a spectrum analyzer measures the power and frequency of each signal that makes up a complex RF signal. Therefore, the frequency measurement uncertainty and the power measurement uncertainty must both be considered.

Frequency Measurement Uncertainty

The frequency measurement uncertainty of the displayed center frequency, span, stop and start frequencies, and marker frequencies depend on the following:

Time since the last annual preventive maintenance
Resolution bandwidth of the measurement
Span of the measurement

The first two uncertainties are usually negligible, compared to those related to the span. For the PSA described in Chapter 9, the displayed frequency uncertainty is 3% of the span. For an RF signal at 1 GHz being observed with the span of 100 MHz, the frequency uncertainty is 3 MHz. The span should be set for the smallest frequency range that contains all significant frequency components of the signal.

Power Measurement Uncertainty

Historically, spectrum analyzers have had large power measurement uncertainties up to plus or minus several dB. However, modern spectrum analyzers, such as the PSA model discussed in Chapter 9, have lower uncertainties, almost as good as provided by power meters. The power measurement uncertainty depends on exactly

TABLE 13.1 A Comparison of Power Measurement Uncertainty

Specification	PSA E4440	8586E
	Absolute Measurement	
Frequency response (0–3 GHz)	±0.4 dBm	±1.8 dBm
	Relative Measurement	
Log fidelity	±0.07 dB	±0.85 dB
Range of log fidelity	Unlimited	100 dB
IF gain uncertainty	None	±1.0 dB
Resolution bandwidth switching	±0.03 dB	±0.5 dB
Calibrator	—	±0.3 dB
Calibrator + log + IF gain + resolution bandwidth switching	±0.24 dB	±2.65 dB

what the spectrum analyzer is required to measure, as will be shown in later examples.

The power measurement uncertainty of the Agilent PSA E4440 discussed in Chapter 9 is compared in Table 13.1 with that of the Agilent 8586E, which was a top of the line spectrum analyzer in the late 1990s.

The major design changes that give the PSA E4440 spectrum analyzer its improved performance are

Flat frequency response

Digital IF section with digital resolution bandwidth, digital log amp, digital video bandwidth, digital detector

Same 50 MHz power reference as used in power meters

Examples of Measurement Uncertainty of PSA 4440E Spectrum Analyzer Under Different Measurement Conditions

1. Measurement of 900 MHz CW signal at −5 dBm
   - Reference level uncertainty: ±0.27 dB
   - Frequency response uncertainty: ±0.38 dB
   - RSS uncertainty: ±0.48 dB
2. Measurement of 10 GHz fundamental with its second harmonic at 20 GHz
   - Frequency response uncertainty at 10 GHz: ±1.0 dB
   - Frequency response uncertainty at 20 GHz: ±2.5 dB
3. Measurement uncertainty due to mismatch:
   - DUT output port: $\rho_1 = 0.167$
   - PSA input port: $\rho_2 = 0.4$ (Because of its wide frequency coverage, the sensor match of the spectrum analyzer is very poor.)
   - Spectrum analyzer input attenuator setting: 0 dB
   - Measurement uncertainity: ±0.59 dB

If the spectrum analyzer attention setting is increased to 10 dB, to reduce its sensor mismatch, the mismatch measurement uncertainty is reduced from ±0.59 to ±0.13 dB.

13.5 MEASUREMENT UNCERTAINTIES OF RATIOED MEASUREMENTS WITH A VNA

The procedure for calibrating a VNA for making ratioed measurements (gain, loss, S-parameters) was discussed in Chapter 8. Causes of the ratioed measurement uncertainties are further discussed in this section.

The causes of ratioed measurement uncertainties are the following:

Directivity of the directional couplers: critical when measuring small mismatches
Test port mismatch: critical when measuring large mismatches
Source mismatch: critical when measuring large mismatches

Directional couplers are not perfect. Some of the incident power leaks into the arm of the coupler that is measuring reflected power. This causes a major problem when small mismatches are being measured, because the reflected power is so small. The difference between the coupling in the correct direction and the coupling (leakage) in the wrong direction is called the "directivity" of the coupler. For example, if the coupling in the correct direction is 10 dB, and the coupling in the wrong direction is 30 dB, the directivity is 20 dB.

A numerical example of the problem of coupler directivity when measuring small mismatches is presented in Figure 13.7. The incident power is 0 dBm. The coupled sample of the incident signal from the port R coupler, which is a sample of the incident power, is −10 dBm. The incident power passes through the port A coupler, which is reversed, and is reflected by the mismatch that has a 20 dB return loss. The reflected signal now passes through the port A directional coupler in the correct direction. The reflected signal to be measured is −30 dBm.

The leakage signal for the incident signal coming out of the port A coupler is also −30 dBm. Therefore, the coupled signal from the mismatch and leakage signal are both −30 dBm. The total signal could be 0, which would represent a perfect match, or it could be −27 dBm or any value in between, depending on the phase. Thus, the measured mismatch expressed as return loss could appear as any value between 17 dB and infinity.

Figure 13.8 shows the uncertainty of measuring the return loss of a mismatch with a coupler that has a directivity of 35 dB. The range of measurement uncertainty is shown as a function of the mismatch being measured. With a mismatch of 20 dB, the uncertainty is between 1.70 and 1.40 dB, a range of over 3 dB. Note that coupler directivity is not a problem if the mismatch is bad, approaching 0 dB, which is a short. In the case of a large mismatch, the leakage power of the forward signal is small compared to the reflected power.

However, when the reflected power is large, two new uncertainty problems arise, as illustrated in Figures 13.9 and 13.10. Figure 13.9 shows the test signal passing

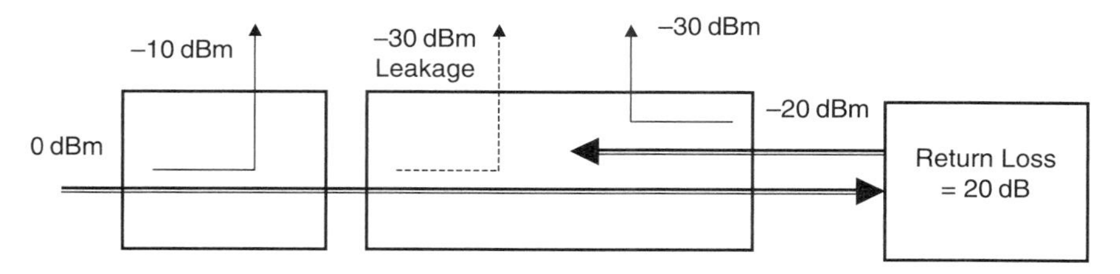

Figure 13.7 Directional coupler directivity measurement errors.

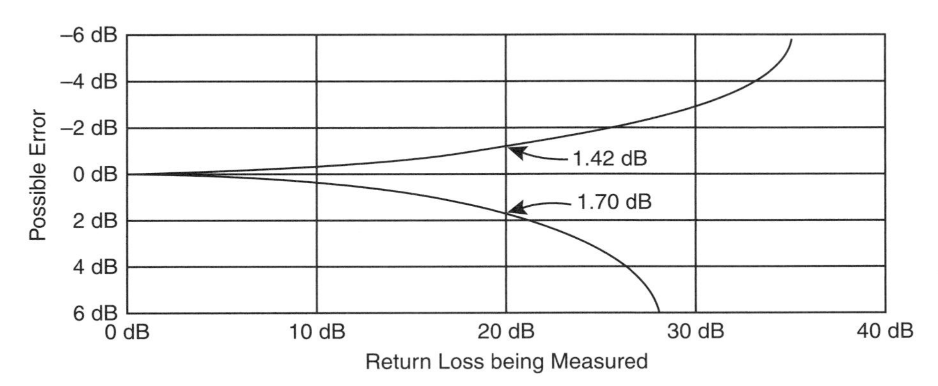

Figure 13.8 Error limits caused by coupler directivity of 35 dB.

through the coupler for port A and being reflected by a short, which is a very bad mismatch. The reflected signal enters the reversed coupler and the sampled signal is E_r. However, because the test port is not perfectly matched, some of the reflected signal from the short is reflected at the coupler test port and returns to the short to be rereflected. As shown by the vector drawing, the measured signal in port R consists of the original signal reflected from the short, which is to be measured, and a false signal attributable to the mismatch looking into the coupler. The total measured signal at the coupler port is greater or less than the actual value, depending on the phase between the two signals.

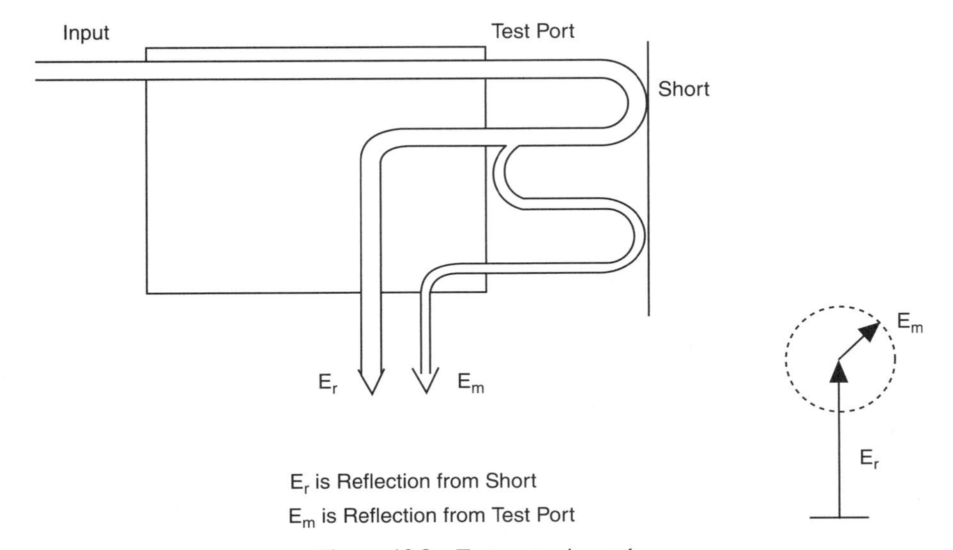

Figure 13.9 Test port mismatch.

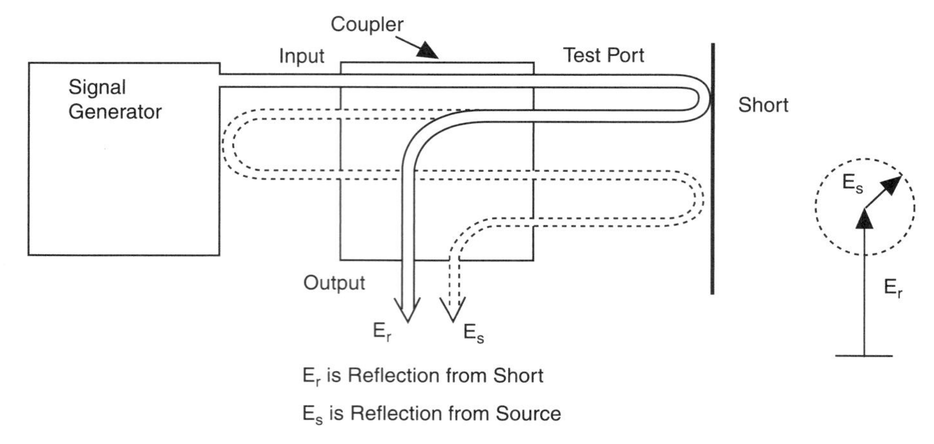

Figure 13.10 Source mismatch.

Figure 13.10 shows a situation similar to that shown in Figure 13.9, except the interfering signal passes through the coupler in the reverse direction and then is reflected off the mismatch of the source being measured.

The solution to all of the above causes of measurement uncertainty in ratioed measurements is to use the SOLT calibration procedure. As discussed in Chapter 8, this calibration procedure can be performed either manually or electronically.

SOLT calibration generates six systematic errors in the forward direction:

Directivity
Source match
Reflection tracking
Load match
Transmission tracking
Isolation

The reverse error model is a mirror image, giving a total of 12 errors for two-port measurements.

Table 13.2 compares the measurement capability of the VNA with and without the use of the correcting technique. The second column shows the hardware capability of the VNA. The third column shows the effective performance of the VNA using the calibration corrections.

Figure 13.11 reveals the uncertainty of S_{21} and S_{11} measurements that can be obtained using the SOLT calibration techniques. The top graph shows the uncertainty of S_{21} as a function of S_{21} from a 10 dB gain to a -90 dB loss. Note that over a power range of S_{21} from $+10$ to -50 dB, and over a frequency range of 0.3 MHz to 3 GHz, the measurement uncertainty is less than ± 0.1 dB.

TABLE 13.2 A Comparison of Corrected and Noncorrected Measurements

Specification	Uncorrected	Corrected
Directivity	30 dB	51 dB
Source match	16 dB	49 dB
Load match	18 dB	51 dB
Reflection tracking	±1.5 dB	±0.005 dB
Transmission tracking	±1.5 dB	±0.009 dB

The bottom graph in Figure 13.11 shows the uncertainty of S_{11} as a function of S_{11} from 0% reflection to 100% reflection. Note that over this mismatch range from an open to a short, and over a frequency range of 0.3 MHz to 3 GHz, the measurement uncertainty of S_{11} is below 0.02 dB.

13.6 NOISE FIGURE MEASUREMENT UNCERTAINTY

Chapter 11 explained the effect of the noise figure of individual components on the overall noise figure of the system. The effect on the noise figure of components that occur later in the chain is negligible if the preceding elements have enough gain.

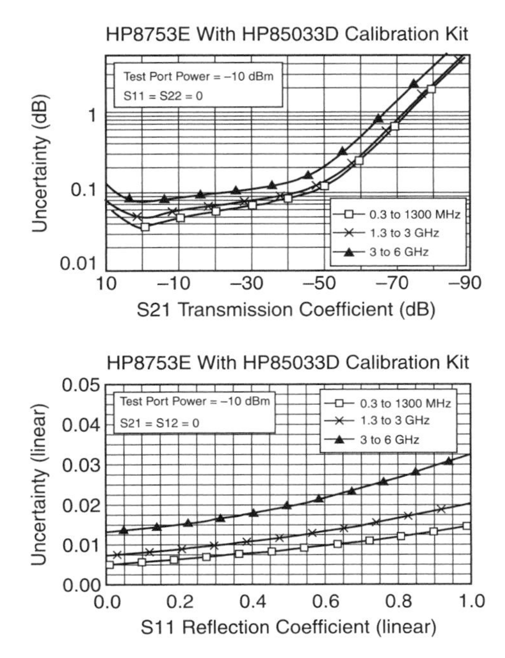

Figure 13.11 Uncertainty of S_{21} and S_{11} corrected measurements.

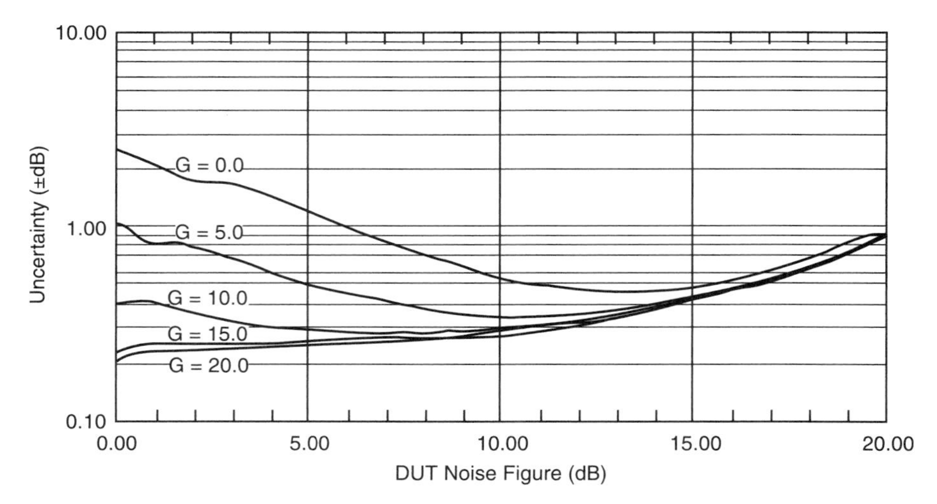

Figure 13.12 Typical noise figure measurement uncertainty.

The uncertainty of noise figure measurements is affected by the gain of the DUT, whose noise figure is being measured, and by the noise figure of the spectrum analyzer that is making the measurement. To convert a spectrum analyzer into a noise figure meter, a low noise preamp must be installed in the spectrum analyzer ahead of the mixer. This has the effect of lowering the noise figure of the spectrum analyzer.

The uncertainty of the noise figure measurement is then determined by the noise figure and the gain of the DUT. If the DUT has enough gain (~20 dB), then the noise figure of the spectrum analyzer has no affect on the measurement. If the DUT has less than 10 dB gain, then the noise figure of the spectrum analyzer does affect the measurement. These results are provided in the graph in Figure 13.12, which shows the uncertainty of the measurement as a function of the noise figure of the device in dB with DUT gain as a parameter of the curves. As shown, if the DUT has 20 dB gain, then the uncertainty of the noise figure measurement is less than ±0.3 dB for any noise figure from 0 to 10 dB. If the gain of the DUT is 5 dB, then the uncertainty of the noise figure measurement varies with the noise figure, being ±1.0 dB at 0 dB noise figure and ±0.5 dB at 5 dB noise figure.

13.7 ANNOTATED BIBLIOGRAPHY

1. Application Note 1449: Fundamentals of RF and Microwave Power Measurements, Parts 3 and 5, Agilent Technologies, Santa Clara, CA.
2. PSA High-Performance Spectrum Analyzer Series Amplitude Accuracy Product Note, Document 5980-3080EN, Agilent Technologies, Santa Clara, CA.
3. Boyd, D., Calculate the uncertainty of NF measurements, Microwaves and RF Magazine, October 1999.

4. Application Note 1287-3: Applying Error Correction to Network Analyzer Measurements, Agilent Technologies, Santa Clara, CA.

Reference 1 provides a thorough discussion of RF measurement uncertainty, beyond what was covered in this book.

Reference 2 discusses measurement uncertainties in spectrum analyzer power measurements.

Reference 3 shows how to determine the uncertainty of noise figure measurements.

Reference 4 provides background information about network analyzer error correction.

CHAPTER 14

COMPONENTS THAT DO NOT HAVE COAXIAL CONNECTORS

All the test equipment types described in the previous chapters have coaxial fittings to which the device to be tested must be connected. However, many of these devices do not have coaxial input and output connectors. In order to make measurements on devices with noncoaxial connectors, the device has to be mounted in a test fixture with transitions between the device connections and the coaxial connectors that can connect to the test equipment.

The drawing in Figure 14.1 shows an RF device without coaxial connectors mounted on such a test fixture. Figure 14.2 is a photograph of such a test fixture. Figure 14.3 shows an RF device being tested with RF probes. In this case the probes themselves are the test fixture.

With these arrangements, the test equipment will be measuring the device plus the test fixture. The measurements must then be corrected to give the characteristics of the device alone. There are four methods of achieving this:

1. using SOLT calibration standards fabricated in microstrip;
2. using through, reflection, and load (TRL) standards that are easier to implement in microstrip, and a different calibration procedure;
3. de-embedding, which uses computer modeling of the microstrip from the coaxial measurement plane to the device plane, and then calculation to remove the effect of these regions from the measured data; and
4. including fixture mismatches as part of the VNA calibration error coefficients.

RF Measurements for Cellular Phones and Wireless Data Systems. By A. W. Scott and R. Frobenius

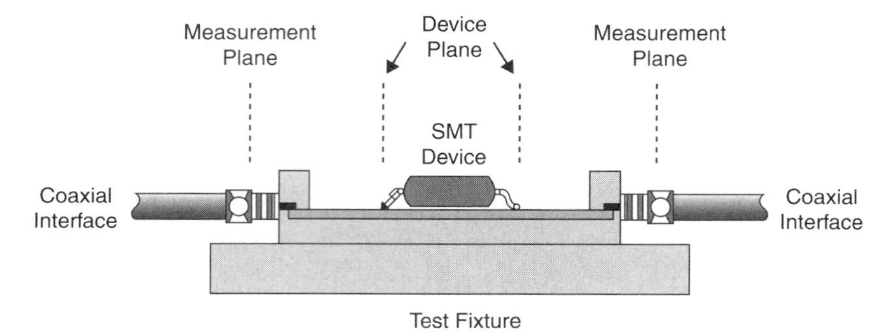

Figure 14.1 VNA measurements on nonpackaged devices.

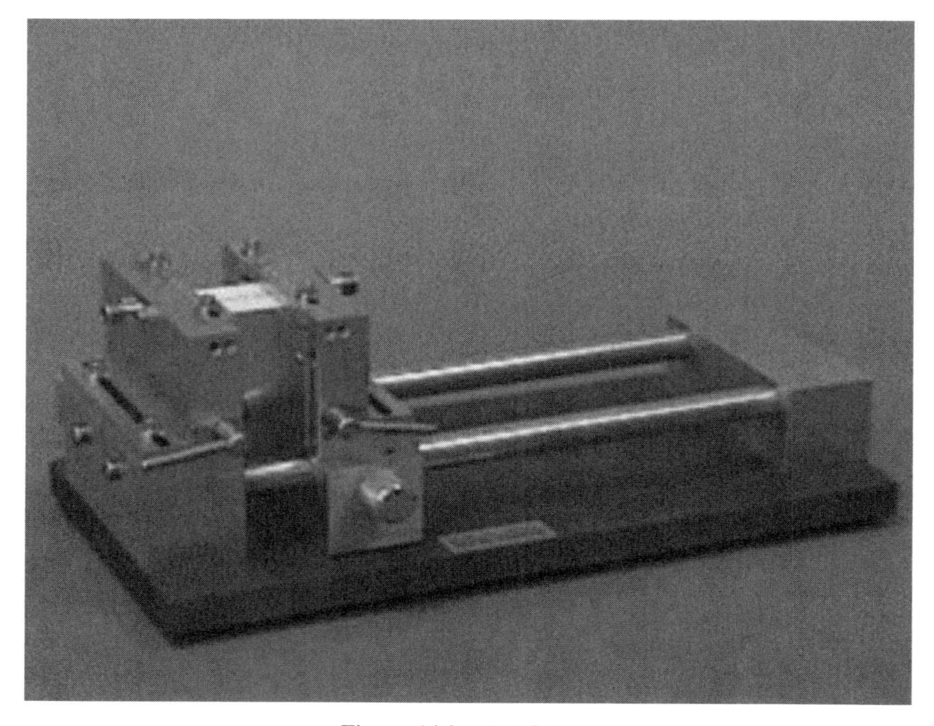

Figure 14.2 Test fixture.

14.1 USING SOLT CALIBRATION STANDARDS FABRICATED IN MICROSTRIP

As described in Chapter 7, the VNA is normally calibrated with a SOLT standard. These standards are easily fabricated in coaxial cable. The 12 equation calibration procedure using SOLT standards is part of the VNA's software. One approach to

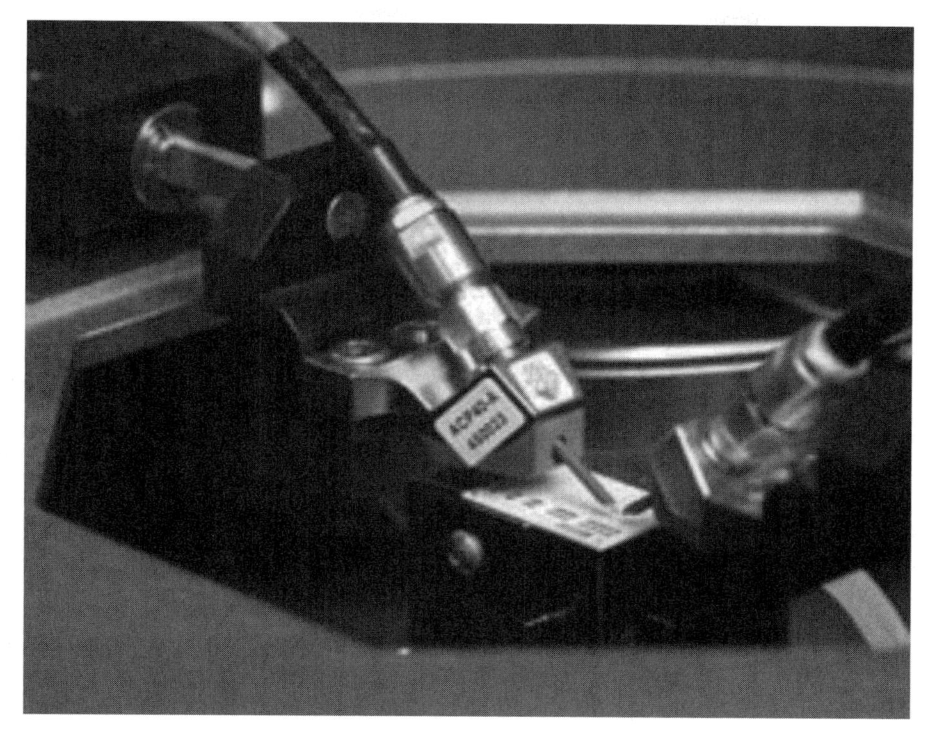

Figure 14.3 Wafer probe.

making measurements on RF components with microstrip outputs is to fabricate the SOLT standards in microstrip. A set of such standards is provided in Figure 14.4. The standards are fabricated on a microstrip board, and the microstrip lines are matched to coaxial connectors for direct connection to the VNA. The upper left microstrip lines are designed to be connected to the DUT.

The four SOLT standards are mounted on the same microstrip board adjacent to the DUT. The through standard is easy to make in a microstrip. The short is achieved by connecting the end of a microstrip line to the ground plane using a

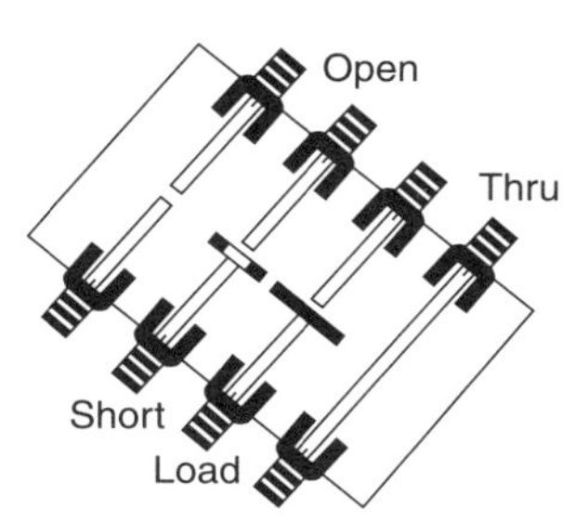

Figure 14.4 SOLT standards in a microstrip.

via. The open is approximated by simply ending the microstrip line, but it needs additional characterization to account for the radiated fields. The 50 Ω termination is the most difficult to achieve because of fringing RF fields. A first step in its design is to use two 100 Ω resistors in parallel that are attached to opposite sides of the microstrip line at its end.

14.2 TRL STANDARDS IN MICROSTRIP

TRL uses standards based on the characteristic impedances of microstrip lines, rather than a set of discrete standards. Different characteristic impedances can be easily obtained in microstrip lines by changing the width of the lines relative to the board thickness. There are no fringing field problems to account for. Different error models are used with the TRL calibration procedure, as compared with the SOLT calibration. Possible models are TRL, LR and match (LRM), and TRM.

A different set of equations must be installed in the VNA software for TRL calibration than the SOLT calibration, because the 12 term calibration procedures are different. Installing the required TRL software into an existing VNA is not a major issue.

However, for maximum accuracy, the VNA using TRL standards should have four receivers. As discussed in Chapter 7, a standard VNA has three receivers to measure incident power with the R receiver, reflected power on the A receiver, and transmitted power on the B receiver. The R receiver is switched from port 1 to port 2 to make measurements in one direction (for S_{11} and S_{21}) or the other (for S_{12} and S_{22}).

Adding a fourth measurement channel to an existing VNA is a significant hardware upgrade. To permit the TRL calibration procedure to be used with existing three receiver VNAs, a modified calibration called TRL* was developed. It is not as accurate as the TRL calibration procedure with four receivers, but it allows reasonably accurate measurements to be made.

14.3 DE-EMBEDDING

An accurate measurement of the device on its own can be achieved by using computer modeling of the microstrip from the coaxial measurement plane to the device measurement plane (as defined in Fig. 14.1) and using the modeled results to remove the effect of these regions from the measured data.

The microstrip region may be represented as a lossless length of transmission line, in which case its only effect is to add a phase shift to the measurement, which can be subtracted easily. Alternately, the microstrip region can be represented as a lossy length of transmission line, and it then adds attenuation and a phase shift to the measurement. The effect of the microstrip region can also be represented as a inductance capacitance network as shown in Figure 14.5. The computer model can be experimentally modeled.

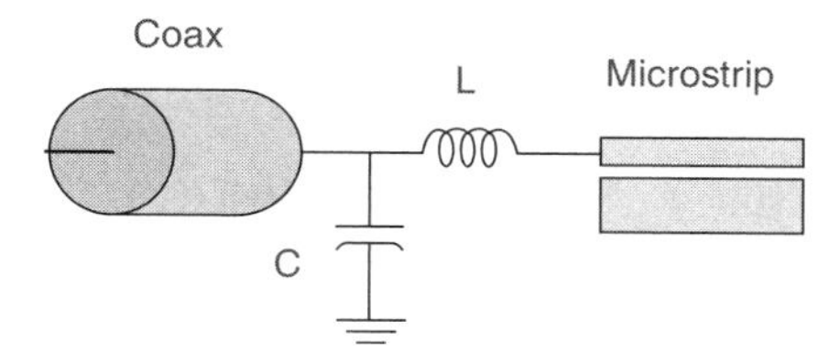

Figure 14.5 De-embedding.

Unfortunately, the de-embedded approach does not allow for real-time feedback to the operator, because the measured data needs to be captured and postprocessed to remove the effects of the fixture. The VNA is not a general purpose computer, so the correction must be made on an auxiliary computer and then sent back to the VNA for display.

14.4 INCLUDING THE FIXTURE EFFECTS AS PART OF THE VNA CALIBRATION

It is possible to perform the de-embedding calculation directly on the VNA by using a different calibration model. If the test fixture effects are included as part of the VNA calibration error coefficients, real-time de-embedded measurements can be displayed directly on the VNA. However, the same test fixture design must be used for all measurements with this particular VNA.

14.5 ANNOTATED BIBLIOGRAPHY

1. Application Note 1364-1: De-Embedding and Embedding S-Parameter Networks Using a Vector Network Analyzer, Agilent Technologies, Santa Clara, CA.
2. Product Note 8510-8A: Network Analysis Applying the 8510 TRL Calibration for Non-Coaxial Measurements, Agilent Technologies, Santa Clara, CA.
3. Application Note 1287-9: In-Fixture Measurements Using Vector Network Analyzers, Agilent Technologies, Santa Clara, CA.
4. Rytting, D. Network analyzer error models and calibration methods, Paper presented at the IEEE MTT/ED Seminar on Calibration and Error Correction Techniques for Network Analysis, OGI Center for Professional Development, September 30, 2004.

References 1–3 provide background information on the techniques for measuring devices that require a test fixture or that do not have coaxial connectors.

Reference 4 provides a thorough mathematical analysis of error sources on the VNA and the basis for SOLT and TRL calibration techniques.

PART III

MEASUREMENT OF INDIVIDUAL RF COMPONENTS

CHAPTER 15

RF COMMUNICATIONS SYSTEM BLOCK DIAGRAM

Figure 15.1 shows a block diagram of an RF communications system. The block diagram is generic. It applies to any type of wireless RF communications system: cellular phone, wireless LAN, satellite communications system, and even a deep space probe. Any RF communications system must contain all of the devices shown. Of course, the performance requirements of each device vary from system to system.

15.1 RF COMMUNICATIONS SYSTEM COMPONENTS

The function of each device in the block diagram is as follows:

PLO: Generates the RF carrier at the correct frequency

Modulator: Varies the frequency, amplitude, or phase of an IF carrier to put information onto it

Upconverter: Shifts the modulated IF signal to RF

Power amplifier: Increases the power level of the modulated RF carrier

TX antenna: Transmits the RF carrier in the direction of the receiver

RX antenna: Collects the transmitted RF signal at the receiver

RF filter: Allows only a specified range of RF frequencies to pass and blocks all other frequencies

LNA: Amplifies the weak received RF carrier

RF Measurements for Cellular Phones and Wireless Data Systems. By A. W. Scott and R. Frobenius

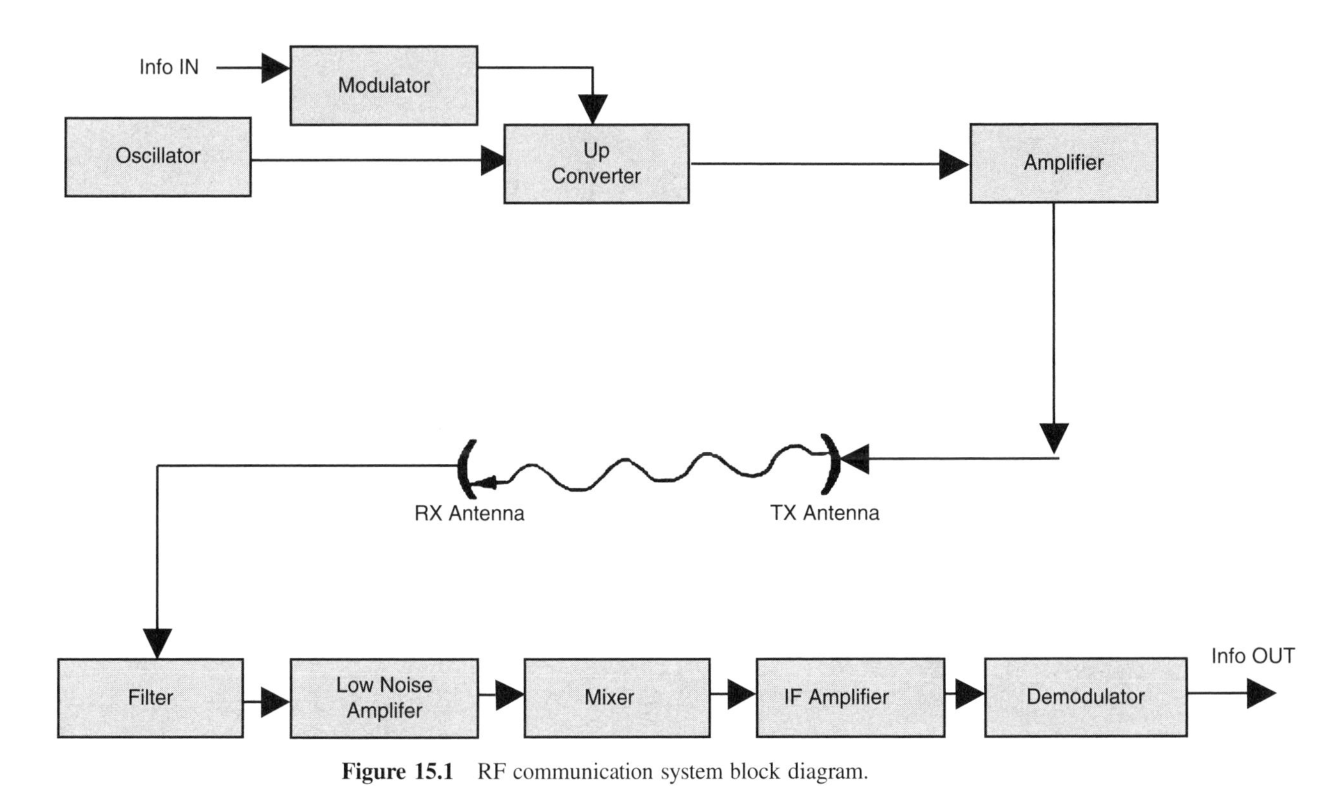

Figure 15.1 RF communication system block diagram.

Mixer and IF amplifier: Shifts the RF carrier to a lower frequency below the RF band and amplifies it to a level where it can be demodulated.

Demodulator: Removes the information from the low frequency carrier

The devices, in the order that they appear starting from the input to the transmitter and ending at the receiver output, are discussed in Chapters 16–24.

Figure 15.2 shows a mobile communication system block diagram, consisting of a mobile unit and a base station. Both contain a transmitter and a receiver. However, the transmitter in the mobile unit does not communicate to the receiver in the mobile unit (as might be inferred incorrectly from Fig. 15.1). The transmitter in the mobile unit connects wirelessly to the receiver in the base station. Correspondingly, the transmitter in the base station does not connect wirelessly to the receiver in the base station. It connects wirelessly to the receiver in the mobile unit.

The RF path from the mobile unit to the base station to the mobile unit is called the uplink. The RF path from the base station to the mobile is called the downlink.

The design of the mobile unit and the base station are very different. The mobile unit is designed to be as simple as possible to reduce its cost and size. The mobile unit transmits the minimum possible RF power to reduce battery size and amplifier heat dissipation. The mobile unit receiver has only moderate low noise performance to reduce cost. Its antenna is as small as possible.

In contrast, the base station is made as complex as necessary to make the overall communication system work. It uses multiple diversity antennas to reduce multipath fading and very low noise amplifiers, sometimes even with cryogenic cooling, to achieve the S/N necessary to achieve the required BER. The base station transmits whatever power is required to achieve the required S/N with the moderate noise level of the mobile unit receiver.

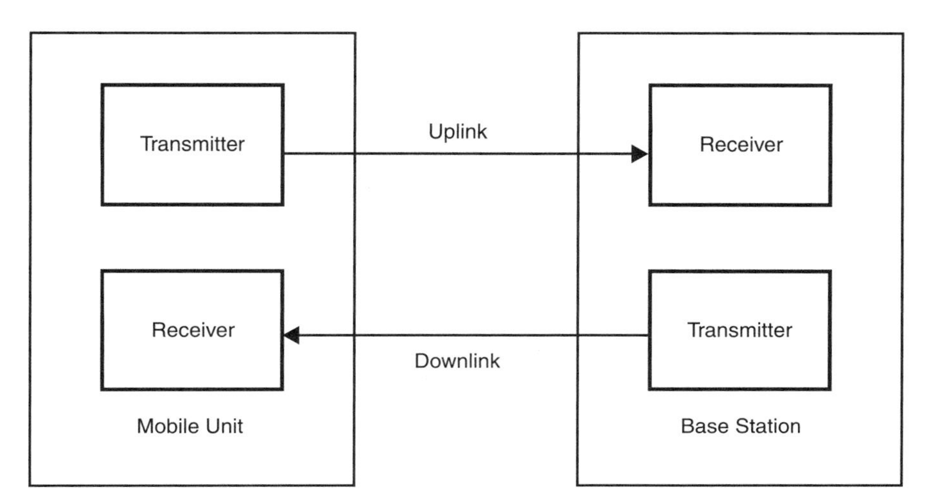

Figure 15.2 Wireless communication system block diagram.

Most base station coverage areas are divided into three sectors per cell, and Figure 15.3 is a block diagram of the RF portion of one sector of a cellular phone base station. Three antennas are used per sector. Figure 15.4 shows the antenna layout looking down toward the top of the base station tower. Two RF receiving antennas are located at the ends of the 10 ft long side of the triangular mounting platform on the top of the base station tower. A single transmitting antenna is mounted midway between the two receiving antennas.

One sample of the received signal, containing the RF signals from many users, is received by one of the antennas and passes through the RF filter, as shown at the top of Figure 15.3. These signals are amplified by the LNAs and divided into 20 identical parts (although only 4 are shown in the figure). Each of the 20 identical signals from the LNA is sent to a separate mixer. Each mixer separates out a single user frequency channel, which is further amplified and detected. This receiving process is repeated on all 20 channels, each of which carries a different user's signal.

As shown in Figure 15.3, an identical receiver channel processes the RF signals from the other receiving antenna. The two signals from the pair of diversity antennas is then compared and the strongest signal is used for further transmission.

All of the base stations in a cellular phone system are connected to a Mobile Switching Office (MSO) by a wire line or a microwave relay system The MSO controls the base stations, assigning the mobiles they will connect to and switching the mobile's communication to another base station as it approaches the edge of a cell. The MSO connects the signals from the mobile users to the Public Switched

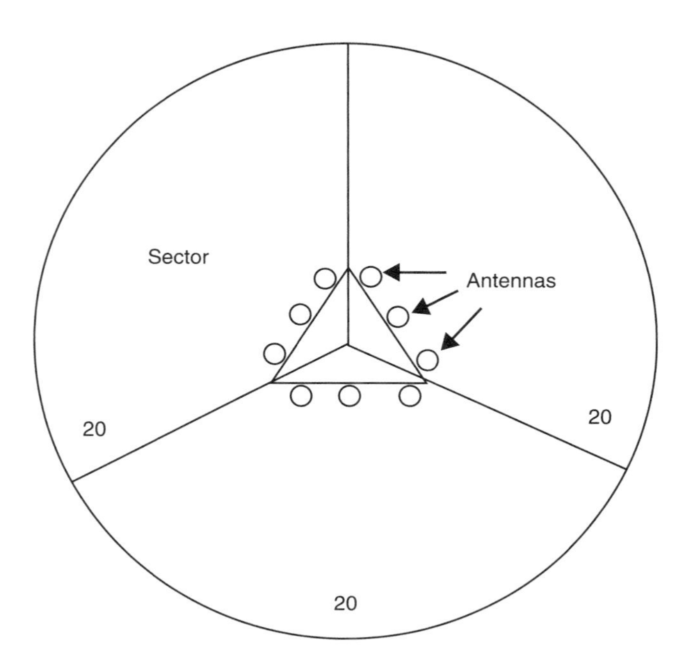

Figure 15.3 Base station antenna layout.

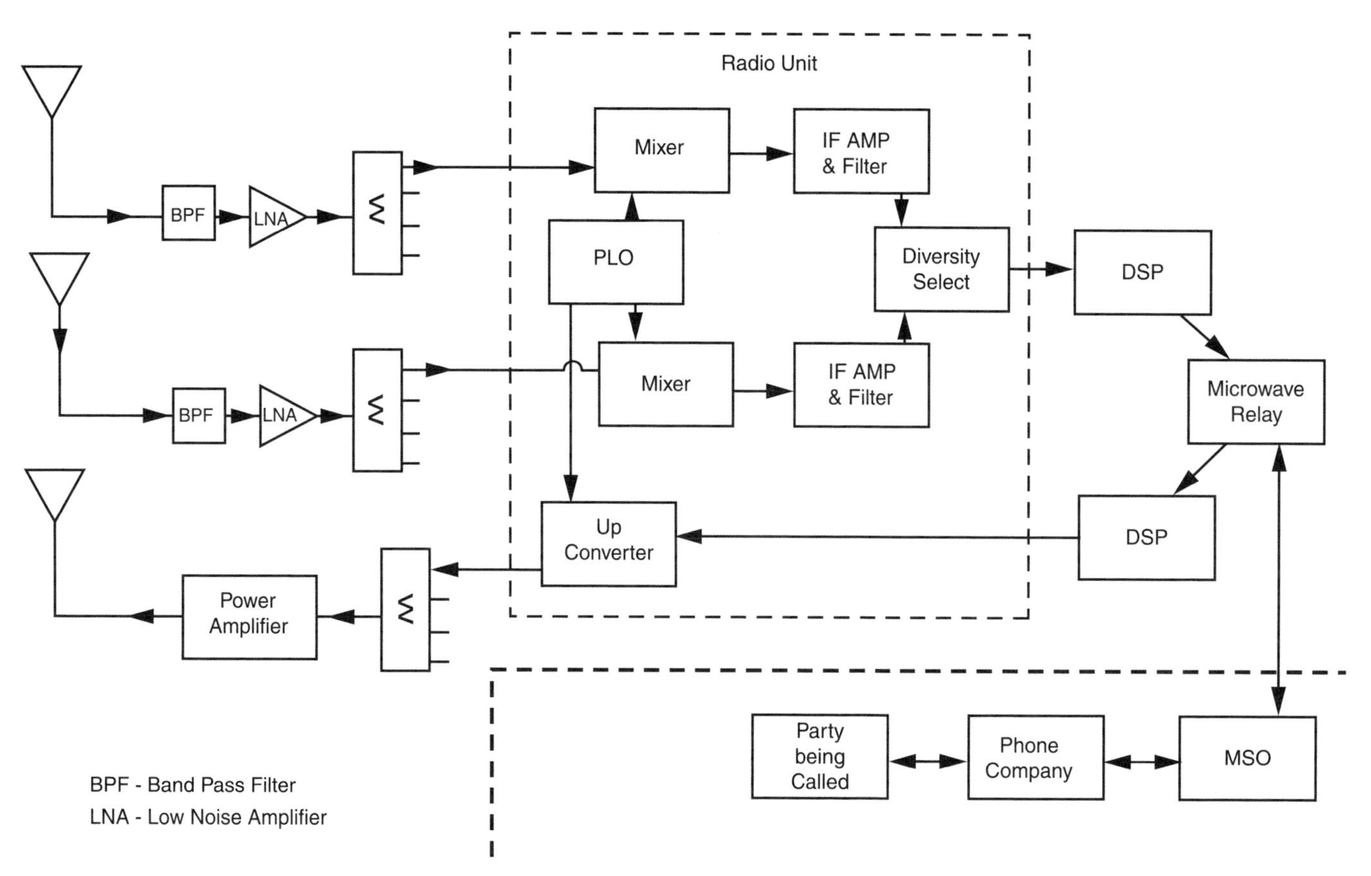

Figure 15.4 Cellular phone base station block diagram.

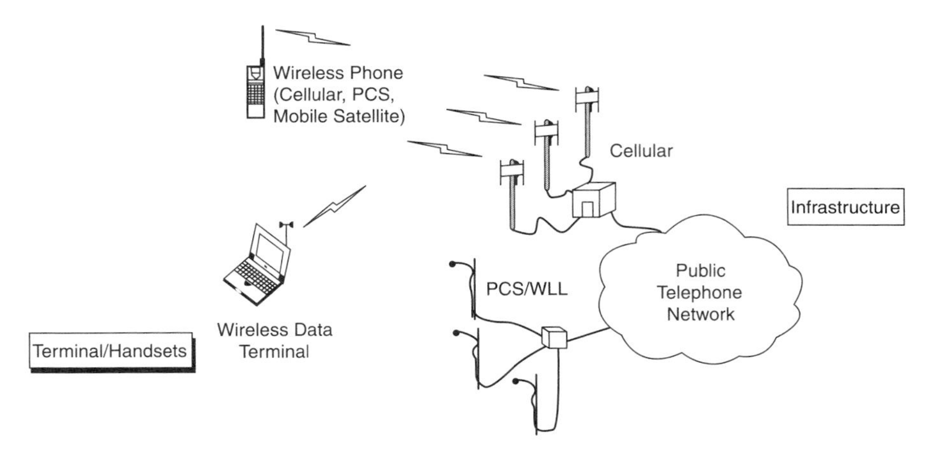

Figure 15.5 Wireless services.

Telephone Network (PSTN). The MSO looks to the PSTN just like a Private Branch Exchange (PBX) of a large business office. The PSTN then connects the signals from the cell phone users to their end user destinations, as shown in Figure 15.5.

The end user then responds to the call and sends an answer message back to the PSTN by a wired telephone line. The PSTN sends the return signal to the MSO. The MSO then sends the return signal to the appropriate base station.

In the transmitting half of the base station, the signals being transferred back to the mobiles are upconverted to the transmitter TX frequency, combined together at high power, and sent wirelessly back to the mobiles.

The names "mobile unit" and "base station" shown in Figure 15.2 are commonly used, but generic. For cell phone systems, the mobile unit is called the "user unit (UE)". The base station is called the "base transceiver station."

High-speed, short-range wireless data transmission systems have a block diagram similar to that shown in Figure 15.2. The mobile unit is correctly called the UE and the base station is correctly called the "access point."

Every wireless system has all of the components shown in Figure 15.1. The specifications of the components may differ from system to system, but they will always be there.

15.2 ANNOTATED BIBLIOGRAPHY

1. Cheah, J. Y. C., Introduction to wireless communications applications and circuit designs, In Larson, L., Ed., RF and Microwave Circuit Design for Wireless Communications, Chap. 2, Artech House, Boston, 1996.

Reference 1 describes RF communication system component requirements in more detail than this book.

CHAPTER 16

SIGNAL CONTROL COMPONENTS

RF signal control components vary the frequency, power, and other characteristics of the RF signal. Because many of these control components use semiconductor diodes for their operation, semiconductor diodes are discussed first.

16.1 RF SEMICONDUCTORS

RF signal control components use PN (or varactor), Schottky, and PIN semiconductor diodes. Semiconductors are useful as electronic devices because their characteristics can be varied by changing the applied electrical control signal.

Semiconductor doping is reviewed in Figure 16.1. A semiconductor has four electrons in its outer shell. The outer shell can hold as many as eight electrons, and the atoms of the semiconductor are bound in a crystal by the outer electrons. For example, in a pure silicon crystal, each atom shares the four electrons in its outer shell with its four neighbors and each neighbor shares one electron with this atom. This sharing of outer shell electrons holds the crystal together. Thus, there are no free electrons that can move through the material under the effect of an applied electric field, so the semiconductor material is an insulator.

If the dopant level is less than 10^{-9}, which is 1 part/billion, the number of unbound electrons is negligible, and the semiconductor material is type I, an insulator. However, the conductivity of the material can be changed to any level simply by adding dopants, for example, phosphorus or arsenic, which have five electrons in their

---

RF Measurements for Cellular Phones and Wireless Data Systems. By A. W. Scott and R. Frobenius

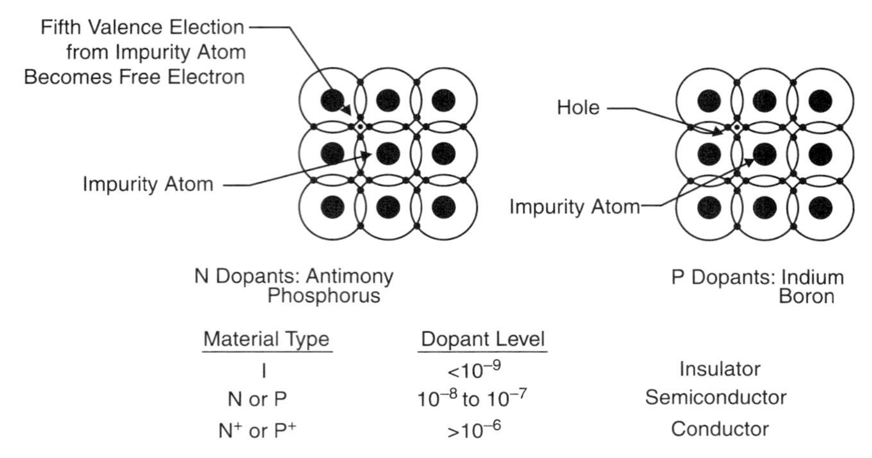

Material Type	Dopant Level	
I	$<10^{-9}$	Insulator
N or P	10^{-8} to 10^{-7}	Semiconductor
N^+ or P^+	$>10^{-6}$	Conductor

Figure 16.1 Semiconductor doping.

outer shell. When a dopant atom replaces a silicon atom in the crystal lattice, four of the outer shell electrons are shared with the neighboring silicon atoms to bind the crystal together. However, the fifth electron is free to move through the material, so the material becomes a semiconductor. This type of doping makes N material, because the material has an excess negative charge.

In contrast, P material has an excess positive charge. In this case, dopants such as gallium or indium are used, which have only three electrons in their outer shell. When the P dopant atom replaces a silicon atom, the three electrons are shared with the neighboring silicon atoms, but an electron is missing from the crystal lattice. This missing electron is called a hole. An electron from a neighboring silicon atom fills this hole, but this leaves a corresponding hole in the atom that supplied the electron. As electrons fill the holes and leave other holes, the hole appears to "move" through the material as a positive charge, hence, the designation P material.

The unique semiconductor action, particularly in regard to forming junctions between the N and P types of material, occurs at doping levels between 10^{-8}, which is 1 part in 100 million, and 10^{-7}, which is 1 part in 10 million. If dopants are added until the doping level exceeds 10^{-6}, which is 1 part in 1 million, then the material is called N+ or P+. The unique semiconductor junctions can no longer be formed when the material is so heavily doped, and N+ and P+ materials are considered to be conductors.

This type of doping makes silicon or germanium, which in their pure form are insulators, into conductors with a controlled level of conductivity. Compound semiconductors like gallium arsenide, whose atoms have three and five electrons, respectively, in their outer shells, can also be formed. These compound semiconductors can also be doped to be P or N. However, all of thse semiconductor materials have no other special properties when doped and appear as resistors with various resistance values.

The unique characteristics of semiconductors, which make them useful for signal generation, amplification, and control, are achieved by forming junctions between P- and N-doped materials. The basic junction is the PN, which is shown in Figure 16.2. The PN junction occurs when P-doped semiconductor material and N-doped semiconductor material are placed together. (The materials are not actually placed together, but the junctions are grown inside the material.) The electrons in the N material are attracted by the holes in the P material and move across the junction to fill the holes. The attraction is from electronic forces in the crystal lattice. As shown in Figure 16.2a, a depletion region is then formed between the N and the P materials, and this region is the junction. The junction no longer has electrons in the N material (because they have moved into the P material to fill the holes) or holes in the P material (because they have been filled with electrons), so it is an insulator.

All of the electrons in the N material do not move over and fill the holes in the P material, because as the electrons leave the N region they leave the N region positively charged. Both N and P semiconductors are electrically neutral. The N material, because of its doping with atoms with five electrons in their outer shells, has free electrons that can move through the material, but the nucleus of each dopant atom has one more positive charge than the silicon atoms, which balances out the extra electron in its outer shell. However, when the electron leaves the N material and enters the P material, it leaves the N material positively charged and correspondingly makes the P material negatively charged. The positively charged N material then exerts a force to draw electrons back into it. When this force balances the electron orbital forces, which pull the electrons into the holes, the flow of electrons into the holes stops. Consequently, the depletion region exists for only a small distance on either side of the junction, partly in the P material and partly in the N material.

The electrical characteristics of a PN junction are shown in Figure 16.2b. The current flowing through the junction is plotted as a function of the bias voltage applied to the junction. (The bias voltage is shown as the voltage applied to the P side relative to the N side.) With a negative bias, no current flows through the junction. The P side already has more electrons than it should, and connecting the

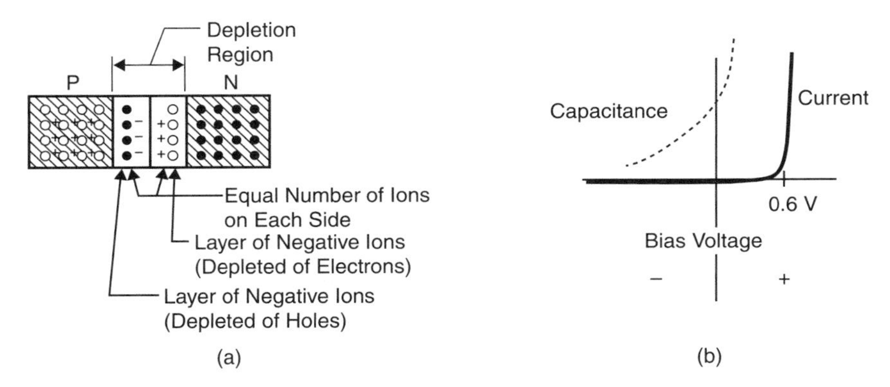

Figure 16.2 PN junctions.

negative terminal of the battery to the P side just makes the situation worse. Hence, the PN junction conducts only a small leakage current, which is usually assumed to be negligible. The diode is thus called reverse biased. The PN junction then appears like a variable capacitance. As the negative bias is increased, the junction thickness becomes greater and the capacitance decreases.

The diode is forward biased when the P material is connected to the positive terminal of the battery. However, current still will not flow until the forward bias is high enough to overcome the internal electronic forces. For silicon at room temperature this occurs when the external bias is 0.6 V. As illustrated in Figure 16.2b, as the voltage increases above 0.6 V, the junction current increases rapidly.

PN junctions are used at low frequencies as rectifier diodes for power supplies. For RF applications, the variable capacitance characteristic of the PN junction is used for electronic tuning of the PLO, as described in Chapter 17.

Related semiconductor junctions used in RF devices are compared in Figure 16.3. The PN junction is shown in Figure 16.3a. Figure 16.3b shows a Schottky junction, formed between a metallic conductor and an N-doped semiconductor. When the metal and N-doped material are placed together, electrons from the N material are attracted to the metal and leave an insulating junction in the N material. The characteristics of the Schottky junction are similar to those of the PN junction, except that the capacitance of the Schottky junction is reduced. By special fabrication techniques, the Schottky junction begins to conduct when the bias voltage just becomes positive.

An ohmic junction is shown in Figure 16.3c. The ohmic junction is used to make electrical connection to the N- or P-doped material. If a metal conductor is connected directly to N-doped semiconductor material, a Schottky junction is created. To make an electrical connection to the N material without any junction characteristics, an N+ region, which is a heavily doped semiconductor, is connected to the N region and the metal is connected to the N+ material. The ohmic junction has no unique characteristics; current flows equally well in either direction with no capacitance effects.

The PIN diode (Fig. 16.3d) is formed from P and N material with a thin layer of undoped or I type material between. The I material reduces the capacitance in the reversed-bias case, and the PIN diode is used as an electronically controlled attenuator or an electronic switch.

16.2 ELECTRONICALLY CONTROLLED ATTENUATORS AND SWITCHES

RF wireless handsets and base stations must operate under a variety of conditions. For example, a single handset must be able to operate in two or more licensed frequency bands. Received or transmitted power may vary over a 90 dB dynamic range because of different transmitter to receiver distances and multipath fading as the mobile moves.

To accommodate all of these changing operating conditions, RF components must be switched into and out of circuits and RF power must be continually adjusted. This dynamic control of RF power is accomplished by PIN diode switches and attenuators.

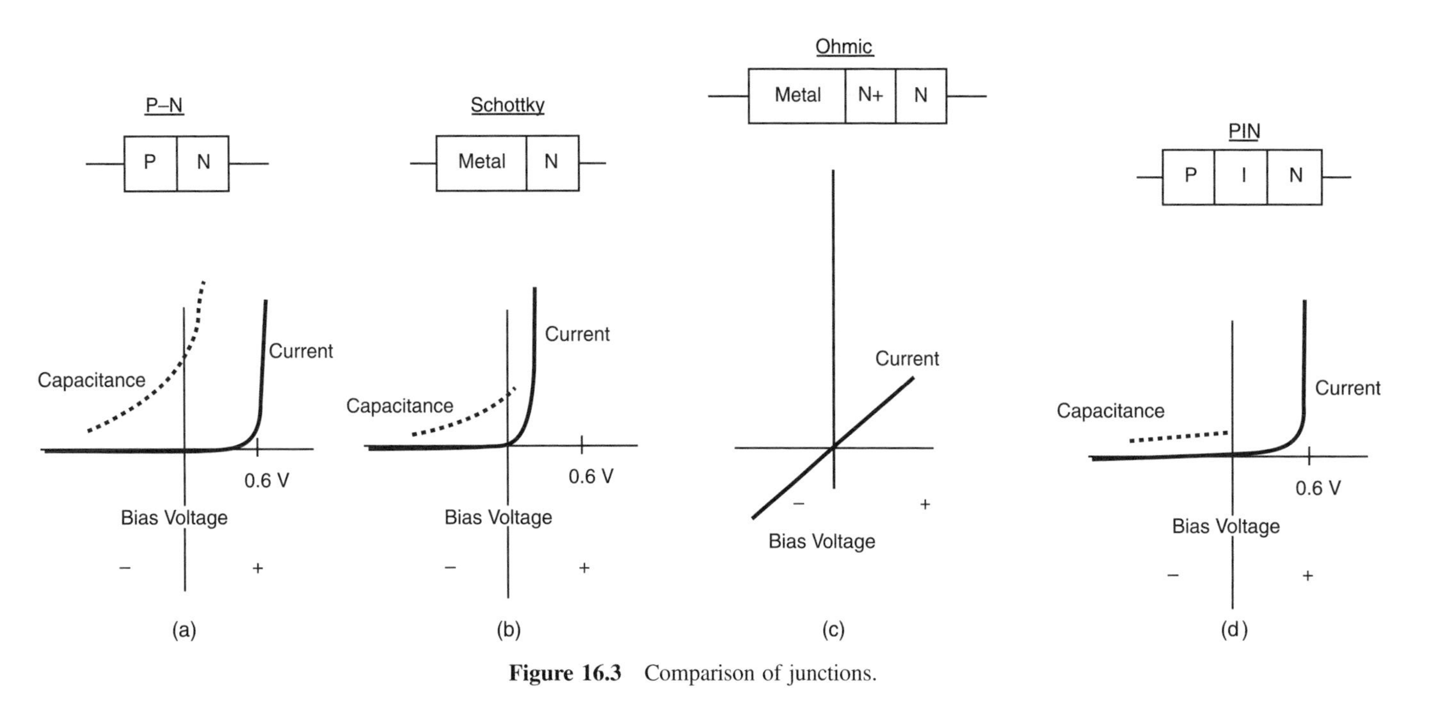

Figure 16.3 Comparison of junctions.

As described in Section 16.1, the PIN diode is a PN junction with a layer of undoped or I material between the P and N regions to reduce the capacitance in the reversed-bias state. Figure 16.4 shows the attenuation, when the PIN diode is placed in a transmission line, as a function of the bias voltage. With a negative or reverse bias, the diode provides no attenuation, because no current flows and its very small capacitance can be turned out with a fixed inductor. When it is forward biased, the diode conducts and attenuates the RF signal. Varying the positive-bias level varies the attenuation.

A reflective PIN diode attenuator is illustrated in Figure 16.5a. It consists of a PIN diode mounted in shunt across a coaxial transmission line. When the diode is reverse biased, the PIN diode has little effect on the RF signal, and the signal travels through the assembly without attenuation. However, when the PIN diode is forward biased, it looks like a resistor and as the bias current increases, the resistance decreases. As this occurs, the PIN diode reflects more and more of the incident signal, so less of it leaves the output. Typical attenuation as a function of the forward-bias current is shown in Figure 16.5b. Because the forward-biased current varies so rapidly with forward-biased voltage, PIN diode characteristics are usually specified in terms of forward-biased current rather than forward-biased voltage. Note that as the bias current is varied from 0.01 to 10 mA, the attenuation varies from a fraction of a dB up to 20 dB. If additional attenuation is required, additional PIN diodes can be added to the attenuator assembly.

When the PIN diode is used in shunt across the transmission line, the power that does not get through the attenuator to the output is reflected back into the input line. This problem can be avoided by using a matched attenuator as shown in Figure 16.6. In this case, a 3 dB quadrature hybrid is used with two PIN diodes. The input signal divides between the diodes; and as the bias level on the diodes is varied, varying but equal amounts of power are reflected from each diode. As this reflected power travels back through the hybrid, because of the phase relationships of the 3 dB quadrature

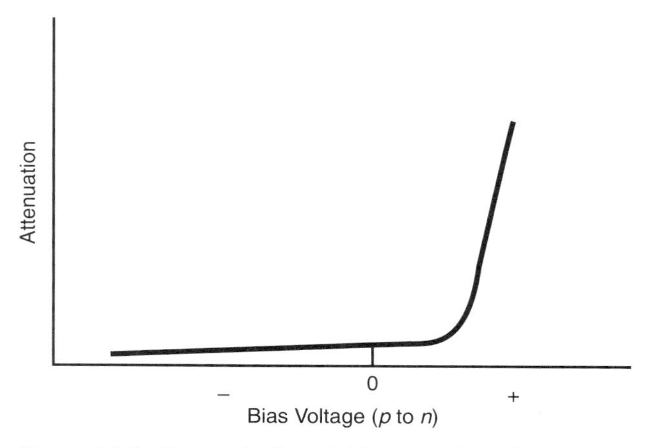

Figure 16.4 Electronically variable attenuation of PIN diode.

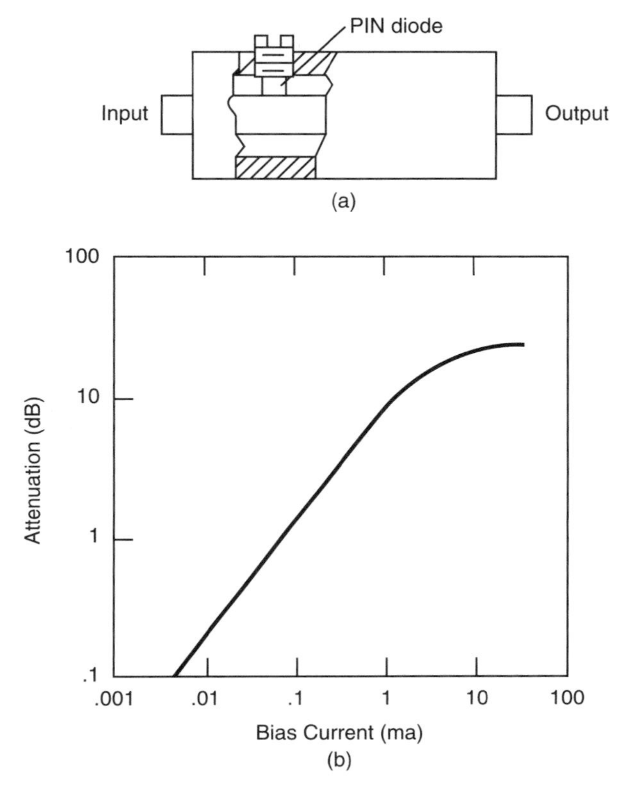

Figure 16.5 Reflective PIN diode attenuator.

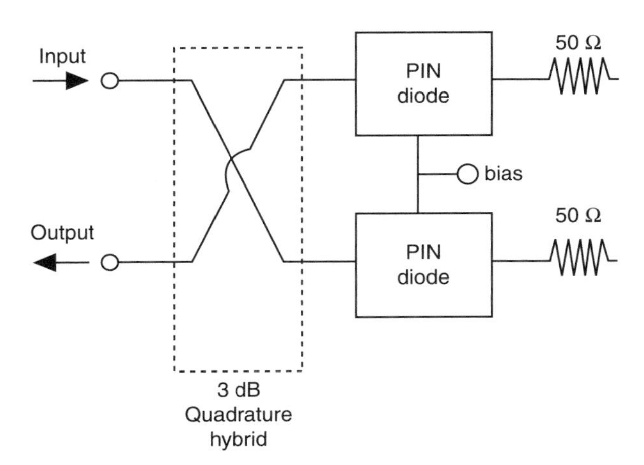

Figure 16.6 Matched PIN diode attenuator.

hybrid the reflected powers add up at the lower port and cancel at the input port. Therefore, the input sees a perfect match, and the amount of power leaving the output is controlled by the PIN diodes. Note that the remaining power that is not reflected passes through the PIN diodes and is absorbed in the 50 Ω internal terminations.

A switch directs RF power from one transmission line to another or turns the RF power on and off. Electronically actuated RF switches use PIN diodes. With a reverse bias, the PIN diode has little effect on the microwave signal. With a large forward bias, the PIN diode completely absorbs the RF. For use in a switch, the PIN diode control voltage is not continuously varied, as in an attenuator, but is switched from a reverse-biased level to a forward-biased level.

16.3 MEASUREMENTS OF PIN DIODE ATTENUATORS AND SWITCHES

Measurements of PIN diode attenuators and switches can be easily made with the VNA described in Chapter 8. The same measurement procedure should be used that was used for measurement of the RF filter, with the exception that measurements must be made at several control voltage settings.

At any control voltage setting, the performance of a PIN diode attenuator or switch can be completely characterized by its four S-parameters.

16.4 ANNOTATED BIBLIOGRAPHY

1. Scott, A. W. Microwave signal control components, In Understanding Microwaves, Chap. 7, Wiley–Interscience, New York, 1993.

Reference 1 covers signal control components in more detail than this book.

CHAPTER 17

PLOs

The function of the PLO in a wireless communication system is to generate the RF signal that will carry digital information wirelessly, in the form of the modulation of amplitude, frequency, or phase, from the transmitter location to the receiver location.

17.1 CHARACTERISTICS AND OPERATION OF A PLO

The most important characteristics of a PLO are its RF frequency stability, its capability of being rapidly tuned from one RF frequency to another, and phase noise.

If the power level of the PLO is too high or too low, it can be easily attenuated or amplified as required. However, if the RF frequency is incorrect, the transmitter will jam other system users and its own system receiver will not receive a signal because it will be tuned to the wrong frequency.

The frequency stability requirements are 10^{-12} or better, which would be a stability of 1 kHz for a 1 GHz RF signal.

At frequencies below the RF band, the transmitter can be stabilized by using a quartz crystal as its resonant circuit. Unfortunately, quartz crystals do not have resonant frequencies in the RF band. Therefore, the RF PLO must divide its frequency to a value below RF, where its divided down frequency can be compared to the reference frequency of a quartz crystal.

Figure 17.1 shows a block diagram of an RF PLO. The PLO consists of two parts: a VCO, shown in the shaded box, and a PLL.

RF Measurements for Cellular Phones and Wireless Data Systems. By A. W. Scott and R. Frobenius

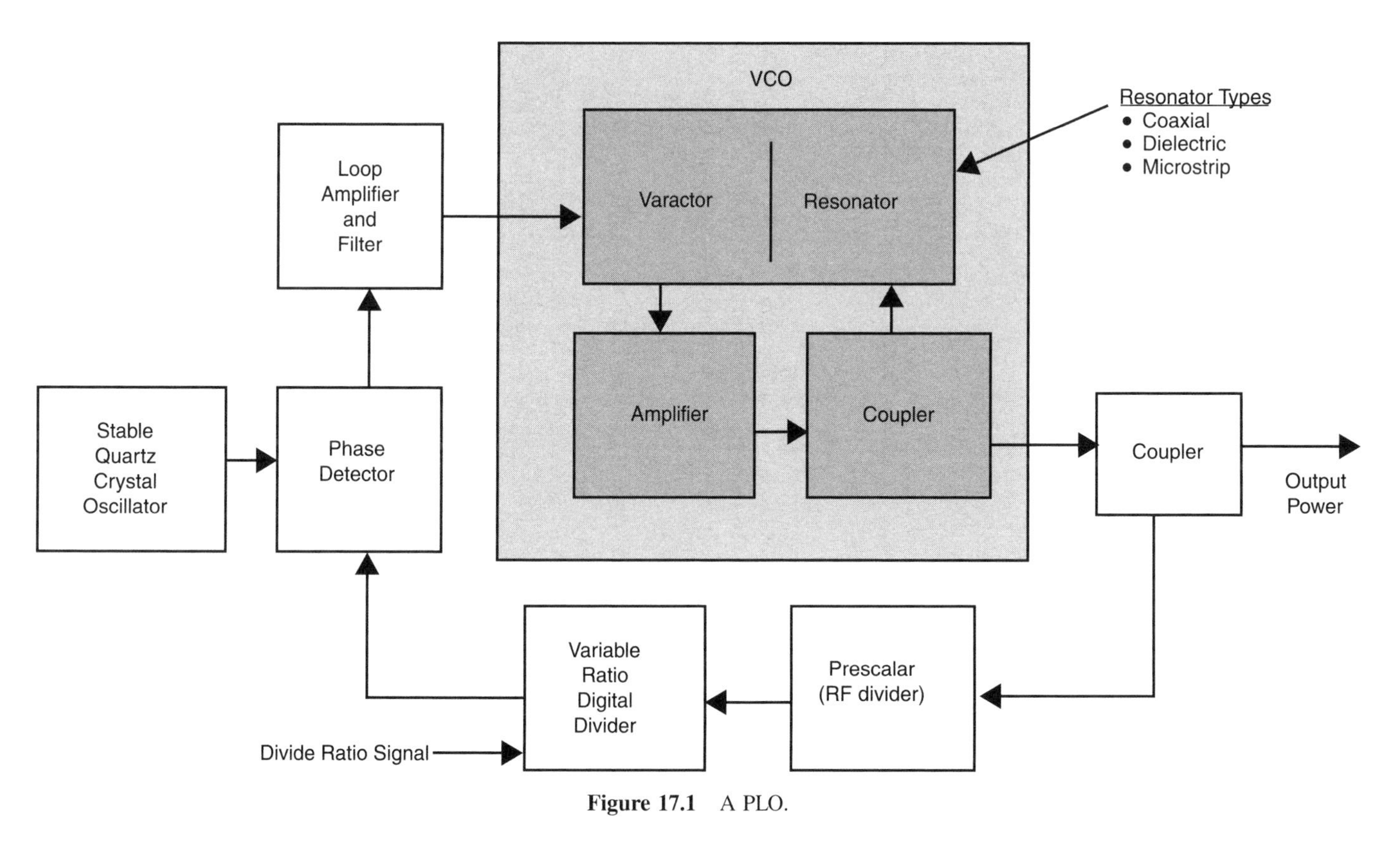

Figure 17.1 A PLO.

The function of the VCO is to generate the RF signal, with provision to adjust its frequency to the required value by the application of a control voltage. The VCO consists of four components:

An RF transistor to generate the RF signal
A directional coupler to take a sample of the RF signal
An RF resonator to control the frequency of the oscillator
A varactor diode to adjust the frequency of the resonator

When the VCO is first turned on, the transistor amplifies broadband noise. A sample of the amplified noise output is coupled by the directional coupler into the RF resonator. Only the amplified noise at the desired frequency is transmitted through the resonator back to the input of the amplifier. The broadband noise and the one frequency that was passed through the resonator are amplified a second time. The resonator again blocks the amplified noise, but the one frequency that originally passed through the resonator is now amplified a second time. This process continues until the one frequency that passes through the resonator is amplified up to the saturation power of the amplifier at around the 0 dBm level.

The PLL next takes a sample of the output power from the VCO through a second directional coupler. It then divides the frequency down by a factor of about 10 times in a prescalar, which is an RF frequency divider. This division reduces the frequency to a value below the RF band where it can be further divided in a standard fractional digital divider IC. The frequency of the divided down RF signal is then compared in a phase detector to a quartz crystal reference signal. If the frequencies are not equal, an error signal is developed. The error signal is amplified and fed to the varactor diode of the VCO, which adjusts the VCO frequency until the phase detector voltage is zero.

If the VCO frequency needs to be changed to another value, for example, when a cellular phone is instructed by the base station to tune to another channel, the fractional digital divider ratio is changed by the control signal.

Figure 17.2 shows the attenuation of the resonator as a function of frequency. At all frequencies except one, the resonator blocks the incident signal with an attenuation of over 40 dB. At the desired resonant frequency, the RF signal passes through the resonator with very little attenuation. Figure 17.3 shows how the resonator is

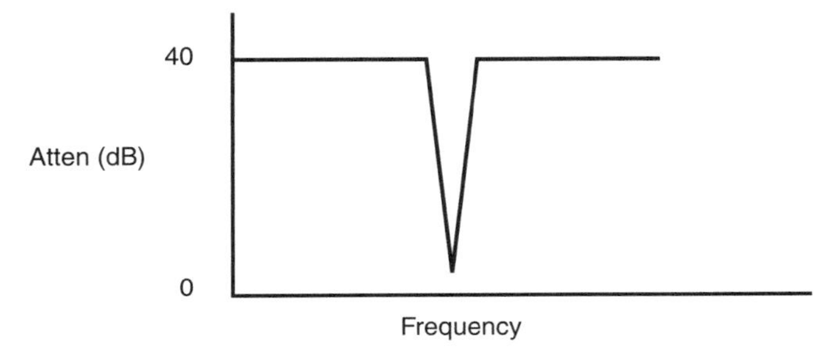

Figure 17.2 RF characteristics of resonators.

(a) RF wave passes through line

(b) RF wave blocked by break in line so very little signal passes through

(c) RF wave blocked except at frequency whose half wavelength is equal to distance between breaks

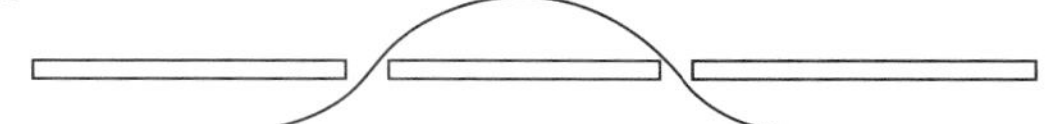

Figure 17.3 Basic resonator fabrication.

made. The resonator is simply a half-wavelength long transmission line. The transmission line can be a waveguide, coaxial cable, or microstrip, depending on the desired size and frequency range. Figure 17.3a shows a continuous length of microstrip line. The RF wave passes through the line. If the line is cut as shown in Figure 17.3b, the RF wave is blocked and very little signal passes through. However, if the line is cut in two places as shown in Figure 17.3c, the RF wave is blocked even more, except at the frequency whose half-wavelength is equal to the distance between the cuts, and at this frequency it passes through the microstrip line.

Other types of half-wavelength transmission lines are the following:

Teflon filled coaxial cable
Ceramic filled coaxial cable
Microstrip spiral
Multilayer cofired ceramic with embedded inductors and capacitors

The half-wavelength Teflon filled coaxial cable is too long for use in the 300 MHz to 5 GHz frequency range, but it is useful at higher frequencies. Coaxial cable can be shortened by using a high dielectric ceramic as the support material. A special material developed for this purpose is barium tetratitanate, which has a dielectric constant of approximately 100, so the wavelength is reduced by 10 times. At 1 GHz such a half-wavelength resonator would be only 0.6 in. long.

A microstrip line on Teflon–fiberglass substrate would be too long for use at 1 GHz just like a standard coaxial cable, but the microstrip line could be wound into a spiral to reduce the size of the resonator to 0.5 in $\times$ 0.5 in.

A resonator could also be built with tiny inductors and capacitors to be resonant at 1 GHz. However, this resonator would be very fragile. A practical way of achieving such a resonator is to screen print spiral inductors and capacitor plates on 3 mil thick unfired ceramic substrates using silver glass ink. Many hundreds of such inductors and capacitors could be screen printed at one time on a large ceramic sheet. Several layers of these sheets could be stacked together to obtain the required resonant characteristics and then fired at 800°C. The 1000 arrays of capacitors and inductors could then be cut apart to serve as the resonators for many PLOs. This process is called low-temperature cofired ceramics (LTCCs).

Resonators built by any of the processes just described have two problems:

1. variation of the resonator length (and therefore frequency) with ambient temperature and
2. tuning to a different frequency channel on demand.

The PLO is required to operate over a temperature range of -20° to $+40^\circ$C so that the wireless system can operate in winter cold and summer heat. Over this temperature range copper expands 1 part/1000 and ceramic expands 1 part/10,000, whereas the resonator frequency must be held to 1 part/million (which is 1 kHz out of 1 GHz). The solution to both problems is achieved by adding a varactor diode to the resonator to permit electronic tuning.

As described in Chapter 16, a varactor diode is a PN semiconductor junction, which is shown in the Figure 17.4. At 0 bias some of the free electrons in the N-doped semiconductor region have moved across the junction to fill the free holes in the P-doped semiconductor region, creating an insulating layer between the P and N regions. Therefore, at 0 bias the varactor diode behaves like a capacitor. As the P side of the junction is biased negative, the insulating junction region increases in width and the junction capacitance decreases. This electronically controlled

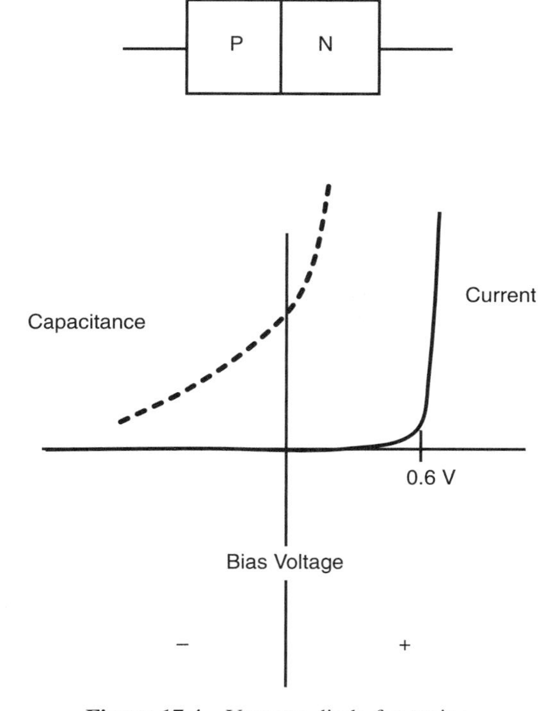

Figure 17.4 Varactor diode for tuning.

capacitance mounted on the resonator allows its resonant frequency to be changed by the application of a control voltage.

The VCO described in the previous paragraph allows the oscillator's RF frequency to be changed by the application of a low frequency voltage to the varactor diode. This control voltage must be accurately controlled to hold the RF frequency constant to the required 1 part/million. The generation of this precise voltage is achieved by the PLL shown surrounding the VCO in Figure 17.1.

The PLL operates as follows: for this example, assume that the PLO is operating at 1000 MHz (1 GHz). A sample of the VCO is obtained from the directional coupler at the VCO output. The 1 GHz signal is applied to a prescalar, which is a simple digital divider that divides the 1 GHz frequency down by a fixed ratio of 10 times. Normally, digital circuits do not work at RF frequencies, but special digital circuits like the prescalar can be made that do simple processes like a fixed division by 10 times. The output frequency of the prescalar is 100 MHz in the example. This sampled and divided down signal from the VCO is now below the RF band, where a wide variety of standard digital dividers are available. By using a variable ratio divider, whose division ratio can be varied by a control signal, the RF signal can be divided down to any value, even by fractional ratios. For the example being discussed, assume a dividing ratio of 10 is chosen, so the signal is divided down to 10 MHz.

This sample of the RF output of the VCO that has been divided down by a factor of $10 \times 10 = 100$ by the prescalar and the variable ratio digital divider is next compared in a phase detector to the frequency of a 10 MHz quartz crystal oscillator. Simple quartz crystal oscillators are readily available at frequencies below the RF band. If a 180° phase shift is applied to the crystal reference signal before it enters the phase detector, the two signals will be out of phase and the output of the phase detector will be 0. The output of the phase detector is then applied through an amplifier to the varactor diode in the VCO. Because the VCO is at the desired frequency and the correction signal is 0, the frequency will be maintained. However, within a short period of time, the VCO frequency will begin to drift. The divided down sample of the VCO frequency will drift out of phase with the quartz crystal reference frequency, and the control voltage to the VCO will no longer be 0. The VCO frequency will then change until its divided down frequency matches that of the quartz crystal reference. The PLL is a servo system controlling the frequency of the oscillator.

If the PLO needs to change to a different frequency, the divide ratio of the variable ratio divider is changed. For example, if the frequency needs to be changed to 900 MHz, the divide ratio is changed from 10 to 9.

17.2 PHASE NOISE AND ITS SIGNIFICANCE IN A DIGITAL RF COMMUNICATIONS SYSTEM

Figure 17.5 shows the amplitude of the oscillator power as a function of frequency across a small frequency range around the carrier. A graph of power from a perfect oscillator as a function of frequency would be a single line, meaning that the oscillator

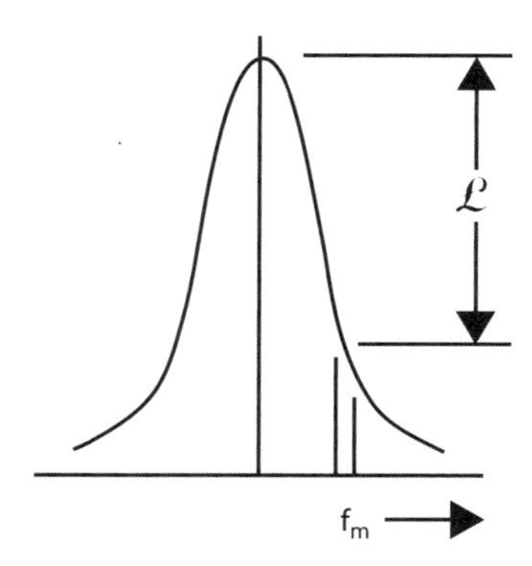

Figure 17.5 Definition of phase noise.

is generating its power by a single frequency. However, because of noise on the power supply voltage, the oscillator frequency is actually being frequency modulated over a very small range around the main oscillator frequency. This frequency variation is called phase noise. The oscillator output can be considered as a frequency spectrum. Phase noise is defined as the power in a 1 Hz bandwidth at a frequency (f_m) from the carrier. The phase noise is measured in decibels below the carrier power (dBc).

Figure 17.6 shows the phase noise of a typical PLO used for cell phones. The graph shows the single side band phase noise around a carrier frequency of 1 GHz. The vertical scale is phase noise in dB below the carrier power/Hertz. The horizontal scale is the frequency offset from the carrier in Hertz. Only the upper side band is

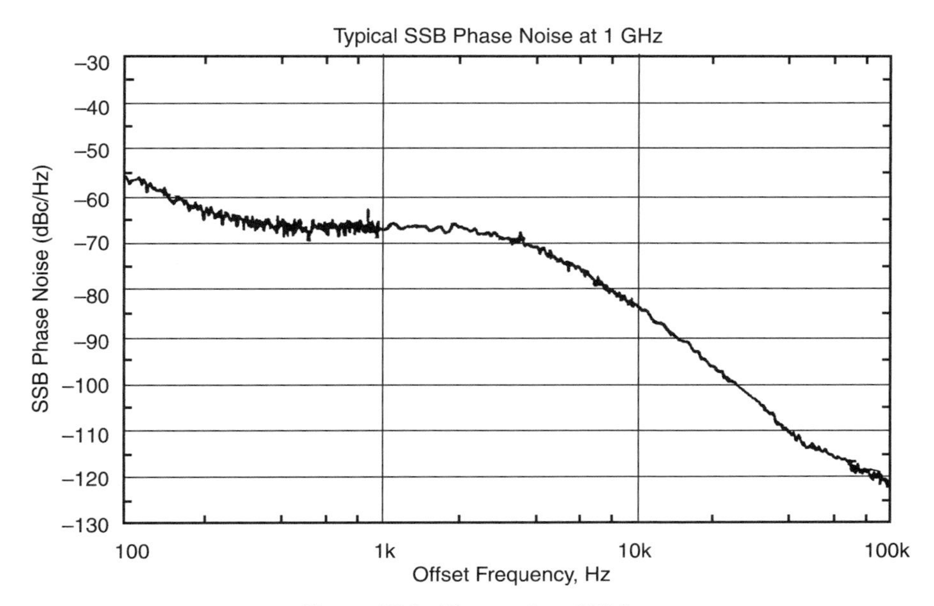

Figure 17.6 Phase noise of PLO.

shown, because the lower side band is identical. For the particular PLO shown, that phase noise is −56 dBc/Hz at an offset frequency of 100 Hz, −68 dBc/Hz at 1 kHz, and −84 dBc/Hz at 10 kHz.

The significance of phase noise in an RF communications system is as follows. Digital information is modulated onto the RF carrier by step phase changes. This phase noise can produce bit errors. The effect of phase noise is determined by the integrated phase noise across the bandwidth of the channel. For example, if the PLO has the phase noise characteristic shown in Figure 17.6, and if the bandwidth of the channel is 20 kHz, the integrated phase noise in decibels below the carrier power is twice the integrated value of the curve from 100 Hz to 10 kHz. (The doubling of the integral is due to the use of the single side band graph.) At frequencies less than 100 Hz, the phase noise is significantly reduced by the PLL and the integration bandwidth is small, so this part of the integration can be ignored.

To not cause a problem, the integrated RMS phase noise must be less than one-tenth of the phase shift between states. The double side band integrated phase noise of Figure 17.6 is about 0.001 radian $\times$ 57°/radian = 0.6° RMS. If the modulation type is QPSK, the separation between states is 90°, and 10% of the separation is 9°. However, if the modulation type is 64PSK, the separation between states would be 5.625° and a phase noise of 0.6° would be more than one-tenth of the separation between states and thus would be a problem.

A PLO is also used as the LO in the receiver mixer. This LO adds an equivalent amount of phase noise to the overall system. Note that the phase noise added by the transmitter and the receiver PLOs is in addition to the thermal noise that enters the receiver from the surroundings and is generated in the receiver LNA, as discussed in Chapter 23.

17.3 CHARACTERISTICS OF PLOs THAT NEED TO BE MEASURED

The characteristics of a PLO that needs to be measured are its frequency, tuning sensitivity, output power, and phase noise. The measurement results shown were made on a PLO with the following characteristics and settings:

Manufacturer	Qualcomm
Model	Q0420
Frequency	2.200 GHz
Output power	7 dBm
Phase noise at 10 kHz offset	−87 dBc

Frequency

The frequency of a PLO is more accurately controlled than that of the reference source in a frequency counter or other frequency measurement instrument. For example, the frequency counter discussed in Chapter 7 is a direct frequency counter up to 325 MHz. At higher frequencies the counter must mix the signal to be measured with an internal source and measure the difference frequency. The frequency of the PLO under test may be more accurate than the reference oscillator in the frequency meter.

Tuning Sensitivity

Sometimes the frequency of an oscillator, such as a VCO, needs to be measured as a function of the applied tuning voltage. This would be an important quantity to know in designing and testing a VCO. Such a measurement can be made by measuring the frequency of the oscillator at different tuning control voltages.

Power

The power of a PLO can be accurately measured with a power meter to ± 0.2 dBm, as discussed in Chapter 6. Care must be taken to use a low pass filter at the PLO output to block any harmonics in the signal. Figure 17.7 shows the output power of a PLO as a function of the supply voltage.

Phase Noise

The phase noise of a PLO can be measured with a spectrum analyzer, either manually or with special signal processing software.

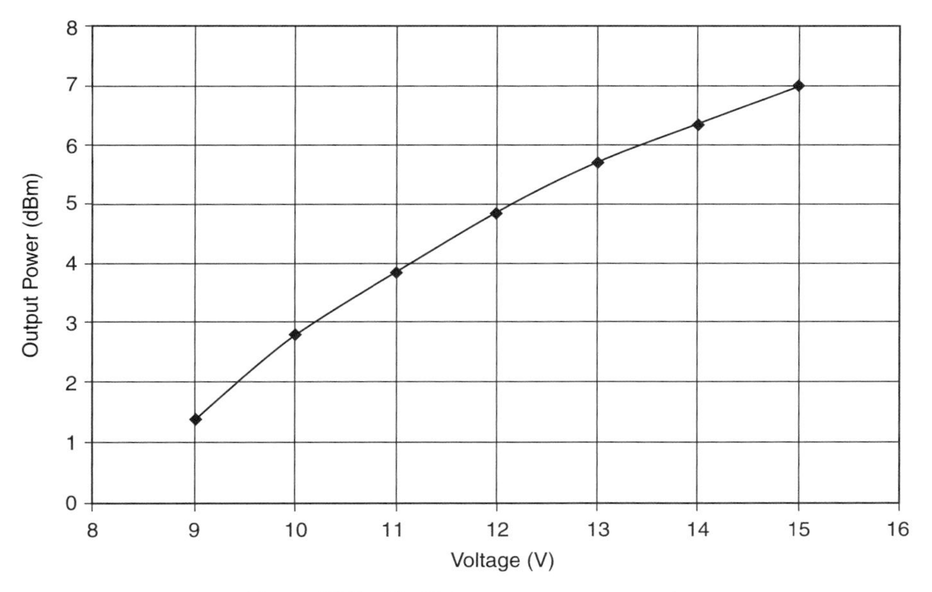

Figure 17.7 Oscillator power measurement.

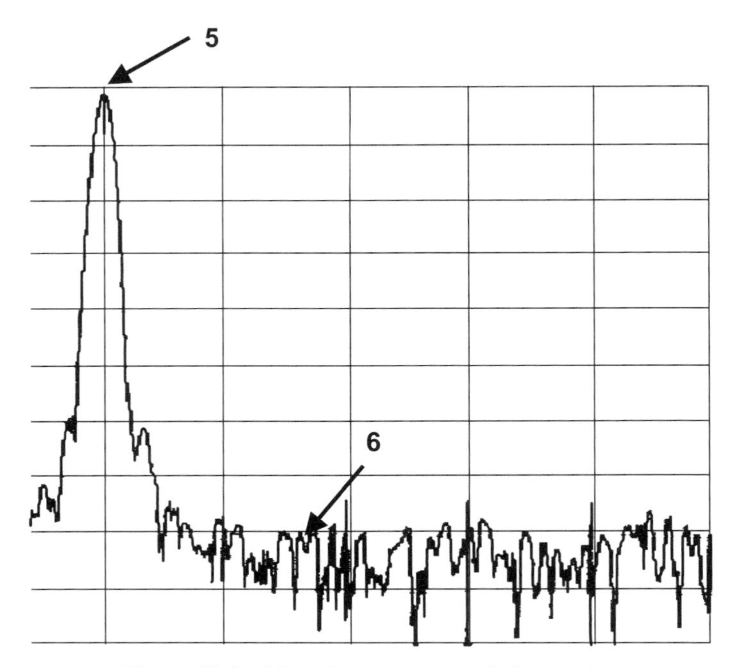

Figure 17.8 Manual measurement of phase noise.

Figure 17.8 shows the spectrum analyzer display to manually measure phase noise. The procedure is as follows:

1. Set the center frequency of the display to the PLO center frequency.
2. Set the peak close to the left-hand edge of the display.
3. Set the span to the range over which phase noise is to be measured.
4. Set the resolution bandwidth to 1 kHz.
5. Use the peak search to set the marker.
6. Activate delta mode on the marker and set the offset frequency to where the phase noise is to be measured (e.g., 10 kHz).
7. Activate the marker noise function, and read phase noise in dB below the carrier power/Hertz.
8. Repeat steps 5–7 to measure phase noise at other offset frequencies.

The PSA spectrum analyzer described in Chapter 9 can be procured with special software that automatically measures phase noise using a procedure similar to the manual procedure described above. When the automatic procedure is installed, it can be accessed by pressing the Measure hard key and then the Phase Noise soft key that is in the Measure Menu. Figure 17.9 shows the automatic phase noise measurement display. The stop and start frequencies and the vertical phase noise scale can be adjusted. The phase noise can be automatically integrated from a start

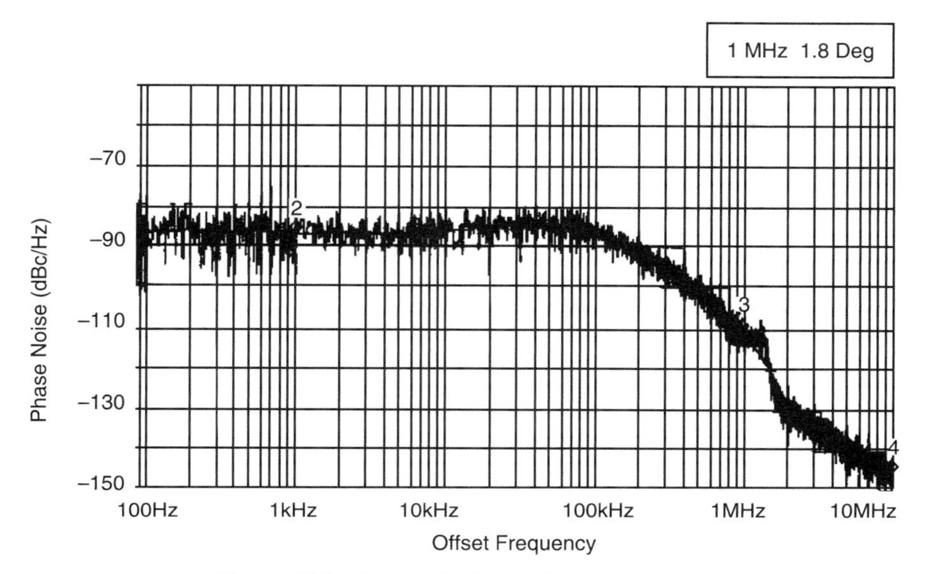

Figure 17.9 Automatic phase noise measurement.

to a stop frequency. The double side band integrated value from 100 Hz to 1 MHz is tabulated at the top right of the display and is 1.8° for this particular PLO.

The minimum phase noise that can be measured is limited by the phase noise in the PLO of the spectrum analyzer making the measurement. Figure 17.10 shows the nominal phase noise floor of the spectrum analyzer described in Chapter 9. A comparison of Figures 17.9 and 17.10 shows that the spectrum analyzer is entirely adequate for measuring the phase noise of the device measured in Figure 17.9 because the spectrum analyzer phase noise floor is 25 dB below the oscillator being measured.

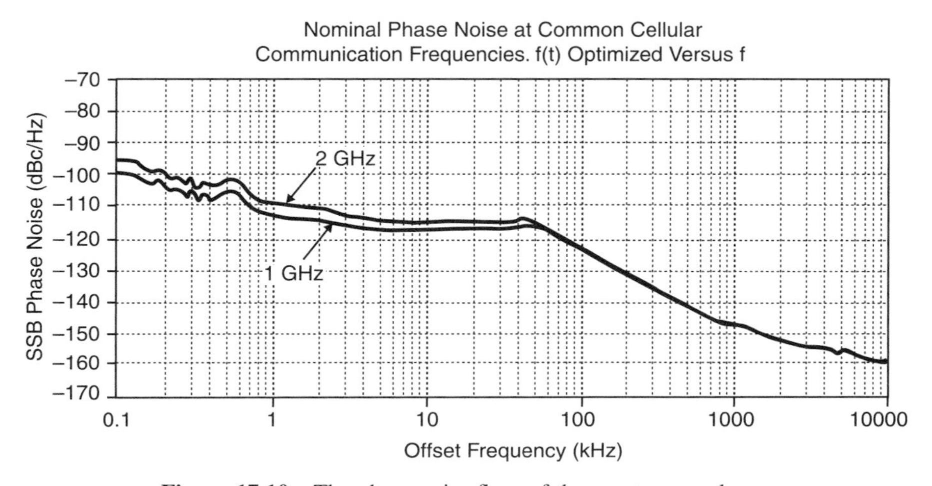

Figure 17.10 The phase noise floor of the spectrum analyzer.

17.4 EXAMPLE PROCEDURE FOR PHASE NOISE MEASUREMENTS OF PLOs

The phase noise of a PLO can be measured using a spectrum analyzer. The Noise Marker function of the spectrum analyzer is used. This function reports the power measured at the marker and normalizes the value to the amount of power in a 1 Hz bandwidth.

Measurements Being Described

PLO phase noise using the noise marker

PLO phase noise using measurement option (if installed)

Specifications of PLO

Manufacturer	Qualcomm
Model	Q0420
Frequency	2.200 GHz
Output power	7 dBm
Phase noise at 10 kHz offset	−87 dBc

Significance to Wireless System Performance

Phase noise will affect modulation performance for phase-dependent modulation schemes as well as affecting the S/N of the received signal.

Generic Procedure

The steps to perform the measurements are listed in the following table. The left column lists the actions to be performed, and the right column lists the motivation for taking each step and provides additional background information in some cases. The instructions are intended to be generic across most spectrum analyzer models.

Procedure	Notes
Preset the spectrum analyzer.	
Set the center frequency to 2.2 GHz, span to 500 kHz, and reference level to 5 dBm.	
Connect the PLO to the input of the spectrum analyzer and observe the signal.	
Adjust the center frequency so that the signal peak is close to the left edge of the display. Set a marker to view the peak.	This step is taken simply to maximize the usable screen space.

(Continued)

Procedure	Notes
Adjust the resolution bandwidth to 10 kHz. (A smaller resolution bandwidth will be needed if measuring close to the carrier.)	A small resolution bandwidth is required to measure the noise floor closer to the signal peak without energy from the carrier tone affecting the measurement. A larger resolution bandwidth will require less averaging to reduce the variance in readings from sweep to sweep, because the noise power is "averaged" over a larger bandwidth.
Change the marker mode to delta mode. Enter a frequency offset for the marker delta such as 20 kHz.	When switching to delta mode, the ESA/PSA series analyzers set the reference marker at the location where the marker was originally set. You can then input a frequency offset from that original position and the marker will report the power level relative to that original position.
Activate the Marker Noise function.	The noise marker takes the average of 32 measurement cells, corrects for the fact that the resolution bandwidth filter is not a perfect rectangle, and corrects for averaging done by the video filter. On the ESA/PSA models the average is taken over about one-half of a division.
Read the value for the phase noise at various offsets from the carrier.	To measure further away from the carrier, you will have to adjust the span. Measuring close to the carrier is limited by the smallest resolution bandwidth of the spectrum analyzer as well as the span settings.
Integrated phase noise can be calculated if enough data points are measured.	
A measurement option is available that automatically plots the phase noise and reports integrated phase noise power and angle.	This is Option 226 for Agilent ESA/PSA spectrum analyzers.

To measure the normalized noise power without using the noise marker function, you can perform some of the normalization calculations yourself. For example, if the resolution bandwidth is set to 10 kHz, then to normalize to 1 Hz you would divide the power reading by 10,000. If you are using a decibel scale, then dividing is performed by subtracting; a factor of 10,000 is 40 dB, so you subtract 40 dB from the measured power reading. If log averaging is used (as in most spectrum analyzers), then you will need to add 2.5 dB to the reading, because the log averaging underreports noise values by 2.5 dB. For example, phase noise is measured at 200 kHz from the carrier using a resolution bandwidth of 10 kHz. Log averaging is used with 100 readings. Using the noise marker function and with a delta marker reference set at the carrier frequency, the normalized noise power reading is −90.5 dBc/Hz. Using a regular marker, the reading is −53.2 dBc. We subtract 40 dB to normalize the reading to 1 Hz and arrive at −93.2 dBc/Hz. Now we must add in the 2.5 dB correction factor for log averaging and arrive at −90.7 dBc/Hz. We are still 0.2 dB off, because we did not take into account the equivalent noise bandwidth of the resolution bandwidth filter, which can be a factor of 0.2–0.5 dB, depending on the spectrum analyzer model.

Averaging will be required to obtain a reasonably consistent reading. The variance of a noiselike signal reduces as the square root of the number of averages; thus, in order to reduce the variance of the readings, a large number of readings must be taken. For example, if 10 averages are not enough to reduce the variance of the readings, then the number of averages should be increased to 100, and so on. Some newer spectrum analyzers allow the selection of RMS power averaging, which does not cause the 2.5 dB underreporting of noise power.

Integrated phase noise can be calculated if enough data points are measured. For high accuracy phase noise measurements, a dedicated phase noise measurement system is used instead of a spectrum analyzer. The most important component of the phase noise measurement system is a signal source with exceptionally low phase noise performance.

17.5 ANNOTATED BIBLIOGRAPHY

1. Larson, L. and Rosenbaum, S., RF and microwave frequency synthesizer design, In Larson, L., Ed., RF and Microwave Circuit Design for Wireless Communications, Chap. 6, Artech House, Boston, 1996.
2. Product Note: Self Guided Demonstration for Phase Noise Measurements, Document 5988-3704EN, Agilent Technologies, Santa Clara, CA.
3. Application Note 1303: Spectrum Analyzer Measurements and Noise, Agilent Technologies, Santa Clara, CA.

Reference 1 describes the design of PLOs in more detail than this book.

Reference 2 describes making phase noise measurements with the Agilent PSA series spectrum analyzer.

Reference 3 provides a detailed explanation of noise statistics, variance, and the effect of log averaging.

CHAPTER 18

UPCONVERTERS

The complicated modulations that are used in cellular phones and wireless data transmission systems are difficult to generate at RF frequencies. Thus, in most RF communication systems, the modulated signals are generated at low frequency with digital processing chips that do not work at RF and then shifted to the desired RF frequency in an upconverter.

18.1 HOW AN UPCONVERTER WORKS

Figure 18.1 shows a block diagram of an upconverter. A stable frequency below the RF range, for example, 100 MHz, is generated with a simple quartz crystal oscillator. Digital information is then modulated onto this low frequency carrier in a digital IC. An RF PLO, such as the one discussed in Chapter 17, is used to generate an RF signal at 900 MHz, 100 MHz below the desired transmitted frequency. The modulated low frequency signal and the RF signal are then added together to form a sum frequency of 1000 MHz that is now carrying the digitally modulated signal. The upconverter also produces a difference frequency at 800 MHz, and this is removed by the bandpass filter.

The key components of an upconverter are a combiner and one or more Schottky diodes. The combiner performs the function of combining the RF signal from the VCO with the digitally modulated IF signal into a common RF transmission line, while minimizing the leakage of the modulated IF signal into the VCO line, and vice versa. The upconversion is accomplished by applying the combined PLO signal and the digitally modulated IF signal to the Schottky diodes.

RF Measurements for Cellular Phones and Wireless Data Systems. By A. W. Scott and R. Frobenius

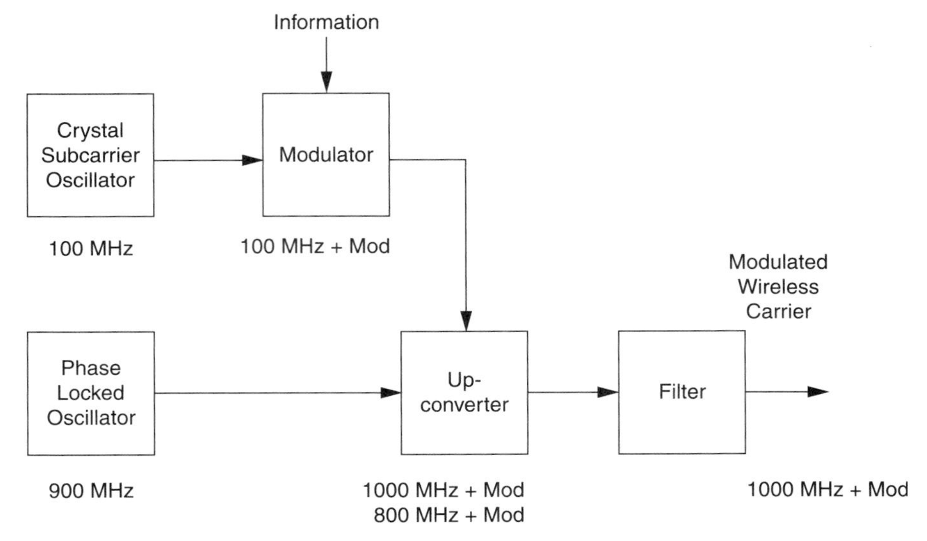

Figure 18.1 Upconverter block diagram.

The fabrication and electrical characteristics of a Schottky diode are provided in Figure 18.2. Schottky diodes were described in detail in Chapter 16. They are also used in power meters, as described in Chapter 6, and as mixers, as will be described in Chapter 24.

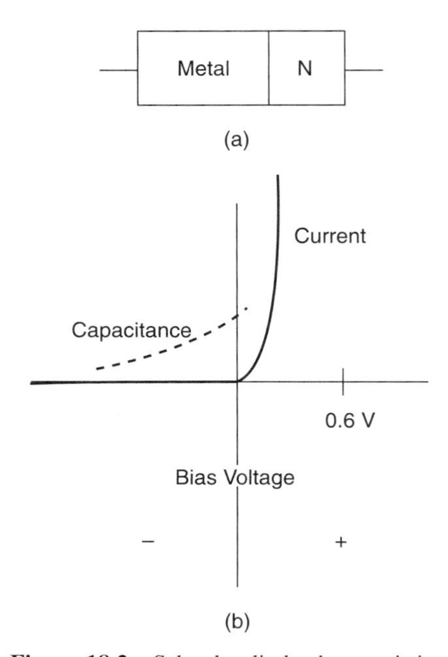

Figure 18.2 Schottky diode characteristics.

Figure 18.2a shows the physical fabrication of a Schottky diode. The diode is formed by the junction of a metal to an N-doped semiconductor. It is similar to the PN diode used as a variable capacitor for tuning the resonator of a VCO, except that a metal region is used instead of the P semiconductor. Figure 18.2b shows the current and capacitance of a Schottky diode as a function of the voltage across the junction. The capacitance in the reverse-biased operation (with the metal region biased negatively) is much lower than in a PN junction. In addition, current begins to flow in a Schottky diode when the voltage of the metal region just becomes positive, instead of at a positive voltage of about 0.6 V as is the case in a PN diode. Most importantly, the current–voltage characteristics are very nonlinear in the conduction region, and this is what creates the upconverting process.

18.2 MATHEMATICAL THEORY OF UPCONVERTER AND MIXER ACTION

The frequency adding in an upconverter (and the frequency subtracting in a mixer) can be understood from the mathematical formulas of Figure 18.3. The I-V characteristics of the Schottky diode in its conducting direction shown in Figure 18.2b can be represented by a power series, which is given by Eq. (1) of Figure 18.3. The first term (k_1) is the constant conductance of the diode, and it gives the component of current that is directly proportional to V. The second term is the second-order nonlinearity, and higher order nonlinearities exist. It is the second-order nonlinearity that creates the upconverted signal.

Equation (2) is the representation of a single RF signal as a function of time at frequency f. Equation (3) describes an RF signal with two frequencies f_1 and f_2.

Equation (4) shows the two frequency signals of Eq. (3) applied to the second-order nonlinearity term of Eq. (1).

A standard algebraic expansion of Eq. (4) leads to Eq. (5). The first and third terms of Eq. (5) vary with time as f_1 and f_2, respectively. The center term of Eq. (5) can be

(1) $i = k_1 V + k_2 V^2 + k_3 V^3 + \cdots$

(2) $V_{RF} = V \sin(2\pi f t)$

(3) $V_{RF} = V_1 \sin(2\pi f_1 t) + V_2 \sin(2\pi f_2 t)$ with 2 signals at f_1 and f_2

(4) Nonlinear output: $k_2[V_1 \sin(2\pi f_1 t) + V_2 \sin(2\pi f_2 t)]^2$

(5) Algebraic Expansion of (4):

$$k_2 \{V_1{}^2 \sin^2(2\pi f_1 t) + 2[V_1 \sin(2\pi f_1 t)]\,[V_2 \sin(2\pi f_2 t)] + V_2{}^2 \sin^2(2\pi f_2 t)\}$$

(6) Trigonometric expansion of center term

$$V_1 V_2\,[\sin 2\pi(f_1 + f_2)t + \sin 2\pi(f_1 - f_2)t]$$

Figure 18.3 Mathematical derivation of frequency mixing in an upconverter.

expanded by a standard trigonometric procedure to give Eq. (6), which shows the creation of sum and difference frequencies.

This adding and subtracting can also be explained by looking at signal waveforms, as will be described in Chapter 24.

18.3 MEASUREMENT OF UPCONVERTER PERFORMANCE

The specifications for the particular upconverter being tested are the following:

Manufacturer	NEC
Model	UPC8106TP
RF output frequency	2.45 GHz
LO frequency	2.20 GHz
LO power	7 dBm
IF input frequency	250 MHz
Conversion gain	1.0 dB
Output at P_1dB	−10 dBm

Figure 18.4 shows a test setup for measuring upconverter performance. RF output power is to be measured as a function of RF input power. The ratio of the two powers is "conversion gain." For the example shown, the IF power has a frequency of 250 MHz, which is obtained from the signal generator described in Chapter 5. The RF signal from the upconverter is measured with an RF spectrum analyzer, such as described in Chapter 9. Although a gain measurement is being made, it cannot be made with a VNA like that in Chapter 8, because the output and input signals are at different frequencies. The PLO described in Chapter 17 generates the LO signal at a frequency of 2200 MHz. The output of the upconverter is 2450 MHz.

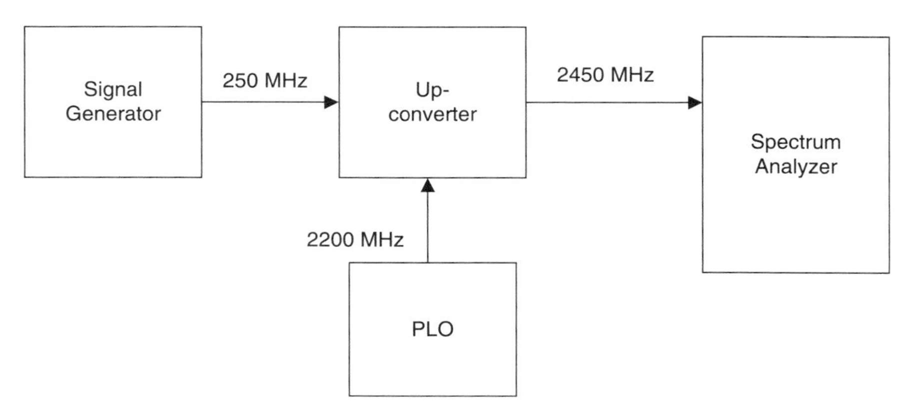

Figure 18.4 Upconverter test setup.

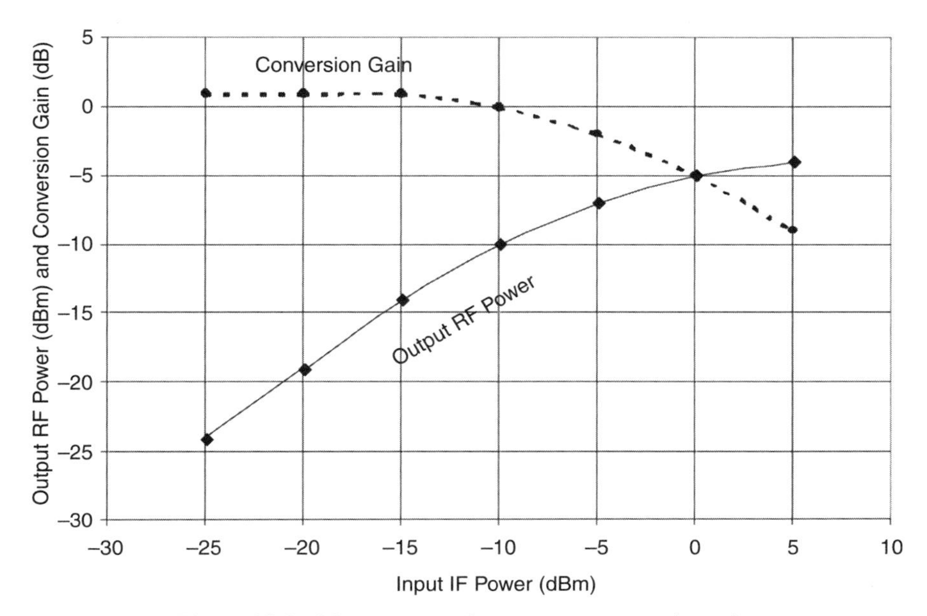

Figure 18.5 Measurement of upconverter conversion gain.

Figure 18.5 graphs the output RF power from the upconverter and the conversion gain as a function of IF input power. When the input power is in the range from −25 to −15 dBm, the conversion gain is constant at 1 dB. At an input power of −10 dBm, the conversion gain drops by 1 dB, and the output power is 0 dBm. Further increases in input power force the upconverter into saturation, and its gain continues to drop. At 0 dBm of input power, the output power from the upconverter is −5 dBm and the conversion gain has dropped to −5 dB.

The purpose of the upconverter is to convert an IF signal that has been modulated to carry information up to the RF transmission frequency. Figure 18.6 compares the spectrum of the modulated IF signal and the upconverted RF signal for the modulator whose conversion gain was shown in Figure 18.5. Analog frequency modulation with a modulation index of 1 is illustrated, because it is easy to understand the modulation waveforms. The left-hand figure shows the IF signal. The right-hand figure shows the upconverted RF signal at 2.45 GHz. Note that the two spectra are identical.

The upconverter generates many other frequencies than just the desired upconverted signal. Figure 18.7 shows the spectrum of the upconverter over a frequency range from 1.2 to 2.9 GHz. A bandpass filter that passes only the frequency range from 2.4 to 2.5 GHz was added to the upconverter output. This filter was measured in Chapter 8. The spectral line marked A is the desired upconverted signal at 2.45 GHz. The spectral line marked B is the LO signal at 2.2 GHz, which has been attenuated by the filter. The spectral line marked C is the lower side band at 1.95 GHz, which has also been attenuated by the filter. The other spectral lines, which are −30 dB below the desired signal, are caused by harmonics of the RF and LO

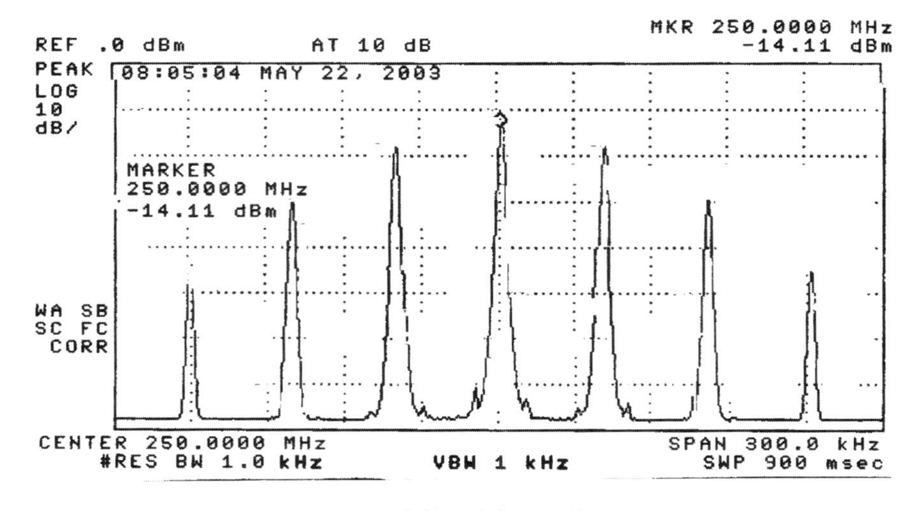

IF Signal @ 250 MHz
with FM Sidedbands
Singal Levels is – 10 dBm

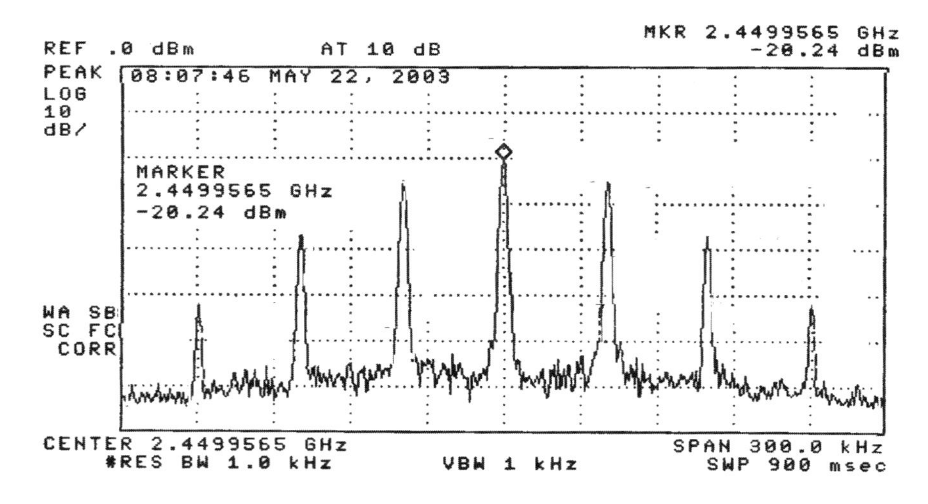

Upconverted Signal @ 2.45 GHz

Figure 18.6 Spectrum of IF and upconverted signal.

mixing together. Additional filtering could be added to reduce these other signals to a lower value if required. The measurement shown in Figure 18.7 was made at the 1 dB compression point of the upconverter, where the IF input power was 0 dBm and the RF output power was −1 dBm. If the IF input power had been increased beyond this point, more unwanted spectral lines would have been generated.

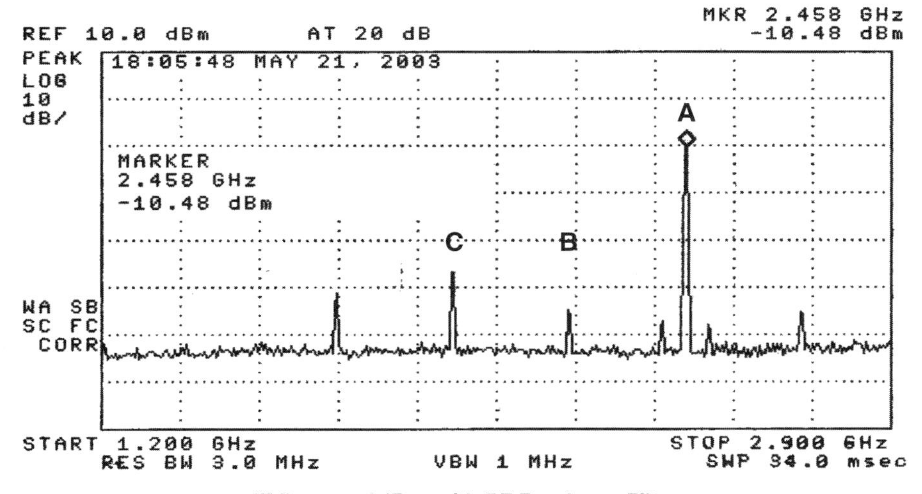

IF Power = 0dBm, with RF Bandpass Filter

Figure 18.7 Spurious signals from the upconverter.

18.4 GENERIC PROCEDURE FOR UPCONVERTER MEASUREMENT

Set up the equipment as shown in Figure 18.4. The steps to perform the measurements are listed in the table following this paragraph. The left column lists the actions to be performed, and the right column lists the motivation for taking each step and provides additional background information in some cases. The instructions are intended to be generic across most spectrum analyzer models.

Procedure	Notes
Preset the spectrum analyzer and signal generator.	
Set the signal generator to generate a signal at 250 MHz and −25 dBm.	This is the IF frequency signal to be input into the upconverter.
Connect the signal generator output directly to the input of the spectrum analyzer.	Observe the signal coming from the signal generator.
Set the spectrum analyzer to a center frequency of 250 MHz, span of 300 kHz, and reference level of 10 dBm.	In this measurement, 10 dBm is the largest signal we expect.
Connect the upconverter as the DUT.	
Set the spectrum analyzer center frequency to 2.45 GHz.	The output of the upconverter should be the LO frequency ± the IF. We could also look at 1.95 GHz.
Observe and record upconverter output at −25, −20, −15, −10, −5, 0, and +5 dBm. Calculate the gain at these input power levels.	Based on these measurements you can determine the 1 dB compression point and the saturation output power.

(Continued)

Procedure	Notes
Apply FM modulation to the signal on the signal generator with an FM rate of 40 kHz and a deviation of 40 kHz.	The goal of using an upconverter is to be able to perform modulation at a lower IF frequency, where it is easier to implement.
View the signal directly on the spectrum analyzer (connect signal generator directly to the spectrum analyzer). Set the spectrum analyzer center frequency to 250 MHz.	Verify the input signal has modulation applied.
Connect the upconverter and view the signal on the spectrum analyzer. Set the spectrum analyzer center frequency to 2.45 GHz. You can adjust the resolution bandwidth to view signal peaks more sharply.	The FM modulated signal should appear at the upconverted frequency. This is the purpose of an upconverter.
Set the spectrum analyzer start frequency to 1.2 GHz and stop frequency to 2.9 GHz.	Observe the 1.95 GHz peak for a lower side band, LO feed through at 2.2 GHz, as well as other products of mixing harmonics.
Adjust the input power to +5 dBm.	Observe additional harmonics and intermodulations being produced.
Adjust the input power to −10 and −15 dBm.	Observe the effect of operating the upconverter in the linear region of its operation versus nonlinear operation.
Set the input power level to the 1 dB compression point for the device.	
Insert a filter at the output of the upconverter.	The filter should reduce the level of the undesired signals. A second filter can be applied as well to further reduce the unwanted signals.

18.5 ANNOTATED BIBLIOGRAPHY

1. Pozar, D. M. Mixers, in Microwave and RF Design of Wireless Systems, John Wiley & Sons New York, 2001.

Reference 1 provides a mathematical analysis of the frequency translation properties of mixers and upconverters, as well as design examples of various mixer/upconverter topologies.

CHAPTER 19

POWER AMPLIFIERS

The purpose of the power amplifier in a wireless communication system is to increase the power level of the RF wave that is carrying the modulation, so that by the time it is broadcast by the transmitter antenna and travels wirelessly to the RF receiver, it is large enough to achieve the required S/N. An example of this path loss was given in Chapter 2.

The power amplifier must

1. have high efficiency to minimize battery drain and
2. not distort the information modulation that the RF wave is carrying.

These two requirements are somewhat mutually exclusive, so amplifier design and testing must determine the operating conditions where both requirements are satisfactory for system performance.

19.1 RF TRANSISTORS

RF transistors use field effect or bipolar designs, just like transistors below the RF band. Like their lower frequency counterparts, the RF signal travels from the input to the output of the transistor by the flow of electrons as a signal is being amplified. The problem with RF transistors is the "transit time effect," which is the delay of the electrons passing through the transistor relative to the RF wave. If this delay is one-half of the RF cycle, the current will arrive 180° out of phase with the applied RF voltage, and no power will be generated. A numerical example will make this clear. The electron velocity in silicon is 1/3000 the velocity of light. This is a very large velocity: 100,000 m/s. The transit time problem arises because at RF frequencies, the period (time) of one RF cycle is very short. The period of a 1 GHz signal is only 1 nanosec (10^{-9} s). Therefore, in one cycle at 1 GHz, the electrons travel 10^5 m/s $\times$ 10^{-9} s $=$ 100 μm. If the transistor is 50 μm long in the direction of RF travel, the electrons will arrive at the transistor output 180° out of phase with the RF wave, and no RF power will be generated. If the transistor is 5 μm long, the electrons will arrive at the transistor output 18° out of phase with the RF wave, and the performance will still be degraded compared to a low frequency transistor. As the RF operating frequency increases, the transit time problem increases directly as the operating frequency.

Solutions to the transit time problem are the following:

1. reduce the spacings between the transistor elements and
2. use semiconductor materials like gallium arsenide (GaAs) and silicon germanium (SiGe) where the electrons travel two or more times faster than they do in silicon.

19.2 SEMICONDUCTOR MATERIALS FOR RF TRANSISTORS

Semiconductor materials in which electrons travel faster than they do in silicon include

silicon germanium (SiGe)
gallium arsenide (GaAs)
gallium nitride (GaN)
silicon carbide (SiC)

The atomic structure of these materials was discussed in Chapter 16.

Silicon germanium is an alloy of silicon and germanium. Both materials have four electrons in their outer shells, and therefore are easy to form into solid materials. The advantages of SiGe are low cost like silicon and electrons travel twice as fast in SiGe as they do in silicon.

The third-order intercept (TOI) of amplifiers made with SiGe is 5 dB higher than amplifiers made with any other semiconductor material. This issue of TOI will be discussed in more detail in Chapter 26. Because of this third advantage, SiGe is now used for almost all low noise amplifiers in wireless equipment. Unfortunately, SiGe has unique nonlinearities at saturation. Consequently, it is not useful in power amplifiers.

Gallium arsenide is a compound semiconductor formed from gallium, which has three electrons in its outer shell, and arsenic, which has five electrons in its outer shell.

The advantage of GaAs is that electrons travel twice as fast through GaAs as they do through silicon.

However, GaAs transistors are two or more times as expensive to fabricate than silicon transistors, because the raw materials are more expensive, and disposing of the process waste materials is more expensive.

In the frequency range up to about 1.5 GHz, silicon is the material of choice for power amplifiers because of its low cost and usually acceptable performance. In the frequency range above 2.5 GHz, gallium arsenide is the material of choice because of its significantly better RF performance. In the 1.5–2.5 GHz frequency range, which includes the PCS and Universal Mobile Telecommunications System (UMTS) cellular phone bands and a wireless LAN band, both materials are useful for power amplifiers.

Gallium nitride and SiC have RF properties comparable to GaAs. In addition, both have higher thermal conductivities, and thus are the materials of choice for high power base station transistor amplifiers. Other semiconductor materials for use at frequencies above 5 GHz are indium gallium phosphide and indium gallium arsenide.

19.3 TRANSISTOR FABRICATION PROCESSES

RF transistor fabrication processes include the following:

Metal–semiconductor FET (MESFET)
Bipolar transistor
High electron mobility transistor (HEMT)
Heterojunction bipolar transistor (HBT)
Laterally diffused metal–oxide–semiconductor (LDMOS)
Pseudomorphic HEMT (PHEMT)

These fabrication processes are used with the materials described in the previous section.

MESFETs

The operation of a FET is shown in Figure 19.1. The FET is fabricated on an insulating substrate, which serves as the transistor support. An epitaxial layer of doped semiconductor material is deposited on top of the substrate, and the FET is built into this layer. The FET has a source, a gate, and a drain. The source is at one end of the transistor, and the drain is at the other. A voltage is connected to the drain, and electrons are drawn from the source to the drain. The gate is placed between the source and drain on the top surface of the epitaxial layer. A small input voltage on the gate controls a large flow of current to the drain at a higher voltage, and this process provides signal amplification. PN junctions do not have to be grown through the material in a FET. All the electrodes are fabricated on the top surface.

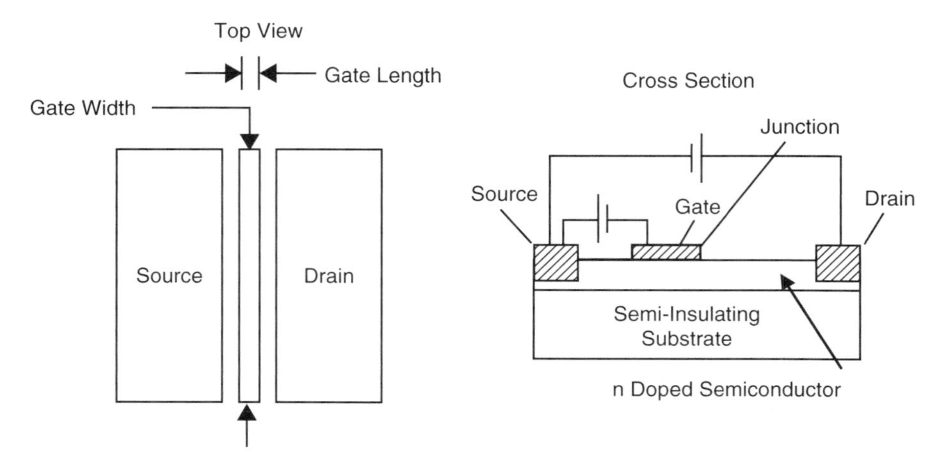

Figure 19.1 FET.

The most commonly used transistor for computer circuitry is a MOSFET. Computer circuitry operates up to 100 MHz rather than at RF frequencies. The MOSFET uses a metal-oxide-semiconductor junction for its gate. The layer of oxide reduces the power consumption, which is important for putting thousands of FET computer gates on a single chip, but the added capacitance of the oxide layer reduces the frequency response. Because frequency response is not a critical parameter, MOSFETs use silicon because of its lower cost.

If the MOSFET is made as described above, using an N-doped semiconductor, the transistor is called an NMOS. If it uses a P-doped semiconductor, it is called PMOS, and the polarities of all voltages are reversed. If N-doped and P-doped regions are fabricated on the same chip, the device is called CMOS. The CMOS design simplifies the connection of thousands of gates on the same chip, so it is the preferred design for computer chips.

RF FETs are made with a metal to semiconductor junction (also called a Schottky junction) at the gate. For this reason the microwave FET is often called a MESFET, meaning that the gate is a MES junction. The semiconductor material used for RF FETs is GaAs and similar compound semiconductors. Silicon could be used, but electrons travel approximately twice as fast in gallium arsenide as they do in silicon, so better high frequency performance is obtained.

Figure 19.2 explains the operation of a MESFET. In Figure 19.2a, the gate of a MESFET is shown operated at the same voltage as the source; that is, the source–gate voltage is 0. In this case electrons move through the entire thickness of the epitaxial layer, and the FET draws the maximum current, called the saturated drain–source current.

As shown in Figure 19.2b, as the gate voltage is made negative, a reversed-biased Schottky junction is formed around the gate. An RF FET is always operated with its gate voltage negative with respect to its source voltage. Electrons are drawn out of the semiconductor into the gate electrode from a region around the gate. This region

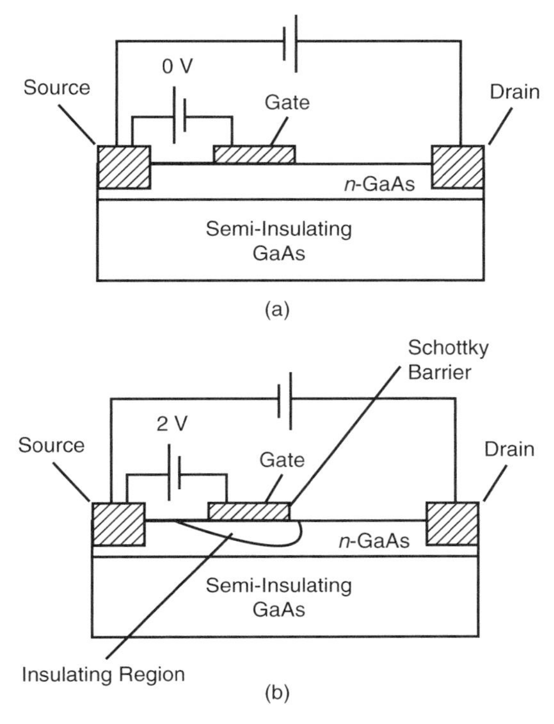

Figure 19.2 MESFET.

becomes an insulator, and the electrons flow from the source to the drain is therefore impeded, because electrons can no longer flow through the Schottky barrier. As the negative voltage on the gate is increased, the size of the insulating barrier region increases and the current flow from the source to the drain is further reduced. If the gate voltage is negative enough, the insulating region around the gate can be made to extend across the entire epitaxial layer and completely cut off the current flow.

The FET achieves amplification because a small voltage applied to the gate controls a large amount of current flowing through the transistor, and this current can be used to generate a large voltage in the output circuit.

Top views of MESFETs are provided in Figure 19.3. The gate length determines the transit time of the MESFET. Typical gate lengths are 0.5 μm. Note that, as defined in Figure 19.3a, the gate length is the short dimension of the gate and the width of the gate is the long dimension. The gate width determines the power capability. With a gate width of 50 μm, the FET provides about 1 mW of power. For higher power levels, parallel gates must be used.

Figure 19.3b shows a top view of a power MESFET. The increased power is obtained by using multiple sources, gates, and drains. The problem with the FET is that all three electrodes are on the top surface of the substrate, so some sort of

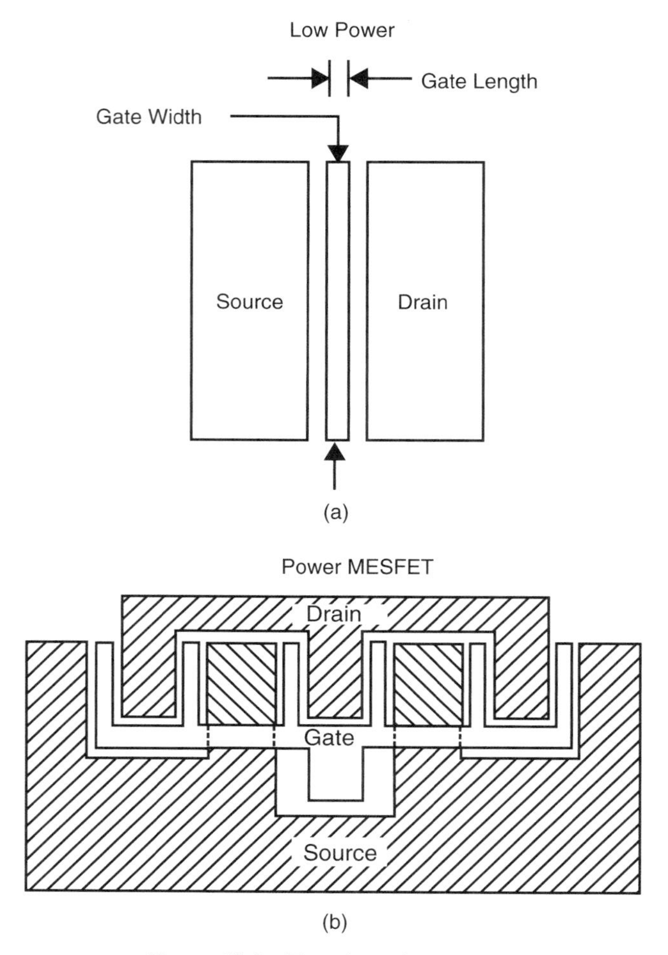

Figure 19.3 Top view of MESFETs.

bridging connection must be used. As shown, the interconnection of the three electrodes, all on the same surface, is accomplished by forming the source and the drain electrodes first as interdigital fingers. After this is done, the surface is covered with an insulating layer and the gate electrode is deposited over it. A special technique called *bridging* is used to etch away the insulating layer where the gate connection passes over the source electrode to reduce capacitance.

Bipolar Transistor

The special fabrication techniques used in a microwave bipolar transistor to reduce the effects of transit time and internal capacitance and resistance are illustrated in Figure 19.4. The major time delays are the charging time of the emitter–base

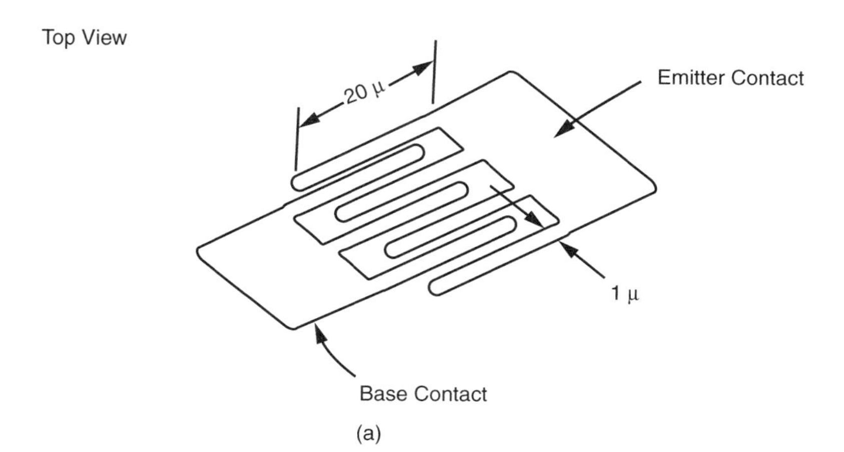

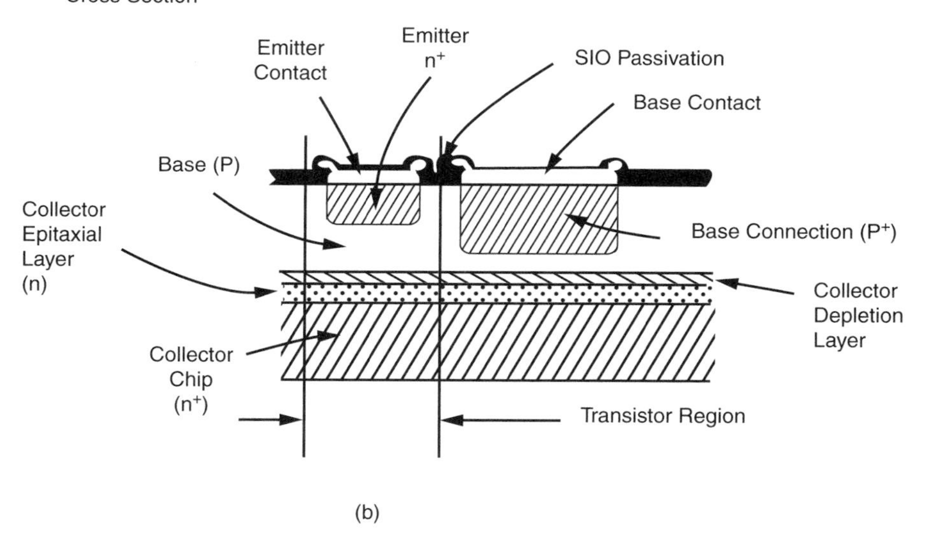

Figure 19.4 Bipolar transistor.

junction, the electron transit time through the base, and the electron transit time through the base–collector junction. The emitter–base charging time is minimized by using interdigital finger geometry for the emitter and the base (Fig. 19.4a). The interdigital finger design provides large areas for current to flow from the emitter to the base around the periphery of the fingers. At the same time, the design minimizes the emitter–base capacitance. This capacitance is determined by the area of the fingers, so they are made as narrow as fabrication allows, approximately 1 μm wide and 20 μm long.

The base transit time is reduced by making the base width as small as possible. No voltage is applied across the base, so only the fabrication process limits the base width. The planar fabrication process permits a base width of a few tenths of a micron. The transit time through the base–collector junction is controlled by the junction thickness. Its thickness cannot be arbitrarily reduced because it must stand off the reverse voltage applied to the transistor. To minimize transit time, the operating voltage of the transistor must be as low as possible, which in turn reduces transistor power capability. Typical delay times for a bipolar transistor such as shown in Figure 19.4 are an emitter–base charging time of 3 ps, a base transit time of 6 ps, and a base–collector transit time of 14 ps, for a total transit time of 25 ps from the emitter to the collector. This may seem very small, but at 4 GHz one RF cycle lasts only 250 ps, so the total delay of a 4 GHz signal in passing through this transistor is one-tenth of an RF cycle.

HEMTs and HBTs

The high frequency performance of FETs and bipolar RF transistors can be improved with special fabrication techniques and materials. An improvement over FET performance is obtained by using the HEMT. A comparison of the FET and HEMT is provided in Figure 19.5a and b. The FET (Fig. 19.5a) is built on a semiinsulating gallium arsenide substrate, and an N-doped layer of gallium arsenide is deposited

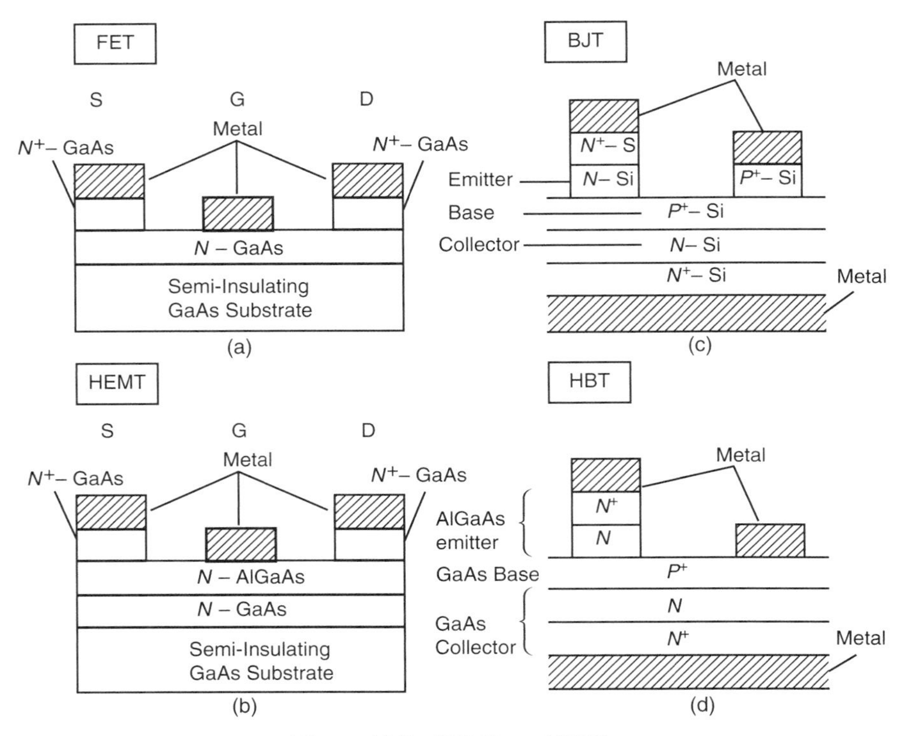

Figure 19.5 HEMTs and HBTs.

onto it. (This figure is the same as Fig. 19.2, but with the vertical scale exaggerated.) An electrical connection is made at each end of the N-doped layer to form the source and drain electrodes. To achieve an electrical connection, a layer of N+-doped material must be between the metal electrode and the N-doped semiconductor. The gate is a MES junction on the N-doped gallium arsenide with no N+ intermediate layer. The electrons flow through the N-doped gallium arsenide channel from the source to the drain, and the thickness of the channel is controlled by the negative voltage applied to the gate.

Like the FET, the HEMT shown in Figure 19.5b is built on a semiinsulating gallium arsenide substrate. Unlike the FET, however, the HEMT consists of multiple layers of different materials, including lightly N-doped aluminum gallium arsenide and N-doped gallium arsenide. This layer of different semiconductor materials is called a heterojunction. The source, drain, and gate electrodes are the same as in the FET. The difference between the FET and the HEMT is that in the FET the electrons travel in the N-doped gallium arsenide channel, whereas in the HEMT the electrons travel in the heterojunction layer between the two materials. The electrons suffer fewer collisions with the ionized doping atoms, which results in a higher electron velocity. This higher velocity of electron flow in the heterojunction channel from the source to the drain gives the HEMT its better high frequency performance and its name. PHEMTs have an even more complex arrangement of layers of semiconductor materials, including layers of gallium arsenide, aluminum gallium arsenide, indium phosphide, and indium gallium arsenide. The purpose of all of these layers is to increase the velocity of the electrons in the channel from the source to the drain, which allows the PHEMT to operate to over 100 GHz, whereas the standard FET can only operate to about 30 GHz.

A greatly improved bipolar transistor called the HBT is compared to the standard silicon bipolar transistor in Figure 19.5c and d. The standard silicon bipolar junction transistor (BJT) is shown in Figure 19.5c. (This is the same drawing as Fig. 19.4 but with the vertical scale exaggerated.) In the standard silicon BJT, the emitter is made of N-doped silicon, the base is P-doped silicon, and the collector is N-doped silicon.

The HBT is shown in Figure 19.5d. Its emitter is made of N-doped aluminum gallium arsenide. The N+ layer allows the emitter electrodes to be connected to the N-doped aluminum gallium arsenide. The base is made of P+-doped gallium arsenide, and the collector is made of N-doped gallium arsenide. Because different materials are used for the emitter and base, the transistor is called an HBT. The use of the emitter base heterojunction allows the base to be heavily doped to reduce its resistance, which significantly reduces the charging time of the base–collector junction. The heavy doping and the use of gallium arsenide instead of silicon allows the HBT to operate up to 50 GHz and provides higher efficiency than either the FET or the HEMT.

LDMOS

A power RF transistor technology for high power base station amplifiers is the LDMOS. This technology can be fabricated in silicon for low cost. The lateral diffusion reduces the effective gate length to less than its physical length to reduce transit time. Recent developments replace the silicon substrate with silicon carbide for better thermal conduction to achieve even higher RF power.

19.4 MODULATION DISTORTION CAUSED BY POWER AMPLIFIER NONLINEARITY

The lower curve of Figure 19.6 shows the RF output power of a gallium arsenide MESFET power amplifier as a function of RF input power. The amplifier is designed to operate in the PCS and UMTS cellular phone bands around 1.9 GHz. At an input power of –10 dBm, the output power is +13 dBm, so the amplifier gain is 23 dB. At an input RF power of –5 dBm, the output is 18 dBm, so the gain is still 23 dB, and the gain remains constant up to an input power of −2 dBm. This range of operation up to an input power of −2 dBm is called the "linear" or "small signal gain" range. As the input power increases above −2 dBm, the RF output power continues to increase, but it is no longer directly proportional to the RF input power. Linear performance is shown by the dotted line in Figure 19.6. As the input power increases further, the output power falls away more and more from the linear gain line. When the RF output power is +3 dBm, the gain has fallen away from the linear line by 1 dB,

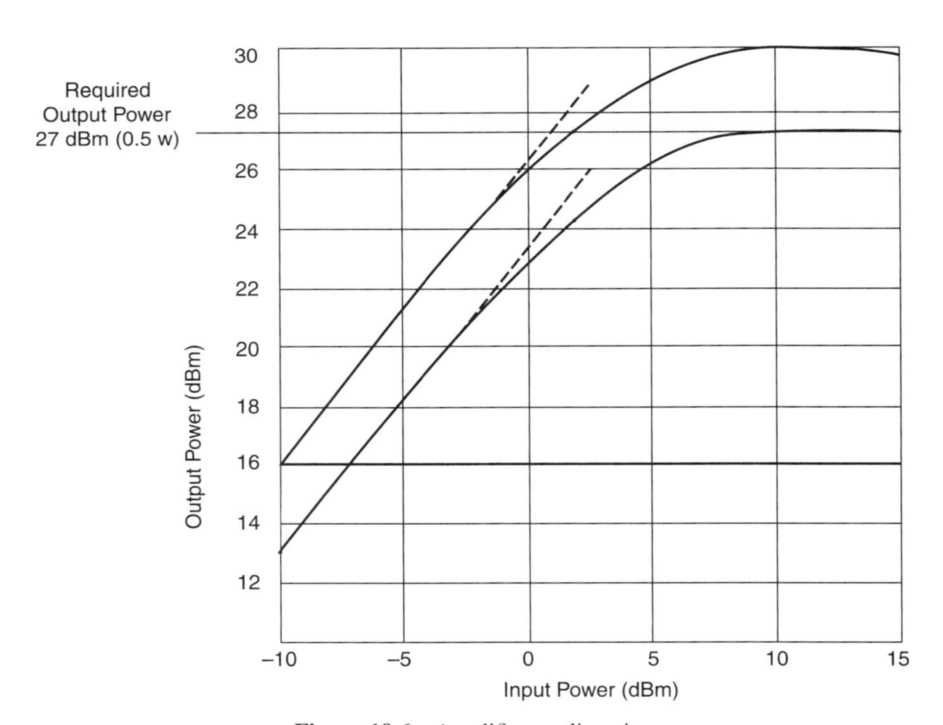

Figure 19.6 Amplifier nonlinearity.

and this point is called the 1 dB compression point. The input power at this point is approximately the maximum power that can be applied to the amplifier without causing distortion to the modulation that the RF is carrying. As the RF input power is further increased to +10 dBm, the amplifier reaches its saturation point, where increases in input RF power produce no increase in output power. At this point the RF output power of the amplifier is +27 dBm, which is 0.5 W. The power supply is supplying 1.0 W of DC power, so the amplifier efficiency at this point is 0.5 W/1.0 W = 50%.

Every RF amplifier, regardless of whether it is a MESFET, a bipolar, or other type of transistor has a similar nonlinear output versus input power characteristic. The reason is that the amplifier cannot provide more RF power than the DC power being supplied to it. The problem with operation at saturation is that any amplitude modulation is corrupted, because if the input RF power is raised or lowered as a result of modulation, there is no change in the RF output power.

Most wireless communication power amplifiers draw the same power from their power supply whether they are operated in the linear range or at saturation. This is because under all output power conditions they draw the same power from their power supply. The DC voltage and current drawn by the amplifier is constant. This is called class A operation. Thus, for any RF power amplifier,

operation near saturation causes distortion and
operation in the linear range causes low efficiency.

One solution to this amplifier nonlinearity problem is to use digital *frequency* modulation. Because all of the modulation is carried as changes in frequency, the amplitude variations of the signal due to amplifier saturation make no difference to the modulation. Unfortunately, as will be shown later, digital frequency modulation uses much more bandwidth than digital phase modulation.

The usual compromise is to use a two times higher power transistor at its 1 dB compression point, where the RF power has dropped by about one-half from the saturated value, and where distortions like spectral regrowth and modulation errors are satisfactory.

The upper curve in Figure 19.6 is derived from the measured performance shown in the lower curve. The amplifier in the upper curve provides exactly two times as much power, and therefore it must be twice as big and requires twice the battery power. For example, to double the power of the lower curve, two of the same transistors that were used to generate the lower curve would have to be operated in parallel. To avoid distortion, this pair of parallel transistors would have to be operated at their 1 dB compression point.

Spectral regrowth, and the ACP associated with it, are illustrated in the next two figures and discussed in more detail in Chapter 32. Modulation error, expressed as EVM, is discussed in Chapter 33.

Figures 19.7 and 19.8 show spectral regrowth of the same transistor as it is operated in the linear range and at saturation, while it is carrying two different types of

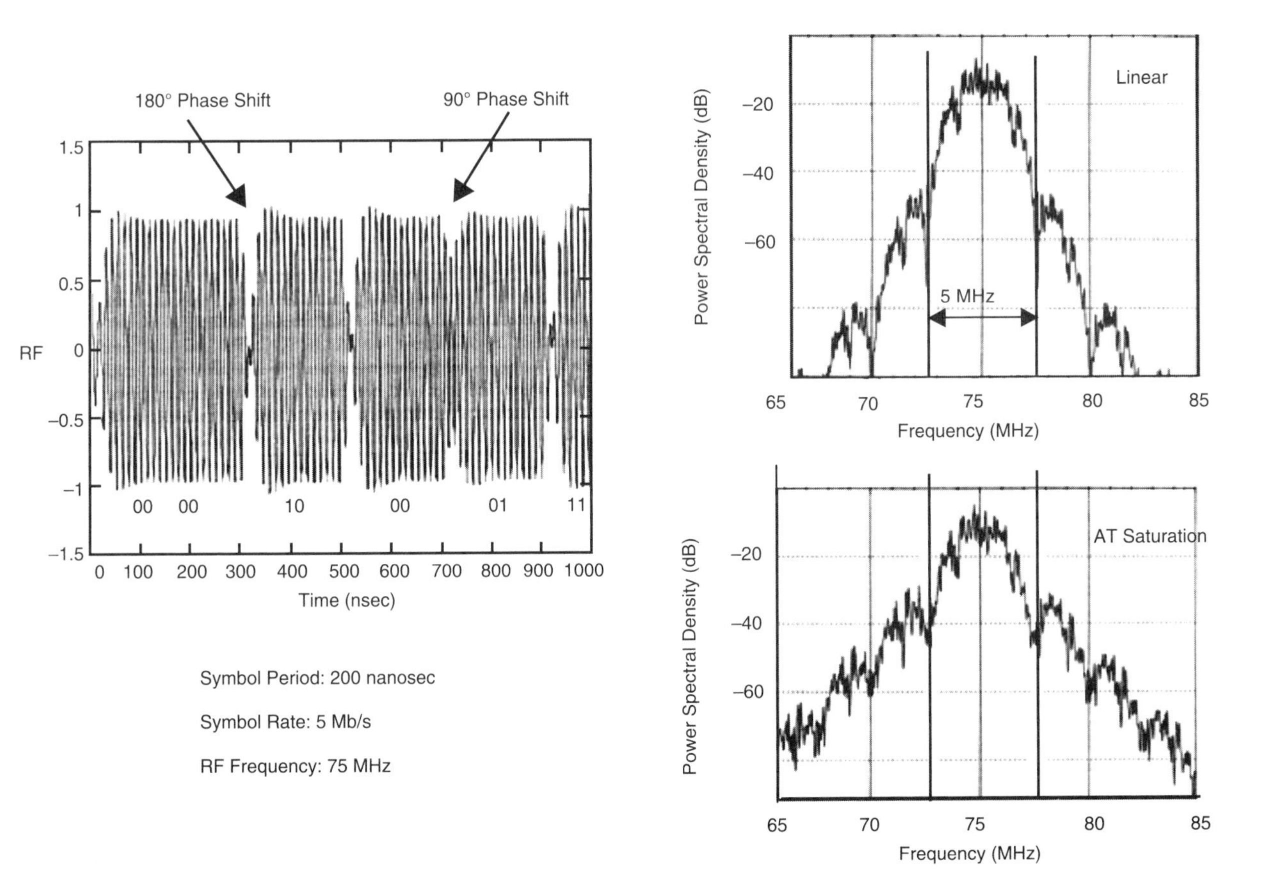

Figure 19.7 Spectral regrowth with QPSK.

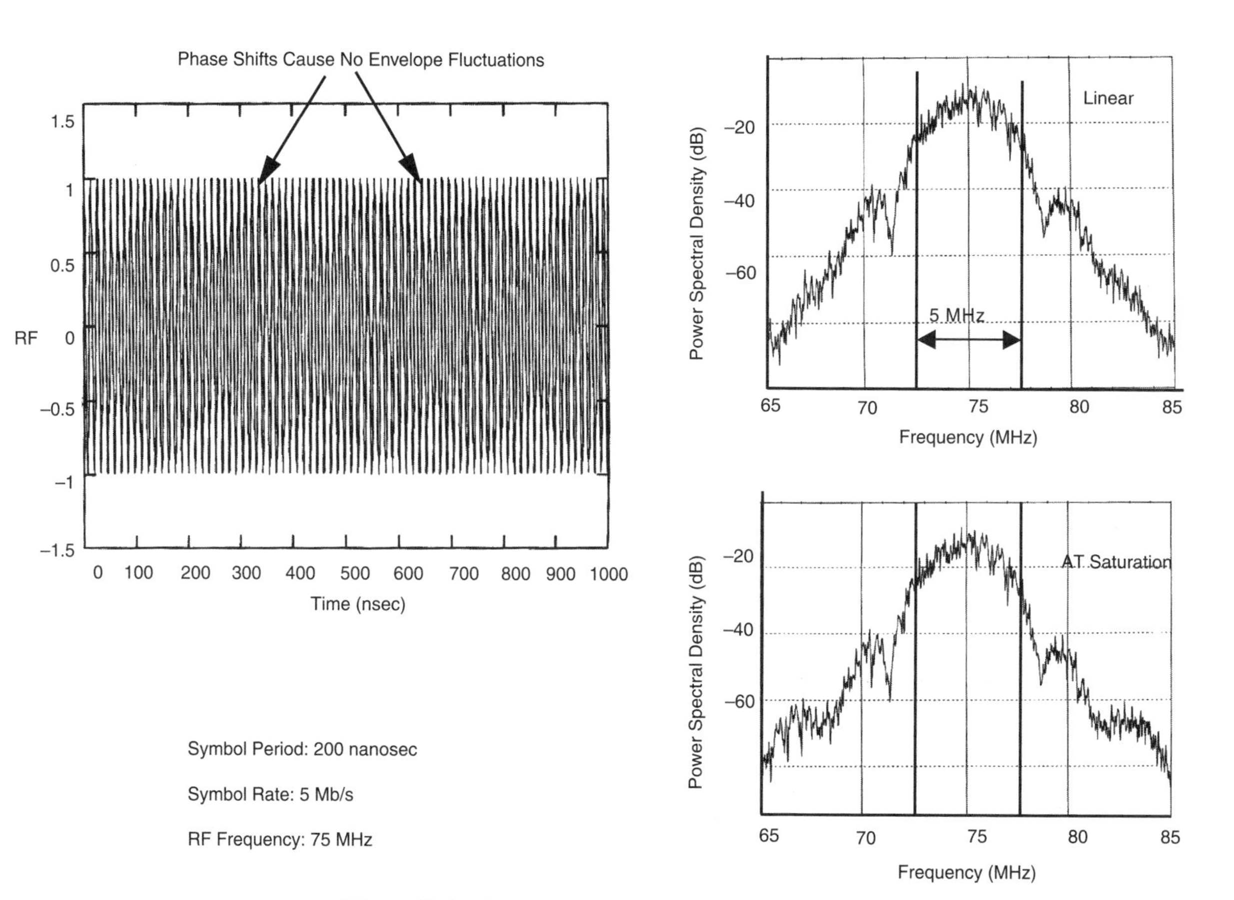

Figure 19.8 Spectral regrowth with MSK.

modulation. The measurements were made below the RF band at 75 MHz, so that the carrier waveforms could be observed with an oscilloscope as well as with an RF spectrum analyzer.

The left-hand drawing of Figure 19.7 shows a signal that has been modulated with QPSK modulation. The symbol rate defines how many times the phase changes per second. With QPSK modulation, 2 digital bits are transmitted at 5 Mb/s. The symbol period is $1/5$ Mb/s $= 200$ ns. In each symbol period there are $75/5 = 15$ RF cycles. The bit pattern shown being transmitted is 00/00/10/00/01/11, corresponding to phase changes of the carrier (relative to a phase reference) of 0°, 0°, 180°, 0°, 90°, and 270°.

The upper right-hand drawing shows the test signal as displayed on a spectrum analyzer for the same signal whose time waveform was shown in the left-hand drawing. The bit rate is 5 Mbps and the signal almost fits into a 5 MHz bandwidth. However, some of the RF power spreads into adjacent channels. This is called "spectral regrowth." The ratio of the integrated power in the lower (or the upper) adjacent channel relative to the integrated power in the desired channel is defined as ACP. The vertical grids are 20 dB apart, so the ACP value of the test signal is approximately -30 dB and it will not degrade the performance of most wireless communication systems. The same ACP spectrum will be obtained if the test signal is amplified by the amplifier whose performance was shown in Figure 19.6, if the amplifier is operated in its linear range. However, if the amplifier is driven to saturation, its spectrum will be as shown in the lower right-hand sketch of Figure 19.7. Note that the power in the adjacent channels is much higher. The ACP in either the lower or upper channel is now about -20 dB, and this is unsuitable for most wireless applications. The solution to this problem is to operate the power amplifier at its 1 dB compression point, where the spectral regrowth, although greater than that of the test signal, is still acceptable.

Figure 19.8 shows the same measurement as Figure 19.7 except that the modulation is a form of digital frequency modulation described in Chapter 4 called MSK, where the RF frequency is stepped the minimum possible amount between a higher RF frequency representing a digital 1 and a lower RF frequency representing a digital 0. Note from the left-hand sketch, which shows the digital modulation on an oscilloscope, that the amplitude of the RF wave is constant. The upper right-hand sketch shows the spectrum of the MSK modulated RF signal. Note that the spectral regrowth is large even before the signal is sent through the amplifier, and a bandwidth much greater than 5 MHz must be used to contain the signal. This is to be expected because frequency modulation is being used.

The lower right-hand curve in Figure 19.8 shows the spectrum when the signal is amplified to saturation. Note that in this case the spectrum is identical to that of the test signal, because the signal contains frequency changes but has constant amplitude.

The power amplifier in a mobile unit needs to be operated at its highest allowable power, which will be at the 1 dB compression point or at saturation, depending on the type of digital modulation used, when the mobile unit is near the outer boundary of a cell. As the mobile unit moves closer to the base station, the received signal at the base station gets as much as 90 dB larger and will saturate the base station receiver, so that weaker signals from other more distant mobile units will be jammed. To avoid this problem, the base station sends a control signal to each mobile telling it to turn its transmitted power up or down so that the received signals from each mobile at the base station will be the same.

When the mobile is located anywhere except at the edges of the cell, it will be operating in a very inefficient mode because its power amplifier draws the same power from the power supply regardless of its power output. To solve this problem, the power amplifier can be designed to operate at stepped power supply voltages, so that it can always be operating near its 1 dB compression point for any output power that is required.

Another solution to the problem of distortion of the digital modulation because of amplifier nonlinearity is to use a "polar modulator." With a polar modulator the phase shift of an RF signal traveling through the amplifier is measured dynamically as a function of input RF power. This is defined as AM to PM conversion, and techniques for measuring it are described in Section 19.6. Using the AM to PM measurement, the phase of the input RF signal is dynamically adjusted to correct for the phase shift of the amplifier output.

A block diagram of one implementation of a polar modulator is shown in Figure 19.9. The double lined arrows show the RF path, and the single lined arrows show the low frequency control signals. The RF output of the power amplifier is sampled by a directional coupler. One sample is directed to a phase detector that generates a phase error signal. This signal is processed and sent to the PLO to cause it to change the phase of its signal. This signal, with its changed phase, is applied to the input of the RF amplifier to compensate for the AM–PM conversion. A second sample from the directional coupler is directed to a power sensor. The output of the power sensor, which represents the power output of the amplifier, is sent to the power supply to change its output voltage as described earlier.

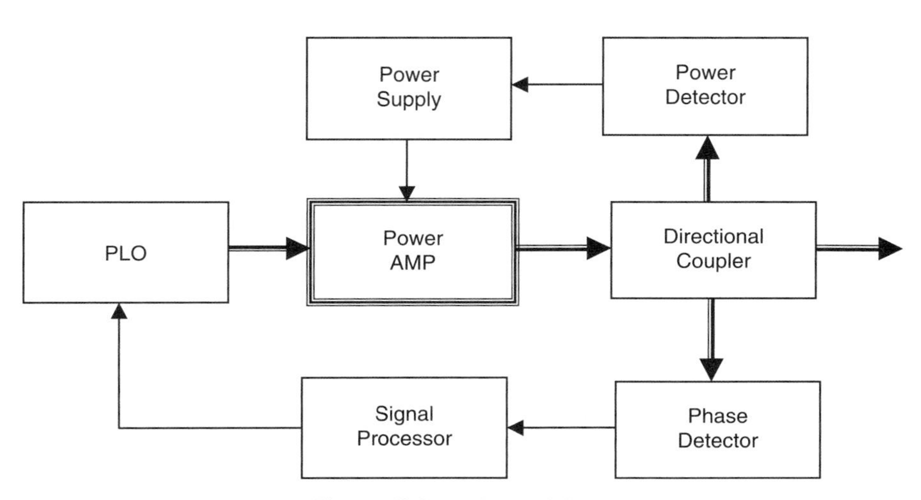

Figure 19.9 Polar modulator.

19.5 MEASUREMENTS TO BE PERFORMED ON RF POWER AMPLIFIERS

The following measurements need to be performed on RF power amplifiers:

1. Swept gain and power versus frequency
2. Swept gain and power versus input power
3. 1 dB compression point
4. Swept phase versus input power
5. AM to PM conversion
6. Harmonic power
7. ACP with digitally modulated signals
8. Required backoff to meet ACP requirements
9. EVM with digitally modulated signals
10. Required backoff to meet EVM requirements

The first five tests are most easily performed on a VNA and are described in Section 19.6 The sixth test, which is the measurement of harmonic power, is performed with a spectrum analyzer and is described in Section 19.7 The remaining measurements depend on the type of modulation of the signal that is being amplified. ACP is measured with a spectrum analyzer and EVM is measured with a VSA as described in detail in Chapters 32 and 33, respectively.

Results of all the above measurements made on a GaAs MESFET operating in the 2.4–2.5 GHz band will now be presented. Manufacturer's specs for this wireless LAN amplifier are as follows:

Manufacturer	Triquint
Model	TQ9207
Frequency	2.450 GHz
Linear gain	13 dB
P1 dB	18 dBm
Psat	21 dBm

19.6 MEASUREMENTS OF AMPLIFIER OUTPUT CHARACTERISTICS VERSUS FREQUENCY AND INPUT POWER

Figure 19.10 shows the test setup for measuring output power, gain, and phase as a function of frequency or input power with a VNA. This setup is the same as the one used to measure passive components whose performance did not change with input power. For high power measurements an attenuator must be added after the amplifier to limit the power to the VNA so that its output detector will not be driven into the nonlinear range to give incorrect readings or be damaged. For very high power measurements, a power sample must be taken from an external directional coupler.

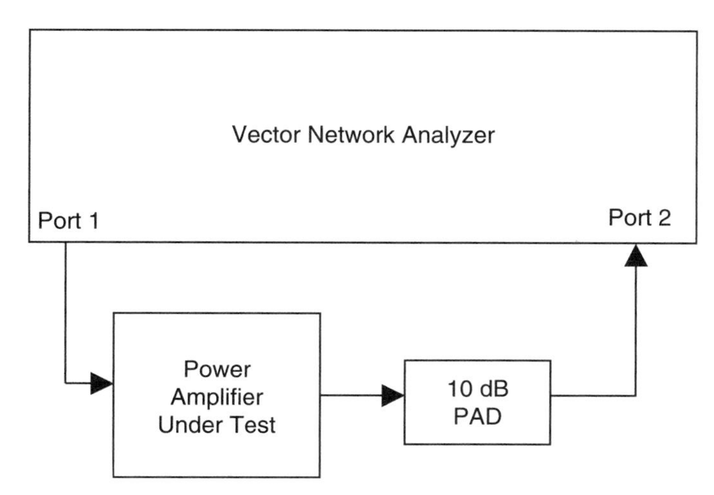

Figure 19.10 Setup for swept power measurements.

The effect of the attenuator and the directional coupler can be removed from the measurement by being in place during the calibration procedure.

Figure 19.11 shows measurements of the power amplifier linear gain on the left graph and power output on the right graph, both as a function of frequency over the 2–3 GHz band. The measurement was made at an input power of 0 dBm. The measurements change with input power level.

With the VNA calibration procedure described in Chapter 8, the gain measurements are accurate to better than ± 0.02 dB. The effect of the attenuator or external directional coupler are canceled out by including them in the calibration procedure.

However, a special calibration using a power meter to calibrate the VNA has to be used to obtain power measurements with an uncertainty of ± 0.2 dBm. This procedure was explained in Chapter 13 on measurement uncertainties. All of the power measurements described in this chapter were made using that calibration procedure.

Figure 19.12 shows the gain (S_{21}) and the output power of the amplifier under test as a function of input power over a 20 dB range from -7 to $+13$ dBm. The upper chart shows gain (S_{21}). A delta marker is set to show the 1 dB compression point (which is defined as the point where the gain has decreased from its linear value by 1 dB). The lower chart shows output power as a function of input power. By adjusting the coupled markers control, the 1 dB compression point can be displayed on the output power curve. The curve shows the same type of measurement as Figure 19.6.

The lower chart of Figure 19.13 shows the same data that was shown in the lower chart of Figure 19.12. However, the upper chart shows the phase shift of the RF power traveling through the amplifier instead of linear gain. At input power levels in the linear range, the phase shift is constant. Note that at the 1 dB compression point, the phase begins to change with power level, and this can cause digital data errors in the phase modulated RF signal that is being amplified. This departure of

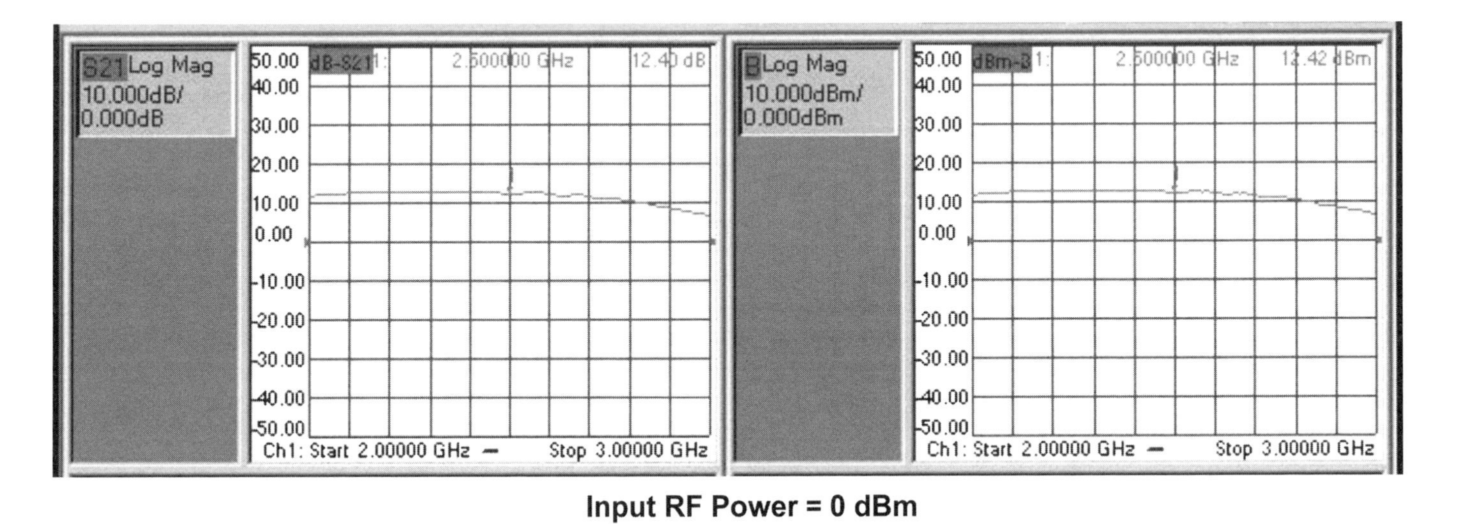

Figure 19.11 Swept power measurements versus frequency.

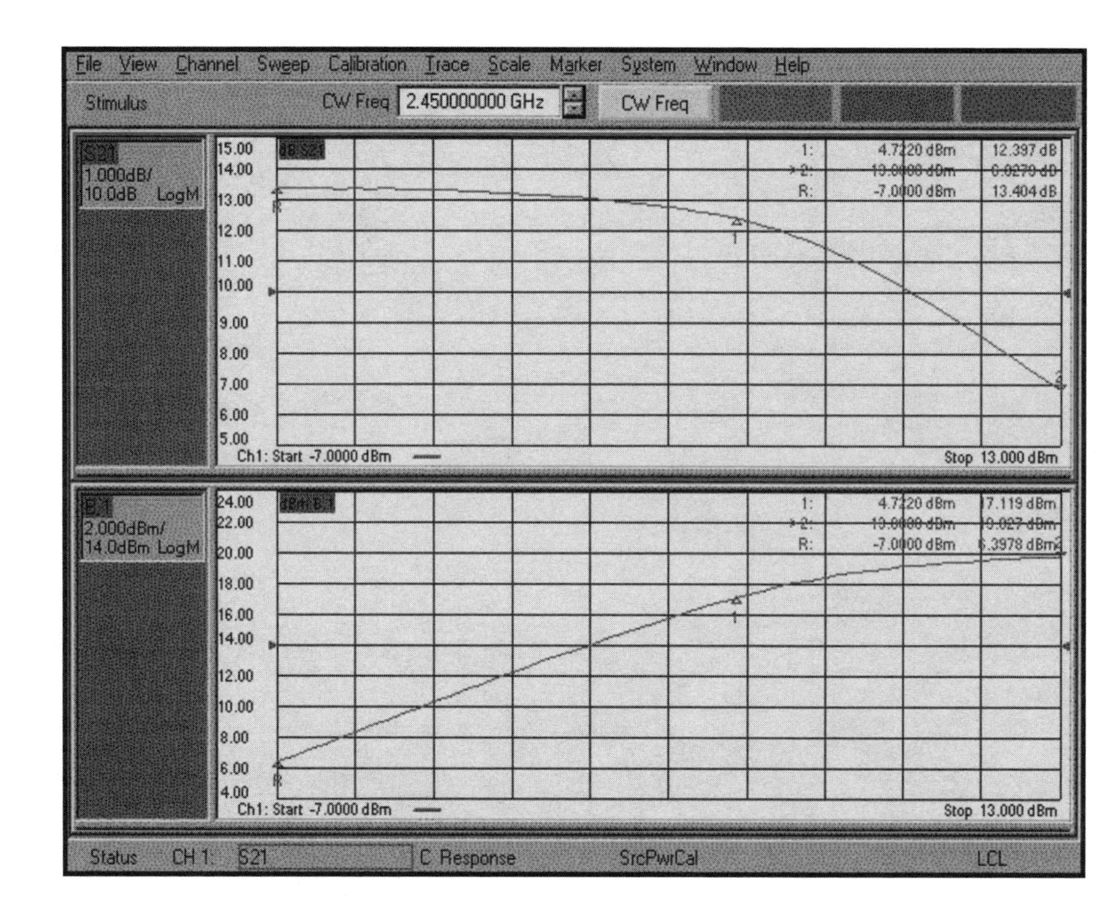

Figure 19.12 Power amplifier swept gain and output power versus input power.

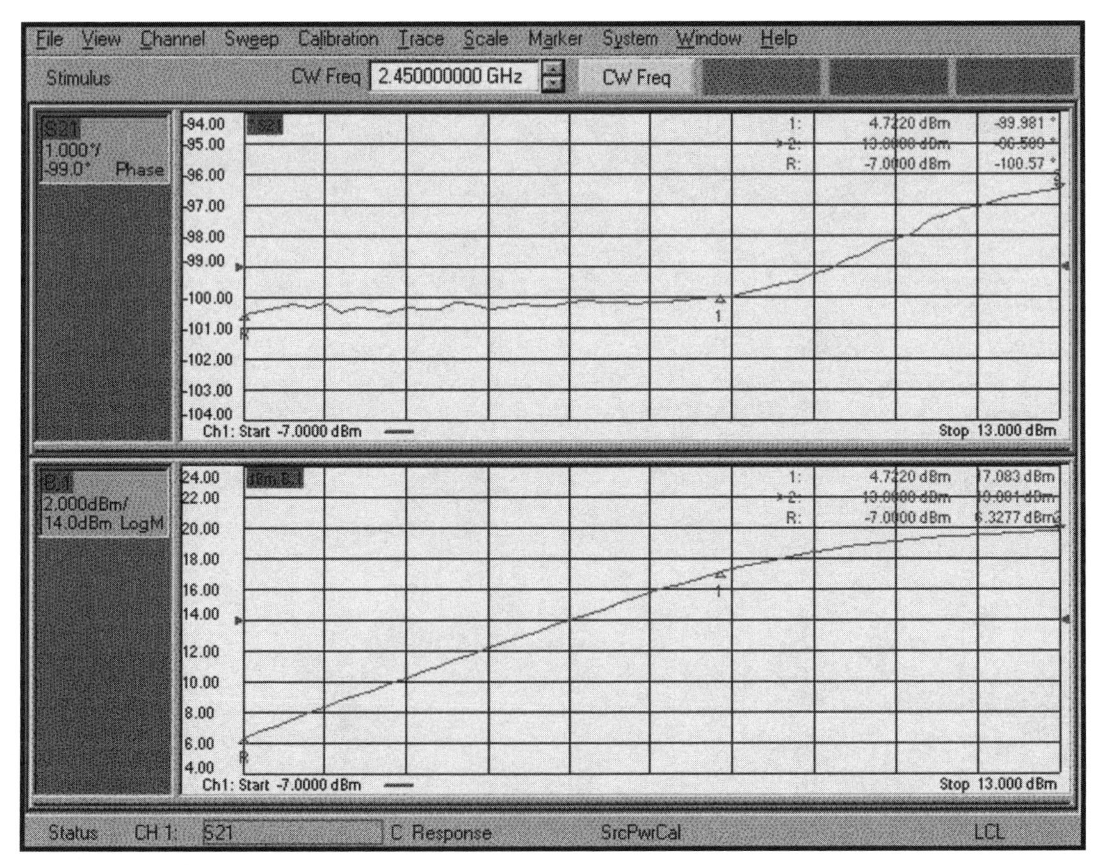

Figure 19.13 Power amplifier swept phase and output power versus input power.

phase shift from its constant value is called AM to PM conversion. For the particular amplifier being measured, the AM–PM is 0.5°/1 dB change of input RF power.

19.7 HARMONIC POWER MEASUREMENTS

Because the power amplifier is distorting the RF signal when it is operating in its nonlinear region, it is producing harmonic outputs at 2, 3, 4, ... times the operating frequency. Government licensing requirements are very stringent in respect to limiting emissions outside the licensed band. They usually specify that any harmonic power must be reduced to −60 dB below the licensed carrier power.

Figure 19.14 shows a test setup for measuring harmonic outputs using the spectrum analyzer that was discussed in Chapter 9. The amplifier is driven with a signal generator operating in its unmodulated mode. Care must be taken to ensure that the signal generator is not generating harmonics. If it is, appropriate filtering must be placed at the signal generator output to eliminate the harmonics. Figure 19.15 shows the harmonic content of the test amplifier when it is operated at saturation. The input frequency is 2.5 GHz. Harmonic outputs are created as follows:

Harmonic	Frequency (GHz)	Power (dBm)	Power Below Fundamental (dB)
Fundamental	2.5	19	—
Second	5.0	−13	−32
Third	7.5	−7	−26
Fourth	10.0	−24	−43
Fifth	12.5	−26	−45
Sixth	15.0	−34	−53

Note that none of the harmonics are below the −60 dB specification.

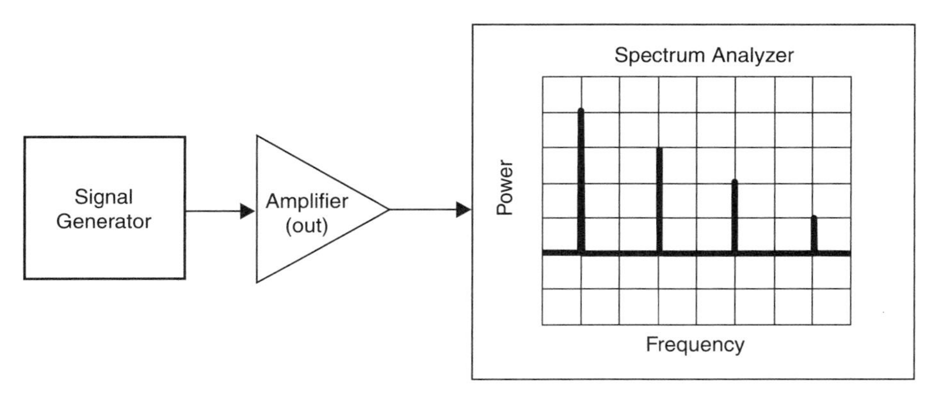

Figure 19.14 Test setup for harmonic power measurement.

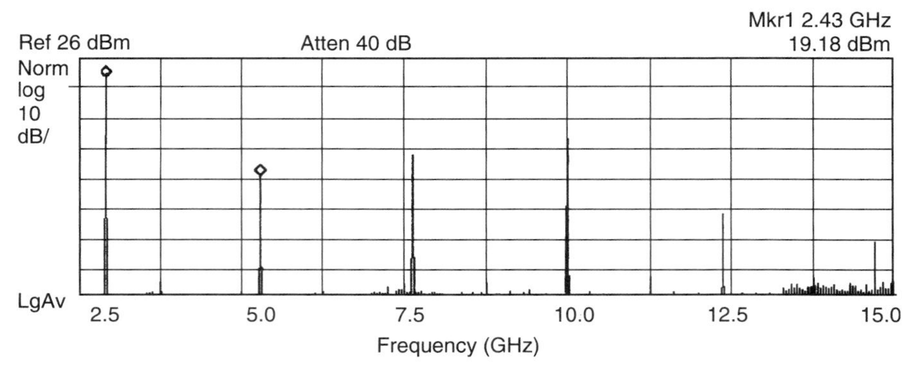

Figure 19.15 Harmonic power measurement.

Figure 19.16 shows the fundamental power at 2.5 GHz, the second harmonic power at 5.0 GHz, and the ratio of the second harmonic to fundamental power all as a function of RF input power. The harmonic power is at the required −60 dB level only when the input power is −4 dBm. The output power at this input is 9 dBm, which is 10 dB below its saturation value. A better solution to meet the harmonic requirement is to use low pass or bandpass filters between the amplifier and the antenna.

19.8 EXAMPLE POWER AMP MEASUREMENTS ON THE VNA

Objective

Our objective is to measure power amplifier gain and determine the 1 dB compression point. First, a frequency sweep is used and the input power is adjusted manually to

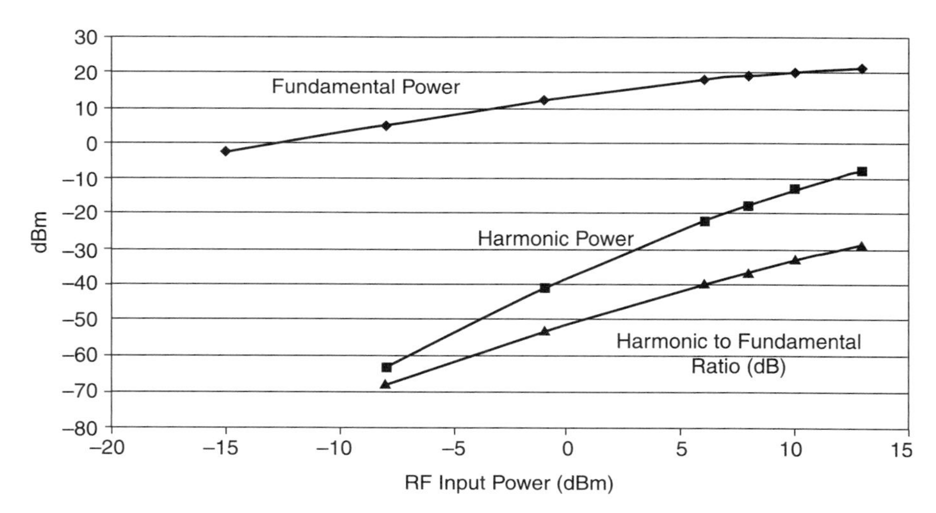

Figure 19.16 Fundamental and second harmonic versus input power.

determine the P1 dB. Second, the VNA is set up to perform a power sweep and the P1 dB can be read directly from the measurement. The output power level of the PA will overdrive the receiver of the network analyzer, so a 10 dB attenuator is connected to the end of the cable on port 2 of the analyzer. The 10 dB pad is left on the cable during the calibration process, so the attenuation is automatically removed from the measurement results that are displayed on the analyzer.

Measurements Being Demonstrated

Gain (S_{21}) versus frequency
Output power versus frequency
Gain (S_{21}) versus input power (swept power)
Output power versus input power
AM–PM versus input power

Specifications of Power Amplifier

Manufacturer	Triquint
Model	TQ9207
Frequency	2.450 GHz
Linear gain	13 dB
P1 dB	18 dBm
Psat	21 dBm

Generic Procedure

Steps to perform the measurements are listed in the following table. The left column lists the actions to be performed, and the right column lists the motivation for taking each step and provides additional background information in some cases.

Procedure for Network Analyzer	Notes
Preset the instrument	We assume that we are starting with one measurement being displayed.
Measurement setup and calibration	
Measure S_{21}.	This will measure the gain of the device.
Set the start and stop frequencies to 2 and 3 GHz, respectively.	
Set the power level of the source to −7 dBm.	
Attach a 10 dB pad to the end of the cable on port 2 of the network analyzer	
Perform a response calibration.	Because we are only looking at S_{21} and not reflection, we decide to skip the full two-port calibration and accept some greater uncertainty in the measurement. In addition,

(*Continued*)

Procedure for Network Analyzer	Notes
	the fact that amplifiers generally have good isolation prevents any reflected signals from traveling from port 2 back to the input, which reduces the mismatch error somewhat.
Activate another trace and set it to measure the input at the B receiver. This trace should share the same "channel" (sweep and source power settings) as the measurement we just set up.	This will measure the power at the output of the power amplifier. This is a nonratioed, power meter style measurement.
Configure and calibrate the power meter for communication with the network analyzer.	Most VNAs allow an external power meter to be configured to communicate with the analyzer for calibration purposes. On newer analyzers, the sensor of the power meter can be zeroed and calibrated from the network analyzer interface prior to calibration.
Perform a source power calibration sweep on the network analyzer with the power sensor attached to the end of the cable from port 1 of the network analyzer.	The first step in performing a power calibration is to ensure that the source inside the network analyzer is outputting an accurate power level across the specified sweep range. This calibration sweep is time consuming because of the communications between the network analyzer and the power meter. The number of sweep points can be reduced to speed the process, at the expense of having less trace data from the measurement.
Perform a receiver calibration with a thru connection between cables.	Once the network analyzer source is outputting power levels with power meter accuracy, the receiver calibration ensures that the power level produced by the source is read as the same power level by the receiver in the VNA.
Connect the power amplifier as the DUT between ports 1 and 2.	
View the gain of the power amplifier on the S_{21} measurement. Set the scale per division for the S_{21} measurement to 1 dB and the reference value to 10 dB. Set a marker at 2.45 GHz. Set a marker at the same frequency on the B measurement. Press the Power key to display the source power setting on the screen. Note that the input power (which you can read from the source power setting) plus the gain (S_{21}) should equal the output power (B).	You can use "coupled" markers so that if you move a marker in one measurement, the marker moves in the other measurement as well.

(*Continued*)

Procedure for Network Analyzer	Notes
Increase the input power until the gain drops by 1 dB. This is the 1 dB compression point. Note the input power level.	
Now we will switch to a power sweep. Select the sweep type menu and change the sweep type to a power sweep. Set the start power to −7 dBm and the stop power to −13 dBm. Set the CW frequency to 2.45 GHz. After setting the CW frequency, check that the calibration is on for both measurements.	On older VNA models the start and stop power levels for the sweep are set with the start and stop hard keys. When you switch from a frequency sweep to a power sweep many VNA models set the default CW frequency to 1.0 GHz. This default frequency is outside of the range where the calibration was performed and causes the calibration to be turned off. The calibration cannot be turned on until the CW frequency is set to 2.45 GHz, which is in the range where we performed the calibrations. Typically, 13 dBm is the largest signal we can send from the built-in source on the VNA. For larger signals, we would have to use booster amps as part of the test setup on port 1, which would also be calibrated by the source power calibration sweep.
With a scale of 1 dB we can simply observe where the S_{21} trace drops by one division. This is the 1 dB compression point. You can also set a marker at the left edge of the trace and then use a marker delta to find where the signal is down by 1 dB.	Note that the output power trace is a familiar curve that appears in textbooks on amplifier design. The swept power curves give a good description of the amplifier's power performance at a single frequency.
Observe AM to PM conversion.	
Maximize the S_{21} plot to use the full screen. Switch the format to phase.	Note that the phase is steady through the linear region of operation. Around the 1 dB compression point, the phase changes as the input power level is changed. This will produce an undesired phase modulation whenever the amplitude of the signal is modulated (which occurs in most modern digital systems).

On newer model VNAs, the swept frequency and swept power measurements can be assigned to two different "channels." These can both be set up and calibrated separately before the DUT is connected.

19.9 ANNOTATED BIBLIOGRAPHY

1. Baringer, C. and Hull, C., Amplifiers for wireless communications, in Larson, L., Editor, RF and Microwave Circuit Design for Wireless Communications, Artech House, Boston, 1996.

2. Cripps, S., RF Power Amplifiers for Wireless Communications, 2nd ed., Artech House, Boston, 2002.
3. Application Note 1408-7: Amplifier Linear and Gain Compression Measurements, Agilent Technologies, Santa Clara, CA.
4. Application Note 1291-B: 10 Hints for Making Better Network Analyzer Measurements, Agilent Technologies, Santa Clara, CA.

Reference 1 explains RF power amplifier circuit diagrams.

Reference 2 gives a thorough explanation of RF power amplifier design and testing.

Reference 3 provides step by step instructions for performing source power calibrations on VNAs.

Reference 4 provides a sample test setup for measuring power amplifiers using an additional amplifier external to the network analyzer for higher power devices.

CHAPTER 20

ANTENNAS

Antennas are used on both the transmitter and the receiving end of every wireless communications system. The transmitter and receiver antennas may be of the same design or a different design. In either case, they perform different functions.

20.1 ANTENNA FUNCTIONS

The purpose of the transmitter antenna is to launch the modulated transmitter power from the power amplifier in the direction of the receiver. The purpose of the receiver antenna is to collect as much of the transmitted power as possible.

The performance specifications of the transmitter antenna are

1. Gain
2. Pattern
3. Beamwidth
4. Polarization
5. Impedance match

Gain

The gain of the antenna can be best understood by discussion of how it is measured. (See Section 20.4 for a detailed discussion of measurement setup and procedures.)

RF Measurements for Cellular Phones and Wireless Data Systems. By A. W. Scott and R. Frobenius

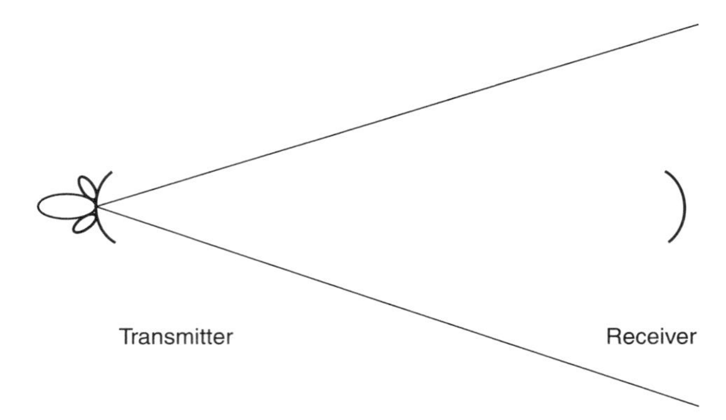

Figure 20.1 Transmitting and receiving antennas.

The antenna measurement is accomplished on a test range with a transmitter at one end and a receiver at the other end, as shown in Figure 20.1.

If the antenna measurement is conducted indoors, the walls of the test room must be covered with RF absorbing material so that none of the transmitted RF power reaches the receiver by reflections from the room walls, floor, or ceiling. Such a room is called an anechoic chamber. (A photograph of such a anechoic chamber is shown later.) If the measurement is made outdoors the direct path between the transmitter and the receiver must be surrounded by vegetation that adsorbs the RF, so that only the direct radiation from the transmitter, and not any reflected power, reaches the receiver.

The receiver used for the antenna test may have any type of antenna and may be any type of receiver tuned to the test RF frequency.

The transmitter may use any convenient power level tuned to the test frequency. The first step is to mount an "isotropic" antenna on the transmitter. An isotropic antenna is an antenna that radiates uniformly in all directions. Such an antenna is difficult to build. The hardware equivalent is called an omnidirectional antenna, which is not perfectly isotropic. The common practical solution is to use a half-wave dipole antenna (described in the next section) as the reference antenna and then add an appropriate correction to the measured gain.

The RF signal radiated from the transmitter with the isotropic or dipole antenna is measured at the receiver and recorded. It is generally small because most of the RF power is radiated in other directions and absorbed by the anechoic chamber walls. The reference antenna is then replaced by the antenna under test, and the measurement is repeated. The received power is much larger when the actual antenna is used. The ratio of the received power with the antenna under test to received power with the reference antenna is recorded and converted to dB.

It is important that the antenna gain be recorded as dB relative to istotropic or dB relative to dipole, depending on the type of reference antenna that is used. Antenna gain is often incorrectly expressed in dB. Failure to follow this rule can lead to critical problems in system performance. The dipole antenna is such a simple structure that its gain in its maximum direction of propagation can be calculated, and it is 2.15 dB greater than the gain of an isotropic antenna. This leads to the following relationships:

Gain (isotropic) = gain (with dipole reference) + 2.15 dB
Gain (isotropic) must be used in path loss calculations
Gain (dipole) is usually measured during antenna tests

Pattern

The gain of any antenna is achieved by reducing the angular width of the beam in both its vertical and horizontal directions. Figure 20.2 shows the pattern of a typical base station antenna in the horizontal plane. The gain in the antenna in any direction from its axis referenced to 0° is shown by the radial scale. The angular scale is in 5° increments. The width of the beam is usually defined as the angle

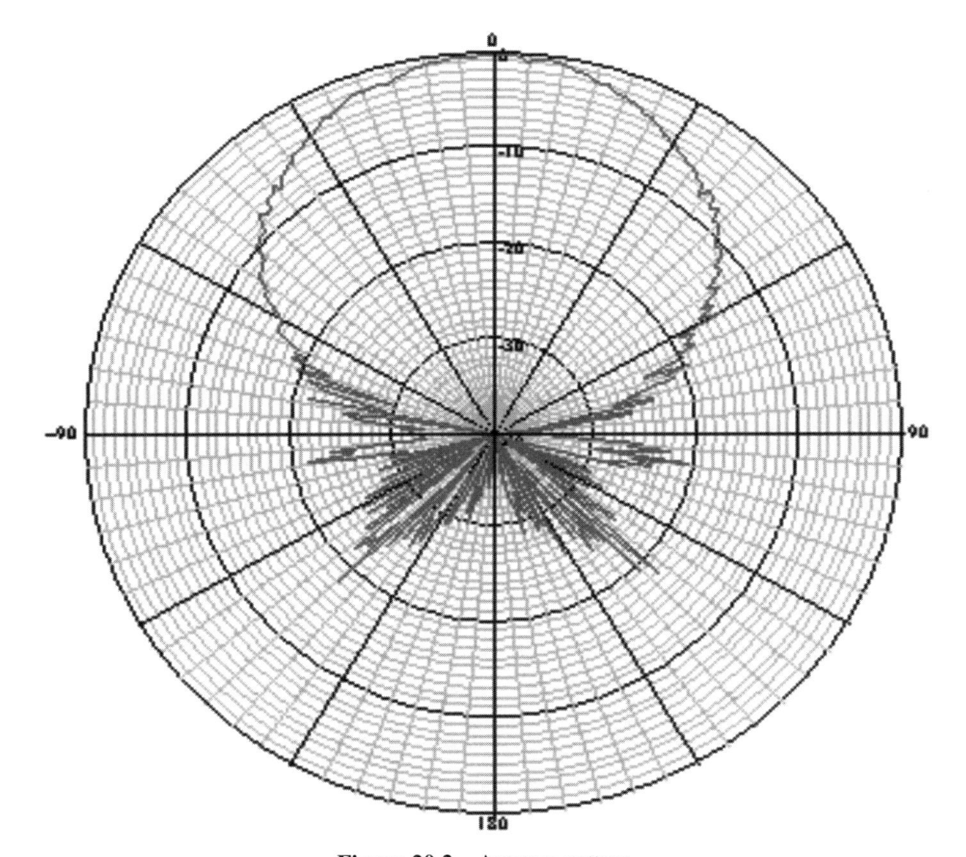

Figure 20.2 Antenna pattern.

when the signal has dropped from its value on the axis by 3 dB. For the antenna pattern shown, the beamwidth is 60°.

In many applications, like microwave relay, satellite communication, and cell phone base stations, where the transmitter knows where the receiver is located, the use of a high gain antenna allows the transmitted power to be concentrated in the direction of the receiver. However, most mobile units have no idea where the base station is located. Therefore, the mobile units must use nondirectional antennas, like half-wave dipoles or patch antennas, which have almost isotropic radiation patterns.

Gain

The gain of a wireless system antenna is completely specified by two antenna patterns such as the one illustrated in Figure 20.2: one in the horizontal plane and one in the vertical plane. However, usually antenna gain is specified as a single number, which is its gain in the direction of maximum transmission or reception.

Side Lobes

Most narrow beam antennas, such as the antenna whose radiation pattern is shown in Figure 20.1, radiate in other directions out the side and back of the antenna. These radiations are called side lobes. Although the radiation from the side lobes is much lower (−20 dB) than that of the main beam, the side lobe radiation can jam other systems that are close to its transmitting antenna. Side lobe control is therefore an important requirement in an area that is crowded with wireless systems.

Polarization

As described in Chapter 2, an RF wave consists of an electric field and a magnetic field, which are at right angles to each other. Polarization describes the direction of the electric field relative to the surface of the earth. The electric field can be at any angle relative to the surface of the earth, depending on the alignment of the antenna. Typically the antenna is aligned so that the electric field is either vertical or horizontal to the earth's surface. By using two antennas, one with vertical polarization and one with horizontal polarization, at both transmitter and receiver locations, a given bandwidth of frequencies can be transmitted twice with different data modulation being carried on each polarization. This dual polarization doubles the data transmission capability of a licensed frequency band. This approach is used by all microwave relay and satellite communication systems.

Unfortunately, this use of polarization diversity to increase data capacity only works with line of site transmission. If the transmission uses reflections from objects in the propagation path, a vertically polarized or a horizontally polarized wave is changed into both polarizations at every reflection, so this approach for increasing bandwidth cannot be used in a cell phone or a wireless LAN system.

Impedance Match

As discussed in Chapter 2, the impedance of an RF wave, which is the ratio of the electric field to the magnetic field, is 377 Ω In free space. The antenna must convert the impedance of the RF wave to some value near 50 Ω in the antenna feed line as it enters or exits the antenna to connect to RF transmitter or RF receiver components. Fortunately, the impedance transformation is easily accomplished by using a half-wavelength dipole antenna.

Antenna Area

Antenna gain is the important characteristic of a transmitting antenna. For a receiving antenna, the equivalent important characteristic is antenna area. The antenna area determines how much of the incoming RF field is collected by the receiving antenna.

The pattern, polarization, and impedance characteristics are the same for a receiving antenna as for a transmitting antenna.

The antenna gain and antenna area are related by

$$G = 4\pi A/\lambda^2$$

where G is the antenna gain expressed as a ratio (not in dB), A is the effective area of the receiving antenna, and λ is the wavelength. Note that the antenna gain is not determined by the antenna area in square meters or square feet or square inches, but by the antenna area in square wavelengths. A numerical example will make this statement clear. Assume a parabolic dish antenna has a diameter of 1 ft. Its area is roughly 1 ft^2. If it is operated at 1 GHz where the RF wavelength is 1 ft, the wavelength squared is 1 ft^2. By the formula above its gain is 4π, which is 11 dB.

If this same piece of hardware is now operated at 10 GHz, the wavelength squared term is 0.01 ft^2, so the gain is 100 times greater or 31 dB.

20.2 TYPES OF ANTENNAS

Half-Wave Dipole Antenna

Figure 20.3 shows a half-wave dipole antenna. The design shows a two wire feed line with the two wires of the antenna extending at right angles to the feed line. The length

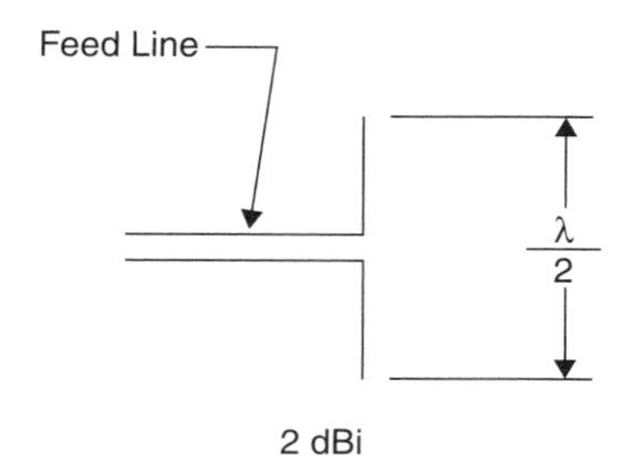

Figure 20.3 Half-wave dipole antenna.

of each antenna wire is 0.25 wavelength, so the length of the antenna is 0.5 wavelength. In actual use the two wire feed lines must be matched to a coaxial or microstrip line at the antenna input.

The radiation pattern of the half-wave dipole antenna is shown in Figure 20.4. It has almost a spherical antenna pattern, except that there is no radiation along the wires of the antenna. Because the half-wave dipole antenna has no radiation along the dipole wires, it has more radiation along a circumference at right angles to the wires. Around the circumference, the gain of the half-wave dipole antenna is 2.15 dB greater than the gain of an isotropic antenna, so the dipole gain is 2.15 dBi.

The direction of polarization is along the length of the wires, as shown in Figure 20.5. If the half-wave dipole is mounted with its wires vertical, the electric field of the radiated RF wave is vertical. If the half-wave dipole is mounted with its wires horizontal, the electric field of the radiated RF wave is horizontal.

The length of the antenna wires of a half -wave dipole has no affect on the radiation pattern, gain, or polarization. The length of the dipole wires controls only the impedance of the antenna at its feed wires. If the dipole is 0.5 wavelength long,

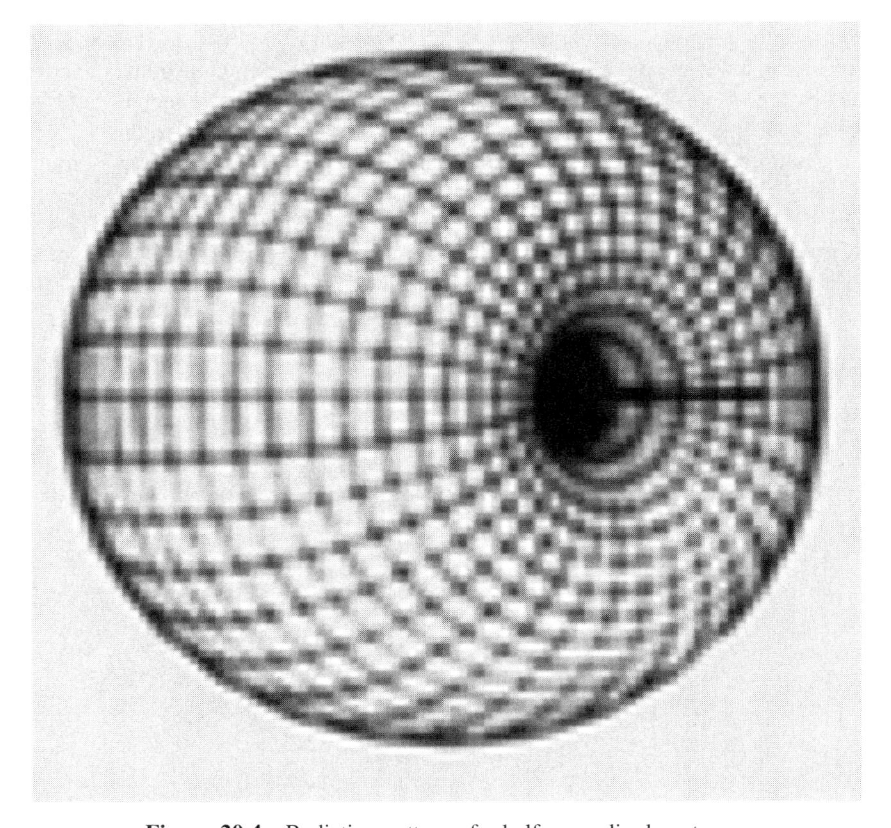

Figure 20.4 Radiation pattern of a half-wave dipole antenna.

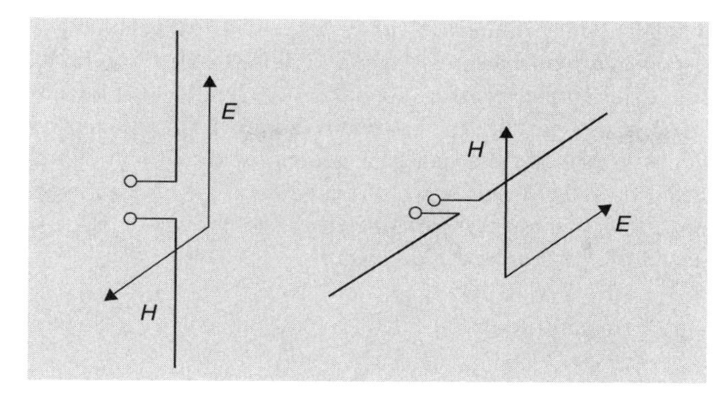

Figure 20.5 Dipole orientation for different polarizations.

the antenna matches the free space impedance of 377 Ω to a pure resistance of 62.5 Ω at the output of the feed wires, which can be easily matched to 50 Ω for RF or 75 Ω for video applications.

If the dipole wires were shortened, the impedance at the output terminals would be reduced as the square of the wire length. If the total length of the antenna wires were one-fourth of the wavelength, the impedance would be 12.5 Ω.

Patch Antenna

A simple RF antenna with the radiation pattern similar to that of the half-wave dipole is shown in Figure 20.6. A rectangle or other shape is etched onto one side of a

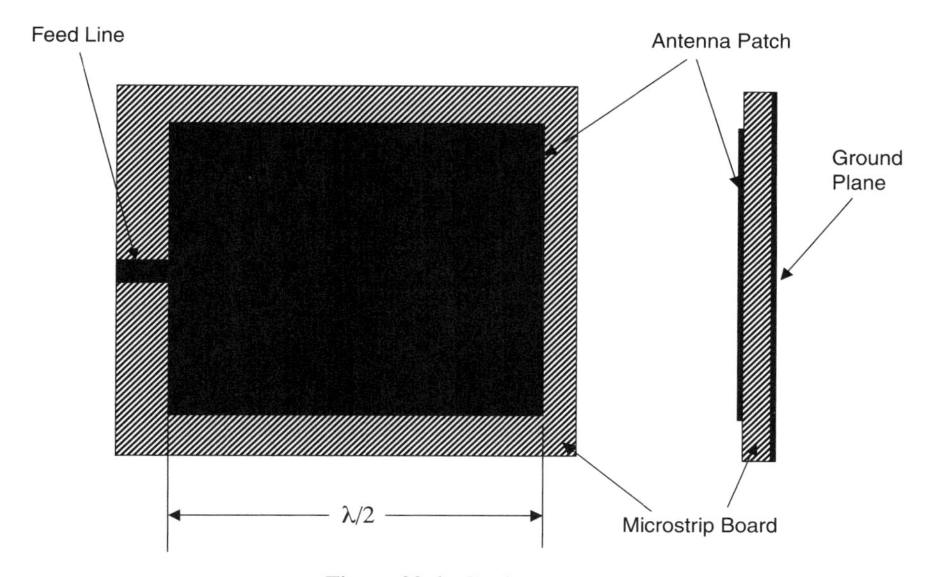

Figure 20.6 Patch antenna.

microstrip board. The long dimension of the patch is 0.5 wavelength, which will be less than one-half of a free space wavelength because of the dielectic loading of the board. If the backside of the microstrip board is not metallized, the gain of the patch antenna will be 2.15 dBi, just like the half-wave dipole. If a ground plane is formed on the back of the substrate, the antenna gain is doubled to 4.3 dBi because all of the power is radiated from the front surface of the substrate. If a high dielectric constant ceramic is used for the microstrip board, the patch antenna can be made small enough to fit inside a cell phone or a laptop computer.

Colinear Dipole Array

Half-wave dipoles or patch antennas can also be used as building blocks to achieve high gain, narrow beam antennas. Figure 20.7 shows a colinear dipole array that is used in most cell phone base stations. A schematic drawing is shown in Figure 20.7a. Four half-wave dipoles are shown mounted one above the other vertically on the antenna mast. The spacing between the dipoles is 1 wavelength so that all are operating in phase. This arrangement reduces the wasted radiation in the upward direction where there are no mobiles and downward to regions inside the base

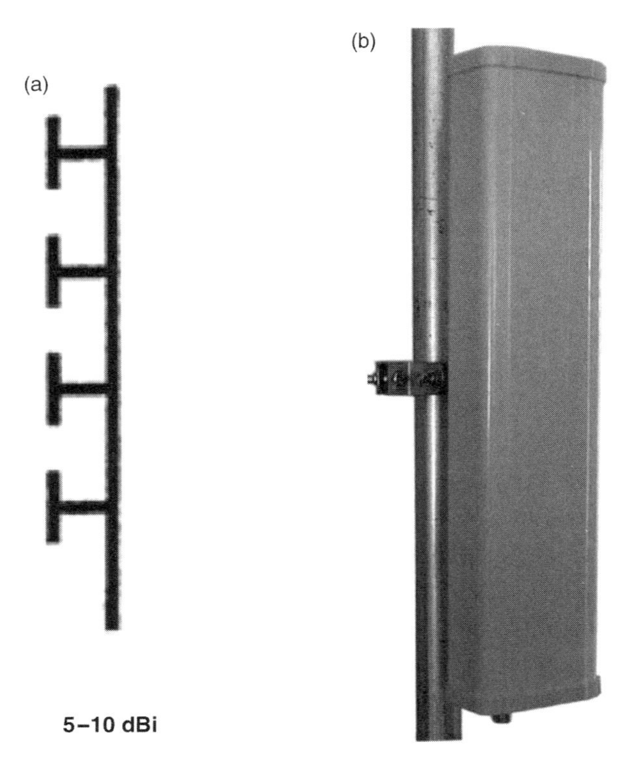

Figure 20.7 Colinear dipole array.

station's fence. The array causes all radiation to be focused into the cell. The resulting gain of the colinear dipole array is equal to the gain of one array element times the number of elements used. For the configuration shown in Figure 20.6a, the gain in each element is about 2.15 dBi; with the four elements shown ($4 = 6\,\text{dB}$), the overall gain is $2.15\,\text{dBi} + 6\,\text{dB} = 8.15\,\text{dBi}$.

Colinear dipole arrays usually have gains between 5 and 10 dBi. Because the gain increases linearly with the number of elements, and tower cost increases more rapidly, economics limits the number of elements to about 6.

Figure 20.7b shows a sectorized colinear dipole array antenna. A reflector is placed behind the array to reflect its power into a 60° sector in the horizontal plane, which increases its gain by 3 times or 5 dB. The gain of a four dipole, sectorized colinear dipole antenna is therefore $2.15\,\text{dBi} + 6\,\text{dB} + 5\,\text{dB} = 13\,\text{dBi}$.

Base station antennas of the type described above are usually enclosed in a plastic housing (Fig. 20.7b) to protect them from environmental conditions. Radiation patterns of such a sectorized colinear dipole array are shown in Figure 20.8. The pattern shown in Figure 20.8a is the pattern in the elevation plane. Radiation directly up is at 90°; radiation directly down is at -90°. Radiation along the horizontal is at 0°. The pattern in the azimuth plane is shown in Figure 20.8b, and it is similar to Figure 20.2.

Parabolic Dish Antennas

One of the most well-known RF antennas is the parabolic dish antenna shown in Figure 20.9. The parabolic dish antenna consists of an antenna feed, which radiates its power into a parabolic shaped reflecting surface. The parabolic reflector concentrates the RF power back along the antenna axis. The parabolic shape has the unique property that all rays from the feed to any point on the parabolic surface travel an equal pathlength from the feed to a plane at right angles to the direction of propagation. This feature gives the parabolic reflecting surface its focusing characteristics and allows it to concentrate the RF power from the feed horn along the antenna axis. The gain, receiving area, and beamwidth are determined by the parabolic reflector. Antenna polarization and bandwidth are determined by the feed. Antenna gains up to 50 dBi are possible.

Patch Antenna Array

The patch antenna array shown in Figure 20.10 achieves the same performance as a small parabolic dish at much less cost. The patch antenna array consists of an array of patches photoetched on a large microstrip board. The feed lines split the incoming RF power into three parts to feed the three vertical columns and split each part into an additional three parts to feed the nine elements of the array. The correct phase of the radiation from each element required to form a radiated beam is achieved by using a specific length of feed line for each element. (These varying sections of line are not shown in the figure.) Compared to achieving the correct phase relationships by using reflections from a parabolic surface, the patch antenna is much

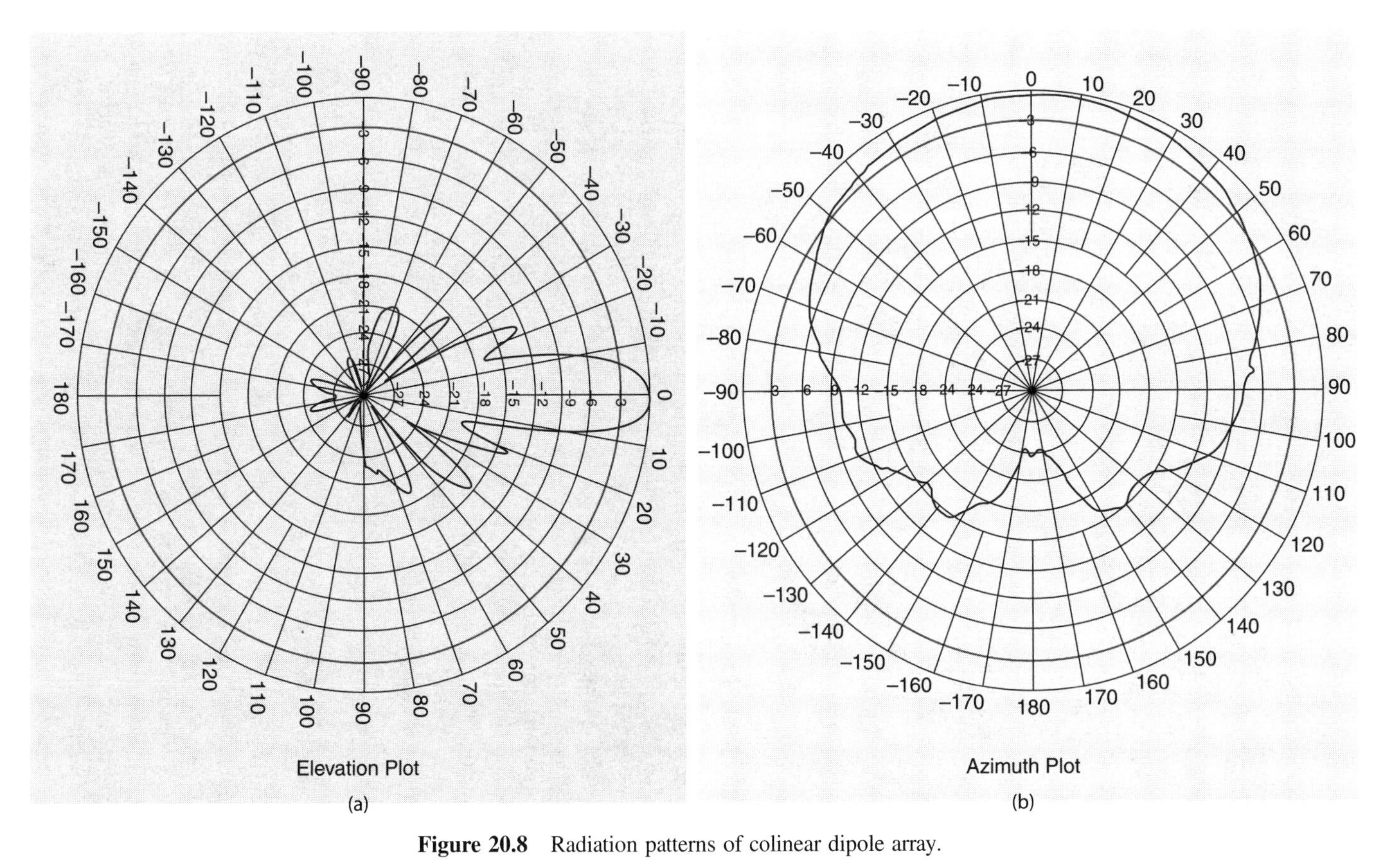

Figure 20.8 Radiation patterns of colinear dipole array.

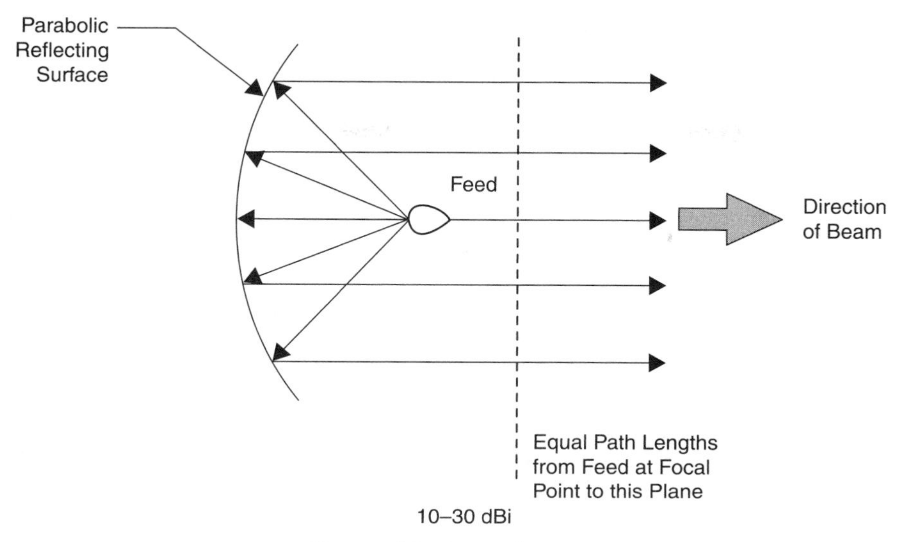

Figure 20.9 Parabolic dish.

simpler. A ground plane on the back surface of the microstrip board makes the antenna unidirectional. Antenna gains between 15 and 25 dBi can be easily obtained.

20.3 MEASUREMENT OF ANTENNAS

An anechoic test room for measuring antennas is shown in Figure 20.11. As described previously, the walls, ceilings, and floors must be covered by RF

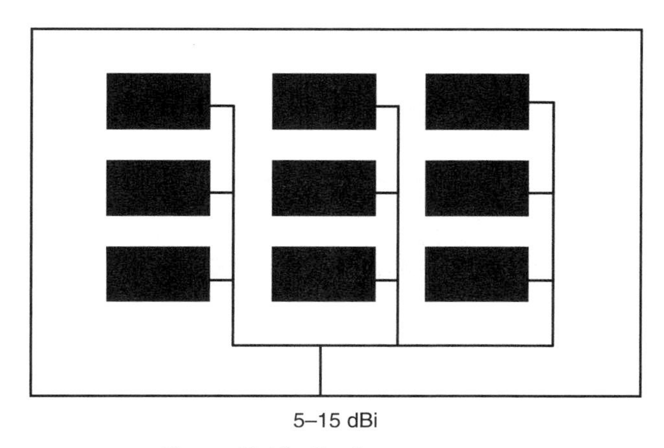

Figure 20.10 Patch antenna array.

Figure 20.11 Anechoic test room.

absorbing materials so that no reflections of the RF signal from the transmitter will get into the receiver. A transmitter with a reference antenna, and the antenna to be tested, must be located at one end of the room and a receiver must be located at the other end.

Figure 20.12 shows a block diagram of an antenna measurement setup. The best way to make the measurement is by using a VNA. The procedure is as follows:

1. Port 1 of the VNA will serve as the transmitter. Port 2 will serve as the receiver. Any type of antenna, like a half-wave dipole, can be used as the receiving antenna and should be connected to port 2 of the VNA. The receiving antenna should be set up at the far end of the antenna range.
2. The VNA should be adjusted to sweep across the frequency range of interest.

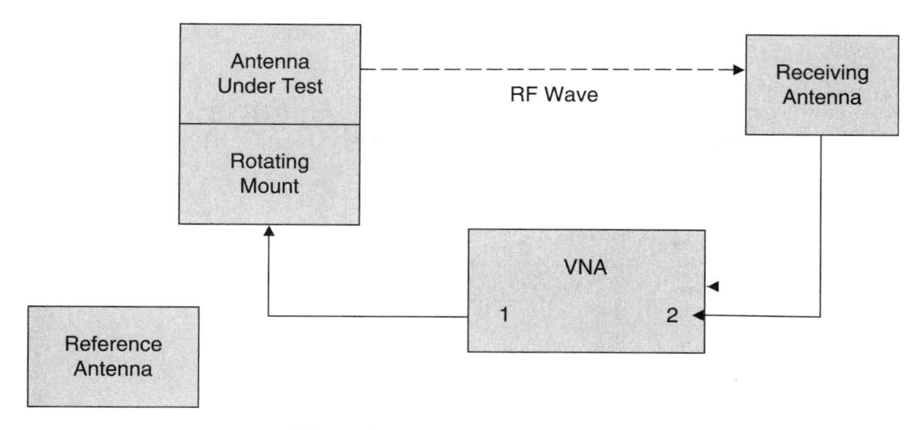

Figure 20.12 Antenna test setup.

3. The VNA is located at the near end of the antenna range, and it is calibrated by connecting an isotropic antenna to Port 1.
4. The VNA is set up to measure S_{21}, at port 1, with the scale set for LogMag (dB). A "through" calibration is made.
5. The isotropic antenna is then replaced with the antenna to be tested, and S_{21} is measured. The measured S_{21} will be the gain of the antenna.
6. If the measurement is made with a half-wave dipole reference antenna, the received signal at port 2 will be 2.15 dB greater than the actual gain. This can be most easily handled by inserting a 2.15 dB attenuator ahead of the half-wave dipole reference antenna during calibration.
7. The antenna under test can be mounted on a rotating joint, and successive measurements at different azimuth angles can be made. This information will provide the antenna pattern in the horizontal plane as shown in Figures 20.2 and 20.8b.
8. The elevation of the antenna can be adjusted to provide the antenna pattern in the elevation plane, as shown in Figure 20.8a.

20.4 DUPLEXERS

For continuous two-way wireless communications, both the mobile unit and the base station must have separate antennas: one for transmission and one for reception. Older walkie-talkie mobile phones used one antenna connected alternately by a manual switch to the transmitter or to the receiver. Having two antennas is not a problem for a base station, but it is inconvenient for a mobile unit.

To achieve a one antenna mobile unit, a duplexer must be used. Two types of duplexers are shown in Figure 20.13. The switch duplexer (Fig. 20.13a) uses a fast acting PIN diode electronic switch (see description in Chap. 16) to alternately

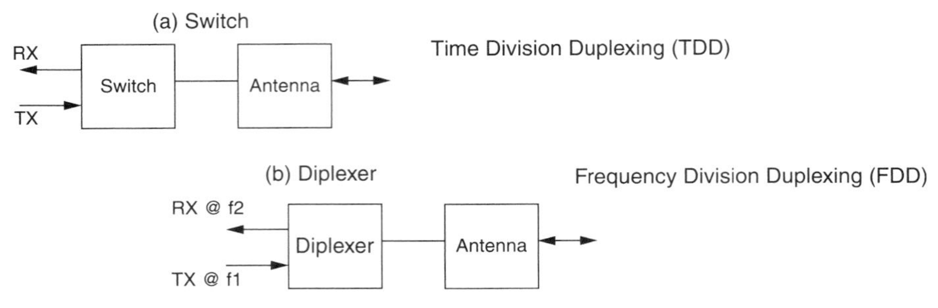

Figure 20.13 Duplexers.

connect the antenna to the transmitter and then to the receiver. A typical connection interval is 10 microsec. The mobile unit generates a digital representation of the talk and listen signals continuously but transmits them in alternate time periods. At the receiving end the signal processing converts the digital signals into two continuous bitstreams and then converts each into an analog voice signal. The listener cannot hear all the processing, and the received signal sounds like a normal voice.

The diplexer (Fig. 20.11b) is used with wireless communications systems that use two different RF frequency bands, one for transmission from the mobile unit to the base station (which is called the uplink or reverse direction) and one for transmission from the base station to the mobile unit (which is called the downlink or forward direction). Almost all cell phone communications systems use this dual frequency assignment. The diplexer contains a separate bandpass filter for the uplink RF band and one for the downlink RF band, and each filter is connected to either the transmitter or the receiver. The uplink and downlink signals are transmitted continuously and are separated by the diplexer. A diplexer for a mobile unit is described in more detail in Chapter 22, which covers RF filters.

Note that both the diplexer and the switch duplexer require the same RF system bandwidth. For example, if the bandwidth for each of the transmitter and receiver channels is 25 MHz, the total bandwidth required for a system using a diplexer is 50 MHz. If the switch duplexer is used, the data signals in either direction must be sent at a two times higher data rate, because each direction is using the channel only half of the time, so that the channel bandwidth must also be 50 MHz.

20.5 ANNOTATED BIBLIOGRAPHY

1. Scott, A. W., Microwave antennas, In Understanding Microwaves, Chap. 13, Wiley–Interscience, New York, 1993.

Reference 1 gives a survey of RF antenna types and performance.

CHAPTER 21

RF RECEIVER REQUIREMENTS

The receiver in a wireless mobile unit must operate under a variety of conditions. When the mobile unit is at the edge of the coverage area, the receiver must have a low noise figure to receive the very low signal from the base station transmitter with a satisfactory S/N for achieving the required BER. When the mobile unit is close to the base station, the received signals intended for all the mobile units in the cell are very large, and two signals may mix in the mobile receiver to form a signal that jams the signal intended for the mobile. This jamming signal is called an intermodulation product.

The received RF signal varies over a 90 dB dynamic range, depending on the transmitter–receiver separation and multipath fading. The receiver must provide adjustable gain, depending on the received signal level, to provide a constant output level of about 0 dBm for demodulation.

Figure 21.1 shows the block diagram of a typical RF receiver in a mobile unit, between the receiving antenna and the demodulator. The receiver is made up of four basic parts, whose types of performance are as follows:

1. RF filter: Allows only a specified range of RF frequencies to pass and blocks all other frequencies
2. LNA: Amplifies the weak received RF carrier
3. Mixer: Shifts the RF frequency to a lower frequency where it can be more easily amplified and filtered
4. IF amplifier: Amplifies the IF frequency to a power level where it can be demodulated

RF Measurements for Cellular Phones and Wireless Data Systems. By A. W. Scott and R. Frobenius

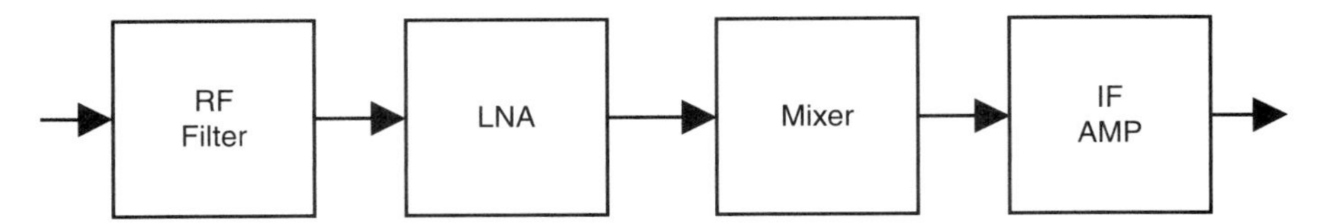

Figure 21.1 RF receiver block diagram.

All components contribute to the gain or loss of the RF signal as it passes through the receiver. The filter and the RF amplifier contribute to the noise figure of the receiver. The LNA and the mixer contribute to the intermodulation products.

Intermodulation products are explained in Figure 21.2 and in more detail in Chapter 23. Figure 21.2a shows three signals applied to the input of an LNA. Figure 21.2b shows the output of the LNA when the inputs are at a low level and shows that each of the three signals are amplified. As the input signals are increased and drive the amplifier near saturation, the output appears as shown in Figure 21.2c. The three input signals are amplified, but many more signals caused by harmonic mixing occur. These harmonic mixing signals are called intermodulation products. Note that one mixing product of the signals at 1.0 and 1.1 GHz occurs at 1.2 GHz and jams the desired signal. As explained in Chapter 23, the higher the IP3 of the

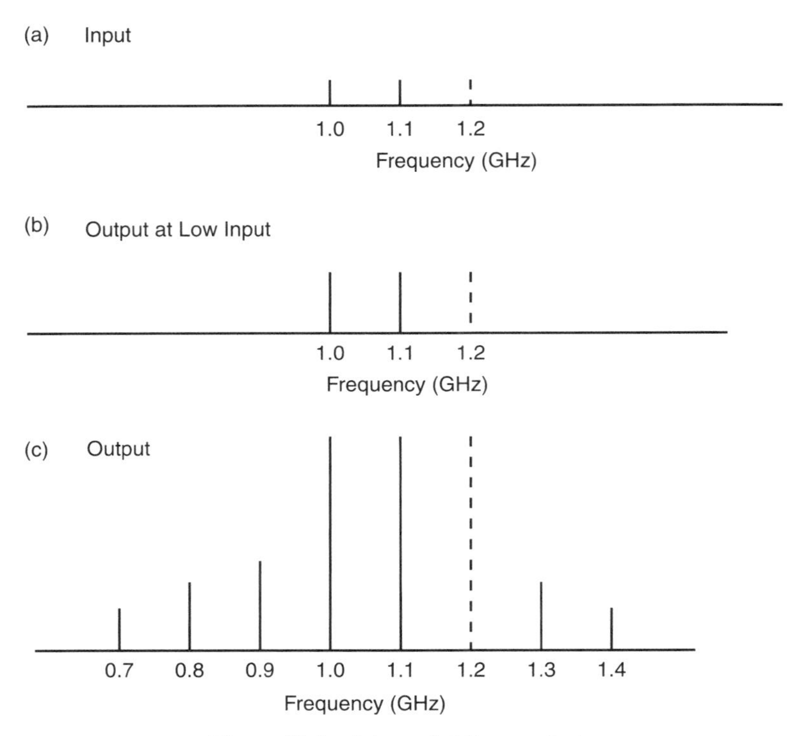

Figure 21.2 Intermodulation products.

component is, the better will be the performance of the mobile receiver when it is close to the base station.

The situation is different for the base station receiver. It must receive signals from all of the mobiles in the cell. If it raises its gain to receive a weak signal from a distant mobile, it will be jammed by a nearby mobile. If it reduces its gain to properly receive the signal from a nearby mobile, it will not be able to receive the signal from a distant mobile. Therefore, the base station tells each mobile using control instructions to continuously raise or lower its transmitter power, so that the signals from all mobiles will be at the same power level in the base station receiver regardless of where they are located in the cell.

Chapters describing the different receiver components are as follows:

Chapter 22 discusses RF filters

Chapter 23 discusses LNAs and defines gain, noise figure, and intermodulation products

Chapter 24 discusses mixers

Chapter 25 discusses measurement of noise figure

Chapter 26 discusses measurement of intermodulation products

Chapter 27 discusses overall system performance

21.1 ANNOTATED BIBLIOGRAPHY

1. Cheah, J. Y. C., Introduction to wireless communications applications and circuit designs, In Larson, L., Ed., RF and Microwave Circuit Design for Wireless Communications, Chap. 2, Artech House, Boston, 1996.

Reference 1 describes RF receiver requirements in more detail than this book.

CHAPTER 22

RF FILTERS

22.1 RF FILTER CHARACTERISTICS

The purpose of RF filters is to allow a range of frequencies to pass and to block unwanted frequencies. Characteristics of RF filters are

Frequency range
Out of band RF loss
In-band RF loss
Sharpness of the filter skirts
Group delay

Figure 22.1 shows an RF front end filter. This filter allows the 25 MHz band of frequencies in the cellular range from 869 to 894 MHz being transmitted from the base station to enter the mobile receiver while blocking all other frequencies. This band includes 832 individual user channels, each of which is 30 kHz wide.

After frequency characteristics, the next most important characteristic is the out of band loss. The filter is being used to block interfering signals of other systems from getting into the cell phone mobile receiver. For the example shown in Figure 22.1, the front end filter has a loss outside the band of the 40 dB. This high loss is required to block out the cell phone transmitted band from 824 to 849 MHz, which is just below the cell phone received band. Below this band are the UHF TV channels. Above the

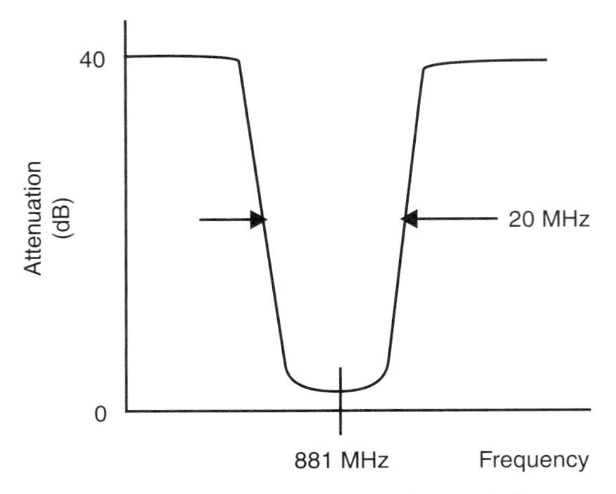

Figure 22.1 Filter frequency characteristics.

cell phone receive band are pager channels at around 930 MHz and air traffic control radar channels at around 1 GHz.

The in-band insertion loss of the filter is important, but it less important than the out of band loss because interfering signals must be blocked out of the cell phone receive channel. However, as described in Chapter 23, the noise figure contribution of the RF front end filter is equal to its insertion loss, so insertion loss is still important.

The skirts of the RF front end filter define the frequency range between the blocking and passing frequencies. If an interfering signal frequency falls into the skirt region, it will not be attenuated effectively. The width of the skirt regions can be reduced by using more filter elements, which makes the filter larger and more expensive.

RF filters are also used throughout wireless transmitters and receivers to eliminate unwanted frequencies from uconverters and mixers and to eliminate image noise, transmitter harmonics, and other unwanted signals.

Filters in the IF frequency range are required to separate out individual voice or data channels. A filter affects the amplitude of the RF signal as well as its phase. The phase variation across the filter passband is defined by "group delay." This is a critical parameter for the individual user channels, which is discussed in detail in Section 22.5.

22.2 RF FILTER DESIGN

All filters from low frequency through the RF, microwave, and millimeter wave bands are made with an array of coupled resonators, as illustrated in Figure 22.2. Coupled resonators allow the RF wave to leak from one resonator to the others in a controlled fashion. Figure 22.2a shows an array of three uncoupled resonators connected in

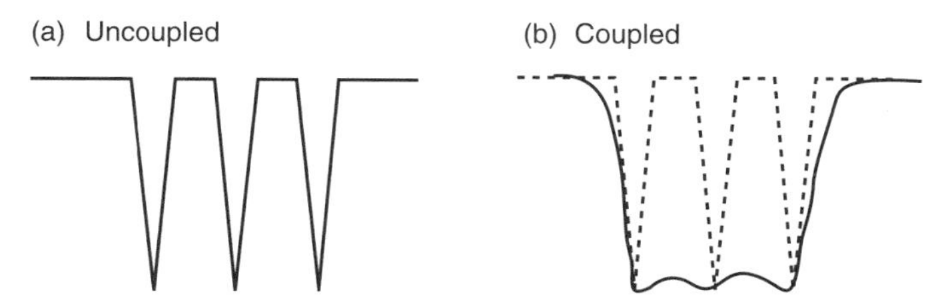

Figure 22.2 How filters are made. Filters are made of an array of coupled resonators (power leaks from one resonator to the next).

parallel to the input and output transmission lines. The RF wave passes through the resonators only at their resonant frequency. However, if coupling is allowed, the passband of the filter fills in between the resonant frequencies across the desired filter bandwidth as shown in Figure 22.2b.

22.3 TYPES OF FILTERS

Microstrip Filter

Figure 22.3 shows a three element filter fabricated on a microstrip board. The filter pattern is etched onto the microstrip board. It consists of three microstrip lines which are 0.5 wavelength long at three different frequencies across the 2.4–2.5 GHz band. Because of their proximity, the RF wave couples from each resonator to the others. The input and output feed lines can be seen at either end of the resonator. This filter is simple to make, but it is large because the wavelength on the microstrip board is 2.5 cm at 2.45 GHz.

Ceramic Block Filter

Filter size can be significantly reduced by fabricating the resonators on a high dielectric ceramic support material. Figure 22.4 shows a two element ceramic block filter fabricated using barium tetratitanate ceramic, which has a dielectric constant of 90.

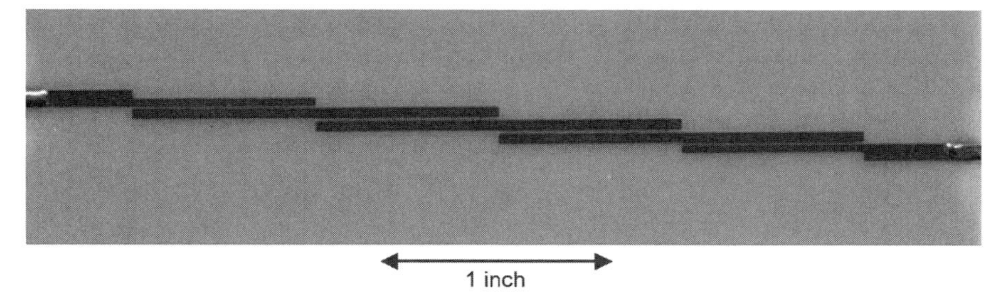

Figure 22.3 Microstrip filter at 2.45 GHz.

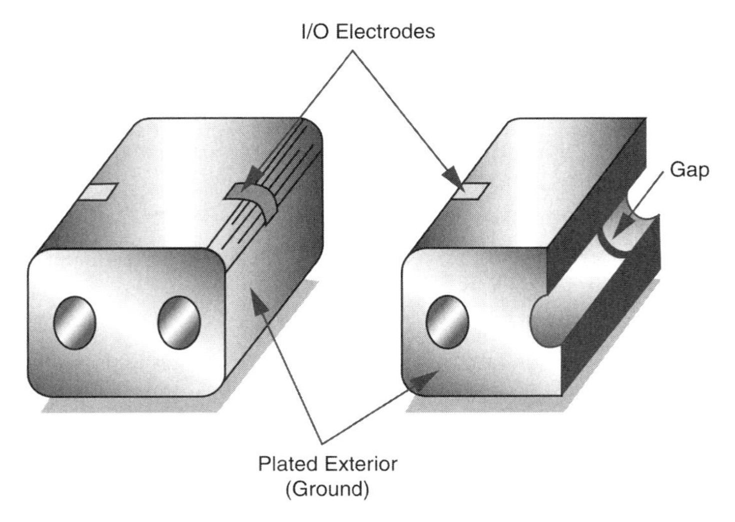

Figure 22.4 Ceramic block filters.

This high dielectric reduces the RF wavelength and consequently the length of the resonator by about 10 times. The resonator consists of two coaxial cables, with a common square outer conductor and cylindrical inner conductors. The inner and outer conductors are metallized onto the ceramic and then copper plated. There is no metallic surface between the two inner conductors, which allows for the coupling, and the amount of coupling is controlled by the spacing between the center conductors. Each resonator has a slightly different length, and thus a slightly different frequency. The ceramic block resonator has approximately the same RF filter characteristics as the microstrip filter shown in Figure 22.3, but it is only 5.7 mm long $\times$ 3.5 mm wide $\times$ 2.5 mm high.

Surface Acoustic Wave (SAW) Filters

The size of RF filters can be reduced by a factor of two in all dimensions (for a volume decrease of eight times) as compared to the ceramic block filter by using a SAW filter. A schematic drawing of a SAW filter is shown in Figure 22.5. It consists of a slab of piezoelectric material mounted in an RF transistor package. A piezoelectric material has the characteristic that when an electrical signal is applied to electrodes on its surface, it vibrates mechanically. Alternately, when it is vibrated mechanically, it generates an electrical signal. This characteristic is used to make loudspeakers for mobile phones. The choice of material determines the frequency range of operation. An IF filter for cellular phones can be made using quartz, which is piezoelectric. This IF filter is discussed in Section 22.5. Lithium niobate can be used to make an RF SAW filter for cell phone and wireless LAN frequency bands.

Referring to Figure 22.5, the piezoelectric lithium niobate substrate is metallized, and two sets of interdigital fingers are photoetched on its surface at opposite ends of

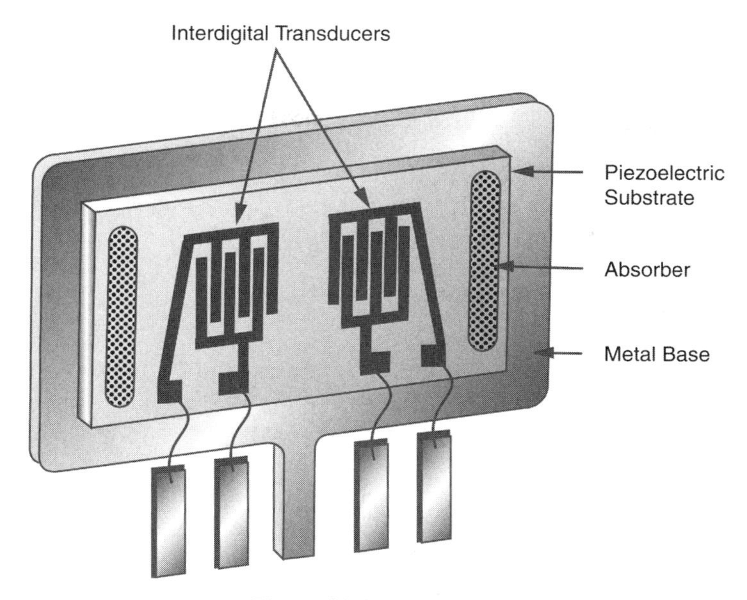

Figure 22.5 SAW filter.

the substrate. The RF signal to be filtered is applied to one pair of the fingers. The fingers are spaced 0.5 wavelength apart, which is in the 1 μm range at frequencies between 800 MHz and 2.4 GHz, because of the low velocity of mechanical wave propagation in the lithium niobate compared to the velocity of RF waves in microstrip materials. The SAW resonators shown in Figure 22.5 have five elements that are defined by the slightly different spacings between each pair of fingers. The RF wave is converted to a mechanical vibration at the RF frequency and travels down the surface of the lithium niobate substrate to the identical resonators at the opposite end. The acoustic wave sets up an RF wave in the output resonator, which is connected to microstrip output lines that are identical to those used at the input. Because not all of the acoustic wave power is coupled into the resonators, absorbers are placed on each end as shown to prevent multiple reflections. Only RF signals in the frequency range of the SAW resonators can pass through the filter. The RF loss of SAW filters is from 1 to 3 dB, about the same as the RF loss of the ceramic block filters.

Film Bulk Acoustic Resonator (FBAR) Filters

The size of RF filters can be reduced by another two times in both the width and length dimensions, as compared to a SAW filter, and by using the FBAR shown in Figure 22.6. The resonator is a thin film of aluminum nitride (AIN) suspended over a silicon substrate. The resonator vibrates mechanically at the RF frequency. The length of the resonator is 0.5 wavelength at its mechanical vibration frequency,

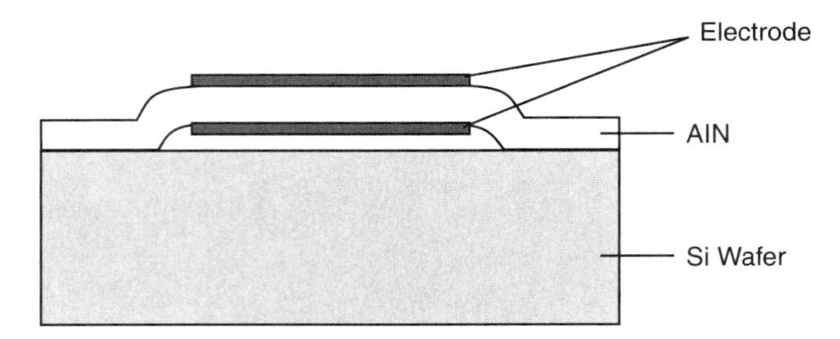

Figure 22.6 FBAR filter.

which is the RF frequency. The FBAR filter is fabricated by RF transistor fabrication techniques on a silicon wafer. The tiny air gap between the aluminum nitride resonator and the silicon wafer is achieved by etching.

Figure 22.7 shows a schematic drawing and an enlarged photograph of an FBAR duplexer. As discussed in Chapter 20, which covered antennas, a duplexer allows the transmitter signal to be routed to the antenna and the received signal to be routed from the antenna into the receiver, and it isolates the transmitter and receiver from each other. The duplexer assembly shown is $3.8 \times 3.8 \times 1.3$ mm high. It can handle 33 dBm of transmitter power. The receiver band performance is

RF frequency band: 2110–2170 MHz,
Blocking of the transmitter signal: 45 dB
Insertion loss: 2.0 dB

The transmitter band performance is

RF frequency band: 1920–1980 MHz
Insertion loss: 1.6 dB

Base Station Filters

Reducing the size and cost of RF components is always desirable. It is essential to do this for mobile handsets. However, the requirement on the base station RF equipment

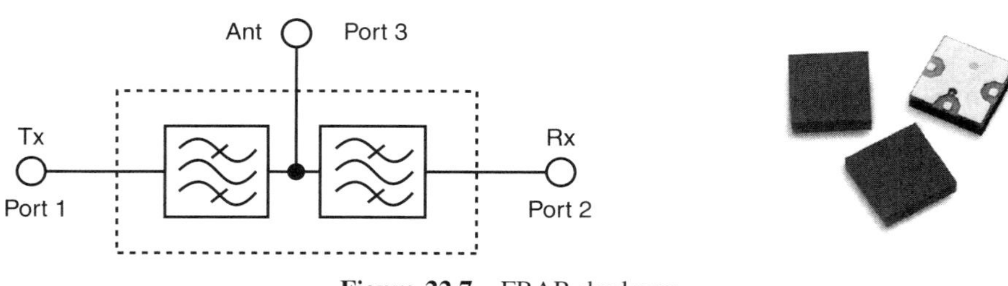

Figure 22.7 FBAR duplexer.

is to optimize RF performance so that overall base station–mobile system works properly. To reduce the loss of the base station front end filters, air filled coaxial cavity resonators are often used. These are 0.5 free space wavelength long. Because they have less than 1 dB insertion loss, the transmitter power requirement from the mobile unit is significantly reduced.

22.4 MEASUREMENT OF RF FILTERS

The performance of RF filters is most easily measured with a VNA. Figure 22.8 shows the test setup. After a two-port calibration is made, the filter under test is connected between port 1 and port 2 of the VNA.

Measurements were made on the following RF filter:

Manufacturer	Toyo
Model	TDF2A-2450T-10
Center frequency	2.45 GHz
Passband width	$\pm$50 MHz
Insertion loss in band	1.5 dB
Attenuation at $\pm$280 MHz	20 dB

Measurements were shown for this filter in Chapter 7, where it was used to demonstrate the operation of the VNA.

Figure 22.9 shows the filter's insertion loss as a function of frequency over the frequency range from 2 to 3 GHz. As discussed previously, the purpose of this filter is to block out interfering signals outside its 2.4–2.5 GHz passband. The specification requires that the attenuation be at least 20 dB at $\pm$280 MHz from the filter's center

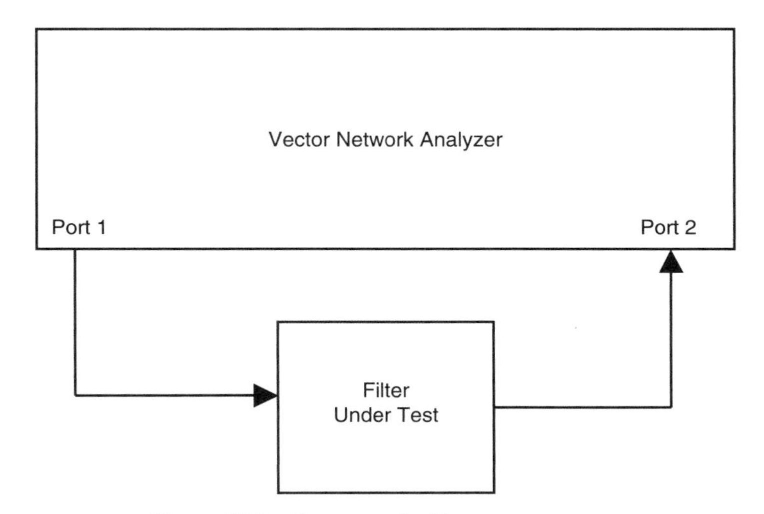

Figure 22.8 Test setup for filter measurements.

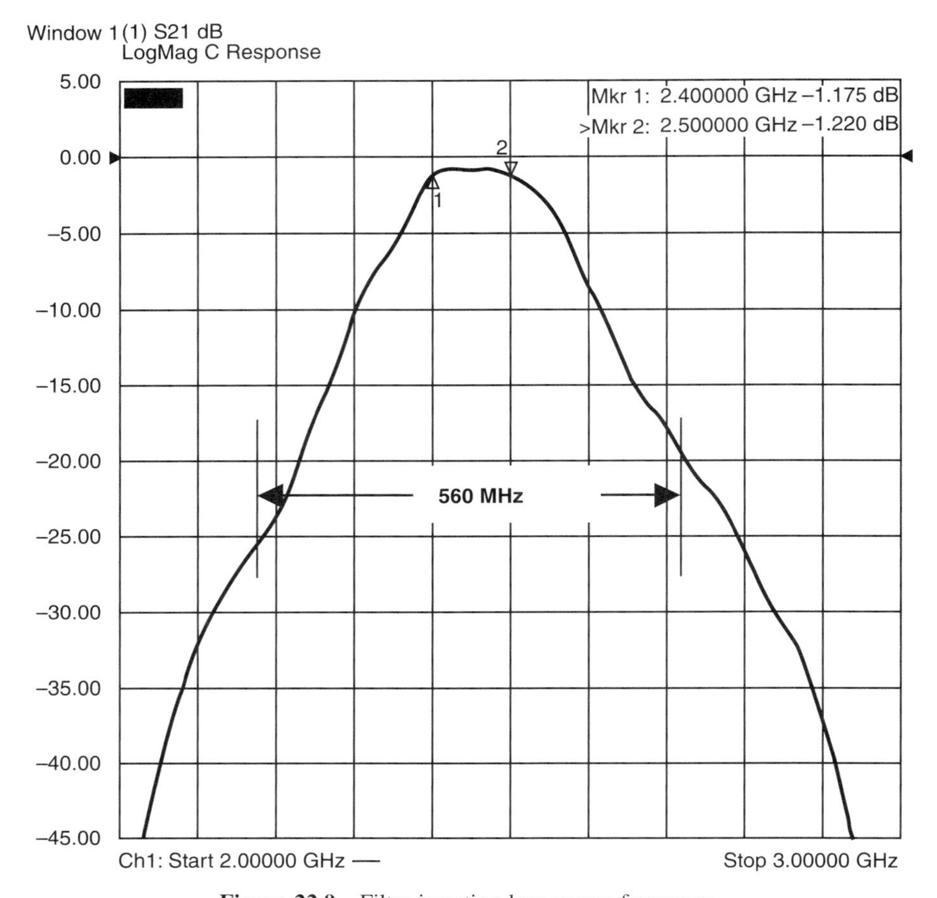

Figure 22.9 Filter insertion loss versus frequency.

frequency of 2.45 GHz. As shown, the filter meets this requirement, being −26 dB at 280 MHz below the center frequency and −21 dB at 280 MHz above the center frequency. The next requirement is that the attenuation be less than 1.5 dB over the operating band; as the markers show, this requirement is met.

Figure 22.10 graphs the mismatch of the filter looking from both the input and the output ends. The mismatches are expressed as S-parameters S_{11} and S_{22}, which are both 0.2, a return loss of 14 dB.

22.5 GROUP DELAY AND ITS MEASUREMENT

The RF filter measured in the previous section allows all frequencies of the particular RF system to pass and then be amplified by the LNA. The individual voice or data channel assigned to a particular user is then selected by the mixer. This selection

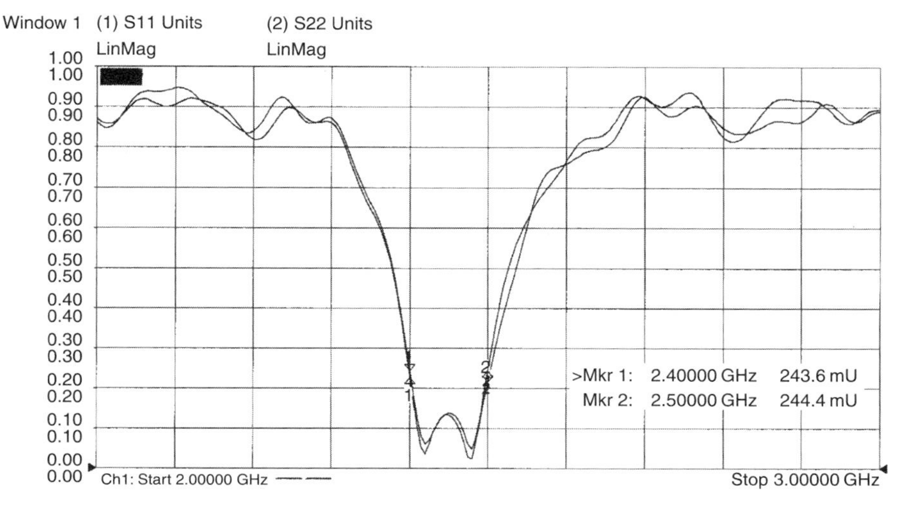

Figure 22.10 Input and output match of filter.

process involves the use of a narrowband IF filter that exactly matches the user channel bandwidth.

A critical performance parameter of the IF filter is its group delay, which is explained in Figures 22.11 and 22.12. Figure 22.11 defines group delay. The graph shows the phase shift through a component-like a filter as a function of frequency. The slope of the phase shift versus frequency line is the group delay. If the component is a coaxial cable, the curve is a straight line that has a constant slope, so the group delay is constant. When an RF signal is modulated to carry information, it forms a set of modulation side bands, each at a different frequency. It is essential that all side bands travel at the same velocity so that all components of the modulated wave will reach the receiving end at the same time. Then, the modulated wave can be

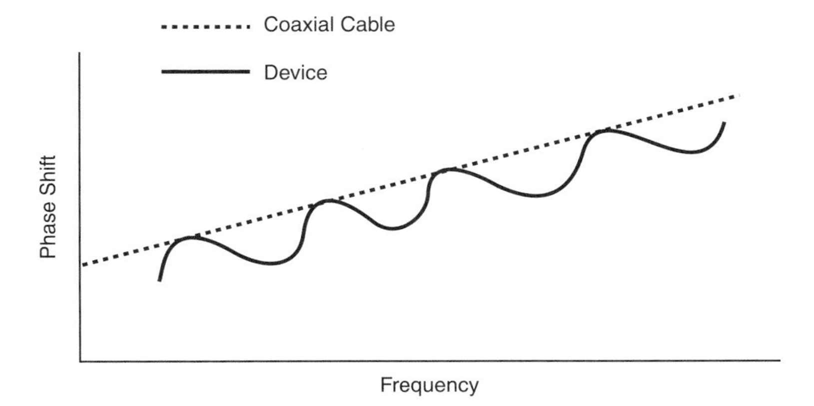

Figure 22.11 Group delay.

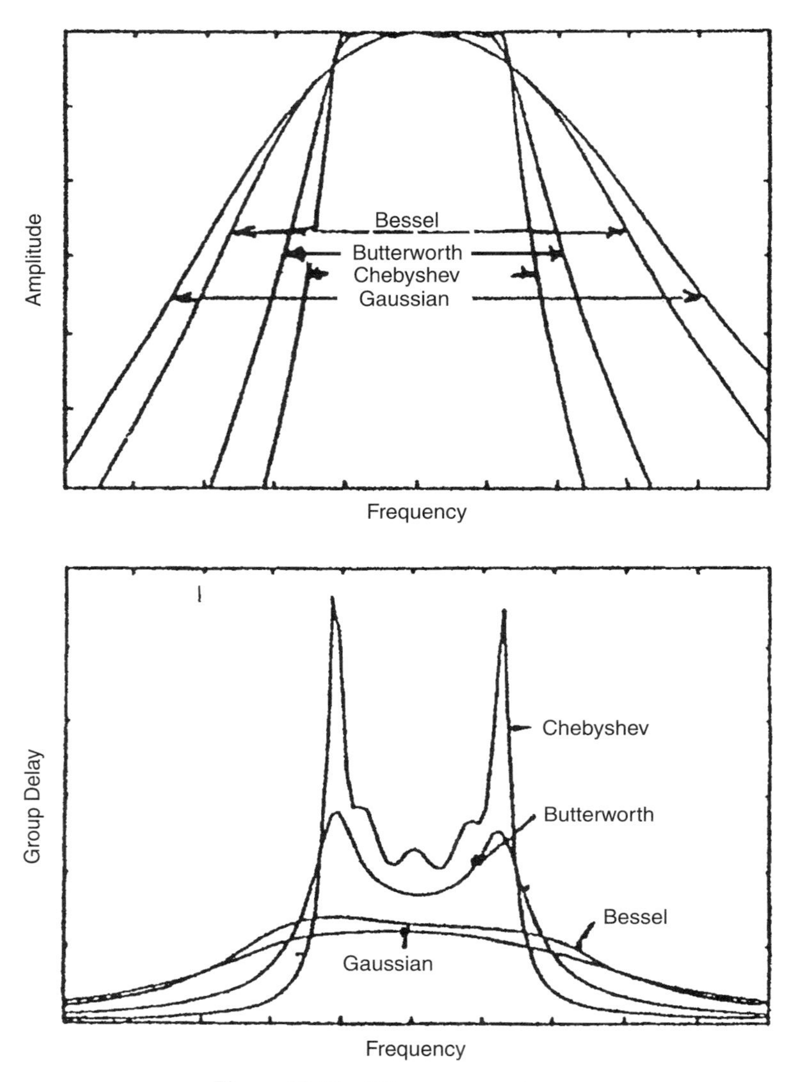

Figure 22.12 Filter response shapes.

properly demodulated to give the information signal. If the transmission system does not have a constant slope like the solid line in Figure 22.11, the modulation side bands will arrive at different times and the modulation will be distorted.

The top graph of Figure 22.12 shows the attenuation characteristic of various filter designs, each using five filter elements. The Gaussian and Bessel filters have broad skirts and do not have sharply defined passbands. The Butterworth filter has a more sharply defined filter passband, and the Chebyshev filter has the sharpest

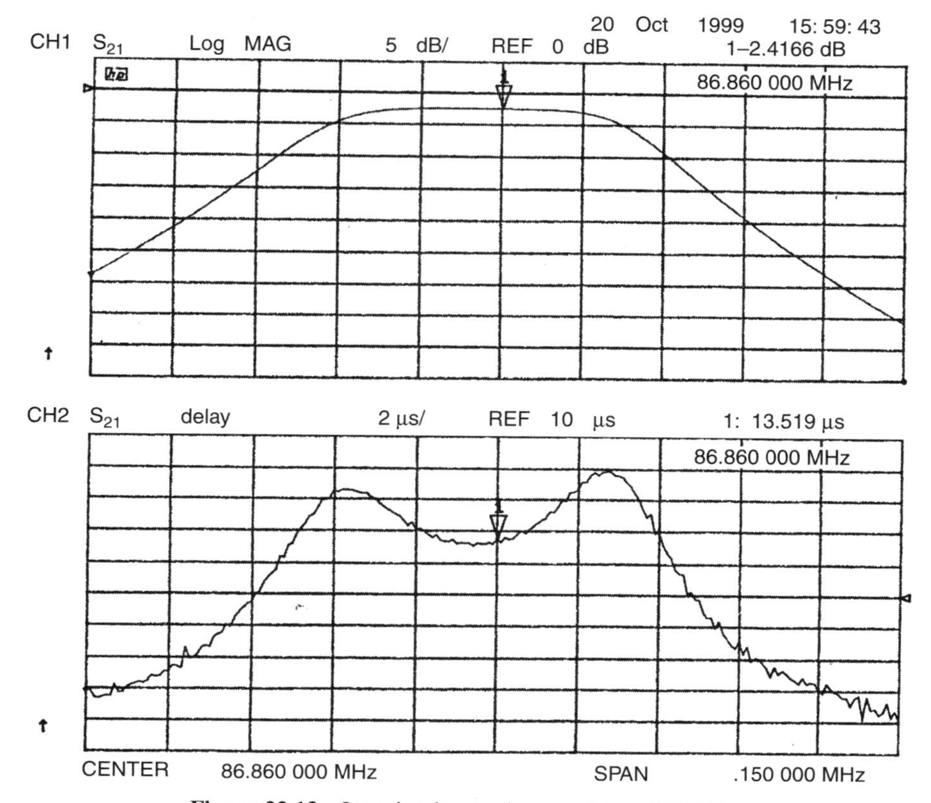

Figure 22.13 Insertion loss and group delay of IF filter.

filter response of all. The bottom graph shows the group delay characteristics of the same four filters. Remember that the requirement for nondistortion of the modulation is that a filter have the same group delay over the bandwidth of the filter. This is approximately achieved with the Gaussian and Bessel filters, but not with the Butterworth and Chebyshev filters.

Figure 22.13 shows the insertion loss and group delay of an IF SAW filter, which has the following characteristics:

Manufacturer	Toyo
Model	SF086WA-001
Center frequency	86.85 MHz
Passband width	± 15 kHz
Insertion loss in band	4 dB
Attenuation at ± 100 kHz	40 dB
Group delay deviation at ± 15 kHz	5 μs
Filter size	$13 \times 6.5 \times 2$ mm

The filter is made of quartz, which when combined with the need to operate at the 86.85 MHz center frequency, accounts for the larger size of the filter compared to an RF SAW filter.

The upper curve of Figure 22.13 shows the attenuation characteristics of this filter over a span of 150 kHz centered at 86.85 MHz. The filter meets the specification requirements of less than 4 dB insertion loss in the band and more than 40 dB attenuation at ± 100 kHz from the center frequency.

The lower curve shows the group delay, which is measured in microseconds. The unit of group delay is time, which can be understood from Figure 22.11, which showed the phase shift through an RF component as a function of frequency. The group delay is the slope of this curve.

$$\text{group delay} = \frac{\text{phase}}{\text{frequency}} = \frac{\text{phase}}{\text{phase/time}} = \text{time}$$

To prevent distortion of the modulation, the group delay must be very constant across the bandwidth of the filter, and the specification allows a variation of $\pm 5\ \mu s$ across a ± 15 kHz frequency range around the center frequency of the filter. The measured group delay is 4 μs, so the filter meets the requirement.

The distortion effect of this filter on a $\pi/4$DQPSK modulated signal, as measured with a VSA, will be shown in Chapter 33.

22.6 EXAMPLE FILTER MEASUREMENT

Objective

Our objective is to demonstrate the out of band rejection of the RF filter. We also want to demonstrate the IF filter and measure group delay. Note that impedance match and S-parameters are not relevant at the IF frequencies (below 100 MHz in this case).

Measurements Being Demonstrated

RF filter insertion loss and out of band rejection
IF filter insertion loss and out of band rejection
IF filter group delay

Specifications of DUT

The specifications for the RF filter are listed in the following table:

Manufacturer	Toyo
Model	TDF2A-2450T-10
Center frequency	2.45 GHz
Passband width	± 50 MHz
Insertion loss in band	1.0 dB
Attenuation at ± 280 MHz	20.0 dB

The specifications for the IF filter are listed in the following table:

Manufacturer	Toyo
Model	SF086WA-001
Center frequency	86.85 MHz
Passband width	±15 kHz
Insertion loss in band	4.0 dB
Attenuation at ±100 kHz	40 dB
Group delay deviation	5 μs

Generic Procedure

Steps to perform the measurements are listed in the upcoming table. The left column lists the actions to be performed, and the right column lists the motivation for taking each step and provides additional background information in some cases.

Procedure	Notes
Preset the network analyzer.	
Set the start and stop frequencies to 2 and 3 GHz, respectively. Set the power to 0 dBm.	
Measure S_{21} and set the reference position to 9 and the reference value to 0 dB.	
Connect the cables and perform a response calibration.	Because we are measuring S_{21}, we elect to skip the full two-port calibration for this demo.
Connect the RF filter as the DUT, and set a marker at 2.45 GHz. Check that out of band rejection meets specification.	You can use marker bandwidth functions to have the VNA report the bandwidth for a specific level of attenuation.
Remove the RF Filter and connect the cables.	
Set the center frequency to 86.85 MHz and the span to 300 kHz (still measuring S_{21}). Perform a response calibration.	
Connect the IF filter as the DUT.	
Set a marker at the center frequency. Measure the bandwidth for out of band rejection. Set markers for the edges of the passband.	
Activate a second display (channel) and measure group delay performance of S_{21}.	Once the measurement is set up, the group delay is simply another format that can be displayed.

22.7 ANNOTATED BIBLIOGRAPHY

1. Matsumoto, M., Ogura, H., and Nishikawa, T., A miniature dielectric monoblock bandpass filter for 800 MHz cordless telephone system, In Microwave Symposium Digest, IEEE MTT-S International, Vol. 1, pp. 249–252, IEEE, New York, 1994.

2. Dubois, M. A., Thin film bulk acoustic wave resonators: A technology overview, Paper presented at MEMSWAVE 2003, Toulouse, France, July 2–4, 2003.
3. Besser, L. and Gilmore, R., Filters and resonant circuits, In Practical RF Circuit Design for Modern Wireless Systems, Vol. I: Passive Circuits and Systems, Artech House, Norwood, MA, 2003.

Reference 1 describes the design and performance of ceramic block filters.

Reference 2 describes the design and performance of bulk acoustic wave filters.

Reference 3 provides a background in RF filter design.

CHAPTER 23

LNAs

The purpose of the LNA in a wireless communication system is to amplify the incoming RF signal well above the thermal noise in the receiver, while adding a minimum of additional noise to the signal. The output of the LNA can then be converted to a lower frequency below the RF frequency in a mixer (as will be described in Chap. 24) where additional amplification is easily obtained and the individual user voice channels or data channels can be separated.

If the receiver is mobile, the received RF signal will vary from an extremely low value when the mobile is at the edge of the coverage area to a very high value when the mobile is close to the base station. At low signal levels the important performance characteristic of the LNA is its noise figure. At high signal levels its important characteristic is intermodulation products, which define how signals from other frequency channels in the same system mix together to form a jamming signal at the frequency of the desired signal.

This chapter is organized into four sections. Thermal noise will be discussed in Section 23.1. Noise figure will be described in Section 23.2. Intermodulation products will be discussed in Section 23.3. The use of S-parameters for estimating the linear performance of an LNA will be explained in Section 23.4.

Measurements of LNA noise figure and intermodulation products will be discussed in Chapters 25 and 26, respectively.

RF Measurements for Cellular Phones and Wireless Data Systems. By A. W. Scott and R. Frobenius

23.1 THERMAL NOISE

The noise at the RF receiver determines the obtainable range of a communication system and the data rate that can be supported. The noise comes from two sources:

1. thermal noise radiating from all objects in the receiving antenna's field of view and
2. noise generated by the first few components in the receiver.

Random electrical signals, called noise, exist in all communication systems. This noise interferes with the information carrying signals and ultimately determines the range of the communication system. At frequencies below the RF band, the major source of noise is man made and is due to electrical motor brushes, auto ignitions, and fluorescent light ballasts. In the RF frequency range these sources of noise are small, because it is difficult to generate RF. In the RF, microwave, and millimeter wavebands the major source of noise is due to the random motion of electrons that occurs in all matter. Every object in the antenna's field of view such as foliage, buildings, vehicles, and people contain electrons, which are moving randomly. This noise is called thermal noise; as its name suggests, its magnitude depends solely on the temperature of the source. This thermal noise is independent of frequency and is equal to kTB, where k is the Boltzmann noise constant, which is a physical constant; T is the absolute temperature of the surroundings, which is $= 0$ K at -273°C; and B is the RF bandwidth of the system. Regardless of the actual temperature of the surroundings, the noise temperature is almost always assumed to be 293 K (20°C) for noise calculations. (The only exception is for satellite and deep space probe calculations where the antenna is looking at the cool sky, and in these cases the temperature is about 15°C cooler.)

At a temperature of 293 K, the noise power density (kT) is -114 dBm/MHz. The noise power density is sometimes quoted as -174 dBm/Hz, which will give the same answer for the total noise power, as long as the system bandwidth is specified in Hertz.

In a 2 MHz bandwidth, the thermal noise is -114 dBm + 3 dB = -111 dBm.

In a 25 MHz bandwidth, the thermal noise is -114 dBm + 14 dB = -100 dBm.

In a 0.1 MHz (100 kHz) bandwidth, the thermal noise is -114 dBm − 10 dB = -124 dBm.

23.2 NOISE FIGURE PRINCIPLES

Noise figure principles were discussed briefly in Chapter 11, which described noise figure meters. An expanded discussion of noise figure principles is presented in this chapter.

An example of the additional noise that is added by RF components is explained in Figure 23.1. The figure shows an LNA, whose purpose is to amplify the weak

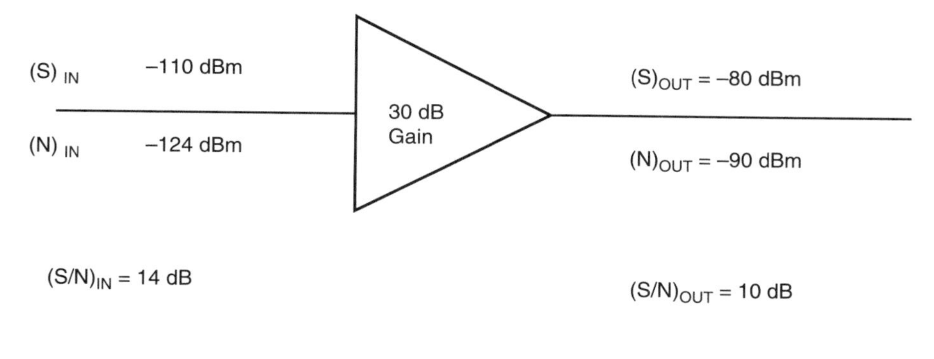

Figure 23.1 Explanation of noise figure.

received signal sufficiently above the RF noise level, so that the noise of the following components in the receiver is not important. For this example, the input signal of −110 dBm would be typical of the received signal when the mobile unit is at the edge of the cell coverage area. The example LNA has a gain of 30 dB, so the weak incoming signal is amplified up to a level at its output of −80 dBm.

Thermal noise is also coming into the amplifier input from the surroundings. In a 100 kHz bandwidth it is −124 dBm, so the S/N at the LNA input is 14 dB. Simple arithmetic would suggest that this input noise will be amplified by the LNA by 30 dB, so that the noise at the output of the LNA would be −94 dBm. However, as Figure 23.1 shows, this is not correct. The actual S/N at the output is 10 dB, which is 4 dB larger than the simple arithmetic suggested. This is because the LNA, being an electronic device with moving electrons inside, adds additional noise. The 4 dB increase in S/N between the input and the output is defined as the noise figure of the amplifier. For calculation purposes it is assumed that the excess noise represented by the noise figure is all added at the input of the amplifier.

Noise figure is defined as the ratio of the S/N going into a device compared to the S/N coming out.

Noise Figure of Passive Components

Passive components like an RF filter or the length of transmission line between the antenna and the LNA also have a noise figure, which is equal to their attenuation. This statement can be understood based on the definition of noise figure. With a lossy component, the RF signal is attenuated in passing through the component. However, the output noise is regenerated in the component, which is at room temperature. The S/N degrades because the noise remains constant through the lossy component but the signal is decreased. Many types of handheld wireless equipment have a total noise figure of about 4 dB. The noise figure of the LNA is about 1 dB and the filter loss is 3 dB. It is therefore essential to minimize the loss of the RF filter and any cabling at the input to a low noise receiver.

Cascaded Noise Figure

By providing enough gain in the first LNA, the noise figure of subsequent components such as amplifiers and mixers will have a negligible effect on the noise figure of the total system. The reason for this is shown graphically in Figure 23.2 and by numerical calculation in Figure 23.3.

Figure 23.2 shows the RF signal and the noise as it passes through the first LNA to the input of the second. The line at the bottom of the chart marked "noise floor" is the *kTB* noise, which is at the input of every component in the equipment. The effective noise at the input to the first amplifier is the noise floor plus the noise figure of the first

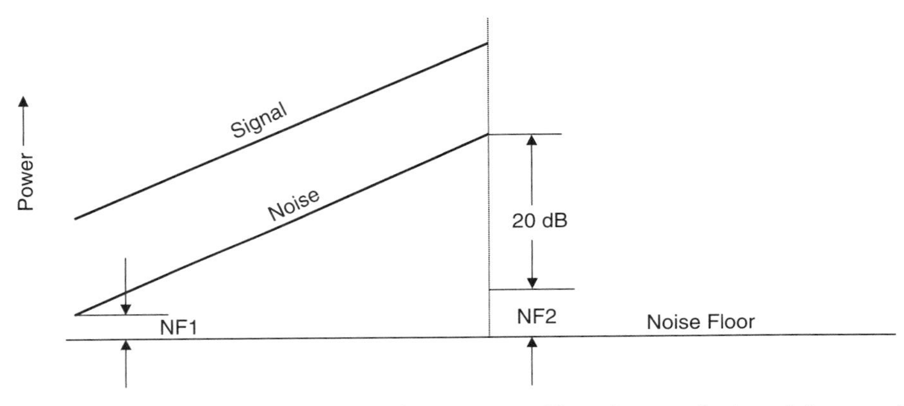

Figure 23.2 Noise figure of a chain of components. The noise contribution of the second stage is reduced by the gain of the first stage.

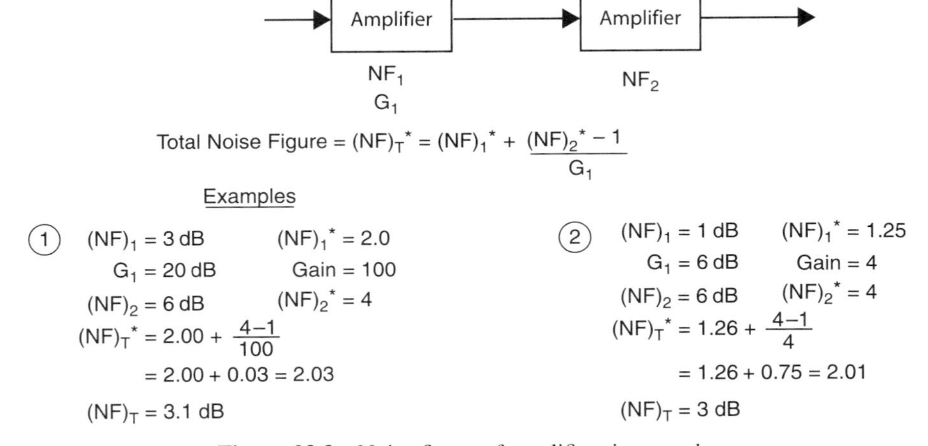

Figure 23.3 Noise figure of amplifiers in cascade.

amplifier (NF_1). As shown by the graph, both the input signal and the input noise are amplified passing through the amplifier and their ratio in decibels remains the same at the amplifier output. Note also that both the signal and the noise are amplified by as much as 20 dB. As the drawing illustrates, the amplified noise coming out of the first stage amplifier is many dB greater than the noise figure of the second stage; consequently, the second stage makes a negligible contribution to the overall noise figure of the system.

Figure 23.3 shows the same effect as discussed in the previous paragraph using a numerical calculation. The sketch at the top of the figure shows a formula for calculating the contribution of the noise figure of a second amplifier to the overall noise figure of the chain. The first amplifier has a noise figure NF_1 and a gain G_1. The second amplifier has a noise figure NF_2. The gain of the second amplifier does not enter into this calculation. The formula for the noise figure of the cascade is shown. Note the asterisks after the NF and G symbols. The formula shows noise powers being added together, so the dB system of units cannot be used in the calculations. Adding decibel values represents the mathematical operation of multiplying. Thus, the decibel values of noise figure and gain must be converted into ratios. That is what the asterisks mean, and this is shown in the calculations.

Example 1 in Figure 23.3 shows the case where the noise figure of the first amplifier is 3 dB and its gain is 20 dB. The equivalent ratios are a noise factor of 2.0 and gain of 100. As the calculation shows, the total noise figure of the cascade is only 0.1 dB greater due to the second stage noise figure than due to the noise figure of the first stage alone, because of the 20 dB (100×) gain of the first stage.

Example 2 in Figure 23.3 shows the same second stage amplifier, but with a first stage amplifier with a noise figure of only 1 dB and a gain of only 6 dB. The cascade noise figure of this combination is 3 dB, only 0.1 dB better than case 1, because even though the noise figure of the first stage is much lower, its gain is so low that the noise figure of the second stage is now having an effect.

Mismatching of the Transistor Input to the Reduce Noise Figure

The noise figure of a transistor can be significantly reduced by deliberately mismatching the transistor at its input terminal to reduce the noise figure. This matching adjusts the phase between the input noise and the noise generated in the transistor, and it results in a partial cancellation of the total noise amplified by the transistor. Figure 23.4 illustrates this effect with contours of constant gain and constant noise figure on a Smith Chart.

If the input of the transistor is matched from the maximum gain point shown on the Smith Chart, the gain would be 14.7 dB. If the input is matched from other points, the gain would be reduced to 13.7 dB, 12.7 dB, and so on down to 6.7 dB as shown by the constant gain circles. However, to achieve the minimum noise figure of 2.5 dB, the input of the transistor must be matched as shown by the 2.5 dB noise figure point. If the input match departed from the optimal noise figure value, the noise figure deteriorates to greater than 3 dB. Achieving the minimum noise figure results in a gain reduction from 14.7 to 10.7 dB for the illustrated transistor.

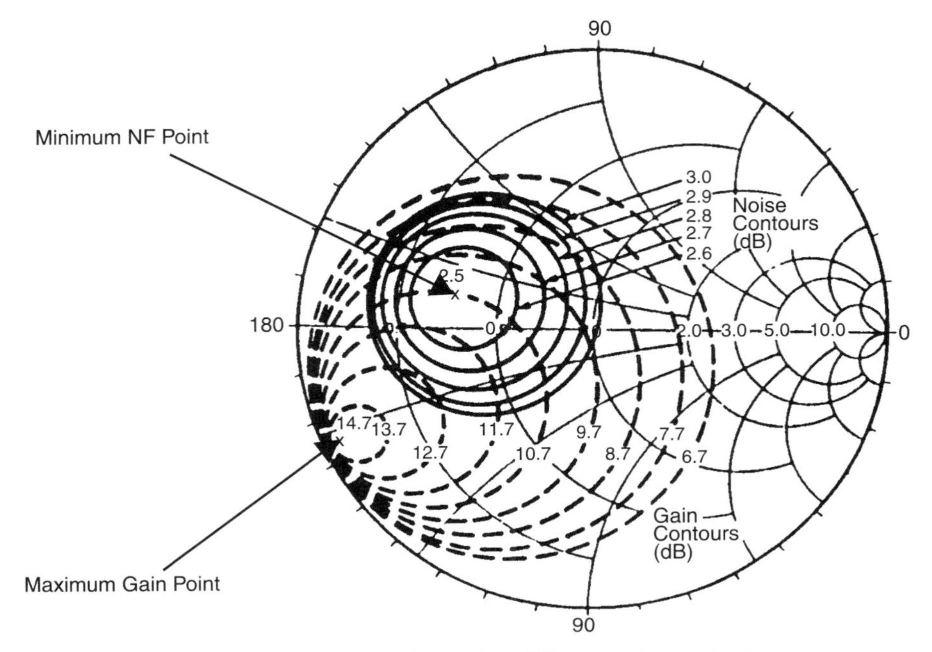

Figure 23.4 Input mismatching of amplifiers to reduce noise figure.

The "associated gain" is the gain that the transistor will have when it has been biased and matched for its lowest noise figure. This gain is less than the gain that could be obtained if the transistor were matched and biased for its highest gain. The most important requirement for the first amplifier stage in a low noise receiver is low noise figure, so often a transistor amplifier is made of two stages. The first stage is matched and biased for minimum noise figure. The noise figure of the second stage is less important because the signal has been amplified above the noise level by the first stage, so the second stage is matched and biased for maximum linear gain.

Figure 23.5 shows the same type of mismatching data for minimum noise figure. Three curves of points are shown on the Smith Chart: one for S_{11}, one for S_{22}, and one for minimum noise figure. Each curve covers the range of frequencies over which the LNA is to be used. The S_{22} points are for matching the transistor output. The S_{11} points are for matching the transistor input for maximum gain. The minimum noise figure points are for matching for minimum noise figure. Note how differently the input must be matched for achieving either maximum gain or minimum noise figure.

The optimum noise figure, maximum available gain, and Smith Chart parameters for matching at different RF frequencies and operating conditions are provided for another transistor type in Figure 23.6. Measurement equipment to determine these matching points and circles is described in Chapter 25.

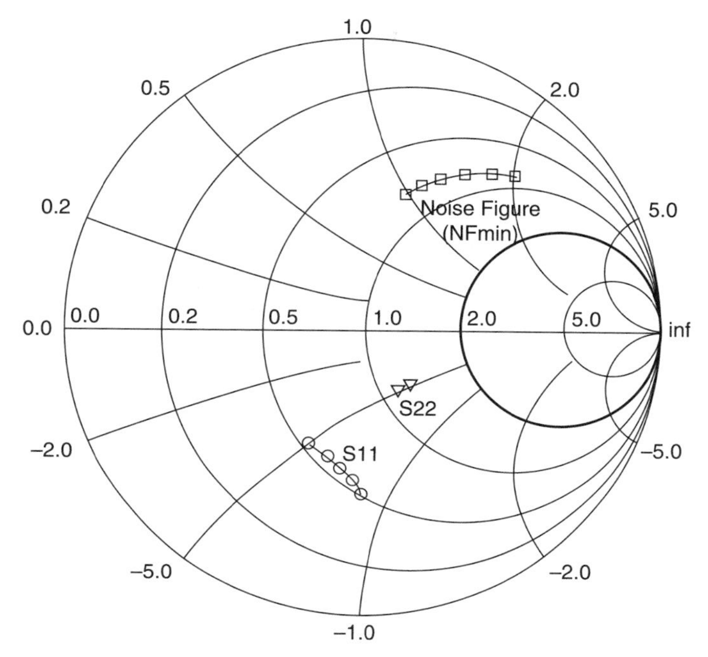

Figure 23.5 Input mismatching of amplifiers to reduce noise figure.

FREQ. (MHz)	NF_{OPT} (dB)	GA (dB)	Γ_{OPT}	
			MAG	ANG
$V_{DS} = 2V$, $I_{DS} = 10$ mA				
900	0.56	20.5	0.76	30
2000	0.63	16.3	0.61	41
2500	0.68	14.1	0.49	51
3000	0.70	13.6	0.39	49
3500	0.76	12.3	0.28	71
4000	0.82	11.6	0.20	80
$V_{DS} = 2V$, $I_{DS} = 30$ mA				
2000	0.60	17.0	0.56	39
2500	0.70	15.3	0.43	46
3000	0.76	14.2	0.32	50

Figure 23.6 Optimized matching for noise figure.

23.3 INTERMODULATION PRODUCTS

In a typical wireless communications system, the mobile unit is continually moving relative to the base station. The base station is sending out constant RF power, so that the received signal at the mobile unit is continually changing over a 90 dB range as the mobile moves. The noise figure of the LNA is the critical factor controlling system performance when the mobile unit is at the edge of the coverage cell, but as the mobile unit moves closer to the base station, the received signal is well above the noise level. When the mobile unit is midway between the edge of the cell and the base station, the S/N is 60 dB. However, as the mobile unit moves closer to the base station, the LNA faces another problem: intermodulation products.

The intermodulation products problem is illustrated by Figure 23.7, which shows the signals generated by the mobile unit's LNA when two signals are sent from the base station at two different frequencies intended for two different users. There are commonly more than two signals, but the problem is usually defined for the two tone case. Figure 23.7 shows the output power of the LNA when two equal signals are applied to its input. The top graph shows the output power when the two input signals are each at a power level of −33 dBm. The gain of the amplifier is about 21 dB, so the output powers for each of the two signals is −12 dBm.

The middle graph in Figure 23.7 shows the output power when each of the two input signals is raised by 10 dB to −23 dBm. Note the two additional frequencies that are generated by harmonic mixing of the two input signals. These additional frequencies are called IP3s, and they are caused by the mixing of the fundamental tone of one of the signals with the second harmonic tone of the other signal. The harmonic signal is generated as the amplifier is driven into saturation.

The bottom graph in Figure 23.7 shows the output power when each of the two input signals is raised another 10 dB to −13 dBm. IP3s have risen to a level that is only 15 dB below the fundamental tones, and many other frequencies are generated by higher order harmonic mixing.

The intermodulation products that occur adjacent to the two desired carrier signals are called IP3s because they are formed by mixing of the fundamental (first order) of one signal with the second harmonic (second order) of the other signal. This gives rise to the name third order because $1 + 2 = 3$. The pair of intermodulation products next to the third order products are called fifth order. They are formed by the mixing of the second harmonic of one fundamental signal with the third harmonic of the other fundamental signal. This gives rise to the name fifth order because $2 + 3 = 5$. Because IP3s are the largest, only they need to be considered, because if they are sufficiently smaller than the desired signal, all higher order intermodulation products will be even smaller.

The simple numerical example of Figure 23.8 shows how the IP3 of two signals can jam a nearby signal. Three different frequencies are shown being transmitted from the base station, which are labeled (1), (2), and (3). The individual voice channels are 30 kHz apart, so the signals are shown as the center frequency of the cell phone channel of 880 MHz +30, +60, and +90 kHz. The second harmonic of signal (2)

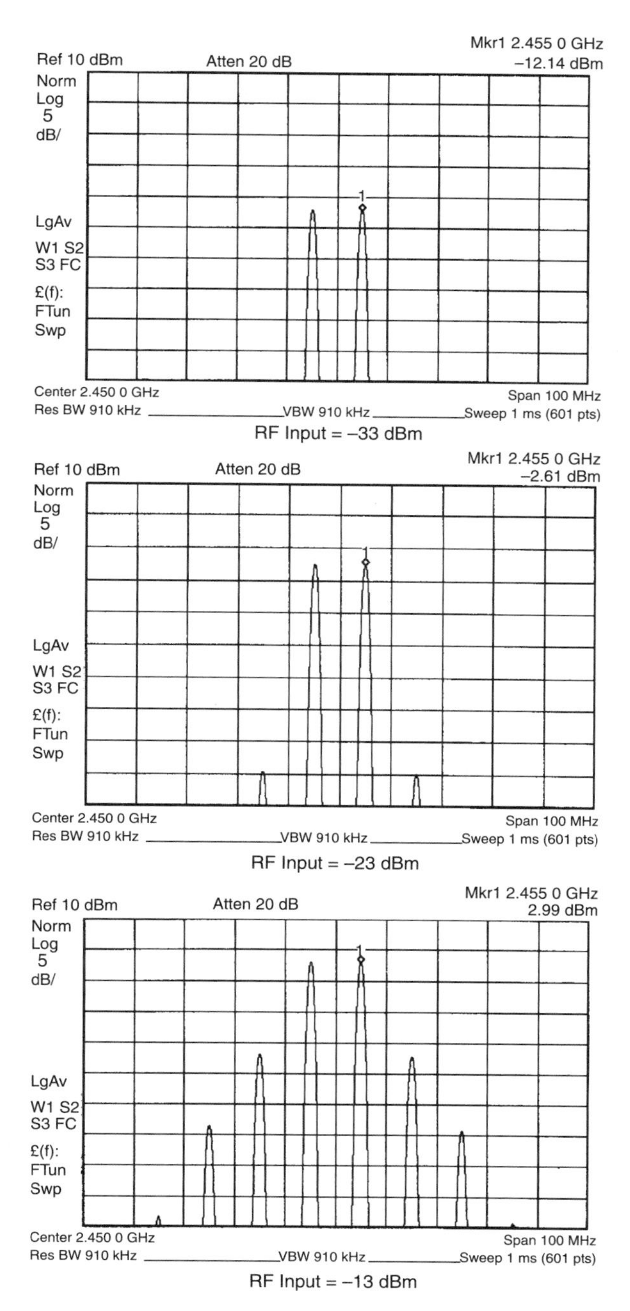

Figure 23.7 Two tone intermodulation products of LNA.

(1) 880 MHz + 30 kHz

(2) 880 MHz + 60 kHz

(3) 880 MHz + 90 kHz

(4)	2 × (880 MHz + 60 kHz)	=	880 MHz + 880 MHz + 120 kHz
	− (880 MHz + 90 kHz)	=	− 880 MHz − 90 kHz
(5)	Intermod Product	=	880 MHz + 30 kHz

Figure 23.8 Third-order intermodulation in a cellular phone receiver.

is shown as signal (4). When signal (3) is subtracted from it, an intermodulation product is created that is at the frequency of signal (1), and thus jams it.

Note that two different phone channels are mixing together to jam a third channel. The intermodulation products are false signals, but they have coherency like a human voice so that they must be 20 dB or further below the desired signal to not be heard.

Mixers also have an intermodulation problem. RF filters do not, because they are passive components that do not generate harmonics at the low power levels of RF receivers. RF communications systems like microwave relay and satellite communication systems, whose transmitters never get close to the receiver, do not have intermodulation product problems either.

The metric defining the intermodulation problem is called IP3. The larger this quantity is, the higher will be the input signal that can be applied to a receiver component without causing excessive intermodulation products.

Figure 23.9 provides a measurement of either one of the two fundamental frequencies and either one of the two IP3s as a function of fundamental RF input power. The fundamental curve has a slope of 1 : 1, whereas the third-order intermodulation curve has a slope of 3 : 1. Both curves show saturation effects, but their linear regions can be extended (dotted curves). The intersection of the linearized fundamental and third-order intermodulation curves is defined as the IP3. The values of this point can be expressed as either Input IP3 (IIP3) or Output IP3 (OIP3). These two IP3 values are related by the linear gain of the amplifier. The usefulness of the IP3 point is that it is a single number that allows the calculation of how far the IP3 power is below the fundamental power.

The OIP3 value can be used to determine how much power the mobile unit can receive before it is jammed by its intermodulation products. The maximum allowed input power formula is

$$P_{\text{in}}(\max) = \text{OIP3} - \frac{1}{2}\,\delta\,(\text{spec}) - \text{gain}$$

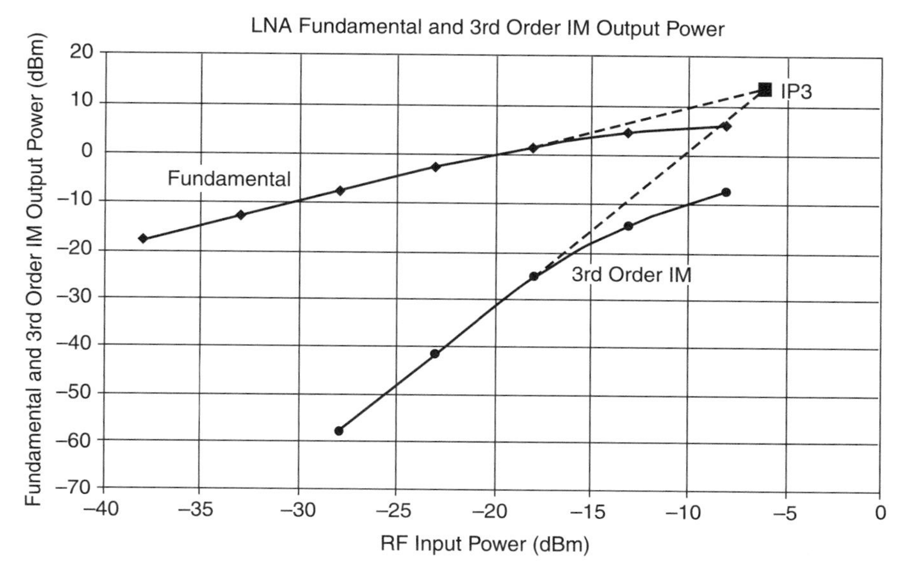

Figure 23.9 Calculation of IP3.

where δ (spec) is the required separation between the fundamental and intermodulation powers. As an example, if δ (spec) = 20 dB, gain = 20 dB, and OIP3 = 14 dBm,

$$P_{\text{in}}(\max) = 14\,\text{dBm} - \frac{1}{2} \times 20\,\text{dB} - 20\,\text{dB} = -16\ \text{dBm}$$

Gallium arsenide or silicon LNA transistors have an IP3 that is approximately 10 dB above their 1 dB compression point. Silicon germanium LNA transistors have an IP3 that is 15 dB above its 1 dB compression point, which makes silicon germanium a popular transistor type for LNAs.

The measurement of IP3s is discussed in Chapter 26.

23.4 S-PARAMETERS AND HOW THEY ARE USED

S-parameters completely describe the RF performance of any component when it is operated in the linear range. The performance of a complete RF system can be determined from the S-parameters of its component parts.

Figure 23.10 defines S-parameters. RF signals going into or coming out of the input port are labeled by a subscript 1. Signals going into or coming out of the output port are labeled by a subscript 2. The electric field of the RF signal going into the component ports is designated a, and that leaving the ports is designated b.

Therefore, S_{11} is the electric field leaving the input divided by the electric field entering the input, under the condition that no signal enters the output. Because b_1 and a_1 are electric fields, their ratio is a reflection coefficient.

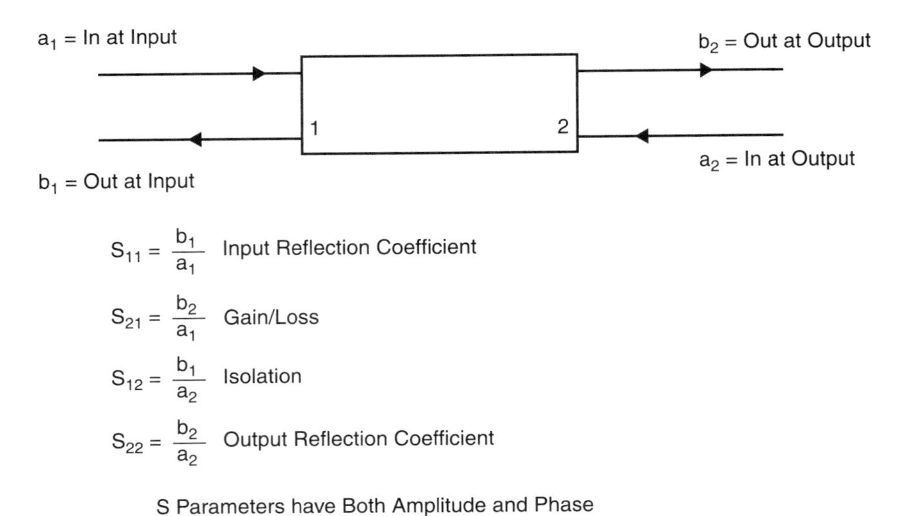

Figure 23.10 S-parameters. The performance specification of an RF component in terms of incident, reflected, and transmitted electric field.

Similarly, S_{21} is the electric field leaving the output divided by the electric field entering the input, when no signal enters the output. Therefore, S_{21} is a transmission coefficient and is related to the insertion loss or the gain of the component.

In like manner, S_{12} is a transmission coefficient related to the isolation of the component. It specifies how much power leaks back through the component in the wrong direction. Coefficient S_{22} is similar to S_{11}, but it is the mismatch looking in the other direction into the component.

It is important to realize that the most important of the S-parameters is S_{21}. It specifies how well the component is doing its intended job. The component was not purchased to reflect RF signals at its input or output ends or to have leakage in the reverse direction. Parameters S_{11}, S_{22}, and S_{12} are just parasitic performance features that tend to degrade performance.

Figure 23.11 presents a tabulation of typical S-parameters of an LNA from a manufacturer's data sheet. These S-parameters apply to operating conditions of $V_{DS} = 2$ V, $I_{DS} = 5$ mA. The data sheet contains several pages of S-parameters at other voltages and currents. The S-parameters are shown at frequencies from 0.5 to 8.0 GHz. Each S-parameter has both amplitude and phase.

The linear gain of a transistor can be calculated exactly from its S-parameters and knowledge of the input and output transmission line impedances. The calculation is complicated because both the amplitude and phase of the S-parameters and the transmission line mismatches must be considered, so the calculation usually requires the use of a CAD program.

However, an estimate of the limits of amplifier gain can be made easily using simplified formulas that involve only the amplitudes of S_{11}, S_{12}, S_{21}, and S_{22}. This calculation is useful to determine if a particular transistor is a good candidate to meet a particular performance requirement, before going through the complexity of a

VDS = 2 V, IDS = 5 mA

Frequency	S11		S21		S12		S22	
(GHz)	MAG	ANG	MAG	ANG	MAG	ANG	MAG	ANG
0.50	0.984	−15.1	4.945	165.0	0.020	80.6	0.807	−7.2
0.60	0.979	−18.0	4.908	162.3	0.023	78.9	0.803	−8.6
0.70	0.973	−21.0	4.899	159.4	0.027	77.0	0.798	−10.0
0.80	0.965	−23.9	4.871	156.7	0.031	75.3	0.793	−11.5
0.90	0.958	−26.8	4.843	153.9	0.034	73.7	0.788	−12.9
1.00	0.949	−29.8	4.825	151.1	0.038	72.1	0.781	−14.4
1.20	0.930	−35.7	4.783	145.6	0.045	68.7	0.767	−17.3
1.40	0.906	−41.5	4.723	140.2	0.052	65.4	0.751	−20.2
1.60	0.881	−47.5	4.660	134.7	0.058	62.2	0.734	−23.1
1.80	0.853	−53.6	4.605	129.3	0.064	59.1	0.715	−26.0
2.00	0.821	−59.8	4.531	123.8	0.070	56.0	0.696	−28.9
2.50	0.737	−76.3	4.332	110.5	0.082	48.2	0.648	−36.0
3.00	0.648	−94.2	4.092	97.6	0.092	41.4	0.600	−42.4
3.50	0.569	−113.6	3.805	85.3	0.098	35.3	0.556	−47.7
4.00	0.512	−133.0	3.516	73.9	0.102	30.5	0.518	−51.8
4.50	0.482	−150.9	3.248	63.8	0.105	27.2	0.480	−54.9
5.00	0.472	−165.2	3.025	54.7	0.108	25.3	0.444	−57.8
5.50	0.468	−175.7	2.846	46.4	0.112	24.5	0.405	−61.0
6.00	0.464	176.0	2.714	38.4	0.118	23.7	0.367	−65.4
6.50	0.456	167.9	2.601	30.5	0.126	22.4	0.331	−71.6
7.00	0.441	158.2	2.505	22.1	0.134	20.2	0.302	−80.8
7.50	0.422	144.3	2.417	13.3	0.142	18.0	0.283	−92.2
8.00	0.411	127.5	2.321	4.0	0.151	15.0	0.281	−105.9

Figure 23.11 LNA S-parameters.

complete vector algebra calculation. These simplified S-parameter formulas are shown in Figure 23.12.

The first equation in Figure 23.12 shows the gain of the unmatched transistor. If the transistor is not matched to the transmission line, some of the input power will not get in and some of the amplified power will not get out, so the gain will be low. The unmatched gain is the lower limit on the amplifier gain.

The second equation in Figure 23.12 shows the gain of the transistor if it is perfectly matched. This is the maximum available gain, and thus represents the upper limit on amplifier gain.

The actual gain will be between the unmatched gain and the maximum available gain, depending on the exact impedance of the input and output match. Note that the amplitude and phase of these impedances, as well as the S-parameters, vary with frequency.

Because some power leaks through the transistor in the reverse direction, as specified by S_{12}, the transistor may become unstable and oscillate as the match is optimized to achieve the maximum available gain. The third equation in Figure 23.12 shows the maximum stable gain, which defines the maximum gain that can be achieved without oscillation. If the maximum stable gain is greater than the maximum available gain, then the maximum available gain can be achieved without oscillation.

If the maximum stable gain is less than the maximum available gain, then the maximum available gain cannot be achieved without oscillation. The maximum gain that can be achieved is the maximum stable gain, and the transistor must be carefully mismatched so that oscillation does not occur.

$$\text{Gain of Unmatched Transistor} = \left| S_{21} \right|^2$$

$$\text{Unilateral Gain with Matched Transistor (Approximate MAG)} = \frac{\left| S_{21} \right|^2}{\left(1 - \left| S_{11} \right|^2\right)\left(1 - \left| S_{22} \right|\right)^2}$$

$$\text{Maximum Stable Gain (MSG)} = \frac{\left| S_{21} \right|}{\left| S_{12} \right|}$$

Example

$\left| S_{11} \right|$	= .60	MAG (Approx)	=	10.0 dB
$\left| S_{21} \right|$	= 2.34	MSG	=	13.7 dB
$\left| S_{12} \right|$	= .10	Unmatched Gain	=	7.38 dB
$\left| S_{22} \right|$	= .39			

Figure 23.12 Formulas for estimating amplifer gain and stability.

All three gains must be calculated at the desired frequency of operation. In addition, the maximum stable gain needs to be calculated over a range of frequencies, because the transistor may oscillate at a frequency other than the desired operating frequency. An example calculation is shown at the bottom of Figure 23.12. The left-hand column shows the amplitudes of the four S-parameters. The right-hand column shows the three gains calculated with these parameters. The unmatched gain is 7.38 dB. With perfect matching, the gain can be increased to 10.0 dB. The transistor will not oscillate until the gain is increased to 13.7 dB, which cannot be achieved. Therefore, the transistor is stable, at least at this frequency.

23.5 EXAMPLE LNA MEASUREMENT ON THE VNA

Objective

Our objective is to measure the performance of the LNA on the VNA (Fig. 23.13). We will determine the 1 dB compression point and view the S-parameters of the amplifier. Note that because the amplifier is on a demo board, we are measuring the S-parameters of the transistor, package, and matching network as well as traces and connectors for the DUT.

We find the 1 dB compression point by two different means. First, we look at gain and output power versus frequency (Fig. 23.14). We manually increase the input power until we observe the gain drop by 1 dB and note the input power level. Second, we set up the network analyzer to sweep the input power at a single, fixed frequency of 2.45 GHz. We can then read the 1 dB compression point from the S_{21}

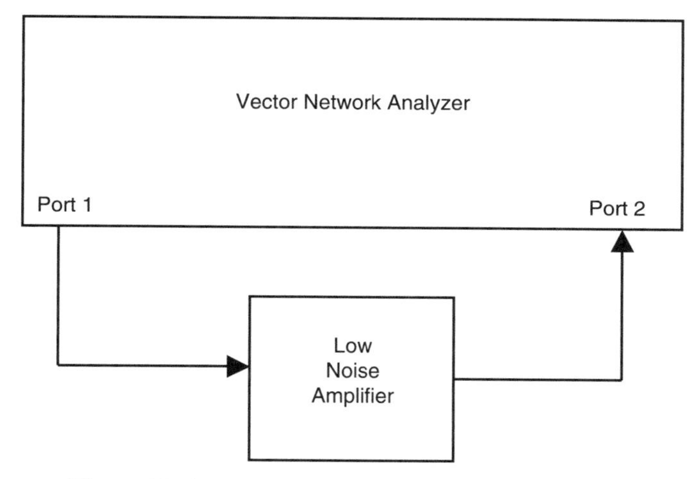

Figure 23.13 Test setup for swept frequency and power.

trace. We finish by looking at the S-parameters in polar form and on the Smith Chart (Figs. 23.15 to 23.19).

Measurements Being Demonstrated

Gain (S_{21}) versus frequency
Output power versus frequency
Gain (S_{21}) versus input power (swept power)
Output power versus input power
S-parameters: S_{11}, S_{22}, S_{21}, S_{12}

Specifications of LNA

The specifications of the LNA to be tested are listed in the following table:

Manufacturer	Triquint
Model	TQ9207
Frequency	2.450 GHz
Linear gain	21 dB
P1 dB	3 dBm
Noise figure	4.5 dBm
OIP3	13 dBm

Generic Procedure

The steps to perform the measurements are listed in the following table. The left column lists the actions to be performed, and the right column lists the

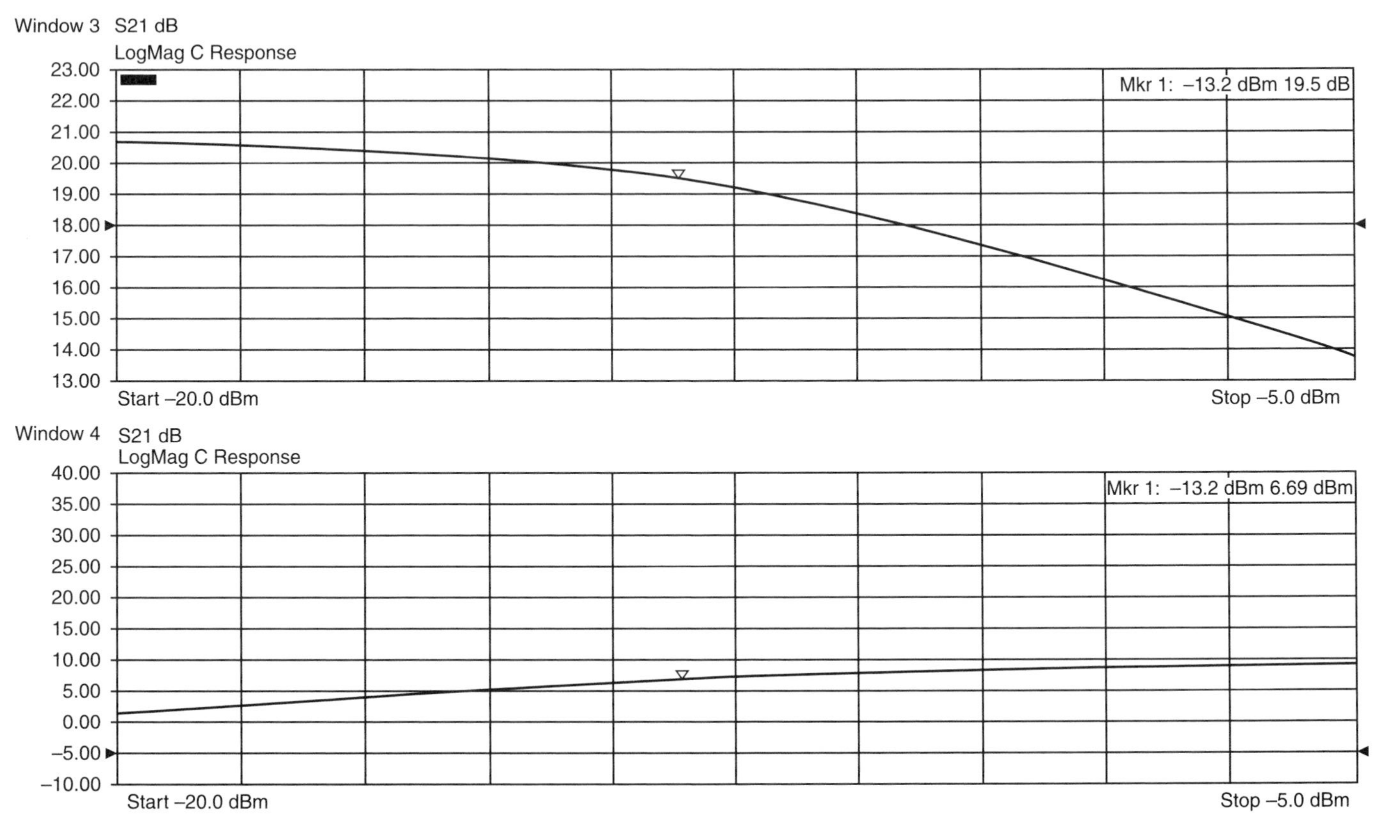

Figure 23.14 LNA gain and power versus input power.

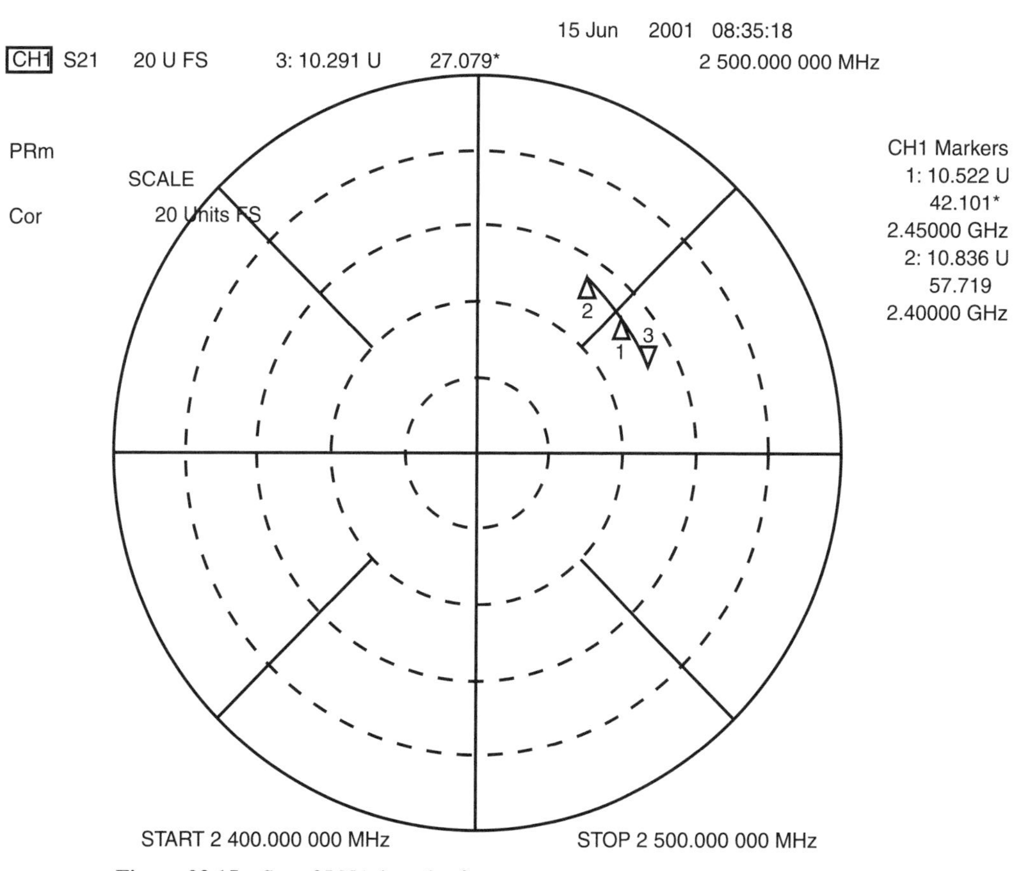

Figure 23.15 S_{21} of LNA in polar format.

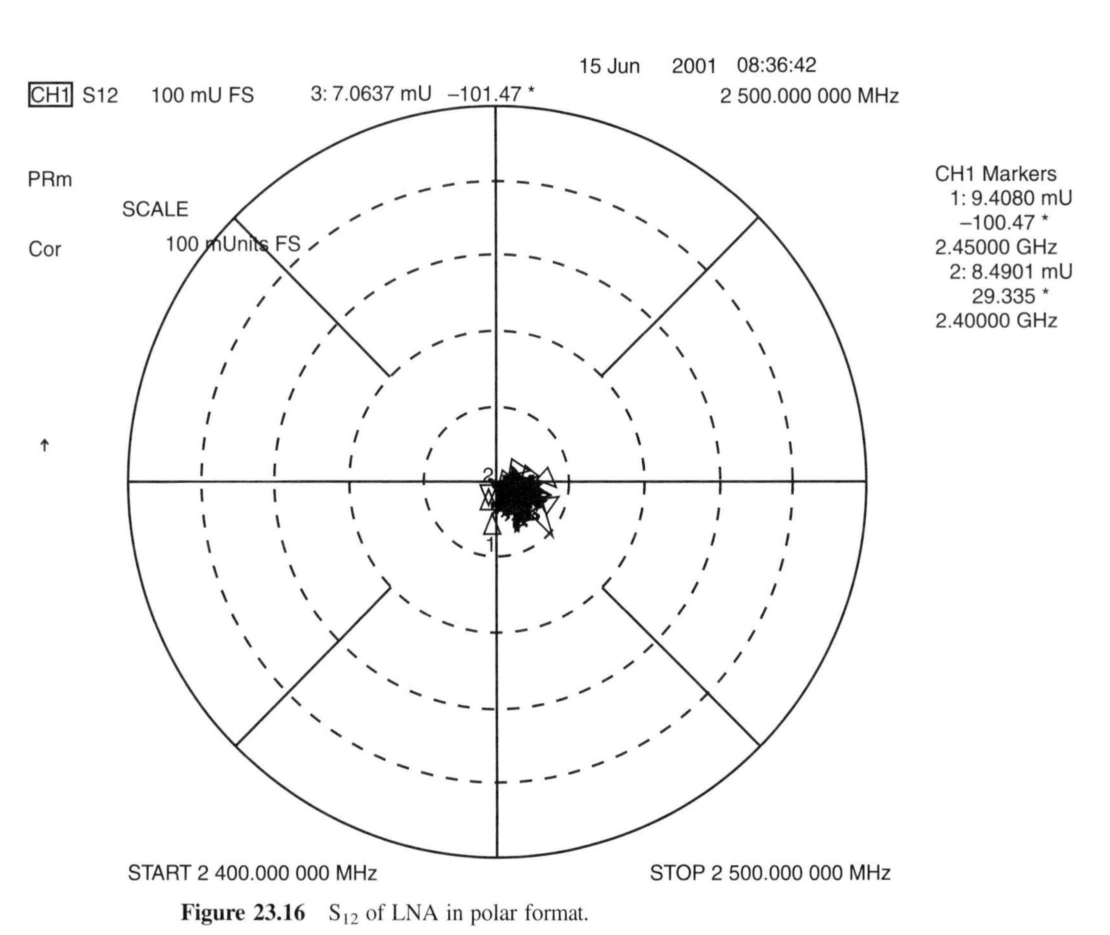

Figure 23.16 S_{12} of LNA in polar format.

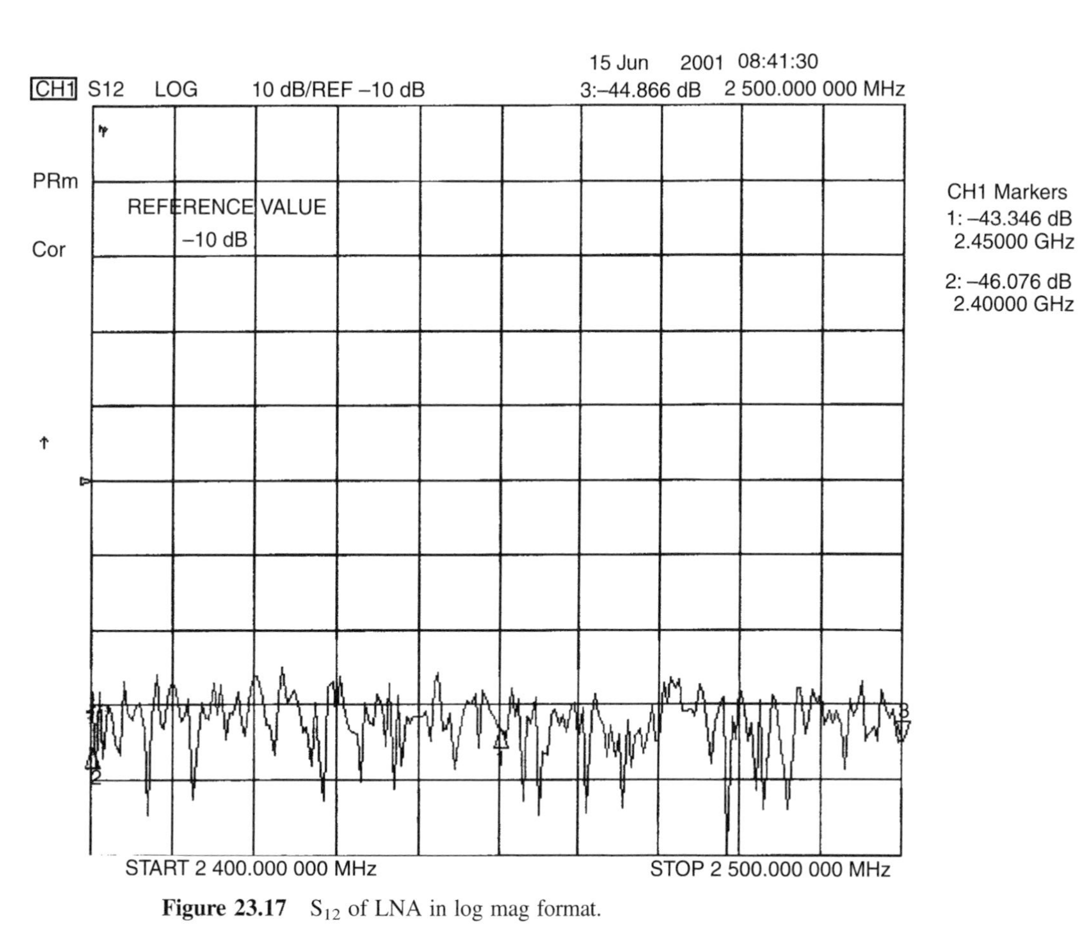

Figure 23.17 S_{12} of LNA in log mag format.

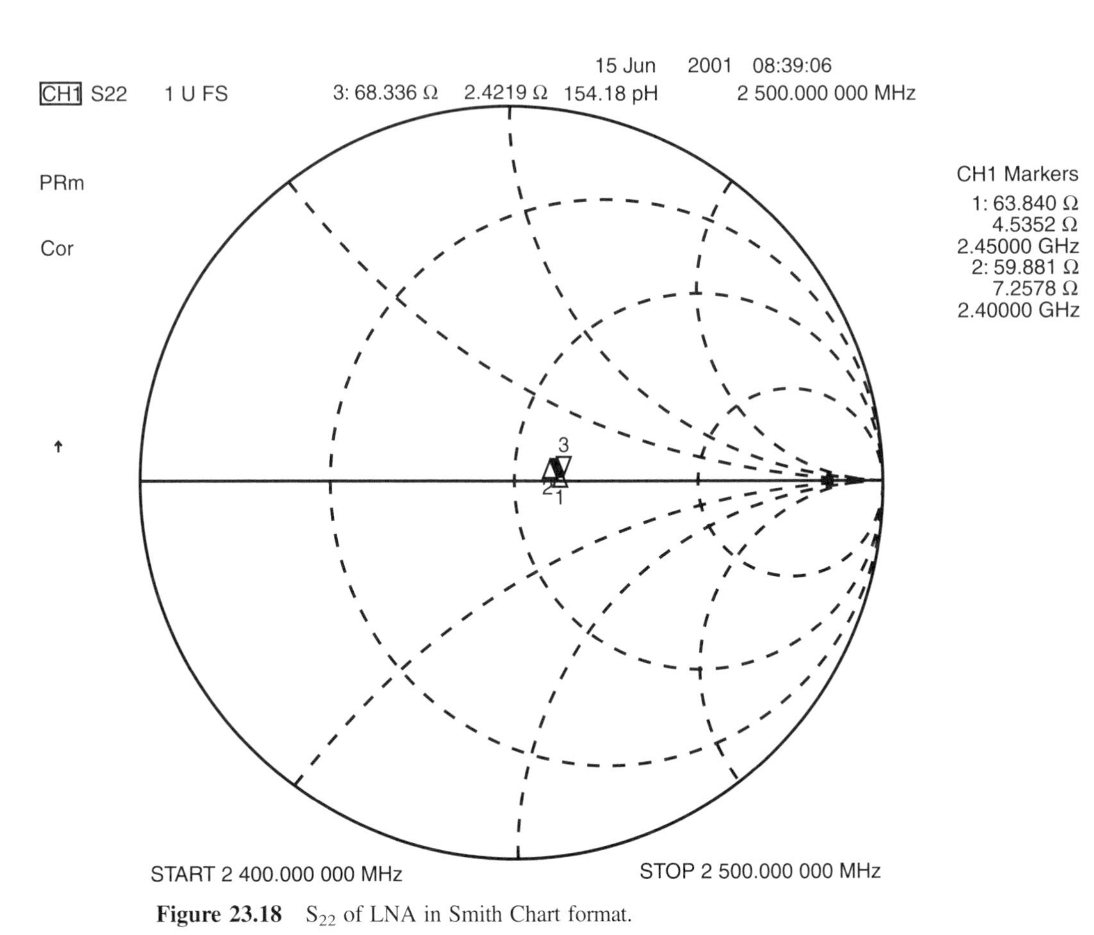

Figure 23.18 S_{22} of LNA in Smith Chart format.

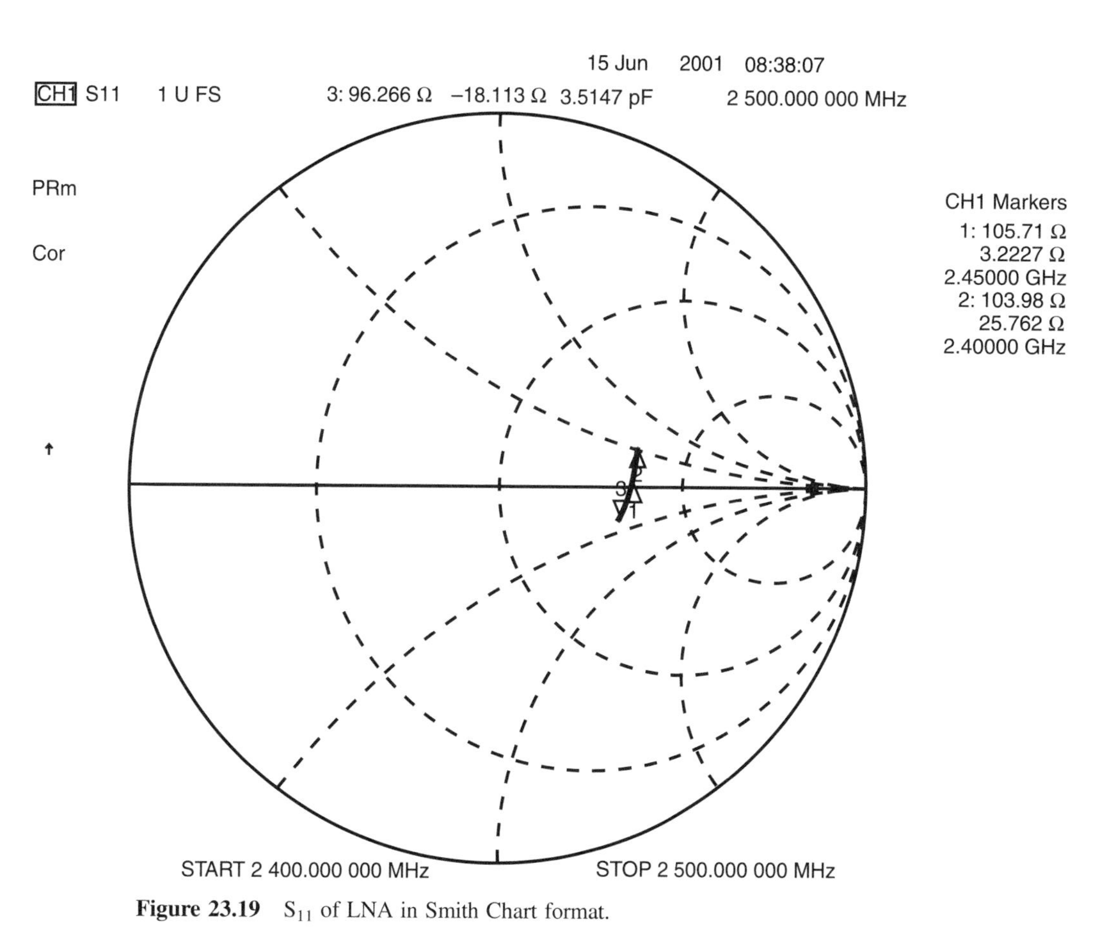

Figure 23.19 S_{11} of LNA in Smith Chart format.

motivation for taking each step and provides additional background information in some cases.

Procedure for Network Analyzer	Notes
Preset the instrument.	
Measurement setup and calibration	
Measure S_{21}.	This will measure the gain of the device.
Set the start and stop frequencies to 2 and 3 GHz, respectively.	
Set the power level of the source to −30 dBm.	
Perform a full two-port calibration.	In order to accurately measure all S-parameters, a full two-port calibration is required.
Verify that the calibration worked. Connect a short or open to the cable on port 1 and measure S_{11}. Repeat on port 2 and measure S_{22}.	The return loss should be zero.
Activate another trace and set it to measure the input at the B receiver. This trace should share the same "channel" (sweep and source power settings) as the measurement that was just set up.	This will measure the power at the output of the LNA. This is a nonratioed, power meter style measurement.
Perform a receiver calibration with a thru connection between cables.	The receiver calibration ensures that the power level produced by the source is read as the same power level by the receiver in the VNA. To achieve the same accuracy as a power meter reading, a source power calibration using a power meter connected to port 1 of the VNA should be performed before this step. This example does not demonstrate power meter accuracy for this measurement.
Connect the LNA as the DUT between ports 1 and 2.	
View the gain of the LNA on the S_{21} measurement. Set the scale per division for the S_{21} measurement to 1 dB and the reference value to 20 dB. Set a marker at 2.45 GHz. Set a marker at the same frequency on the B measurement. Press the Power key to display the source power setting on the screen. Note that the input power (which you can read from the source power setting) plus the gain (S_{21}) should equal the output power (B).	"Coupled" markers move together so that if a marker is adjusted in one measurement, the marker moves in the other measurements as well.

(*Continued*)

Procedure for Network Analyzer	Notes
Increase the input power until the gain drops by 1 dB. This is the 1 dB compression point. Note the input power level.	
Switch to a power sweep. Select the sweep type menu and change the sweep type to a power sweep. Set the start power to −30 dBm and the stop power to −5 dBm. Set the CW frequency to 2.45 GHz. After setting the CW frequency, check that the calibration is on for both measurements. On newer models set the attenuation to 10 dB.	On older VNA models the start and stop power levels for the sweep are set with the start and stop hard keys. When switching from a frequency sweep to a power sweep, many VNA models set the default CW frequency to 1.0 GHz. This default frequency is outside of the range where the calibration was performed and causes the calibration to be turned off. The calibration cannot be turned on until the CW frequency is set to 2.45 GHz, which is in the range where the calibrations were performed. When the VNA source is set to produce a very small signal, the VNA actually produces a larger signal and then attenuates it just before sending it out of port 1. This is done so that the reference receiver can operate at a level with good sensitivity. The range of power that can be swept is limited by reference receiver sensitivity on the low end and maximum source output power on the high end. Recommended power ranges are provided either directly in the power menu on older models or in the online help on newer models.
With a scale of 1 dB we can simply observe where the S_{21} trace drops by one division. This is the 1 dB compression point. You can also set a marker at the left edge of the trace and then use a marker delta to find where the signal is down by 1 dB.	The swept power curves give a good description of the amplifier's power performance at a single frequency.
Observe the S-parameters.	
Maximize the S_{21} plot to use the full screen. Switch the sweep type back to a frequency sweep from 2 to 3 GHz. Switch the format to a polar display.	A polar display will provide magnitude and angle information about the measured S-parameter. The information is reported either as magnitude and angle or real and imaginary components. For active devices the S_{21} magnitude should be >1.
Switch to measure S_{12}.	The reverse transmission will likely be very small. If you autoscale the measurement,

(Continued)

Procedure for Network Analyzer	Notes
	the signal will appear very noisy because of the small magnitude.
Measure S_{11}. Change the format to Smith Chart display.	Note that an LNA is designed with an intentional mismatch on the input. The S_{11} parameter will not be as near the center of the Smith Chart as is the S_{22} parameter.
Measure S_{22} on the Smith Chart.	

On newer model VNAs, the swept frequency and swept power measurements are assigned to two different "channels." These are both set up and calibrated separately before the DUT is connected.

23.6 ANNOTATED BIBLIOGRAPHY

1. Agilent Application Note 1354: Practical Noise Figure Measurement and Analysis for Low-Noise Amplifier Designs, Agilent Technologies, Santa Clara, CA.
2. Besser, L. and Gilmore, R., Linear and low noise RF amplifiers, In Practical RF Circuit Design for Modern Wireless Systems, Vol. II: Active Circuits, Artech House, Norwood, MA, 2003.

Reference 1 describes matching of the LNA for minimum noise figure or maximum small signal gain.

Reference 2 provides detailed background on the procedure to design LNAs.

CHAPTER 24

MIXERS

This chapter is organized into five sections. Basic mixer performance is described in Section 24.1. The selection of individual voice or data channels is discussed in Section 24.2. The removal of image noise is covered in Section 24.3. ZIF receivers are explained in Section 24.4. Finally, mixer measurements are described in Section 24.5.

24.1 BASIC MIXER PERFORMANCE

Every RF receiver uses a mixer to shift the received carrier and its modulation side bands out of the RF band to a lower frequency, where it is easier and more economical to achieve the required gain, and to select individual voice or data channels.

A block diagram of a mixer is provided in Figure 24.1. The RF signal is at 1.0 GHz and it has a 10 MHz bandwidth (upper left sketch), which means that the signal and its modulation side bands are in the frequency range from 0.995 to 1.005 GHz. The modulation side bands carry the information being transmitted by the RF carrier. RF signals at other frequencies may have entered the antenna along with the desired signal, but these other signals have been filtered out before the signal reaches the mixer. The RF signal enters a combiner that contains Schottky mixer diodes that were described in Chapter 16.

A Local Oscillator (LO) signal, generated in the receiver, also enters the combiner. It is offset in frequency from the input RF signal by the center frequency of the IF amplifier. In this example, 70 MHz has been chosen as the IF frequency. The LO frequency is thus at 1.07 GHz.

RF Measurements for Cellular Phones and Wireless Data Systems. By A. W. Scott and R. Frobenius

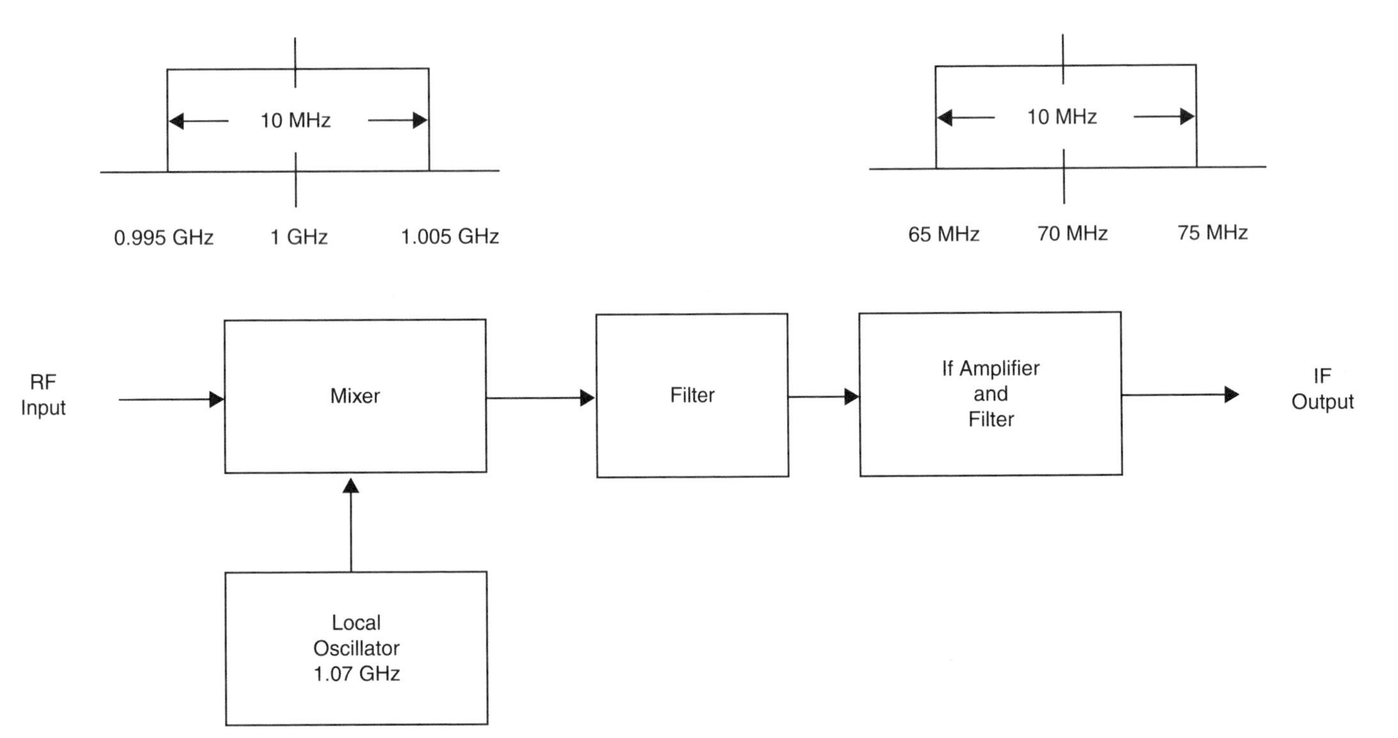

Figure 24.1 Mixers convert the RF signal to a lower frequency below the RF band where it can be more easily amplified and filtered.

The input RF signal enters one arm of the combiner, and the LO signal enters through another arm. These signals are combined and appear at the other arms of the combiner, but they are still at separate frequencies. The mixer diodes are placed in these other arms and mix the input RF and LO signals. The frequencies leaving the mixer include the original RF input signal; the LO signal; a frequency that is the sum of the RF and LO signals, which in this example is 2.07 GHz; and the IF, which is the difference between the RF and LO signal, which in this example is 0.07 GHz or 70 MHz.

All signals except the IF are filtered out, and the IF signal is then amplified by a 70 MHz IF amplifier. As noted, it is much easier to design a 70 MHz amplifier than a 1 GHz amplifier, which is why the carrier signal is converted from 10 GHz to 70 MHz. In the upper right sketch of Figure 24.1 the 10 MHz bandwidth containing the modulation side bands is shifted into the 70 MHz band. The side band of the modulated RF carrier at 0.995 GHz would be shifted to 75 MHz, the side band of the modulated RF carrier at 1.005 GHz would be shifted to 65 MHz, and so on.

Mixer action is explained in Figure 24.2. Different numerical values are used in this example to make the waveform easier to draw. The LO signal (LO), received

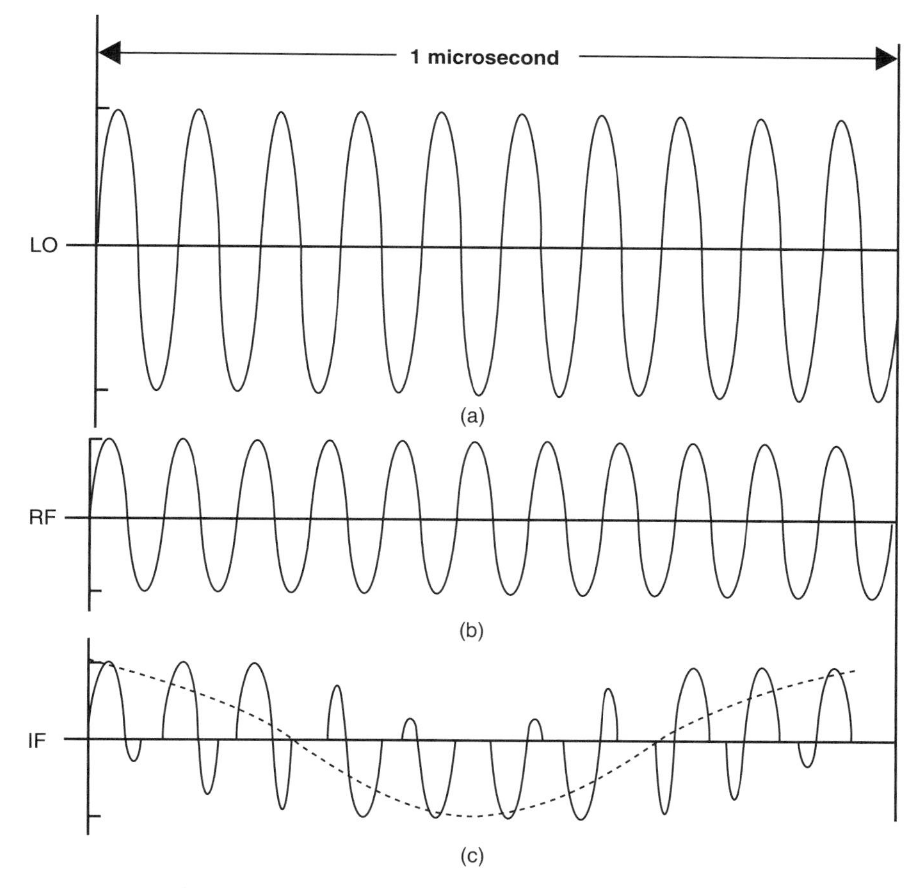

Figure 24.2 How mixers generate the difference frequency.

RF signal (RF), and output of the mixer (IF) are plotted as functions of time during an interval of 1 μs. The LO amplitude is normally more than 10 times greater than the amplitude of the RF and IF output signals, but to make the figure easier to read the amplitudes are not drawn in this way in Figure 24.2.

The LO frequency is 10 MHz, so there are 10 LO cycles during the 1 μs time interval. The RF frequency is 11 MHz, so there are 11 RF cycles during the 1 μs time interval.

The LO signal is large, and when it is applied to the mixer diodes it biases the diodes off during its negative cycle and turns the diodes on during its positive cycle. Therefore, the RF signal can pass through the diodes when the LO signal is positive, but it cannot pass through the diodes when the LO is negative. As shown on the left-hand edge of the figure, all of the positive half of the RF cycle and about one-tenth of the negative cycle of the first RF signal passes through the diode. As time progresses, the RF and LO curves get more and more out of time synchronism, so less of the positive half of the RF wave and more of the negative half gets through the diode. The output of the mixer (Fig. 24.2c) is a complicated waveform containing many frequency components. However, as shown by the dashed envelope (Fig. 24.2c), one frequency component is one cycle in the 1 μs time period, and so is 1 MHz. This frequency is the difference between the 10 MHz LO and 11 MHz RF signals. The output from the mixer is passed through a low pass filter that removes the high frequency components and leaves only the 1 MHz IF signal.

Schematics of various mixer types are shown in Figure 24.3. The major difference between the types is the combiner that is used and the number of diodes. The single-ended mixer is the simplest type, but its performance is poor in most specifications.

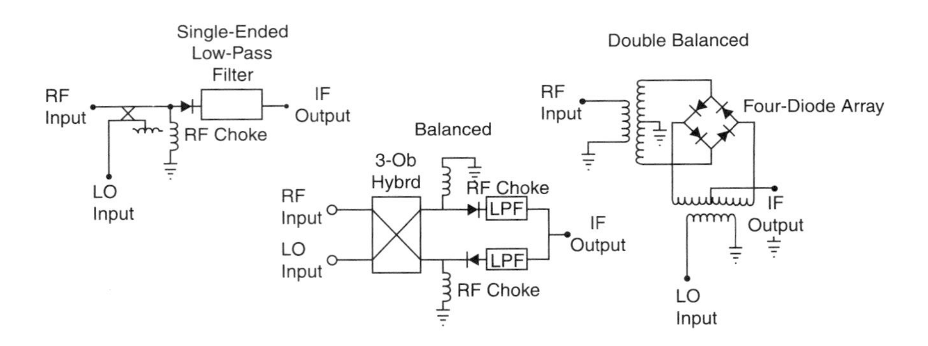

Mixer Type	Conversion Loss	VSWR LO, RF	LO/RF Isolation	Spurious Rejection	Harmonic Suppression
Single-Ended	GOOD	GOOD, POOR	FAIR	POOR	POOR
Balanced (90°)	GOOD	GOOD, GOOD	POOR	FAIR	FAIR
Balanced (180°)	GOOD	FAIR, FAIR	VERY GOOD	FAIR	GOOD
Double-Balanced	VERY GOOD	POOR, POOR	VERY GOOD	GOOD	VERY GOOD

Figure 24.3 Mixer types.

The combiner used in the single-ended mixer is a directional coupler, with the RF input being applied to the main arm of the coupler and the LO input entering through the coupled port. The combined signals are then mixed in a single diode, and a low pass filter removes the LO, signal, and sum frequency, allowing only the IF output to go to the IF amplifier.

The balanced mixer uses a 3 dB hybrid to supply the input RF and LO power to two mixer diodes. A 90° or a 180° hybrid can be used, and different performance trade-offs are accomplished, depending on which hybrid is used. The double balanced mixer is conceptually two balanced mixers. It uses a four diode array.

In the lower part of the RF band from 300 MHz to 3 GHz, the combiners are made of miniature wire wound transformers.

The table in the lower part of Figure 24.3 compares the performance of the different mixer types. Note that the double balanced mixer with its four diodes has the best conversion loss, LO/RF isolation, spurious suppression, and harmonic suppression of any mixer type.

24.2 SELECTION OF INDIVIDUAL VOICE AND DATA CHANNELS

Most wireless communications systems (with the exception of CDMA systems described in Chap. 30) serve multiple users by assigning a separate range of RF frequencies within their total licensed bandwidth to each user. The individual user channel is separated out of the total licensed RF frequency band by the mixer.

Figure 24.4a is a block diagram of the complete RF receiver. The numerical values shown are for the original U.S. analog cellular phone system. (Later this system was upgraded to become the IS-136 digital system using TDMA.) The downlink half of the system is shown for transmission from the base station to the mobile units. Each

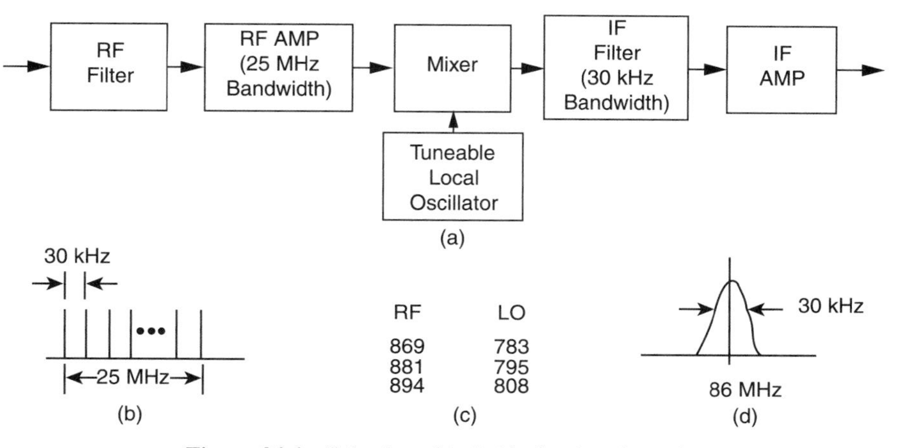

Figure 24.4 Selection of individual voice channels.

user RF channel is 30 kHz wide. The downlink RF bandwidth is 869–894 MHz, which means that 832 analog users can be accommodated (Fig. 24.4b).

The 832 received user channels are conducted through the RF filter and the LNA. There is no practical way to select one 30 kHz wide user channel out of the 869–894 MHz frequency range, especially considering that a different 30 kHz wide channel is selected for each call.

Figure 24.4c shows three user channels: one at the lower end of the downlink frequency range (869 MHz), one at the center of the downlink band (881 MHz), and one at the upper end of the downlink band (894 MHz). When a call is being sent out, the base station notifies the mobile unit which of the 832 channels should be used.

Assume that the base station notifies the mobile that the incoming call should be on channel 1 at 869 MHz. Automatically the mobile unit sets its LO frequency to 783 MHz. The difference frequency coming out of the mixer is therefore 86 MHz, which passes through the 30 kHz bandwidth of the IF filter. The frequency response of this IF SAW filter is shown in Figure 24.3d. All of the other 831 RF channels intended for the other users are shifted down to IF by the mixer, but only the desired signal is shifted down to the correct 30 kHz channel centered at 869 MHz and then amplified by the IF amplifier, which is tuned to a 30 kHz bandwidth centered at 86 MHz. When the call is completed and the mobile unit is disconnected, this 30 kHz wide channel at 869 MHz is assigned to another user.

When the original user wants to make another call, he is assigned another available frequency, for example, 881 MHz in the center of the band. Automatically his LO is set for 795 MHz, and the new incoming RF signal is shifted to 86 MHz and passes through the 86 MHz IF filter.

24.3 THE REMOVAL OF IMAGE NOISE

Figure 24.5 shows the problem of image noise and its solution by the use of an image noise filter. In this example the signal is at 1 GHz and the LO is at 1.1 GHz. The RF signal and the LO signal mix, and the difference frequency is at the desired IF of 0.1 GHz = 100 MHz. If an additional unwanted signal at 1.2 GHz enters the

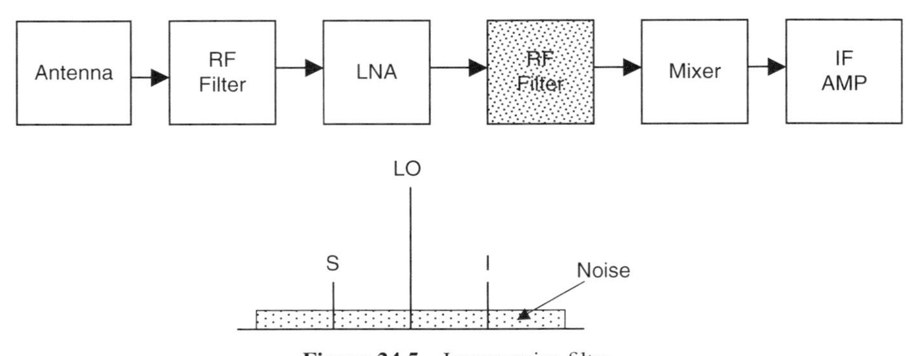

Figure 24.5 Image noise filter.

mixer, it also mixes with the LO to form an IF signal at 100 MHz. If the receiver using the mixer covers only a narrow RF frequency range, the image frequency would be eliminated by the input RF filter of the receiver and would not be a problem.

Note, however, that there is noise at all frequencies in the receiver and this noise is not affected by the RF filter that is at the receiver input. This noise is shown by the shaded bar in Figure 24.5. This image noise is shifted down to the IF frequency and doubles the noise in the IF band that competes with the signal.

The solution to the image noise problem is to add another RF filter, which is similar to the input RF filter, between the LNA and the mixer, as shown.

24.4 ZIF MIXER

The current trend in RF receivers for all mobile units and many base stations is to use a ZIF mixer. The ZIF mixer is also called a ZeroIF or direct conversion mixer. In a ZIF mixer, the LO frequency is equal to the RF frequency, so the difference frequency is approximately DC. Filtering of the desired channel is done with digital processing. A ZIF mixer eliminates the need for an expensive SAW filter and a high gain IF amplifier, and thus reduces component cost and parts count. Figure 24.6 is a photograph of a conventional Bluetooth receiver compared to a ZIF receiver for the same application. The advantages of smaller size and parts count are clearly evident.

In a wireless voice or data communications system, modulation is represented by the phase and amplitude of the RF wave, as illustrated in Figure 24.7. The phase information would be lost if the RF signal was mixed to zero frequency. To solve this problem, the RF signal vector representing the phase and amplitude is analyzed into its in phase (I) component and its out of phase (Q) component, and then both signals are mixed separately to near zero frequency. The I and Q vectors at near DC frequency are then recombined to get the original phase and amplitude vector

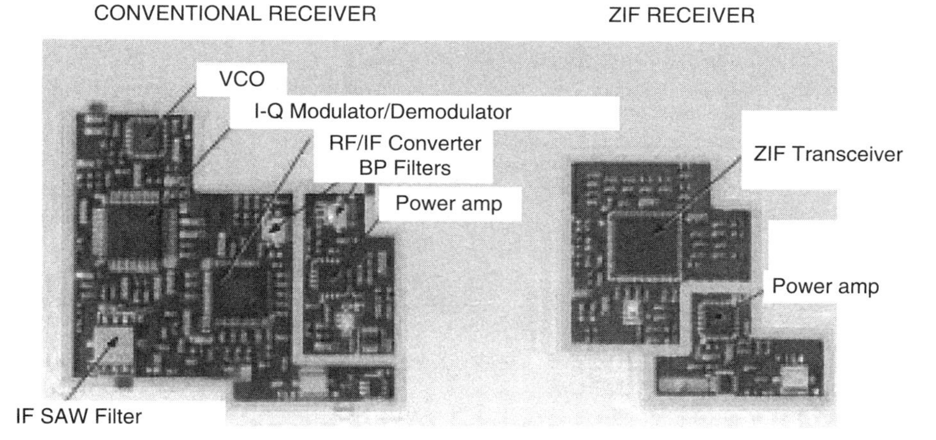

Figure 24.6 ZIF receiver size comparison.

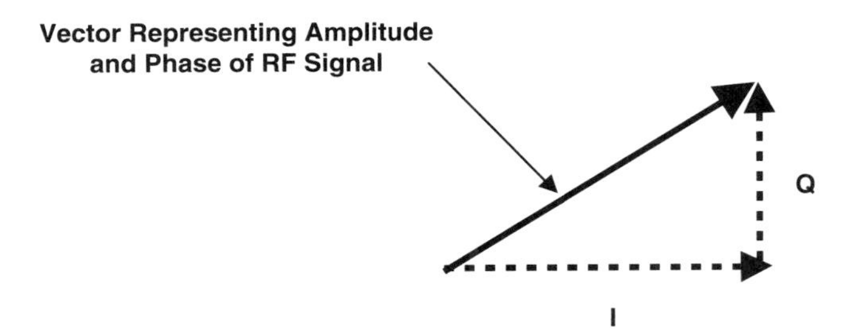

Figure 24.7 I and Q channels for preservation of phase information.

information. Care must be taken to ensure that the gain in the I and Q channels are exactly equal.

ZIF mixers have been used for over 25 years in AM and FM radios and in television receivers. ZIF mixers for RF receivers are more challenging. The main two problems for RF ZIF mixers are I–Q imbalance and DC offset. As shown in Figure 24.7, the phase and amplitude of the digital modulation is carried by the amplitude of the I and Q channels. If the channel gain is not identical between the two channels, the phase and amplitude information will be distorted. This problem is solved by careful circuit design and manufacture.

A much more difficult problem is that of DC offset. In a ZIF receiver, the LO signal is at the same frequency as the RF, and it can be as much as 20 dB larger. Any LO signal that leaks out of the mixer and gets into the RF path can be mistaken for an RF signal. Even a small leakage of the LO signal will be greater than the small RF signal.

Figure 24.8 shows the many mechanisms for LO leakage. Figure 24.8a shows LO leakage paths inside the receiver. The LO signal can be conducted to the RF input port because the semiconductor substrate used to build the mixer is not a perfect insulator. The LO signal at the mixer input can be reflected back into the mixer by the output mismatch of the LNA (S_{22}); because it is at the same RF frequency as the input RF, it gets mixed down to zero IF.

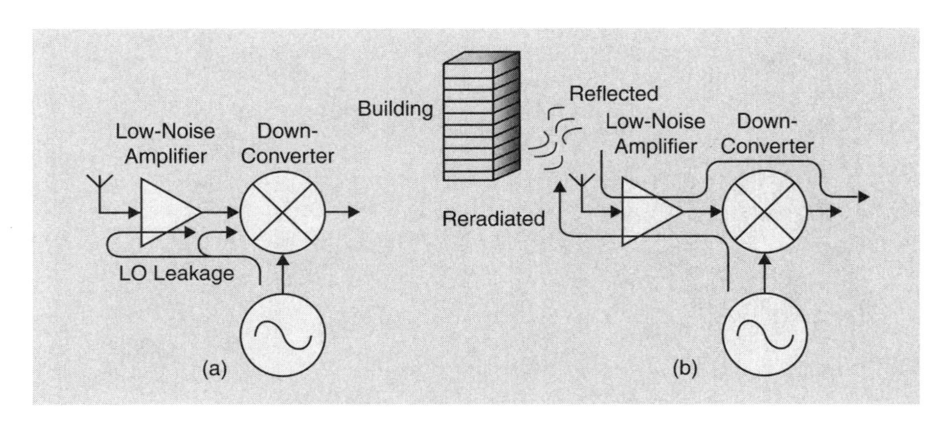

Figure 24.8 DC offset mechanisms.

Alternately, the LO leakage can travel through the LNA in the reverse direction. The amount that gets through is controlled by the S_{12} of the LNA. The LO signal can be reflected at the LNA input as determined by the S_{11} of the LNA and then amplified as it travels back through the LNA. In addition, the LO signal that gets out of the LNA input can be reflected by the duplexer or the mobile unit antenna mismatch and then amplified as it comes back into the LNA.

Another problem (Fig. 24.8b) is when the LO leaks back out through the receiving antenna and is reflected off a large surface (like a nearby building) back into the mobile unit.

All of these reflections cause a false signal that competes with the received RF signal. They can easily be much larger than the received RF signal, which may have traveled a long way to the mobile receiver.

All of the above LO reflections occur in a conventional heterodyning mixer, but they do not cause a problem because the reflected LO power is not mistaken for the RF signal because it is at a different frequency.

Ways to minimize the DC offset mechanisms are

Careful shielding and layout techniques
Selection of a high directivity combiner
Minimizing mismatches of the LNA input and output
Reduction of the LO signal when the RF input signal is small

Other problems with ZIF mixers are

Second-order distortion effects, which occur when large signals at any frequency mix with themselves and form a ZIF frequency
Desensitization of the receiver by large interfering signals
$1/f$ noise, which is noise from man-made effects at frequencies below the normal RF thermal noise

In spite of all the complexities of designing and manufacturing ZIF mixers, they have become the major type of mixer used in wireless RF receivers of the future, because of their advantages of low cost and small size.

24.5 MIXER MEASUREMENTS

Measurements that need to be performed on mixers are

Linear gain versus frequency
1 dB compression point
Input and output mismatch
Noise figure
IP3

The specifications for the particular mixer being tested are

Manufacturer	Triquint
Model	TQ5M31
RF frequency	2.450 GHz
IF frequency	250 MHz
LO power	7 dBm
Conversion gain	2.0 dB
P1 dB	1 dBm
Noise figure	7.0 dB
OIP3	10.0 dBm

This mixer is a conventional mixer with a 250 MHz IF.

The linear gain of a mixer is usually low, only a few decibels. The purpose of the mixer is to change the frequency of the information carrying signal from RF to IF, where achieving the desired gain and selectivity is easy. However, knowledge of the mixer gain is necessary so that the gain requirements of the IF amplifier can be determined.

An important RF characteristic of a mixer is its 1 dB compression point, where significant nonlinear effects begin to occur. These nonlinear effects create many spurious signals that should be avoided.

In an amplifier, the linear gain and 1 dB compression can be measured easily over a range of frequencies with a VNA. This measurement is not possible on a mixer with a standard VNA, because the input and output signals are at different frequencies. (Some VNAs have special modifications to make swept mixer measurements over a limited range of input power.)

The easiest way to measure the linear gain and 1 dB compression point of a mixer is to use a signal generator and a spectrum analyzer. A test setup for making this measurement is provided in Figure 24.9. The same setup was used for the upconverter measurement of Chapter 18 except that the input signal is an RF signal in mixer

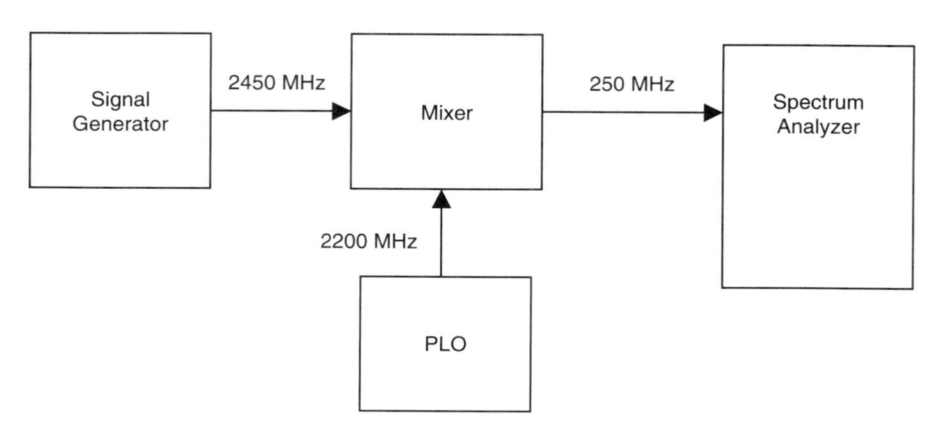

Figure 24.9 Mixer test setup.

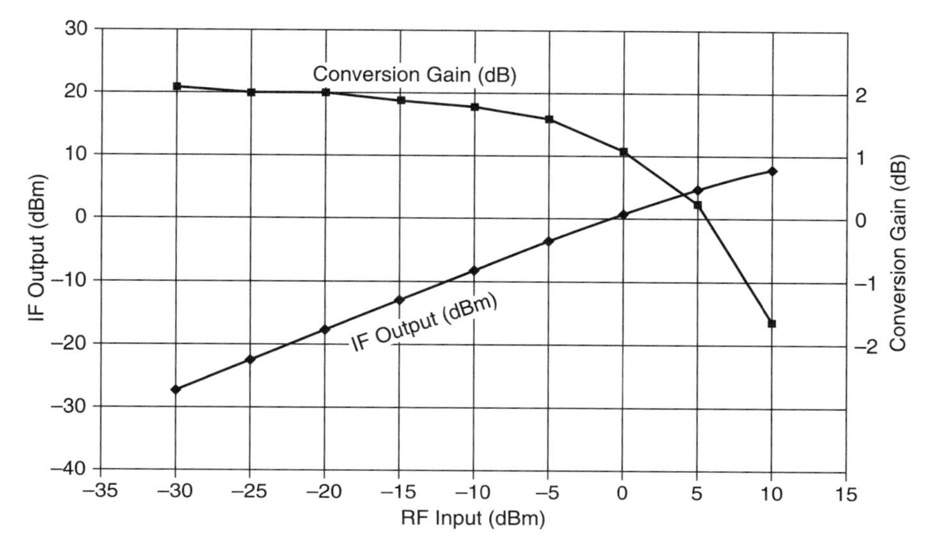

Figure 24.10 Mixer IF output power and gain versus RF input power.

measurement and the output is at IF. The opposite arrangement is used in the upconverter measurement.

Figure 24.10 shows the measured results. Measurements must be made manually, one frequency at a time. The RF input power is varied using the signal generator controls from a level in the linear range up to the point of saturation. The gain and gain compression are easy to measure on the spectrum analyzer. The 1 dB compression point can be easily determined. If required, additional graphs like Figure 24.10 can be generated at different frequencies.

The RF match can be measured using a VNA. The output match is at a frequency below the RF band, and it is not critical. The noise figure of a mixer is comparatively high, from 5 to 10 dB. A receiver is usually designed with enough gain in its LNA stages so that the mixer noise figure can be neglected. Noise figure measurements using special spectrum analyzer hardware and software are described in Chapter 25. The software allows a noise figure measurement to be made on a mixer, even though its input and output signals are at different frequencies.

Intermodulation products generated by a mixer have a major affect on the overall intermodulation performance of the receiver. The IP3 of the mixer determines the largest input signal that the receiver can handle without causing intermodulation product interference. Details of making intermodulation product measurements on both the LNA and the mixer are described in Chapter 26.

24.6 ANNOTATED BIBLIOGRAPHY

1. Moss, S., Mixers for wireless applications, In Larson, L., Ed., RF and Microwave Circuit Design for Wireless Communications, Artech House Boston, 1996.

2. Besser, L. and Gilmore, R., Mixers and frequency multipliers, In Practical RF Circuit Design for Modern Wireless Systems, Artech House Norwood, MA, 2003.
3. Loke, A. and Ali, F., Direct conversion radio for digital phones—Design issues, status and trends, IEEE Transactions on Microwave Theory and Techniques, **50**, 2422–2435, 2002.

References 1 and 2 provide a comprehensive background for mixer design.

Reference 3 provides specific issues related to ZIF mixers.

CHAPTER 25

NOISE FIGURE MEASUREMENT

25.1 NOISE FIGURE MEASUREMENT SETUP AND PROCEDURE

A typical noise figure measurement setup is shown in Figure 25.1. The setup consists of a noise diode and either a noise figure meter or a spectrum analyzer with noise figure measurement hardware and software. In either case, the user interface and setup are generally the same. The noise diode acts as the signal source for the DUT. A 28 V control signal from the instrument turns the noise diode on to take the hot measurement and off to take the room temperature measurement. The ENR is either printed on the side of the diode and has to be manually input into the measurement setup or automatically downloaded from a ROM chip when the diode is first connected to the instrument.

A preamplifier with good noise figure and gain characteristics is inserted before the input of the spectrum analyzer. The gain from this amplifier will mask the noise figure of the instrument components that follow. On newer instruments, the preamplifier is built in and turned on by default when performing noise figure measurements.

The setup of the measurement consists of setting the start and stop frequencies, and, if necessary, inputting the corresponding ENR values. Next, calibration is performed by taking a sweep with the diode connected directly to the instrument input (or external preamplifier input on older instruments). This measures the noise figure of the instrument itself, which can then be calculated out of the final measurement of the DUT by using the formulas for cascaded noise figure.

RF Measurements for Cellular Phones and Wireless Data Systems. By A. W. Scott and R. Frobenius

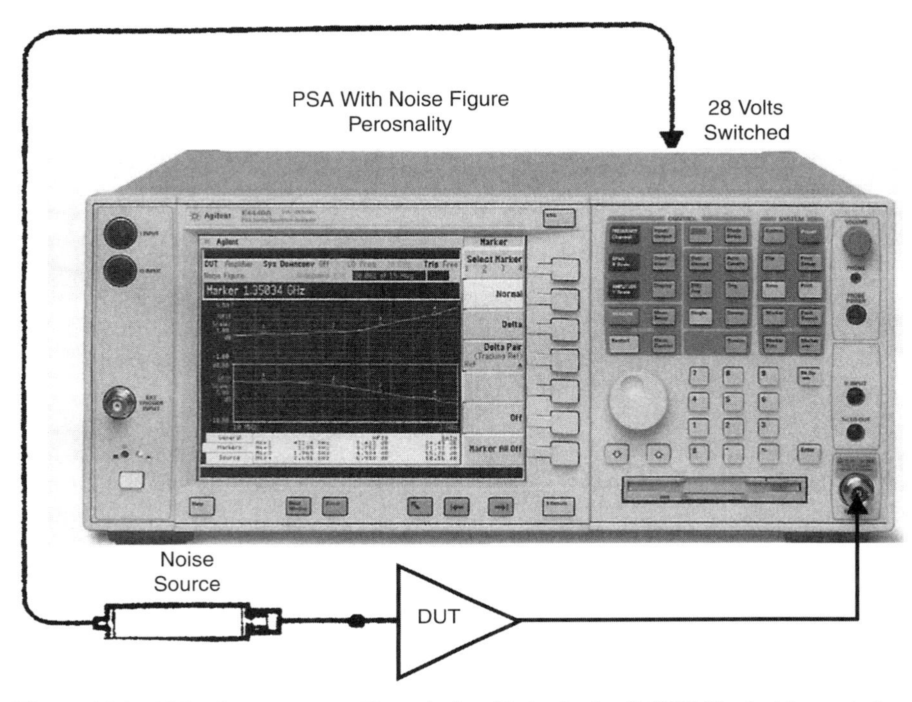

Figure 25.1 Noise figure test setup. Photo Agilent Technologies © 2008. Used with permission.

Once the calibration has been performed, the DUT can be connected and the noise figure and gain will be displayed. The setup, calibration, and measurement procedures are simple and straightforward. However, because we are measuring signal levels near the minimum signal level that can be measured, any interference can be a major source of concern. High accuracy and repeatable noise figure measurements require great attention to detail when eliminating possible sources of interference, such as other systems or even fluorescent lights.

25.2 MEASUREMENT OF THE NOISE FIGURE AND GAIN OF LNAs, FILTERS, AND MIXERS

Figure 25.2 shows a simplified receiver block diagram. At the input, the RF filter rejects signals from other systems. Because it is a passive device, its noise figure is

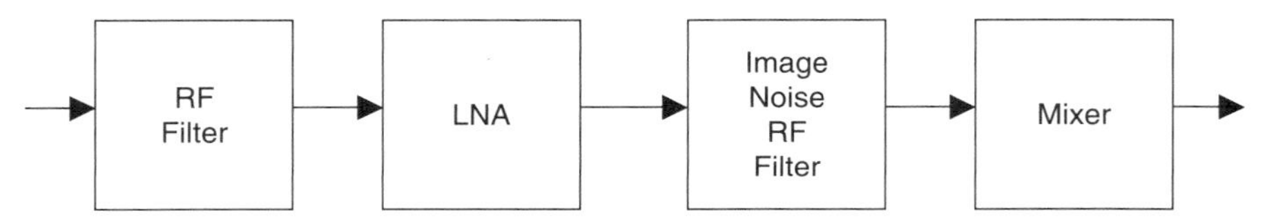

Figure 25.2 Complete RF receiver.

TABLE 25.1 Noise Figure and Gain of Individual Components and Complete Receiver at 2.45 GHz

Component Noise	Figure (dB)	Gain (dB)
RF filter	0.7	−0.7
LNA	4.4	20.5
RF filter + LNA	5.4	19.5
Mixer	8.9	0.5
Complete system	6.5	16.9
Complete system w/out noise image filter	10.9	19.4

equal to its insertion loss. The next component is the LNA, which, as previously described, amplifies the desired signal and the noise so that the noise figure of the following components is not as important. The amplifier also amplifies noise at the image frequency, even though there is no desired signal there. This image frequency is the same distance away from the LO as the desired signal, so the amplified noise at this frequency gets converted down to the IF output just as the desired signal does. As a result, the noise will be doubled at the mixer unless we filter out the amplified noise using an image filter. Next, the mixer downconverts the signal.

The measured noise figure and gain of these components, measured with the test setup shown in Figure 25.1, are provided in Table 25.1. The measured noise figure and gain of the LNA are presented in Figure 25.3. The gain is 20.4 dB and the

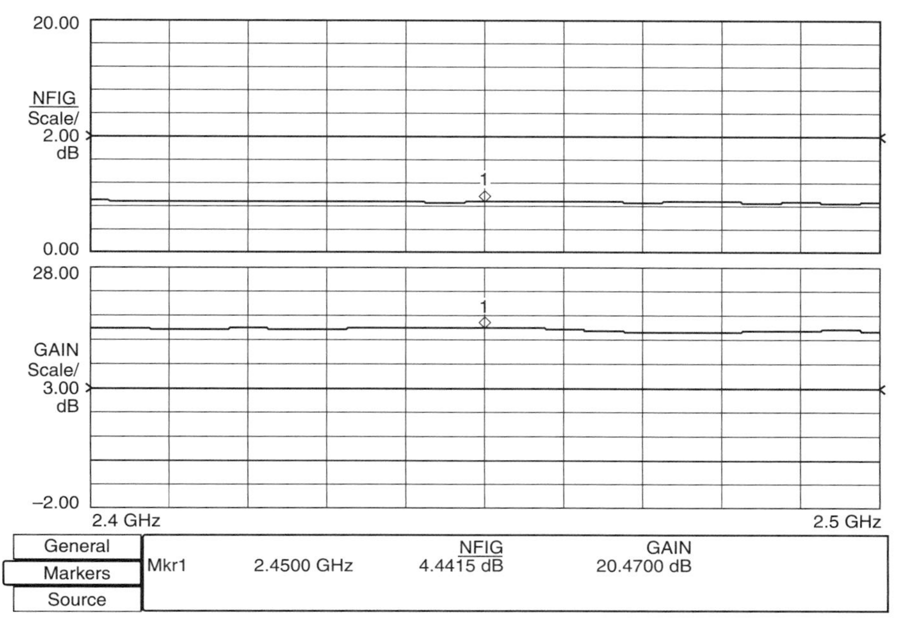

Figure 25.3 Noise figure and gain of LNA.

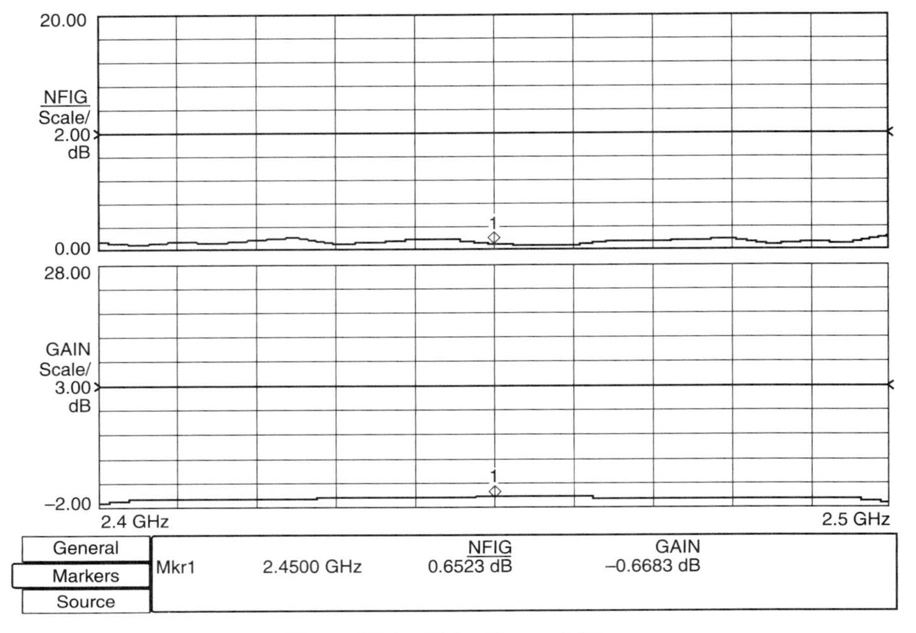

Figure 25.4 Noise figure of filter.

noise figure is 4.4 dB. The noise figure performance of this device is not particularly impressive, because noise figures of about 1 dB can be found in current devices.

Figure 25.4 graphs the results for the filter. The results show that the noise figure does equal its insertion loss, as stated earlier.

Figure 25.5 contains the noise figure and gain of the filter and LNA together. As expected, the noise figure increases by about 1 dB and the gain decreases by about 1 dB, compared to the LNA alone.

Figure 25.6 shows the noise figure of the mixer, which is high and equal to 8.9 dB. The mixer gain is low. Another possible noise measurement pitfall when measuring mixers is the possibility of injecting noise from the LO source. If the LO signal comes from a source that has a broadband amplifier as its final output stage, that amplifier can inject amplified noise to the measurement setup, which can also be downconverted to the final measurement output. This is true for bench-top signal generators, which are designed to output signals at a broad range of frequencies. They are therefore equipped with broadband amplifiers at their last output stage to boost the signal. If such a signal source is used as the LO signal, then a narrow bandpass filter should be applied prior to the input to the mixer to filter out the amplified noise at unwanted frequencies.

Figure 25.7 provides the noise figure and gain of the complete system, including the noise image filter. The mixer and noise image filter add about 1 dB to the noise figure of the complete system, because of the high gain of the LNA.

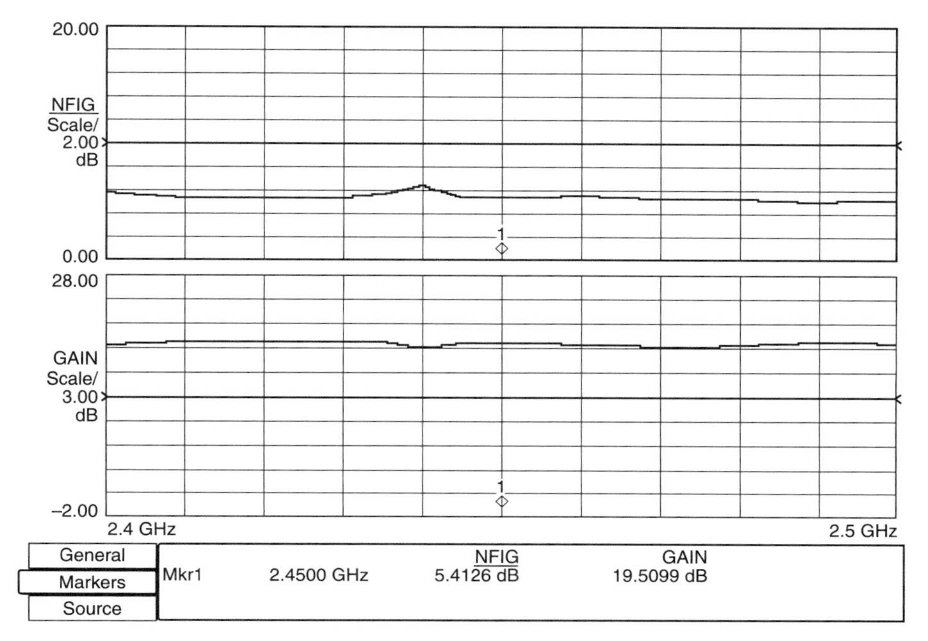

Figure 25.5 Noise figure and gain of filter and LNA.

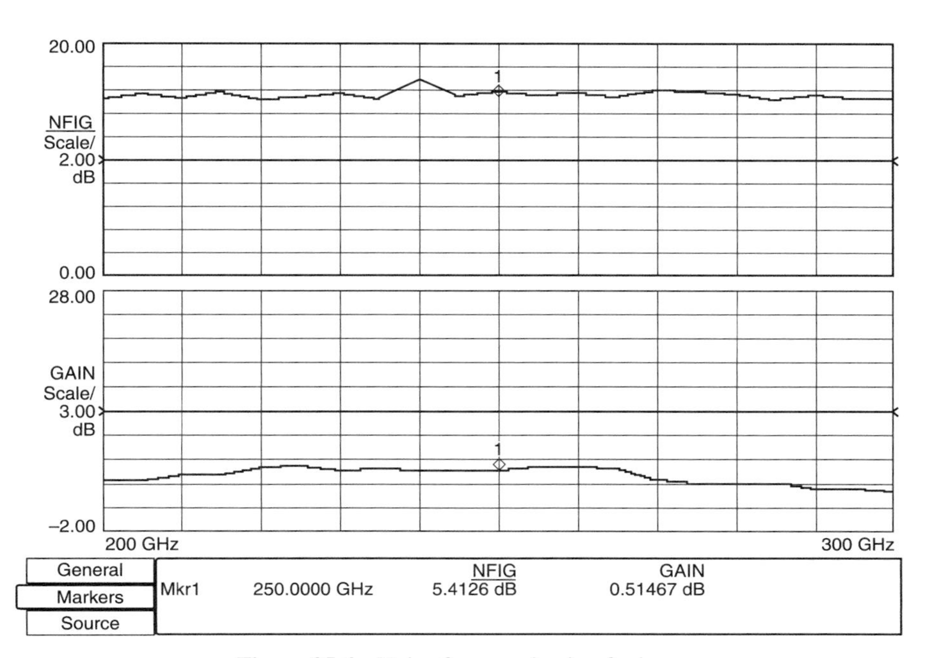

Figure 25.6 Noise figure and gain of mixer.

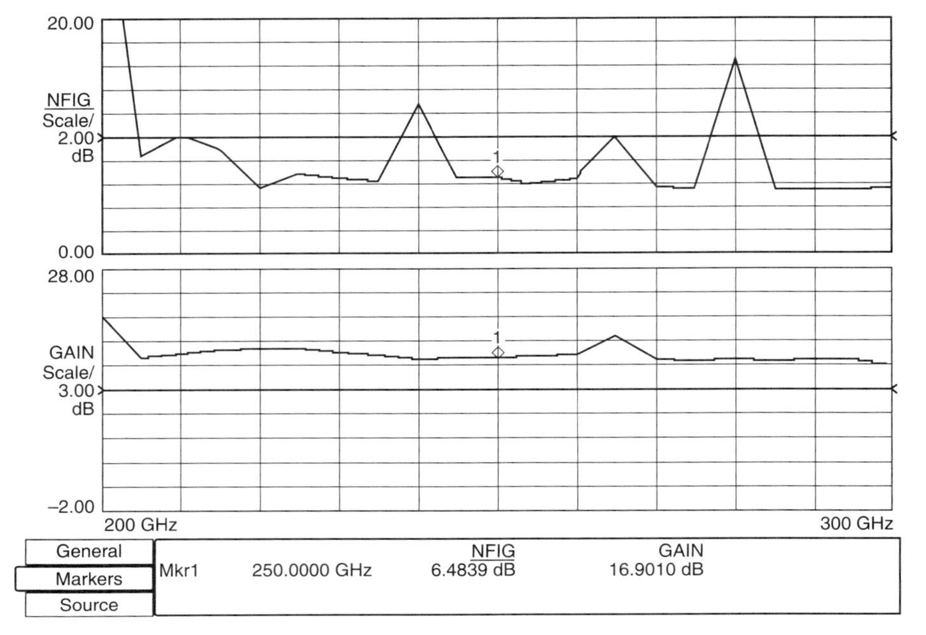

Figure 25.7 Noise figure and gain of complete system.

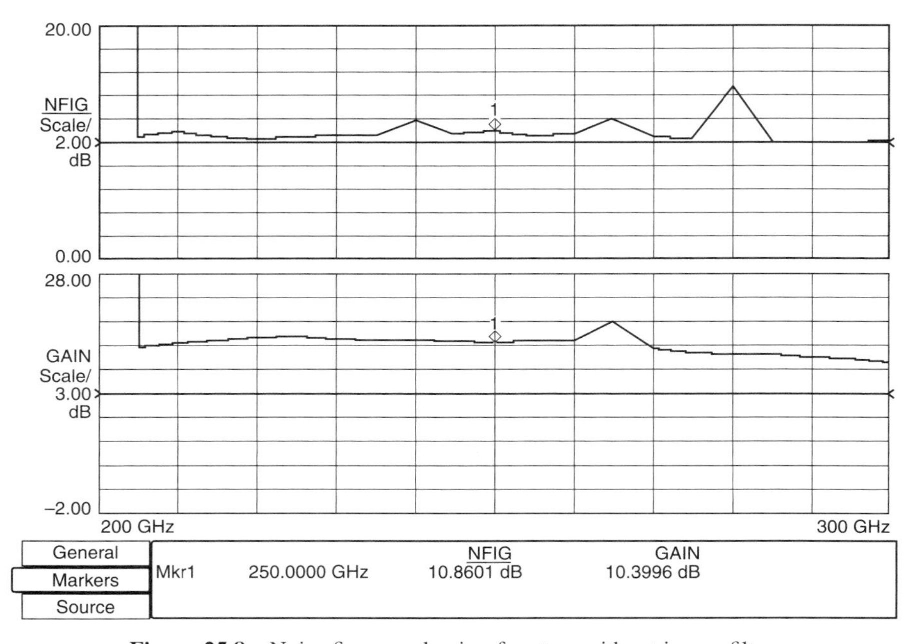

Figure 25.8 Noise figure and gain of system without image filter.

Figure 25.8 charts the noise figure and gain of the complete system with the noise image filter omitted. The noise figure increases by 4.4 dB, and the gain also goes up 2.5 dB.

These measurements were made in an open room, and the equipment was operated in the unlicensed wireless LAN band. This accounts for the unexpected peaks in the measurements. Noise figure measurements should be made in a shielded room.

25.3 APPROXIMATE MEASUREMENTS OF NOISE FIGURE WITHOUT THE NF HARDWARE AND SOFTWARE

The measurements above all relied on the Y-factor method using a noise diode and instrumentation software to control the experiment and calculate the values. If you do not have access to a noise figure measurement setup, you can still make a basic noise figure measurement using just a spectrum analyzer. Note that the Y-factor measurement relied on taking two measurement points to determine the slope of the line (which represents the gain of the device) and y-intercept value. However, if you measure the gain of the device separately (e.g., on a VNA), you can take a single measurement at room temperature and use the gain information and the point-slope equation of a line to determine the y intercept. This is referred to as the "cold source" technique for measuring noise figure, and it is implemented in the newest network analyzer systems that also measure noise figure. The general procedure is described in Figure 25.9.

Figure 25.9 contains the results using this technique. This method gives a 4.2 dB noise figure for the LNA. The measured value with the noise figure meter was 4.4 dB,

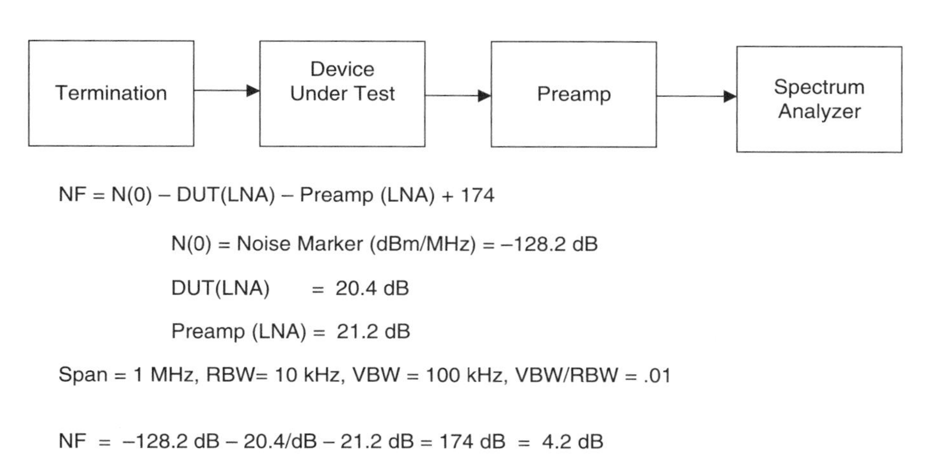

Figure 25.9 Measurement without noise figure hardware and software.

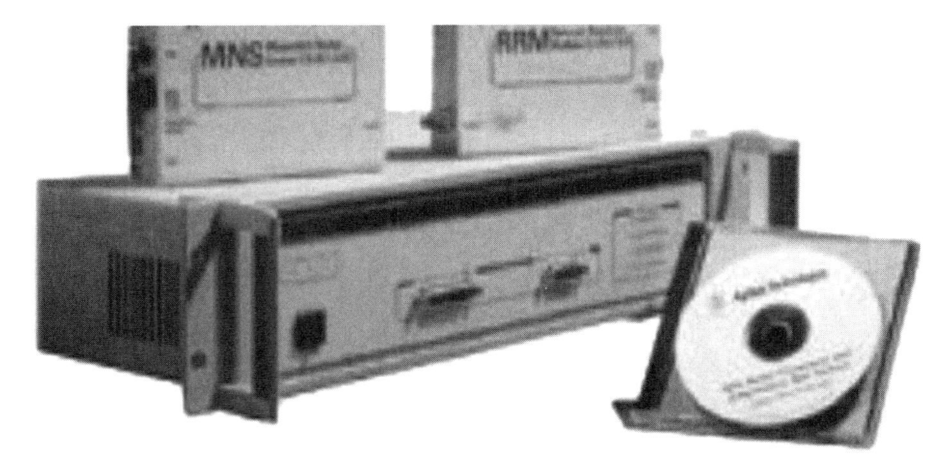

Figure 25.10 Noise parameter measurement system. Courtesy Maury Microwave Corp.

which is fairly good agreement. The latest network analyzers that employ this technique also perform several extra calibration measurements as part of the measurement setup to enhance the accuracy of the procedure.

25.4 MEASUREMENT OF NOISE FIGURE CONTOURS ON THE SMITH CHART

Figures 25.4 and 25.5 of Chapter 23 showed the location of minimum noise figure matching points and constant noise figure circles on a Smith Chart. These data were collected using an NP5 Series Noise Parameter Measurement System manufactured by Maury Microwaye.

The system consists of an electronically controlled tuner, noise figure measuring equipment, and a VNA. The tuner changes the input match to the LNA electronically and measures the noise figure of the DUT at each mismatch point.

The tuner can be controlled to find the minimum noise figure location on the Smith Chart and to plot constant noise figure circles. The electronically controlled tuner is shown in Figure 25.10.

25.5 ANNOTATED BIBLIOGRAPHY

1. Application Note 57.1: Fundamentals of RF and Microwave Noise Figure Measurements, Agilent Technologies, Santa Clara, CA.
2. NP5 Series Noise Parameter Measurement System, Brochure 4T-080, Maury Microwave Corporation, Ontario, CA.
3. Application Note: Noise Figure Scorpion Option 4, Anritsu Corporation, San Diego, CA.

4. Podell, A., Technical Feature: Mixer Noise Measurements, BesserBits Technical Article, Besser Associates in collaboration with Agilent Technologies, http://www.bessernet.com/articles/mixerNoise/mixerNoise.htm.

Reference 1 is a thorough discussion of noise figure in more detail than is covered in this book.

Reference 2 describes an electronic tuning system to measure the optimum match for minimum noise figure.

Reference 3 provides a thorough explanation of the Y-factor technique and the related computational terms.

Reference 4 discusses the challenges of measuring mixer noise figure performance and unexpected sources of noise that may enter the measurement setup.

CHAPTER 26

INTERMODULATION PRODUCT MEASUREMENT

26.1 INTERMODULATION PRODUCTS

When describing receiver performance, the problem associated with detecting weak signals is very intuitive. In our own experience, it is difficult to hear someone speaking softly when we are in a noisy environment. Noise interferes with our ability to detect faint signals. At the other end of the amplitude scale, receivers also experience significant problems when presented with signals that are too strong. Specifically, intermodulation products limit the strength of the signal that can be presented at the input of the receiver without causing excessive distortion.

Thus far, most of the measurements that have been described in this text have relied on passing a single tone (i.e., at a single frequency) through a device and measuring the output response. In a typical wireless communication system, with one base station or access point and several mobile units, signals from the base station to *all* mobiles come into *each* mobile receiver. There is no practical way to filter out a single user channel dynamically at RF. Received signal strength is the same for every signal. In other words, the front-end receiver components in most wireless systems have to be able to handle multiple tones at their inputs.

We have seen that when active devices are driven into nonlinear operation, unwanted harmonics are produced at their output. Furthermore, if multiple tones are present at the input, then unwanted mixing can also occur among the tones and their harmonics. Intermodulation products are the result of this unwanted mixing

RF Measurements for Cellular Phones and Wireless Data Systems. By A. W. Scott and R. Frobenius

action. The resulting outputs occur within our operating band, right on top of our desired signal, and therefore cannot be filtered out.

For example, suppose that we had three signals at the input of our device at 1.0, 1.1, and 1.2 GHz. If these signals are strong enough, the device will become nonlinear and produce harmonics of those signals at 2.0, 2.2, and 2.4 GHz, respectively. In addition, the second harmonic at 2.2 GHz will mix with the fundamental at 1.0 GHz to produce signals at 3.2 and 1.2 GHz. The signal at 3.2 GHz can be filtered out because it is likely outside of our system bandwidth, but the signal at 1.2 GHz lands right on top of our desired signal there. This mixing product of the second harmonic mixing with the fundamental, or first harmonic, is referred to at the IP3 $(2 + 1 = 3)$. The other second harmonics in this example mix with the fundamental tones to produce additional IP3s. That is, the second harmonic at 2.2 GHz mixes with the fundamental tone at 1.2 GHz to produce a third-order product at 1.0 GHz, and so forth.

Figure 26.1a shows two tones presented to the input of an amplifier at 1.0, 1.1, and 1.2 GHz. Suppose that the signal that we are trying to receive is at 1.2 GHz. Figure 26.1b shows what the output spectrum would look like when the input signals are at low levels; no problems are occurring because there are no measurable intermodulation products. In Figure 26.1c we can see the effect of intermodulation products. Note that the third-order product appears right at 1.2 GHz, which is where we are trying to receive one of our signals. As the signal power increases at

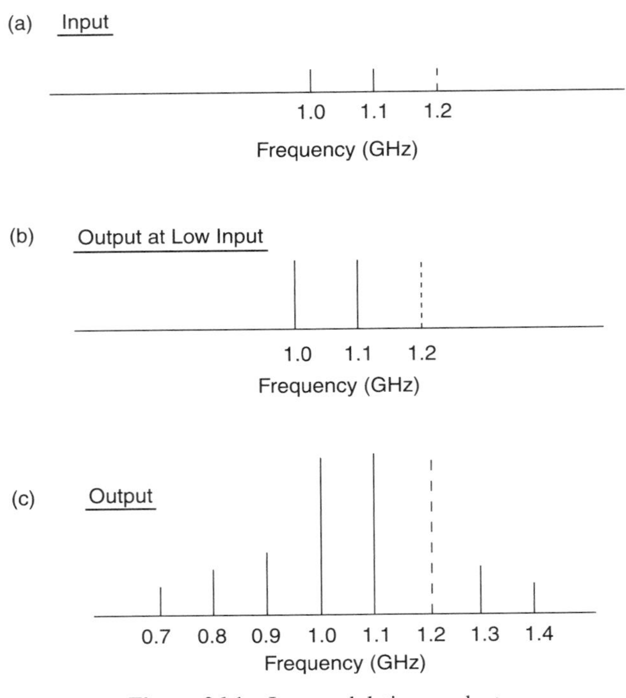

Figure 26.1 Intermodulation products.

the input, additional odd-order products are produced. For example, the third harmonic of the 1.1 GHz tone, which is at 3.3 GHz, can mix with the second harmonic of the 1.0 GHz tone, which is at 2.0 GHz, to produce a fifth-order modulation product at 1.3 GHz. Similar calculations can predict where the 7th, 9th, and 11th tones will appear and so forth.

26.2 THIRD-ORDER INTERCEPT POINT

If we plot the output power versus input power of the fundamental tones along with the third-order products as shown in Figure 26.2, we can observe that the fundamental tones have a slope of 1 : 1. For each 1 dB increase in input power, the output power increases by 1 dB. Because the third-order products are based on mixing between the fundamental tones and second harmonics, their slope is 3 : 1 for each 1 dB increase in input power. As we increase the input power, the DUT eventually starts to compress and these slopes no longer apply. However, if we extrapolated the lines from the linear region, the point at which they intersect is defined as the third-order intercept point. From the standpoint of performance, the higher the IP3 is the better, because this indicates how strong of a signal the device can handle before intermodulation distortion becomes a problem.

A simple, albeit tedious, method to determine the IP3 is to supply a device with two tones at the input and measure the output power of the fundamentals and third-order products versus input power over the operating range of the device. Then, plot the data as shown in Figure 26.2, extrapolate the lines from the

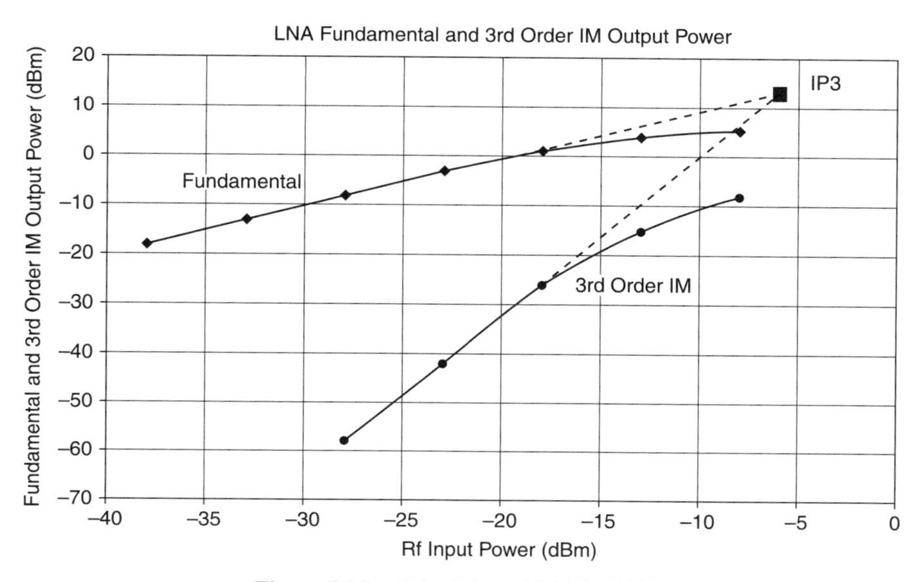

Figure 26.2 Calculation of IP3 for LNA.

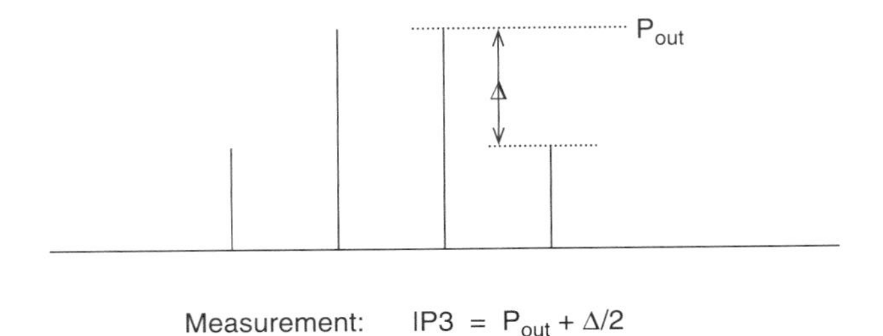

Figure 26.3 Calculation of TOI.

linear region, and determine the IP3 from the graph. By definition, the IP3 is the point where the line from the fundamental tones intersects with the line from the third-order products. We can read off the value either relative to the vertical (output power level) axis, which would be referred to as the OIP3, or we can read the value on the horizontal axis, which would be the IIP3. Note that the OIP3 and IIP3 are related to each other by the gain of the device, that is, OIP3 − gain = IIP3.

If we take into account that we know the slopes of the two lines as 1 : 1 and 3 : 1, respectively, we can use the point–slope equation for a line and simply use one measured data point to construct the lines. Furthermore, if we algebraically set the two point–slope equations for the lines equal to each other, we will find that the equation for the intersection of the lines is

$$\mathrm{IP3} = P_{\mathrm{out}} + \Delta/2 \tag{26.1}$$

as shown in Figure 26.3. Thus, instead of having to plot out the entire input/output power curves, we can measure the IP3 simply by measuring the fundamental output power and the third-order products with two tones as the input to our device. Note that we need to make sure when we are taking the measurement that we are operating the device at a power level where we still have linear gain (i.e., below the 1 dB compression point). Otherwise, the assumption about the slopes of the output curves being 1 : 1 and 3 : 1 is no longer valid and the equation above would not produce the correct result. Most spectrum analyzers and even some network analyzers (models that have two built-in sources) have a one-button measurement for IP3. In some cases manufacturers also refer to this measurement as TOI.

26.3 CALCULATION OF MAXIMUM INPUT POWER

Now that we know where the IP3 comes from, it would be useful to build more intuition about how the IP3 specification is used. As mentioned earlier, the IP3 determines the maximum power handling capability of the DUT before the intermodulation distortion becomes problematic. Specifically, we can rearrange Eq. (26.1) to calculate the maximum power level our device can accept at the input before the third-order products interfere with our desired signal.

$$P_{\text{in}}(\text{max}) = \text{IP3} - \delta(\text{spec})/2 - \text{gain} \tag{26.2}$$

In most cases we will be given a specification [δ(spec)] for how far down the intermodulation products need to be for the system to perform properly. For example, suppose that our system requires that intermodulation products should be 20 dB lower than our fundamental signal. Furthermore, the gain of our device is 20 dB and the IP3 was measured to be 14 dBm. What is the strongest signal that we can present to the input? Using Eq. (26.2), we find that the answer is

$$14\,\text{dBm} - 10\,\text{dB} - 20\,\text{dB} = -16\,\text{dBm}$$

Equation (26.2) is also incorporated into the formula for calculating "dynamic range," which is described later in Chapter 27 under the discussion of overall system performance.

26.4 CAUTIONS WHEN MEASURING DISTORTION PRODUCTS

Whenever we are measuring distortion products, we must remember that the instrumentation also contains analog components that are subject to producing the exact same symptoms that we are trying to measure. The block diagram of a spectrum analyzer, for example, contains analog amplifiers and mixers, both of which will produce harmonics and intermodulation products when they are driven into their nonlinear regions of operation. These internally generated distortion products will add to the measured response from our device and give us an incorrect reading.

A simple step that we can take to check whether the instrument is being overdriven is to manually increase the input attenuation and observe the levels of the intermodulation products. If the intermodulation products do not change, then the instrument was not being overdriven and we can restore the attenuation to the previous setting. If the instrument was generating intermodulation products before we raised the attenuation, then we would observe that the intermodulation products are reduced after we change the attenuation. Measurements should be taken with the attenuation set so that the intermodulation products do not change when we increase it.

26.5 EXAMPLE MEASUREMENT FOR INTERMODULATION PRODUCTS

Objective

Our objectives are the following:

Plot the output spectrum of the LNA with two tones at the input versus power level.

Determine the third-order intercept point from the plot.

Demonstrate the measurement of IP3 using just one input power level.

Measure IP3 for the mixer, as well as the LNA plus mixer in cascade.

Measurements Being Demonstrated

Output spectrum of LNA with two tones at input versus several input power levels using the spectrum analyzer

Measurement/calculation of IP3 using one input power level

Note that some VNAs also provide a built-in application to measure IP3.

Generic Procedure

Steps to perform the measurements are listed in the following table. The left column lists the actions to be performed, and the right column lists the motivation for taking each step and provides additional background information in some cases.

Procedure	Notes
Preset the spectrum analyzer and signal generator.	
Generate two tones, 2455 and 2445 MHz, at a −33 dBm power level for each peak.	Signal generators with multitone capability can be used, or two signal generators can be used with a power combiner.
Set the spectrum analyzer to a 2.45 GHz center frequency, a 100 MHz span, and a 5 dBm reference level.	Depending on the signal sources, we may observe additional signals that are a result of the multitone generation technique, such as carrier feedthrough and so forth.
Connect the LNA as the DUT.	
Measure the lower and upper fundamental tones and the lower and upper third-order products for the following input power levels: −33, −27, −25, −23, −20, −17, −13, and −7 dBm.	Plot the results, perhaps in a spreadsheet.
Manually adjust the attenuation on the spectrum analyzer to make sure that distortion products are not being generated inside the instrument.	If the level of the third-order products is lowered when we increase the input attenuation, then the spectrum analyzer's analog components are producing distortion products that are affecting the measurement results.

(*Continued*)

Procedure	Notes
Plot the output versus input curves. Extrapolate the straight line from the linear region of operation for the fundamental tones (slope = 1 : 1) and the third-order products (slope = 3 : 1). The intersection of these two lines is the third-order intercept point.	If we read the value with respect to the output power level, it is referred to as OIP3. Likewise IIP3 is read from the input power axis. The two values are related by the linear gain of the device.
Reduce the input power to −25 dBm and calculate IP3 from the one measurement data point. Use the one-button measurement if it is available.	Because the slopes of the two lines are known, we can determine the IP3 by taking one measurement in the linear region of operation and calculating the intersection. Some spectrum analyzers feature a built-in calculation measurement.
Disconnect the LNA and connect the mixer as the DUT. Set the spectrum analyzer center frequency to 250 MHz. Set the input power level to −5 dBm.	
Measure/calculate the IP3 for the mixer.	
Connect the LNA output to the input of the mixer. Set the input level to −23 dBm. Measure/calculate the IP3 for the two components connected in series.	Given the individual IP3 values of two components, we can use a formula to calculate the overall IP3 of the two components connected in series. Later in the book we will compare the calculated performance to the measured performance with the two devices connected together.

Some VNAs feature two signal sources as well as the ability to perform IP3 measurements using built in software.

26.6 ANNOTATED BIBLIOGRAPHY

1. Besser, L. and Gilmore, R., The radio as typical RF system, In Practical RF Circuit Design for Modern Wireless Systems Vol. I: Passive Circuits and Systems, Artech House, Norwood, MA, 2003.
2. Pozar, D., Noise and distortion in microwave systems, In Microwave and RF Design of Wireless Systems, John Wiley & Sons, New York, 2001.

References 1 and 2 provide additional background and derivation of the intermodulation product terms.

CHAPTER 27

OVERALL RECEIVER PERFORMANCE

As described in Chapter 21, the receiver in a wireless mobile unit must operate under a variety of conditions. When the mobile unit is at the edge of the coverage area, the receiver must have a low noise figure to receive the very low signal from the base station transmitter with a satisfactory S/N for achieving the required BER. When the mobile unit is close to the base station, the received signals intended for all of the mobile units in the cell are very large, and two signals may mix in a mobile receiver to form a signal that jams the signal intended for the mobile. This jamming signal is called an intermodulation product.

The received RF signal varies over a 90 dB dynamic range, depending on the transmitter-receiver separation and multipath fading. The receiver must provide adjustable gain, depending on the received signal level, to provide a constant output level of about 0 dBm for demodulation.

As has been discussed, the RF receiver is made up of four basic parts, whose functions are as follows:

1. RF filter: Allows only a specified range of RF frequencies to pass and blocks all other frequencies
2. LNA: Amplifies the weak received RF carrier
3. Mixer: Shifts the RF frequency to a lower frequency where it can be more easily amplified and filtered
4. IF amplifier: Amplifies the IF frequency to a power level where it can be demodulated

---

RF Measurements for Cellular Phones and Wireless Data Systems. By A. W. Scott and R. Frobenius

This chapter will show how the various performance characteristics of the four basic components are combined to give the overall performance characteristics of the receiver.

The performance of a typical RF receiver is discussed in Section 27.1. Formulas for combining the gain, noise figure, and IP3 of the various receiver components are presented in Section 27.2.

The use of the noise figure value and intermodulation product value of the individual components to give the values of these characteristics for the overall receiver involves some tedious arithmetic. This problem is handled by using one of a variety of available software programs. One such software program is available from Agilent Technologies. It is easy to use and can be downloaded from the Internet at no charge.

Section 27.3 will show an example of the use of the Agilent software program for calculating the overall performance of a simple receiver consisting of the RF filter, LNA, and mixer, whose performance characteristics were measured in Chapters 22–26.

Mobile phones and data systems must operate over a range of temperatures from -30°C under winter conditions to $+60^{\circ}$C under summer desert conditions. The RF characteristics of the receiver components vary over this temperature range. Section 27.4 shows calculations using the Agilent software program and values of the temperature sensitivity of the individual components to calculate the temperature sensitivity of the overall receiver.

Because a mobile receiver must operate over such a wide range of received power, improved performance can be obtained by electronically switching LNAs in and out of the receiver as operating conditions change. This procedure will be described in Section 27.5.

27.1 OVERALL PERFORMANCE OF A TYPICAL RF RECEIVER

Figure 27.1 shows the overall performance of a typical RF receiver. IF output power from the mixer is shown as a function of RF input power coming into the antenna.

When the receiver is first turned on, with no RF signal applied, there is an IF output signal of -116 dBm. This is the "noise floor." It is the RF thermal noise of the receiver that has been downconverted to IF.

When the RF input power is then turned on at a power level of -140 dBm (left-hand edge, Fig. 27.1), there appears to be no IF signal. This is because the downconverted signal is below the downconverted noise floor.

As the RF input power is increased to -122 dBm, the downconverted IF signal just breaks through the noise floor. The RF signal at this point is defined as the "minimum detectable signal" (MDS). The receiver is not useful for communication purposes at this point, because the S/N is 0 and therefore a signal cannot be demodulated.

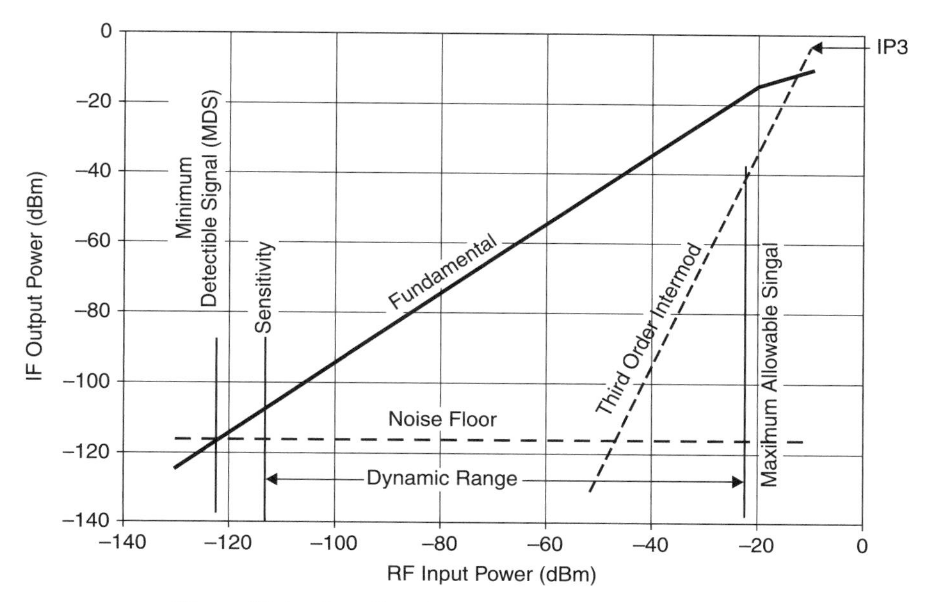

Figure 27.1 Overall performance of a typical RF receiver.

As the input RF signal increases above the MDS value, the IF signal increases in direct proportion and finally reaches a value where the IF S/N is adequate for detection. The RF signal at this point defines the "sensitivity" of the receiver. For example, if QPSK modulation were being used, which at minimum requires a 7 dB S/N (as described in Chap. 4), the RF sensitivity would be −112 dBm. However, using 16PSK modulation with the same receiver, which requires a 16 dB S/N for a 10^{-3} BER, the sensitivity would be −103 dBm. Even though the same receiver is being used, its sensitivity with the higher order modulation is 9 dB less.

As the RF input power increases above the sensitivity point, the IF S/N increases. At an input RF signal of −60 dBm (which occurs when a mobile phone is approximately midway between the edge of the cell and the base station location at the center of the cell), the IF S/N is 60 dB. Noise is definitely not a problem at this point.

As the RF input signal increases to −48 dBm, the IF noise level suddenly begins to increase. This is the beginning of intermodulation product interference. The intermodulation products increase with RF input power with a slope of 3 : 1. For every 1 dB increase in input RF power, the intermodulation product interference increases by 3 dB. The linear intermodulation products are shown by the dotted curve that extends from the OIP3 downward with decreasing RF input power at a slope of 3 : 1.

When the RF input power is at −22 dBm, the intermodulation product power is about 20 dB below the fundamental IF power. This is therefore the maximum RF input power that can be applied to the amplifier without causing intermodulation product interference. It is therefore the "maximum allowable signal."

The RF power range between the sensitivity level, defined by the noise figure, and the maximum allowable signal level, defined by the intermodulation products, is the "dynamic range" of the receiver.

Three things have happened to the incoming signal at the sensitivity point:

1. The RF signal has been raised to a sufficiently high level so that the noise of the following RF components will not be important.
2. The signal has been converted into a narrowband IF channel that contains only one user channel, so there will be no further generation of intermodulation products.
3. The carrier frequency has been shifted to a lower frequency (either into the megahertz range or near zero in a ZIF receiver) where it is easy to amplify the signal up to 0 dBm where it can be easily demodulated.

As is evident from Figure 27.1, the only remaining problem is that of varying the gain of the IF amplifier to get the signal to 0 dBm for demodulation at all RF input power levels from −112 dBm at the sensitivity point to −22 dBm at the maximum allowable signal point. This is a range of 90 dB, but it is easily achieved because the signal to be amplified has been shifted to IF.

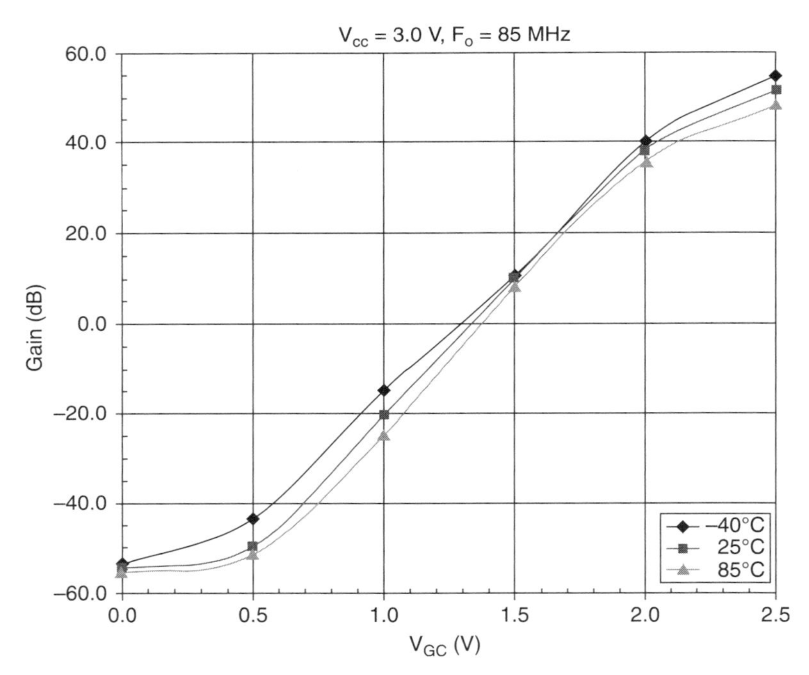

Figure 27.2 Variable gain IF amplifier.

Figure 27.2 shows a variable gain IF amplifier. By adjusting the gain control voltage, the amplifier gain can be varied from −50 to +50 dB. By adding a fixed gain IF amplifier with 50 dB gain, the combination can vary the gain from 0 to 100 dB.

27.2 FORMULAS FOR COMBINING GAIN, NOISE FIGURE, AND OIP3 OF THE RECEIVER COMPONENTS

Formulas for using the gain, noise figure, and OIP3 of the receiver components to calculate their cascaded value for the complete receiver are provided in Figure 27.3. The gain of the overall receiver is calculated in dB simply by adding the gains of each component, expressed in dB, as shown by Eq. (1).

The noise figure of the overall receiver is calculated by Eq. (2) in Figure 27.3. The noise figure and gain of each component must be converted from dB into a ratio for this calculation. Adding noise figures was shown in Chapter 23. Equation (2) is just a generalization of the two LNA cases discussed previously. The noise effect of components further down the chain from the input is reduced by the gain of the preceding components. By careful design with enough gain in the first LNA, only its noise figure and the RF losses in the RF filter and the connecting transmission lines from the antenna have any significant affect on the overall system noise figure.

The OIP3 should be as high as possible, so that the receiver can handle large signals without intermodulation product interference. As Eq. (3) in Figure 27.3 shows, the OIP3 decreases as more components are added into the receiver chain. The contribution of each component is modified by the gain of each preceding component. The important parameter describing the total system intermodulation performance is the maximum input power that can be applied to the overall receiver and still have intermodulation products be a specification value δ below the desired signal power. This topic was discussed in Chapter 23.

The calculation of overall receiver gain, noise figure, and OIP3 is facilitated by the use of simple software, as described in the next section.

27.3 SOFTWARE FOR CALCULATION OF OVERALL RECEIVER PERFORMANCE

Figure 27.4 shows the worksheet for a software program that solves the equations of the previous section to calculate overall receiver performance. This free program can be downloaded from hp.woodshot.com. The software provides many other useful RF calculation programs. The receiver calculation software is accessed by selecting Signals-System with the mouse on the left-hand side of the main page menu and then selecting Noise Calc from the Signals-System menu.

The overall receiver calculation has many useful features, which are accessible from the toolbar. One feature is a Help file, which gives detailed instructions for using the program. Another feature shows three application examples. These become available every time the program is turned on.

1. GAIN

Gain (total) = $G_1 + G_2 + G_3 + \ldots$

Gain expressed as dB

2. NOISE FIGURE

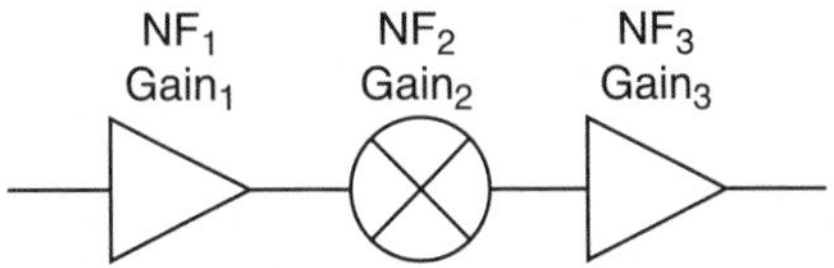

NF_1 $Gain_1$ — NF_2 $Gain_2$ — NF_3 $Gain_3$

The cascaded noise figure for the above is calculated using the equation

$$NF_{TOT} = NF_1 + \frac{NF_2 - 1}{Gain_1} + \frac{NF_3 - 1}{Gain_2 \cdot Gain_1} + \frac{NF_4 - 1}{Gain_3 \cdot Gain_2 \cdot Gain_1} + \ldots$$

3. IP3

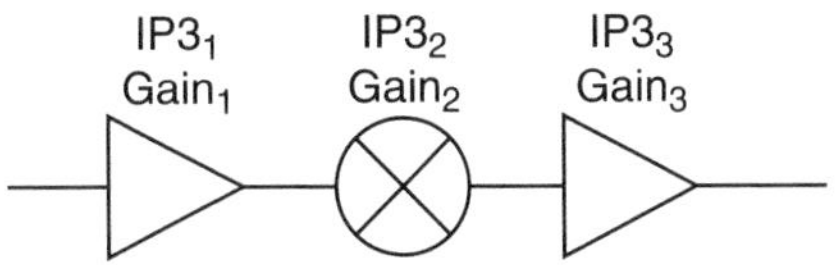

$IP3_1$ $Gain_1$ — $IP3_2$ $Gain_2$ — $IP3_3$ $Gain_3$

The cascaded IP3 for the above is calculated using the equation

$$IP_{TOT} = \frac{1}{\frac{1}{IP3_1} + \frac{Gain_1}{IP3_2} + \frac{Gain_1 \cdot Gain_2}{IP3_3} + \ldots}$$

Figure 27.3 Formulas for cascaded gain, noise figure, and IP3.

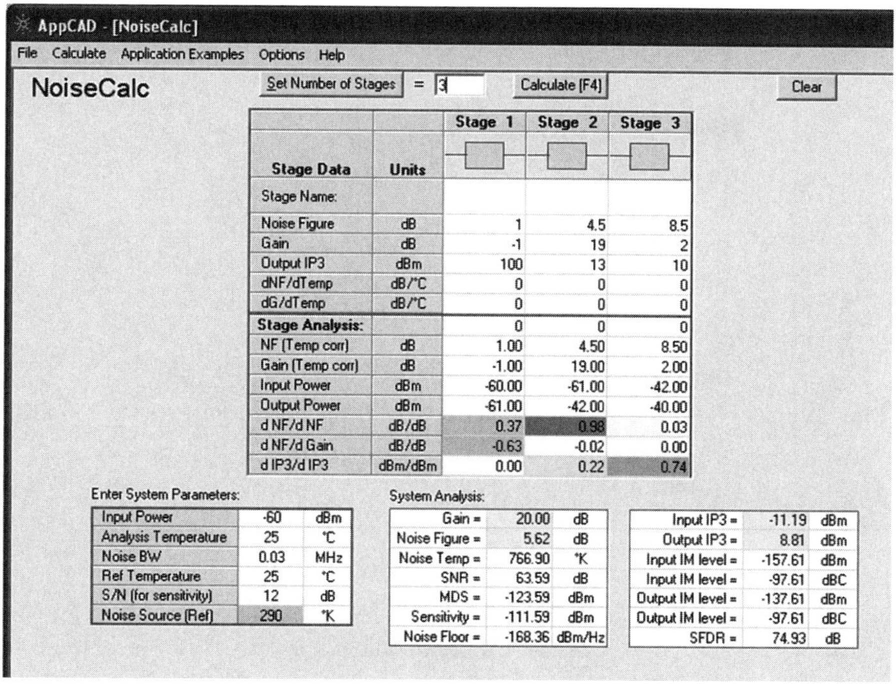

AppCAD - [NoiseCalc]

File Calculate Application Examples Options Help

NoiseCalc Set Number of Stages = 3 Calculate [F4] Clear

Stage Data	Units	Stage 1	Stage 2	Stage 3
Stage Name:				
Noise Figure	dB	1	4.5	8.5
Gain	dB	-1	19	2
Output IP3	dBm	100	13	10
dNF/dTemp	dB/°C	0	0	0
dG/dTemp	dB/°C	0	0	0
Stage Analysis:		0	0	0
NF (Temp corr)	dB	1.00	4.50	8.50
Gain (Temp corr)	dB	-1.00	19.00	2.00
Input Power	dBm	-60.00	-61.00	-42.00
Output Power	dBm	-61.00	-42.00	-40.00
d NF/d NF	dB/dB	0.37	0.98	0.03
d NF/d Gain	dB/dB	-0.63	-0.02	0.00
d IP3/d IP3	dBm/dBm	0.00	0.22	0.74

Enter System Parameters:

Input Power	-60	dBm
Analysis Temperature	25	°C
Noise BW	0.03	MHz
Ref Temperature	25	°C
S/N (for sensitivity)	12	dB
Noise Source (Ref)	290	°K

System Analysis:

Gain =	20.00	dB
Noise Figure =	5.62	dB
Noise Temp =	766.90	°K
SNR =	63.59	dB
MDS =	-123.59	dBm
Sensitivity =	-111.59	dBm
Noise Floor =	-168.36	dBm/Hz

Input IP3 =	-11.19	dBm
Output IP3 =	8.81	dBm
Input IM level =	-157.61	dBm
Input IM level =	-97.61	dBC
Output IM level =	-137.61	dBm
Output IM level =	-97.61	dBC
SFDR =	74.93	dB

Figure 27.4 Software for calculating receiver performance. Agilent Technologies © 2008. Used with permission.

To set up a program, the mouse is used to access the Clear button. Then, the number of stages of the receiver can be entered.

The example shown in Figure 27.4 is for a simple receiver with a filter, an LNA, and a mixer. The sample calculation utilizes the measured performance of the components used as measurement examples in Chapters 22–26.

The combining of the intermodulation products must be done using their OIP3. The software program does not use this nomenclature, assuming that the user understands this. As discussed in Chapter 23, the important intermodulation characteristic is IIP3. IIP3 is equal to OIP3 minus the receiver gain.

An RF filter does not have an OIP3, because it is a linear device. To eliminate the filter from the OIP3 calculation, it is assigned a very high fictitious value. Remember, the higher the OIP3 is, the better. In the example calculation of Figure 27.4, it is given a value of 100 dBm, which is 10 MW. Of course, the filter will never see this much power, but using this value takes it out of the calculation.

The gain, noise figure, and OIP3 for each component are entered in the upper window. The two additional lines in this window are used when calculations are to be made as a function of part temperatures. These calculations will be discussed in Section 27.5, so they are left blank for now.

a. Receiver Performance

- Noise Figure: 5.6 dB
- Gain: 20 dB
- System Bandwidth: 30 kHz
- Required S/N: 12 dB
- MDS Signal: –124 dBm
- Sensitivity: –112 dBm
- Output IP3: 8.8 dBm
- Input IP3: –11.2 dBm

b. Performance Various Input Power Levels

Pin (dBm)	Pout (dBm)	S/N (dB)	OIM3 (dBm)	ORR3 (dB)
–112	–92	12	–294	202
–60	–40	64	–138	98
–20	–1	103	–18.5	18.1

Figure 27.5 Calculation of receiver performance.

The system analysis box in the lower left-hand corner shows the operation conditions for the receiver. In the example of Figure 27.4, the system is operated under the following conditions:

Input power	−60 dBm
Analysis temperature	25°C (part temperature)
Noise bandwidth	0.03 MHz (bandwidth of a single user channel)
Reference temperature	25°C (noise temperature of the surroundings)
S/N (for sensitivity)	12 dB (required for a given BER with a particular modulation)

When these values have been entered, the Calculate button on the tool bar is pressed, and all the values shown in Figure 27.4 are displayed.

Figure 27.5a shows the performance of this overall receiver. The performance parameters are independent of RF power level.

Figure 27.5b shows the receiver performance at various RF input powers. At the sensitivity point, which is an RF input power of −112 dBm, the S/N is 12 dB, which is just adequate for achieving the required BER. The output third-order intermodulation power (OIM3) is −294 dBm. The ratio of the RF output power to the OIM3 (which is defined as ORR3) is 202 dB. Clearly, intermodulation products are not a problem at the sensitivity point.

At an input RF power of −60 dBm, the S/N is 64 dB and the ORR3 is 98 dB. Neither the noise figure nor the intermodulation products are a problem at this input power level.

At an input RF power of −20 dBm, the S/N is not a problem, but the intermodulation products are. Note that they should be 20 dB below the fundamental RF power; but as the ORR3 shows, they are only 18 dB below.

27.4 CALCULATION OF OVERALL RECEIVER PERFORMANCE AS A FUNCTION OF PART TEMPERATURE

The RF performance of RF filters, LNAs, and mixers all vary with part temperature. The amount of each component's performance variation with temperature is given in the part specifications. The Agilent software allows this variation to be taken into account to show the overall receiver performance as a function of temperature.

Figures 27.6 and 27.7 show the performance of the simple receiver, whose performance at 25°C was shown in Figure 27.4, at −30° and +60°C, respectively.

The results of these calculations are shown in the following table:

Temperature (°C)	Gain (dB)	NF (dB)	OIP3 (dBm)
−30	22.8	4.66	9.11
25	20.0	5.62	8.81
60	18.3	6.23	8.57

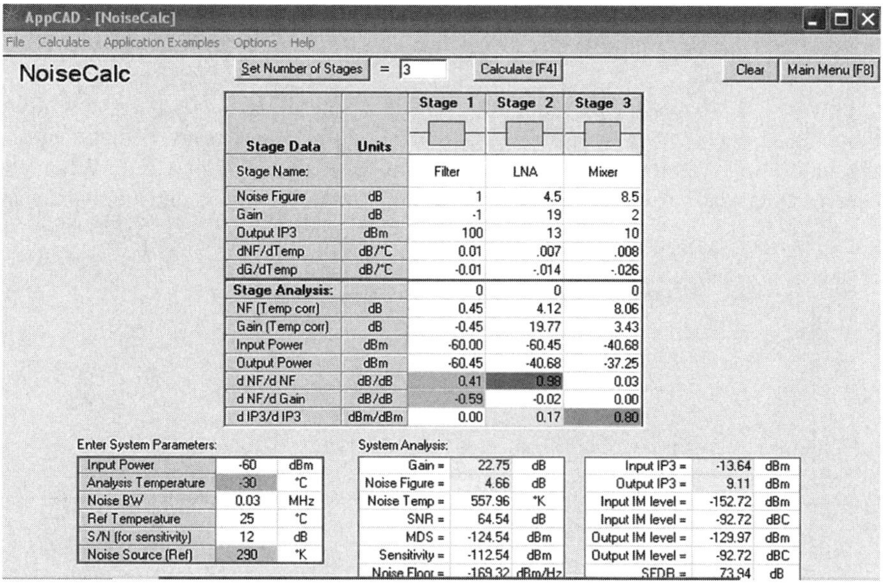

Figure 27.6 Change of receiver performance at −30°C. Agilent Technologies © 2008. Used with permission.

File Calculate Application Examples Options Help

NoiseCalc

Set Number of Stages = 3 | Calculate [F4] | Clear | Main Menu [F8]

Stage Data	Units	Stage 1	Stage 2	Stage 3
Stage Name:		Filter	LNA	Mixer
Noise Figure	dB	1	4.5	8.5
Gain	dB	-1	19	2
Output IP3	dBm	100	13	10
dNF/dTemp	dB/°C	0.01	.007	.008
dG/dTemp	dB/°C	-0.01	-.014	-.026
Stage Analysis:		0	0	0
NF (Temp corr)	dB	1.35	4.75	8.78
Gain (Temp corr)	dB	-1.35	18.51	1.09
Input Power	dBm	-60.00	-61.35	-42.84
Output Power	dBm	-61.35	-42.84	-41.75
d NF/d NF	dB/dB	0.35	0.97	0.04
d NF/d Gain	dB/dB	-0.65	-0.03	0.00
d IP3/d IP3	dBm/dBm	0.00	0.26	0.70

Enter System Parameters:

Parameter	Value	Unit
Input Power	-60	dBm
Analysis Temperature	60	°C
Noise BW	0.03	MHz
Ref Temperature	25	°C
S/N (for sensitivity)	12	dB
Noise Source (Ref)	290	°K

System Analysis:

Parameter	Value	Unit
Gain =	18.25	dB
Noise Figure =	6.23	dB
Noise Temp =	926.58	°K
SNR =	62.98	dB
MDS =	-122.98	dBm
Sensitivity =	-110.98	dBm
Noise Floor =	-167.75	dBm/Hz

Parameter	Value	Unit
Input IP3 =	-9.68	dBm
Output IP3 =	8.57	dBm
Input IM level =	-160.64	dBm
Input IM level =	-100.64	dBC
Output IM level =	-142.39	dBm
Output IM level =	-100.64	dBC
SFDR =	75.53	dB

Figure 27.7 Change of receiver performance at 60°C. Agilent Technologies © 2008. Used with permission.

27.5 SWITCHING THE LNA INTO AND OUT OF THE OVERALL RECEIVER

The performance of an overall RF receiver that must operate over a wide range of input powers can be improved by electronically switching the LNA into and out of the receiver circuit.

Figure 27.8 shows a block diagram of such a setup. The LNA has a low noise figure, high gain, but a low OIP3. In contrast, a mixer has a noise figure that is several dB higher than that of the LNA, low gain, but a high OIP3. When the receiver is operating near its sensitivity point, low noise figure and high gain are

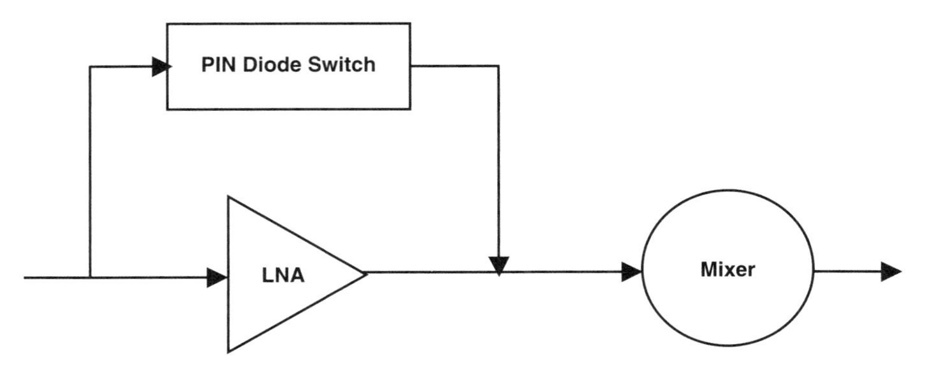

Figure 27.8 Receiver with switched LNA.

very important, but OIP3 is not. When the receiver is operating at its maximum allowable signal point, noise figure and gain is no longer important, but OIP3 is, so a PIN diode switch can be used as shown to bypass the LNA in the signal path.

27.6 ANNOTATED BIBLIOGRAPHY

1. APP CAD software program, hp.woodshot.com, Agilent Technologies Santa Clara, CA.

Reference 1 is a software program that allows the gain, noise figure, and IP3 characteristics of receiver components to be combined to give the overall receiver performance.

CHAPTER 28

RFICs

In previous sections of this book, each RF component has been discussed separately. However, modern RF equipment usually combines many components on a single semiconductor substrate. These parts are called RF Integrated Circuits. RFICs are available that contain all of the required components of the RF transmitter and receiver on a single semiconductor chip. Because of the complex nature of generating, frequency shifting, amplifying, and detecting RF, more often the complete RF systems are made on two chips, using different semiconductor materials on each. For example, gallium arsenide is often used for the power amplifier and LNA on one chip and silicon germanium is used for the VCO, upconverter, mixer, and detector on a second chip.

Because of filtering and shielding requirements, the semiconductor parts are often combined with the filters and shielding inside a module. Typical RFIC chip sizes are $4 \times 4 \times 1.5$ mm high, and modules are typically $10 \times 10 \times 1.5$ mm high.

All wireless communications systems consist of the RF components, whose design and testing has been discussed in detail into the previous chapters of Part III, and the digital signal processing parts that will be discussed in Part IV. Sometimes for relatively simple RF communications systems, all of the RF and digital circuits can be combined on a single chip. The semiconductor material then has to be silicon. The resulting performance will be poorer than if the optimum materials were used, but the total system will be significantly smaller and less expensive. These systems are called System-On a Chip (SOC).

RF Measurements for Cellular Phones and Wireless Data Systems. By A. W. Scott and R. Frobenius

Two representative RFIC systems, both using two RF chips and a few external components, are described in this chapter. The first is a wireless LAN transceiver operating in the 2.4–2.5 GHz band. The second is a cellular phone transceiver operating as a GSM, General Packet Radio Service (GPRS), or EDGE system in any one of the 800, 900, 1800, or 2100 MHz cellular phone bands. (These different kinds of cell phone systems are discussed in Chapters 30 and 31.)

28.1 WIRELESS LAN

Figure 28.1 shows a two chip wireless LAN mounted on an evaluation board. A block diagram is also shown in the figure. This RFIC operates in the 2.40–2.48 GHz Industrial, Scientific, and Medical (ISM) band, and it was manufactured by Triquint.

Looking at the photograph of Figure 28.1, the power amplifier/LNA chip is shown on the right-hand side of the evaluation board and it connects through a ceramic block

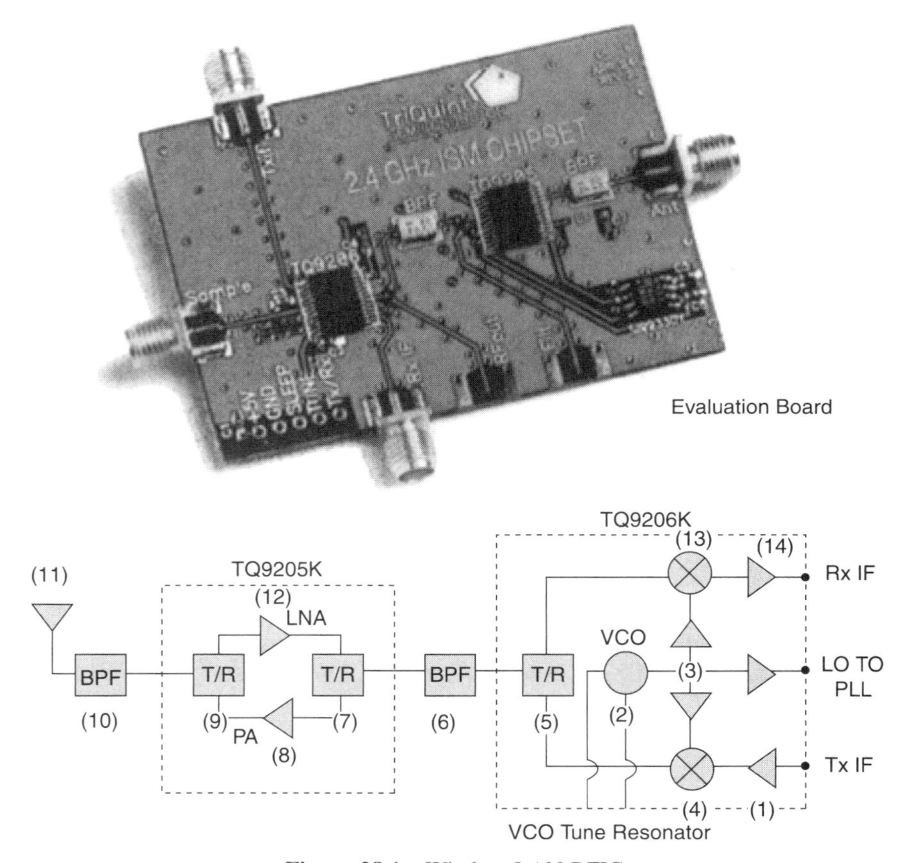

Figure 28.1 Wireless LAN RFIC.

filter to the antenna. The PLO, upconverter, and mixer chip is shown on the left. The block diagram of the two parts is reversed in the figure, with the power amplifier/LNA chip on the left.

The power amplifier used as the DUT in Chapter 19 and the LNA used as the DUT in Chapter 23 are contained in the left-hand chip of the block diagram of Figure 28.1.

In the block diagram in Figure 28.1, the modulated IF signal enters from the lower right-hand side of the chip. It is amplified by an on chip IF amplifier (1) and is transmitted to one port of the upconverter (4). A VCO (2) generates the RF signal for both the TX and RX. The transmitter and receiver signals are at the same frequency, because time division duplexing (TDD) is being used.

The VCO signal is divided into three parts (Fig. 28.1), and each part is amplified (3) on chip. One part is sent off chip to the PLL to generate the phase control signal, which controls the varactor diode in the VCO. The second part of the VCO signal is sent to the receiver mixer (13). The third part is sent to the transmitter upconverter (4).

The three transmit/receive PIN diode switches (5), (7), and (9) allow time division duplexing (Fig. 28.1). During the transmit interval of 10 ms, the upconverted modulated RF signal is transmitted through a bandpass filter (6) to the power amplifier (8). The bandpass filter removes the unwanted sideband of the upconverter. During transmission T/R switches (7) and (9) in the power amplifier/LNA chip are switched to allow the transmitted signal to be amplified by the power amplifier (8) and directed through the system bandpass filter (10) to the antenna (11) for transmission.

During the receiving cycle, the received signal coming into the antenna is routed through the bandpass filter (10), whose function is to allow only frequencies in the band from 2.40 to 2.48 GHz to enter the receiver (Fig. 28.1). The received signal is then routed through the T/R switch to the LNA (12). There is no problem with transmitter signal leakage interfering with the LNA, because they are not on at the same time. The amplified received signal from the LNA then passes through the T/R switch (7) back through the bandpass filter (6) and through the T/R switch (5) to the mixer (13). The mixer shifts the received signal down to an IF frequency where it is amplified in an IF amplifier (14) up to a power level where it can be easily demodulated.

28.2 FOUR BAND GSM, GPRS, EDGE HANDSET

A four band GSM, GPRS, EDGE cellular phone handset block diagram is shown in Figure 28.2. It is manufactured by RF Micro Devices and consists of two RF chips: RF3178 and RF6026.

Because different RF frequencies at 800, 900, 1900, and 2100 MHz are licensed to cellular phone service providers in different parts of the world, this mobile phone is designed to operate in any one of four different frequency bands. In addition, the phone can operate in any of these frequency bands using any of three modulation techniques: digital frequency modulation for a single digitized voice signal in the GSM mode, digital frequency modulation of a data signal using up to eight voice

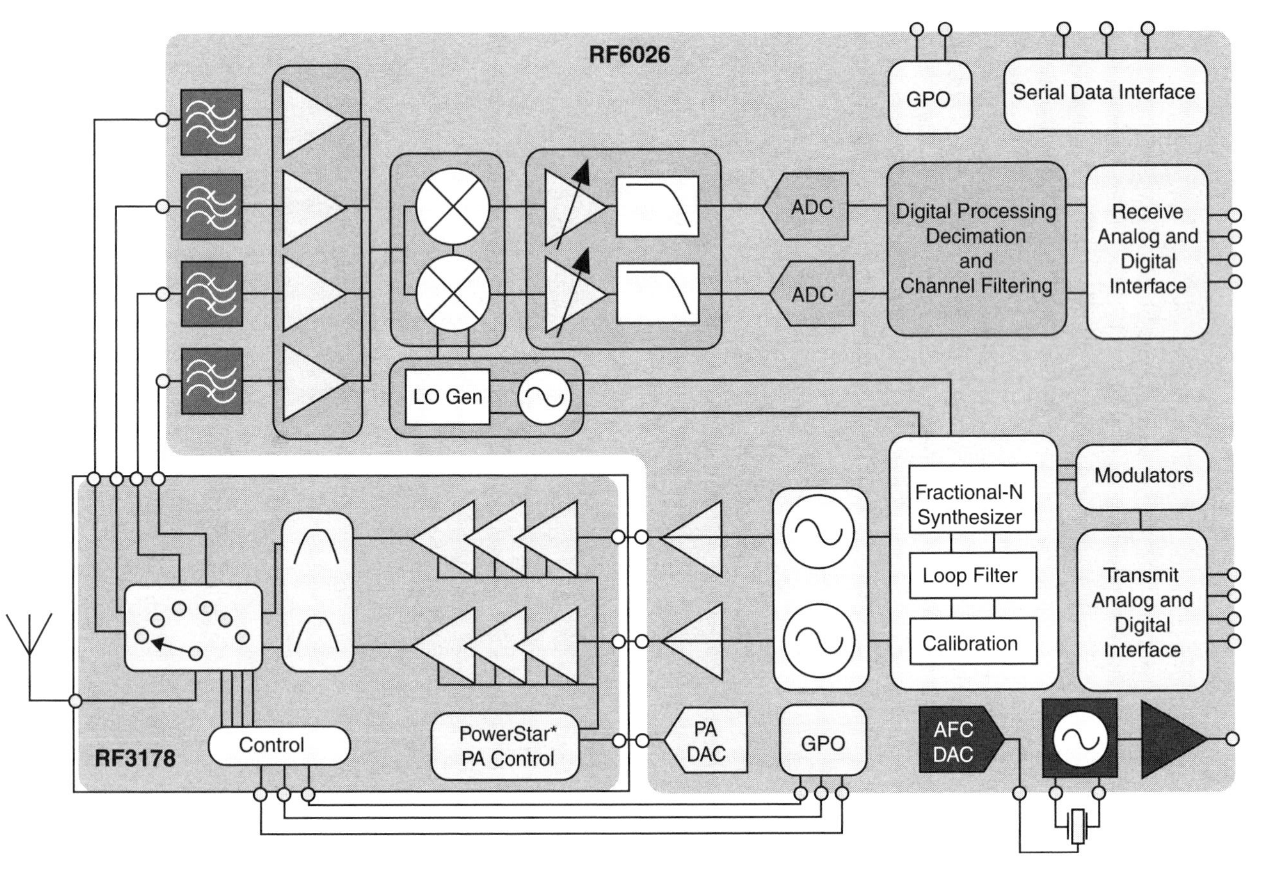

Figure 28.2 Four band GSM, GPRS, EDGE handset.

channels in the GPRS mode, and 8PSK modulation for higher speed data in the EDGE mode.

The RF3178 chip is the power transmit module shown at the bottom left block of the diagram. It contains two amplifier chains: one amplifying the 800 and 900 MHz frequency bands; the other amplifying the 1900 and 2100 MHz bands. The signal from each amplifier is routed through an RF filter to the PIN diode switches that route the RF output from one or the other amplifier to the antenna during the transmit time period. During the receive cycle, the switch routes the received signal from the antenna to the receiver chip.

The power amplifier chip also contains a polar modulator, which allows the amplifier to be operated near saturation without distortion even with 8PSK modulation. This chip uses HBT technology with a GaAs substrate.

Figure 28.3 shows a microphotograph of the RF3178 module. The power amplifier, shielded switch matrix, and control circuits can be seen. The complete module is 7 × 8 × 1.4 mm high.

The RF3178 quad band transmitter module has the following performance:

- >41% system efficiency for GSM, equivalent 51% power amplifier power-added efficiency (PAE)
- >37% system efficiency for Digitial Cellular System (DCS), equivalent 47% power amplifier PAE
- 2–6 dBm drive level, >50 dB dynamic range
- Controlled harmonic levels < −35 dBm
- Low insertion loss in the receive path
- +33.5 dBm GSM output power at 3.5 V
- +31.0 dBm DCS/PCS output power at 3.5 V

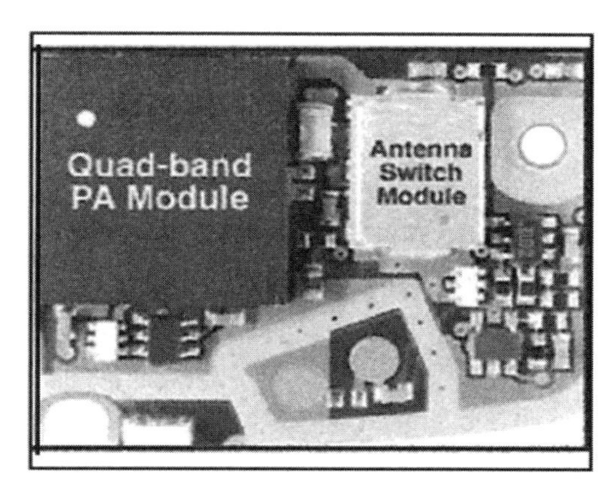

RF3178
7mm x 8mm

Figure 28.3 RF transmitter module.

The rest of the block diagram of Figure 28.2 shows the RF6026 receiver chip. Received signals in any of four RF frequency bands are routed by the antenna switch module in the RF transmitter chip to one of four RF SAW filters in the receiver module. The desired signal is amplified by one of the four LNAs. The lower two RF bands are mixed by one of the mixers and the two upper RF frequency bands are mixed by the other mixer. The mixers can be used in a very low IF or direct conversion (ZIF) mode. The signal from each mixer is passed through a low pass filter and detected.

The RF6026 chip contains several VCOs and PLOs and various processing and control circuits. The entire receiver circuitry is fabricated on a silicon chip. One region of the chip is alloyed with germanium to provide a silicon germanium region for the LNAs. The remainder of the chip is silicon for the mixers, VCOs, and processing circuitry. The total receiver package is $10 \times 10 \times 1.5$ mm.

The RF6026 quad band receiver module has the following performance:

Very low IF and direct conversion receiver architectures

EDGE transmit capability with digital polar modulator and use of near saturated power amplifier for maximum PAE

Highly integrated: Includes all VCOs, loop filters, SAW filters, power amplifier, ramp digital to analog converter (DAC), automatic frequency control (AFC) DAC, and passives

Digital GMSK modulator

Selectable digital channel filters from 80 to 135 kHz

28.3 ANNOTATED BIBLIOGRAPHY

1. POLARIS™ 2 TOTAL RADIO™ Module Brochure, http://www.rfmd.com/pdfs/Polaris2RadioModule.pdf, RF Micro Devices, Inc., Greensboro, NC.

Reference 1 provides the data for the discussion of the quad band cellular phone example in the text.

PART IV

TESTING OF DEVICES WITH DIGITALLY MODULATED SIGNALS

CHAPTER 29

WIRELESS COMMUNICATION SYSTEMS

The complex growth of the cellular phone and wireless data communication industry was described in Chapter 1. There are several major types of cellular phone systems and short-range wireless data systems. The reason for this diversity is attributable to business conditions as much as to advancing technology. Modern full feature mobile units have a manufacturing cost of around $200 and are paid for by the user as part of the monthly service charge. They are usually replaced every 2 years.

The base station equipment is a different issue. It is very complex to allow the wireless phone and data systems to work with the simplest and least expensive hardware in the mobile unit. Because of its high cost, the base station equipment cannot be rapidly updated to incorporate the latest technology advances. In addition, in about the middle of the 30 year history of cell phones, the CDMA technique was demonstrated to provide about twice the user capacity in a given licensed bandwidth compared to systems using FDMA/TDMA techniques. By that time, 80% of the world's mobile phones were FDMA/TDMA and could not be immediately converted to CDMA because of the large base station costs involved. FDMA, TDMA, and CDMA techniques are described in Chapter 30.

However, in spite of the cost of converting to CDMA, it is expected that most of the existing cellular phone systems in the world will be CDMA by 2010, but this will be an evolutionary process.

Techniques for optimally designing the CDMA channels for data and then using Voice over Internet Protocol (VoIP) packet data techniques for voice will further

RF Measurements for Cellular Phones and Wireless Data Systems. By A. W. Scott and R. Frobenius

increase capacity. Again this will involve significant expenditures for base station equipment.

An even higher data rate cellular phone system called Wireless Interoperability for Microwave Access (WiMax) has been developed by using an OFDMA technique similar to the OFDM multiple access system used by the WiFi short-range, high data rate system. OFDM and OFDMA are discussed in Chapter 31. The WiMax system will probably not immediately replace CDMA cell phone systems in North America, Europe, or Japan because service providers already have such a large investment in TDMA and CDMA base station equipment. However, WiMax cellular phone systems are already being implemented in countries that do not as yet have major cell phone installations.

This chapter is organized as follow. A generalized block diagram of a complete wireless communication system is discussed in Section 29.1. Analog voice and video signals are described in Section 29.2. The digitizing of analog signals is explained in Section 29.3. Digital data are described in Section 29.4. Compression of digital voice and video signals is discussed in Section 29.5. Error correction is discussed in Section 29.6. Typical data rates for voice, video, and data systems are tabulated in Section 29.7. Packet switching systems for voice and video, as well as data, are described in Section 29.8.

29.1 BLOCK DIAGRAM OF THE COMPLETE WIRELESS COMMUNICATION SYSTEM

In spite of the technical differences in individual systems, all wireless phone and data systems share the common block diagram shown in Figure 29.1. The RF transmitter and RF receiver are shown on the right-hand side of the block diagram. RF systems were described in Chapter 15, and the design and measurement of the individual RF components making up the RF system were described in Chapters 16–28.

The remaining components in the overall system block diagram are digital processors to convert the analog voice and video signal into digital signals, to apply compression techniques to reduce the transmitted bit rate, to apply error correction bits, and to optimize the number of digital channels that can be transmitted within the RF system bandwidth. These components will be discussed in this and the following two chapters.

The analog signals from the voice microphone and the video camera are shown entering the transmitter part of the communication system from the left-hand side of Figure 29.1. The signals are digitized, and the digital signals are then compressed to reduce their bandwidth requirements. The digital signal representing data cannot be compressed.

The digital signals representing voice, video, and data are then sent to an error coder, which adds extra bits to allow transmission errors to be corrected.

The digital signals to be transmitted from an individual user occupy a small fraction of the service provider's licensed bandwidth. To accommodate many users the digital signals from each user are assigned a specific frequency slot and/or time

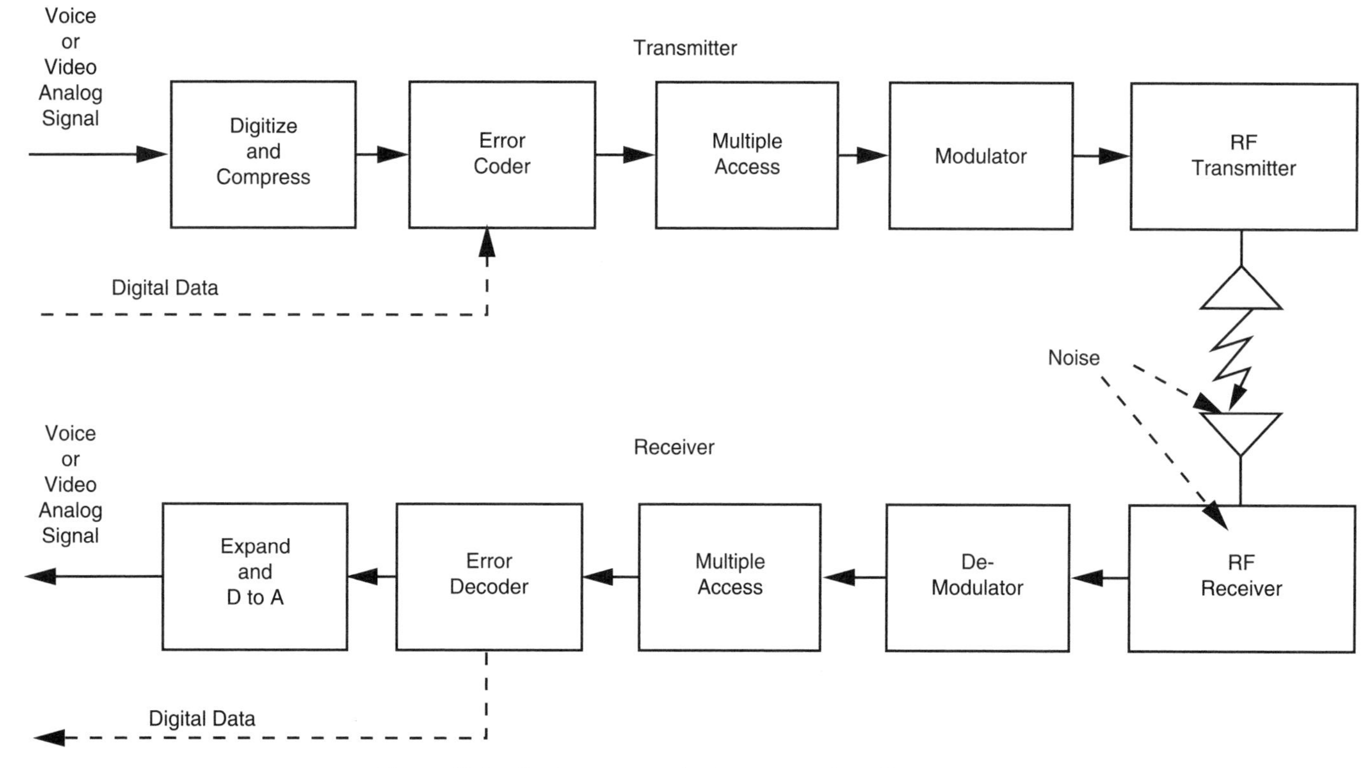

Figure 29.1 Wireless communication system block diagram.

slot, or a coding sequence. This is accomplished in the multiple access block of the overall system.

As the digital signal leaves the multiple access block, it is just a series of digital bits (which may represent voice, video, or data) and system control signals. These bits are next modulated onto an IF subcarrier, which is at a frequency below the RF band where the modulation can easily be accomplished with a digital processor. The modulated IF subcarrier is then shifted to the RF frequency in the upconverter block of the RF transmitter.

In the communication system receiver shown in the lower row of boxes in Figure 29.1, the demodulator, multiple access, error decoder, expander, and DAC simply perform the inverse functions of their counterparts in the transmitter.

29.2 ANALOG VOICE AND VIDEO SIGNALS

It was explained in Chapter 5 that any electronic signal that repeats its waveform in time can be analyzed as a sum of single frequency signals. This procedure is called spectrum analysis.

Figure 29.2 shows the characteristics of the audio signal from a microphone or telephone. In Figure 29.2a, the electronic signal of a human voice from a microphone is shown as a function of time. The major peaks are approximately 1 ms apart, which corresponds to a frequency of 1 kHz. The frequency spectrum of this voice signal is shown in Figure 29.2b. The frequencies in the human voice extend from 30 Hz at the low end to 6 kHz at the high end, with the largest signals around 1 kHz. These characteristics of the audio signal of a human voice are determined by the human vocal chords. A human cannot speak at frequencies below 30 Hz or above 6 kHz.

The frequency components of the audio signal from musical instruments extend over a much wider frequency range. Musical instruments can generate frequencies

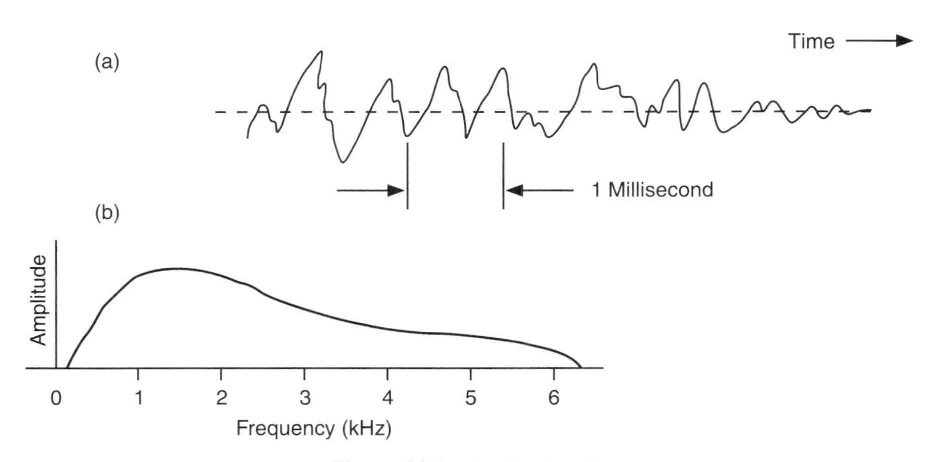

Figure 29.2 Audio signals.

below 30 Hz and well above 6 kHz. However, the average human ear cannot hear audio signals below 30 Hz or above 15 kHz. Therefore, the frequency range of music is taken to be 30 Hz to 15 kHz.

A telephone signal extends from 300 Hz to 3.4 kHz. The telephone set deliberately filters out frequencies from the human voice below 300 Hz and above 3.4 kHz, which is why a person's voice over a telephone can be recognized but the telephone signal does not sound exactly like the person. The telephone set does this to reduce the signal bandwidth so that more telephone calls can be fitted into a given communication system. The bandwidth of a telephone call is conventionally taken to be from 0 to 4 kHz. The actual audio signal is 300 Hz to 3.4 kHz, and this signal is fitted into the 0–4 kHz telephone channel. The narrower frequency range of the actual signal compared with the telephone channel gives guard frequency bands so that the signal from one channel does not overlap into another channel.

The electronic signal of conventional analog broadcast television is shown in Figure 29.3. High definition TV (HDTV) is similar, but it uses a wider bandwidth. The television camera converts a visual image into an electronic signal by projecting the visual image onto the face of the TV camera. An electron beam scans the back of the camera face one line at a time from top to bottom. When the scanning beam reaches the bottom of the tube face, it returns to the top and repeats the scanning process. A simple three-line scanning pattern is shown in Figure 29.3a. If a given spot on the image is bright, many electrons are emitted from the camera face as the electron beam hits that spot during its scan. At this instant of time, the TV camera produces a large output signal.

Four bright areas between the dark areas are shown in the figure, and the resulting electrical signal as the electron beam scans across the face of the camera is shown in Figure 29.3c. Note the four higher amplitude regions of the signal during each scan

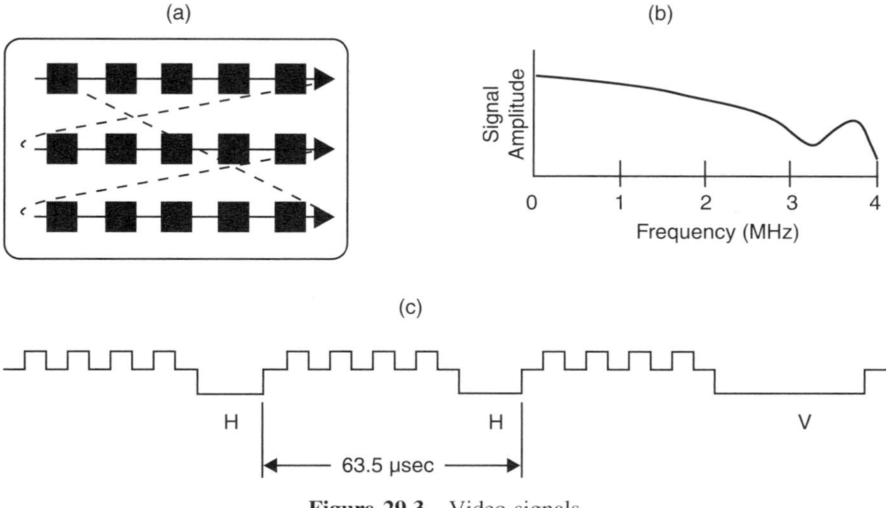

Figure 29.3 Video signals.

corresponding to the four brighter spots and the five lower amplitude regions corresponding to the dark spots.

A broadcast television camera actually has 525 scan lines. The scan must be accomplished so fast that the human eye cannot see it. The 525 lines must be scanned in 0.33 s, and the beam must return to the top of the camera face and repeat the scanning process. Therefore, 63.5 μs are allowed for each line to be scanned across the camera. An interval of 12.5 μs of this time (H regions, Fig. 29.3c) is used for the scanning beam to return to scan the next line. The beam returns from the bottom to the top of the screen to repeat the complete scanning process in the interval marked V in the figure. Therefore, each line of the picture is scanned in 50 μs. Broadcast television has a resolution capability of 400 light and dark spots across one line. This must be scanned in 50 μs, so the highest frequency component, if the finest detail in the picture is used, is 4 MHz. In a normal broadcast television display, part of the screen contains large objects where the signal does not change from bright to dark over most of its scan. These regions give lower frequency components below 4 MHz, so the TV signal contains, (Fig. 29.3b) frequency components extending from 0 to 4 MHz. The concentration of signals at 3.5 MHz is the color information.

The video signal for broadcast television requires 4 MHz of bandwidth. Recall that 1000 telephone calls can be multiplexed together into this same bandwidth. This is easy to remember by noting that "one picture is worth a thousand words." The 525 lines could be scanned at a slower rate than 0.016 s, but then the picture would not seem to be continuously moving. However, a slower scan speed reduces the bandwidth requirements of the transmission system; this is used in video conferencing, where a series of views, rather than a continuously moving picture, are presented. An extreme example of slower scanning is the TV pictures sent back from the deep space planetary probes. One complete picture requires 5 min to transmit, because the signals are sent one line at a time over a 5 min interval. This must be done to reduce the bandwidth requirements in order to reduce noise. At the opposite extreme, more scan lines are used to obtain HDTV. The bandwidth requirements are greater than 4 MHz.

At the receiving end, the process is reversed and the electrical signal is reconverted to a visual image in the TV monitor. The intensity of the electron beam as it scans the monitor is high when the signal is strong and low when the signal is weak. The stronger the electron beam is, the brighter the spot on the monitor. The picture is therefore reproduced as a series of bright and dark spots on the monitor face.

Whether standard broadcast television or HDTV is considered, the video signals require a large bandwidth for their transmission, because they contain many more higher frequency components than a single telephone call does.

29.3 THE DIGITIZING OF ANALOG SIGNALS

Audio and video signals are in analog form. The signal is a voltage that varies with time, and its amplitude is proportional to the amplitude of the audio or video information. These analog signals are often coded into digital signals before transmission to make the signal immune to noise interference.

The advantage of digitizing analog signals to eliminate noise interference is illustrated in Figure 29.4. Part a shows the analog signal (such as an electronic signal from a telephone microphone or a television camera) that is to be transmitted as a voltage varying with time. If noise is added during transmission, the signal is distorted (Fig. 29.4b). However, digitizing the signal in the same high noise environment eliminates the distortion (Fig. 29.4c and d). To do this, the analog signal is not transmitted directly. Instead, the analog signal is measured before transmission. Figure 29.4c shows a measurement at one point in time. Rather than transmitting the 5 V signal at this instant of time, the 5 is coded into a digital signal, represented as 1001110, which is transmitted as a series of positive and negative pulses. When the digital signals, which represent the value of the analog signal at a given instant of time, are transmitted and received they are badly distorted as shown in Figure 29.4d. Note that some of the 1s are large and some are small, but there is no question which digits are 1s and which are 0s. Although a great deal of noise has been added to the digital signal, there is no question that this sequence of digits is 1001110, which represents the value 5. Therefore, the value of the original analog signal is known at the receiving end of the system at this instant in time. At several instants of time, the value of the analog signal to be transmitted must be measured, converted into a digital signal that represents the value, and transmitted. At the receiving end, the original analog signal must be reconstructed from the digital code. Note that the reconstructed signal is nearly a perfect replica of the original signal.

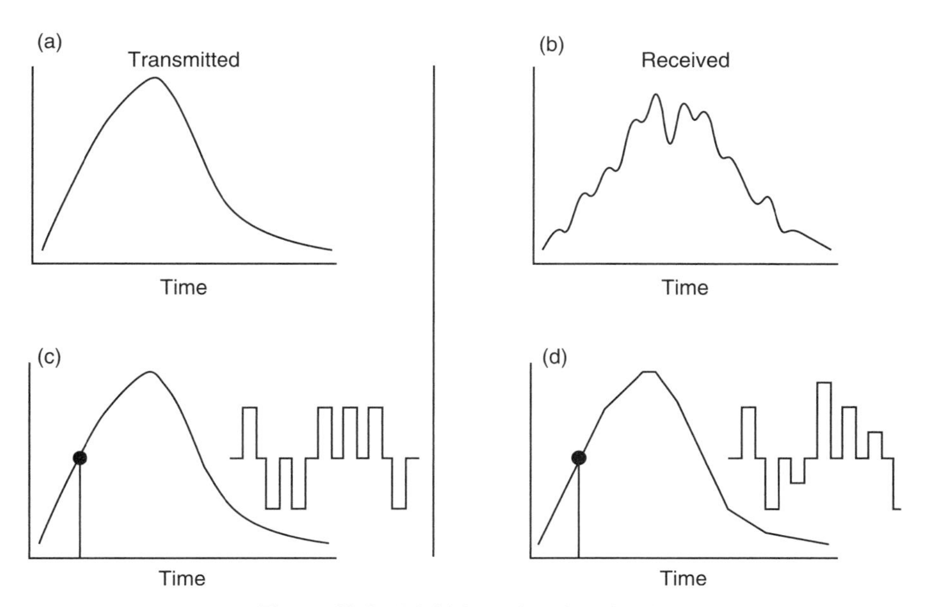

Figure 29.4 Digitizing of analog signals.

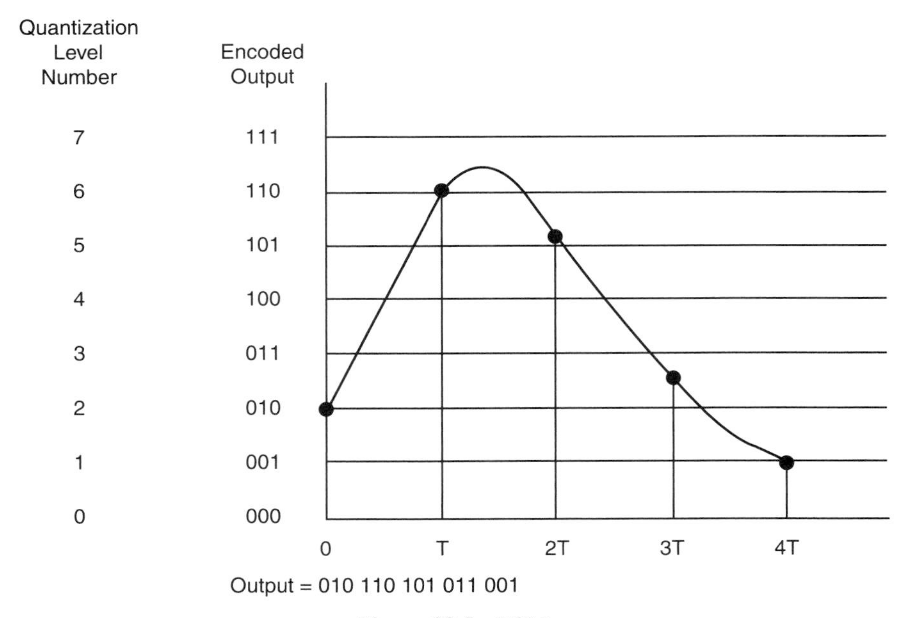

Figure 29.5 PCM.

Figure 29.5 shows an analog signal to be transmitted as a function of time. The signal varies over the amplitude range from 0 to 7, and a three digit representation of the numbers from 0 to 7 is shown. At various sampling times, the signal is measured, represented by its digital code, and the digits are transmitted. Therefore, the transmitted output is 010 110 101 011 001. The 1s and 0s are transmitted and noise adds to them, so upon reception the amplitudes of the 1 and 0 pulses are distorted but there is no question as to which is a 1 and which is a 0.

However, the reconstructed signal is not perfect, because of "quantization" error. The quantization error is illustrated by the analog signal measured at time 3T in Figure 29.5. At this sampling time, the signal is between 2 and 3. Because the digital code only has the numbers 2 and 3, the analog signal is quantized and given the value 3. Thus, there is an error. No error is added by the signal transmission, but the error is inserted when the signal is coded; this error is called the quantization error. When the digital signal is reconstructed into analog form at the receiver, the analog signal is given the value of 3, which is not the true value that the original signal had.

The quantization error can be reduced to an arbitrary small value by making the step size smaller, which requires more digits. If eight digits are used, 256 quantizing values are produced instead of the eight values in the example. The signal/quantization noise ratio would then be 50 dB. Consequently, the error would meet the S/N requirements for telephone transmission. Therefore, an eight digit representation of the analog signal is used in most telephone systems.

The signal must be sampled at twice the highest frequency present in the analog signal to obtain both the amplitude and phase information in the signal. Sampling at a lower rate does not provide a faithful time representation of the signal. Sampling at a higher rate requires more data to be transmitted than is necessary.

It might appear that if digitizing is used, the received S/N is no longer important. However, if the S/N is bad enough, so much noise is added that a 1 could be mistaken for a 0 and vice versa. The likelihood of this happening is a function of the S/N and the type of modulation being used, which was discussed in Chapter 4.

29.4 DATA SIGNALS

Information can be transmitted in far less bandwidth by using a text message than by using a telephone call. Figure 29.6 shows the digital representation of the letters U and F, the number 3, and the carriage return (CR) control signal. These digital representations are taken from ASCII code, the basic digital code used for data transmission. The letter U is represented by 1010101. It can be represented as an electronic signal where the positive voltage represents a 1 and the negative voltage represents a O. The digital electronic signal representing the digits for F, 3, and CR (or enter key) are also shown. The ASCII code has seven digits, so the number of characters is 2^7, or 128 characters. These characters include capital letters, small letters, numbers, symbols, and control signals and represent the keys on a standard computer keyboard.

An error in any of the seven digits would give a completely different meaning to the symbol. For example, if the last digit of the letter U, which is supposed to be a 1, was 0, the digital code would represent an entirely different letter. To eliminate this type of error, most data transmission systems add a parity digit (Fig. 29.6b), so that

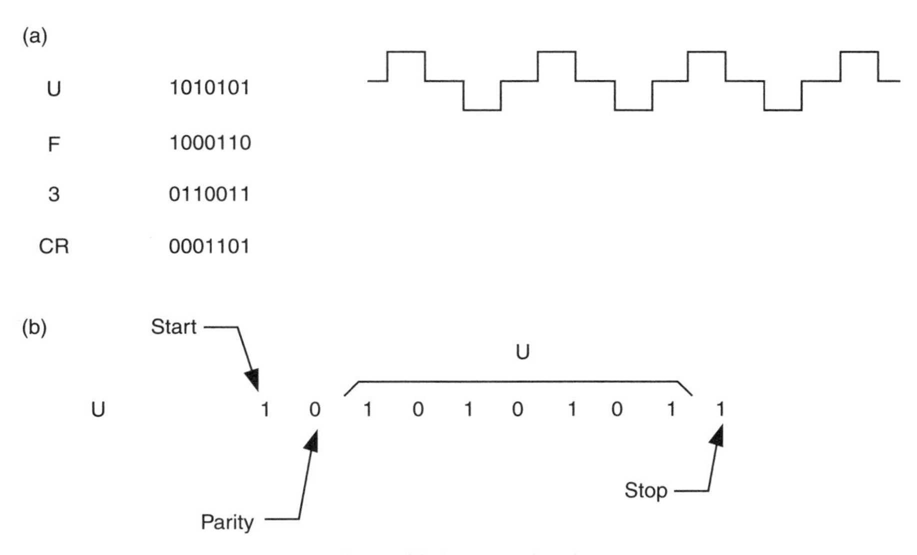

Figure 29.6 Data signals.

the total number of 1s in the symbol is always even. The letter U already has an even number of 1s, so the parity digit is 0. The letter F has an odd number of 1s, so the parity digit is 1. The number of digits is counted when the data signal is received; if the number of 1s is not even, then the receiver knows there has been a transmission error. The receiver does not know which digit is wrong, but it knows that the transmission of that symbol is incorrect and requires it to be retransmitted. The parity digit increases the number of digits from 7 to 8.

Finally, a start and a stop digit are required in most data transmission systems to distinguish between one symbol and the next. The total number of digits then required for transmission of a single symbol (a letter or number) is 10.

For the transmission of 1 word/min, using the standard ASCII code with parity and a stop and start digit, with each word consisting of six letters, and each letter consisting of 10 bits, 1 word/min requires 1 bps, which requires a 1 Hz bandwidth. Therefore, a 120 word/min message, which is the rate of human speech, can be fitted into a 120 Hz bandwidth. Note that a single analog telephone call requires a 4 kHz bandwidth, and a single television channel requires a 4 MHz bandwidth. The bandwidth advantage for the transmission of data is clearly evident.

29.5 COMPRESSION OF DIGITAL VOICE AND DATA SIGNALS

The number of digits in a digital voice or video signal can be reduced to minimize the bandwidth required for transmission. This is called "signal compression." Voice compression is fairly complicated. Video compression is much easier to achieve.

Compression of Voice Signals

The standard digitized telephone voice signal is 64 kbps. This bit rate is determined as follows: the highest frequency in an analog telephone channel is 4 kHz. When a telephone signal is digitized, it must be sampled at twice the highest frequency to get both amplitude and phase information. This sampling rate is therefore $2 \times 4000 = 8$ kbps. Each sampling of the signal must use 8 digital bits to achieve the required 50 dB S/N. Therefore, the bit rate of a digitized phone signal is 64 kbps, which requires 16 times more transmission bandwidth than an analog voice signal.

The bit rate required for understandable telephone conversation can be compressed in various ways, as shown in Figure 29.7. The horizontal axis shows the bit rate of the particular type of compression scheme. The vertical axis shows the quality rating of the compression scheme, as evaluated by typical phone users. The users are asked to rate the compressed telephone signal on a scale of 0–5. A rating of 5 would be a speech signal that did not use a telephone and had all frequency components of the human voice up to 6 kHz. An analog speech signal transmitted over a standard analog telephone system has a bandwidth of only 4 kHz, and it is rated as 4.2 by the users. The same rating is obtained for digitized phone speech using 64 kbps, which is shown by the data point labeled PCM. The digitized phone sounds exactly the same as an analog phone signal if 64 kbps is used.

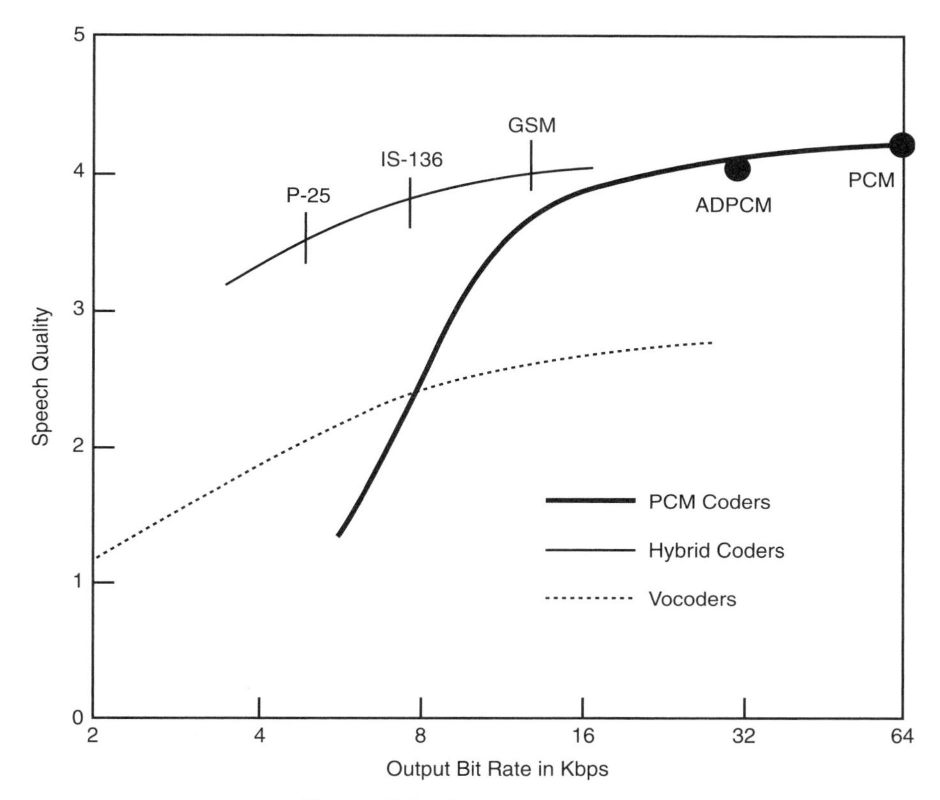

Figure 29.7 Speech compressors.

One simple type of phone voice compression is called ADPCM. ADPCM reduces the bit rate to 32 and 16 kbps, which receive ratings of 4.0 and 3.7, respectively, by typical phone users. With ADPCM, each sample of the analog signal is digitized and compared to the previously digitized sample. Only the difference signal is transmitted. If the samples are obtained frequently enough, a human's vocal cords, tongue, and lips change very little from one sample to the next, so the change in sound is very small. However, below a sampling rate of 16 kbps, ADPCM performance degrades rapidly.

The performance of voice synthesizers or vocoders is shown by the dotted curve in Figure 29.7. Voice synthesizers analyze the speech into its amplitude, frequency (pitch), and filter characteristics matching the voice characteristics created by the human diaphragm, vocal cords, and tongue. Intelligible speech is created, but it sounds to the user like a robot is speaking. Consequently, it is rated as poor in quality.

All cellular phone digitized communication systems use hybrid coders. A block diagram of a hybrid coder is shown in Figure 29.8. In a hybrid coder the voice is first synthesized using 2 kbps. These synthesized coefficients are transmitted wirelessly to the receiver. The synthesized coefficients are also used in the transmitter

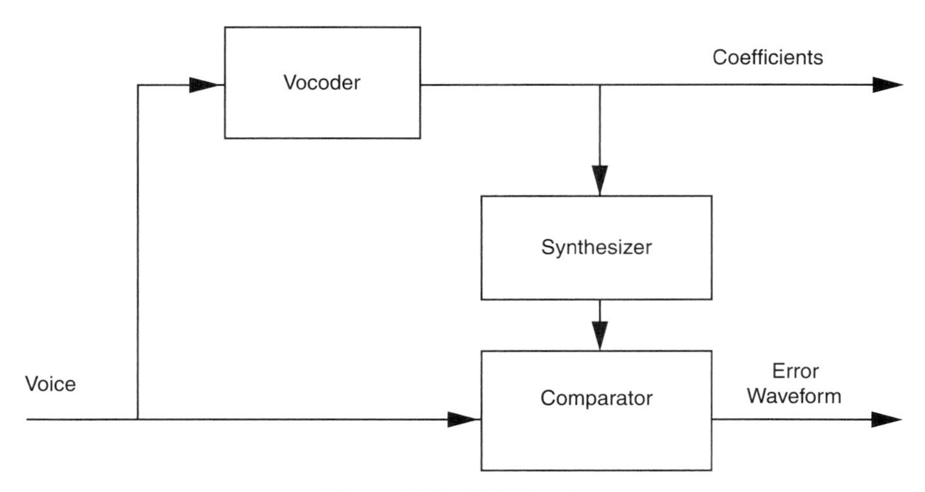

Figure 29.8 Hybrid coder.

to generate the synthesized voice. The bits representing the synthesized voice are compared to the bits representing the actual speech, and the resulting error bits are sent to the receiver to correct the synthesized voice. The more bits used in the error waveform, the better the speech sounds to the listener.

The performance of three different cell phone hybrid coders are shown in Figure 29.7. The GSM coder was designed to digitize the sounds of all world languages, and it does an excellent job. As shown, it requires 13 kbps for the digitized, compressed voice signal. Two kilobits/second are used for the synthesized voice signal, and 11 kbps are used for voice sound correction.

The IS-136 coder was used for the original digital cellular phone system in North America. It was carefully designed to be backward compatible with the original analog phone system, so that three digitized voice channels could be fitted into the existing 30 kHz wide analog channels. Consequently, the available bits for voice sound correction was only 6 kbps, which, when added to the 2 kbps synthesizer, gave a total bit rate of 8 kbps. The IS-136 coder makes every user have an American speech accent. Its advantage is that it uses one-half the bandwidth of the GSM hybrid coder. The P-25 speech compressor is designed for emergency police and fire services, and it is backward compatible with the existing public safety analog mobile phone system. It makes the compressed speech sound like a real human, but not the particular human who was talking. The capability of making the talkers identity known is not a necessity for public safety communication.

Compression of Video Signals

If analog video signals were digitized with PCM, the data rate of a single video channel would be 4 MHz $\times$ 2 $\times$ 7 digits = 56 Mbps. (Note from Section 29.2 that only 7 bits are required to obtain the 40 dB quantization S/N required for TV.)

This high bit rate would require 56 MHz of RF bandwidth, which would require almost the entire VHF band.

Fortunately, unlike voice compression, which is difficult, TV compression is easy. The reason is that most of the TV digital information is redundant. For example, during the evening news broadcast, most of the time nothing in the picture changes from one frame to the next except for the announcer's lips. There was no way to take advantage of this with original analog broadcast TV signals. However, with digital TV, only the bits in certain columns and rows change and new values for these pixels can be corrected for each frame, with all the other pixels kept constant.

The specification for digital broadcast TV that implements this compression is MPEG-2. The only challenge for MPEG-2 is how to handle the situation when the entire scene changes. This is accomplished by delaying transmission of the TV signal until every pixel in the new scene can be changed.

MPEG-2 also has the provision to vary the data rate of each broadcast video program, depending on its action content. Programs with little movement, like a panel discussion of political issues, require a data rate of only 1.5 Mbps. A musical with dancing requires 3 Mbps. Sporting events require from 3 to 6 Mbps, depending on the action content. For example, golf would be at the low end of the bit requirement and hockey would be at the high end. Action movies require 7 Mbps. The bit rate for each presentation is not varied as the action changes but is constant throughout the entire presentation, based on its general action content.

For video presentations on small cell phone screens, the high resolution capability of broadcast TV is not required, so fewer pixels are used.

Various compression techniques have been devised for HDTV. One popular HDTV format in the United States uses 17 Mbps.

29.6 ERROR CORRECTION

Extra bits are added to the information bits of every wireless communication system to allow for the correction of transmission errors. Two types of error correction are shown in Figure 29.9. Figure 29.9a shows Automatic Repeat Request (ARQ). This error correction technique was already explained in Section 29.4, where the digital transmission of data was discussed. With ARQ, the data being transmitted is always an odd number of digits (3, 5, 7, 9, . . .). The number of 1s in the data signal is first determined. An extra parity bit is added so that the total number of 1s will always be even. As shown, the total number of information bits in this example is 5, and the total number of bits in the transmitted signal is 6, because of the parity bit. The group of bits is transmitted, and at the receiver end, the first operation is to count the number of 1s in the symbol. If the number of 1s in the signal is even, the parity bit is dropped and the set of 5 data bits is accepted as good. However, if the number of 1s is odd, one of the bits is wrong. In this case the symbol is rejected

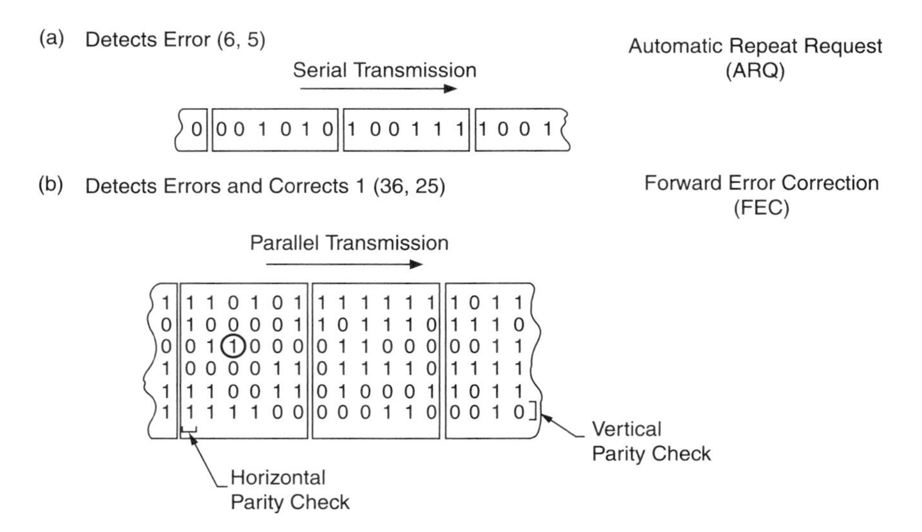

Figure 29.9 Error correction.

and the transmitter is notified to resend the symbol. The parity error correction scheme does not correct errors; it simply detects them.

A more sophisticated technique called Forword Error Correction (FEC) is shown in Figure 29.9b. With this technique a group of five symbols, each with its parity bit, are transmitted as a group. The parity bits for each row of data bits are shown as "horizontal parity check." In addition, a separate parity check can be performed on each column, which is shown as "vertical parity check." With a row check and a column parity check, the particular bit that is incorrect can be detected at the receiver and corrected, so the symbol does not need to be transmitted. For example, suppose the encircled bit in Figure 29.9b was 0 instead of 1. The horizontal parity check would indicate that one of the bits in line three was wrong. The vertical parity check would indicate that one of the bits in column three was wrong. Using both pieces of information would allow the circled bit to be changed to the correct value of 1 at the receiver.

The FEC technique explained above is a simplified version used to illustrate the principle. Figure 29.10 shows a realistic FEC design. The horizontal axis is the BER of the transmission system before error correction. The vertical axis is the BER of the transmission after the FEC is applied. The particular FEC system shown uses a bitstream of 64 information bits and 64 error correction bits. Note that a transmission BER of 10^{-2} can be corrected by 5 orders of magnitude to 10^{-7}. Referring to Figure 3.2 of Chapter 3, to obtain a BER of 10^{-7} by simply increasing the RF transmitted power requires a S/N increase from 4 to 11 dB. This would require 7 dB (5 times) more received RF power. The technical cost of using FEC is that two times more bandwidth is used. All wireless voice and data systems use about two times extra bits for error correction.

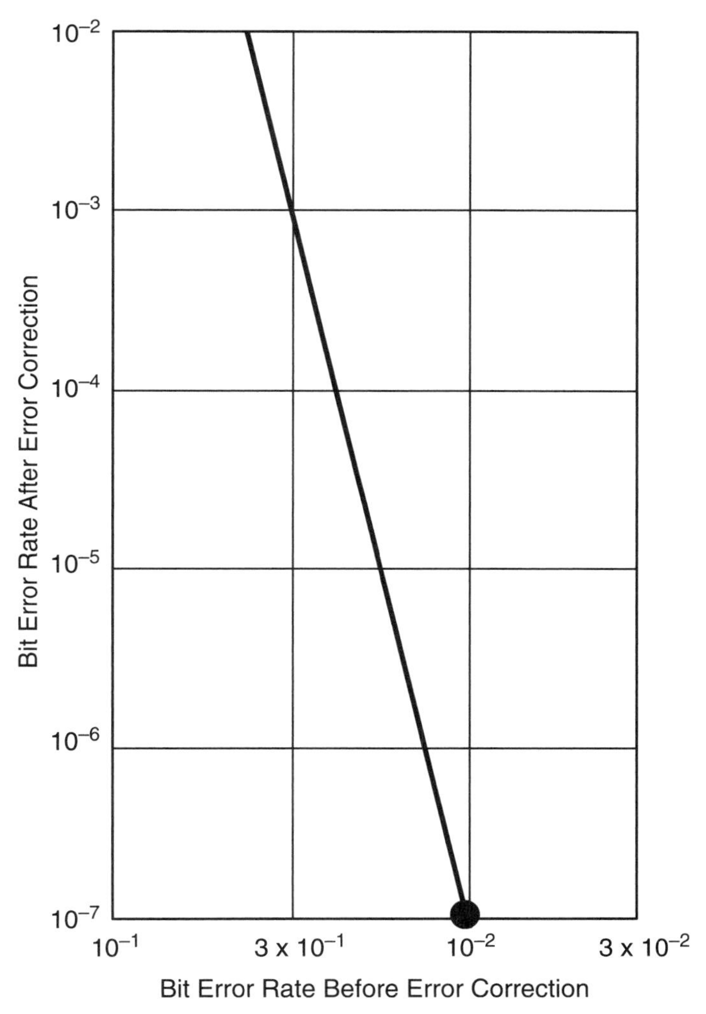

Figure 29.10 Decoder output versus input BER.

29.7 TYPICAL BIT RATES OF COMMUNICATIONS SYSTEMS

Figure 29.11 shows digital data rates in wireless communications systems, with digital data rates for some wired systems for comparison. The left-hand column shows various telephone voice signals. The transmitted digitized phone signal is about twice the actual digitized voice signal when the necessary error correction bits and system control bits are added to the signal. Because CDMA can adjust the bandwidth of the user transmission channels between high usage times (e.g., 5 PM)

Voice	
Analog Telephone Voice	4 kHz
PCM Digitized Telephone Voice (Wireline)	64 kb/s
ADPCM	16 kb/s
Compressed Digitized Telephone Voice	
IS-136 (NADC)	8 kb/s
IS-95 (CDMA)	9.6 – 13 kb/s
GSM	13 kb/s
Digital Wireless Phone with Error Correction	
IS-136	16 kb/s
IS-95	19.7 kb/s
GSM	34 kb/s

Data	
T1 Line (24 PCM Voice)	1.5 Mb/s
ISDN	384 kb/s
Dial-up Telephone Modem	9.6 kb/s
	28 kb/s
	56 kb/s
ADSL	1 Mb/s-350 kb/s
Video	
Analog Broadcast TV	4 MHz
400 h x 300 v pixels	
30 frames/sec	
MPEG-2 Compressed Broadcast Video (30 frames/sec)	1.5-7 Mb/s
MPEG-4 Compressed Video (10 frames/sec)	64 kb/s

Figure 29.11 Bit rates of communication systems.

and low usage times (e.g., 2 AM), a range of bits from 13 to 9 kbps per user can be used. With 13 kbps the telephone voice signal sounds more like a wired phone.

The upper half of the right-hand column in Figure 29.11 shows the data rates of wired digital data systems. The lower half of the right-hand column shows the data rates of analog and digitized compressed TV signals.

The number of voice channels and video channels or data channels that can be transmitted in any given wireless system depends on the type of multiple access techniques used and the RF bandwidth available. These issues are discussed in Chapters 30 and 31.

29.8 PACKET SWITCHING

The original analog and even newer digital cellular phone systems use circuit switching techniques for connecting phone channels between users. This is the same switching technique used for wired phones, and it is wasteful of precious RF bandwidth.

With circuit switching, there is a delay of tens of seconds while the connection is being made between the calling party and the receiving party, including the dial-up time. There is additional idle time on every connection because each party is talking only 40% of the time.

From the very beginning of its use, the transmission of data from one computer to another was accomplished using packet transmission. All communication terminals were permanently connected to a common transmission line as shown in Figure 29.12a. The data being sent were broken up into short packets, and the

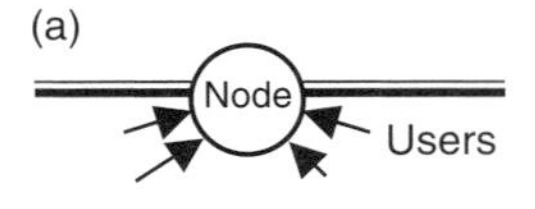

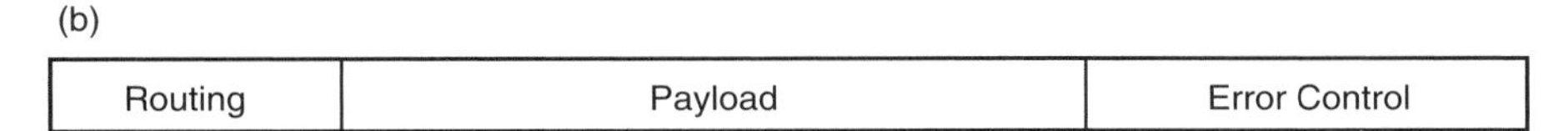

Figure 29.12 Packet switching principles.

packets were transmitted when a channel was idle. Various protocols were developed to ensure that every terminal was given a fair access to the transmission line. The data to be transmitted were broken up into packets, as shown in Figure 29.12b. The packets contain the data payload, extra routing bits at the beginning to indicate where the packets are to be sent, and error correction bits at the end. Representative data transmission protocols are Ethernet and Internet.

Originally, digital voice telephone messages could not be sent over packet data systems, because of the delay between the packet transmissions. However, as wireless transmission speeds have increased, it is now possible to transmit digital wireless voice signals at a sufficiently fast rate so that the human ear cannot hear the delays. Work is currently underway by all cellular phone service providers to convert all their digital wireless voice systems to packet data systems. All cellular phone systems will then be data transmission systems, and the data will represent written text, digitized pictures, or telephone voice.

29.9 ANNOTATED BIBLIOGRAPHY

1. Rappaport, T., Wireless Communications Principles and Practice, Chaps. 7 and 9, Prentice Hall, Upper Saddle River, NJ, 1996.
2. Sklar, B., Digital Communications, Fundamentals and Applications (2nd ed.), Prentice Hall, Upper Saddle River, NJ, 2001.

Reference 1 covers speech coding and wireless networking in more detail than this book does. Reference 2 is an entire text dedicated to the subject of coding and modulation techniques.

CHAPTER 30

MULTIPLE ACCESS TECHNIQUES: FDMA, TDMA, AND CDMA

The need for higher and higher data rates in a given cellular phone licensed bandwidth continues to grow. The reasons are so that the system can handle more voice channels and higher data rate video, Internet, and data transmission requirements.

Techniques for increasing data rate capacity by using multilevel digital modulation techniques were discussed in Chapter 4.

Multiple access techniques FDMA, TDMA, and CDMA will be discussed in this chapter. Data only techniques for CDMA (HSDPA and EV-DO) are also discussed. New multiple access techniques cannot be immediately implemented because of the high infrastructure costs; instead, they must be introduced in an evolutionary fashion. The GSM and cdma2000 evolutionary plans will therefore also be presented.

OFDM multiple access and OFDMA techniques used for 802.11 high-speed, short-range data systems and Wi-MAX will be discussed in Chapter 31.

30.1 FREQUENCY DIVISION MULTIPLE ACCESS (FDMA)

Figure 30.1 shows the concept of FDMA. Each user channel is assigned a small fraction of the total RF bandwidth licensed for use in the particular radio system. FDMA has been used since 1920 for AM radio, FM radio, VHF and UHF television, mobile phone, cellular phone, and other wireless communications systems. For example, the original analog cellular phone system had a total RF frequency range of 25 MHz for both the downlink and uplink transmission paths. The downlink

RF Measurements for Cellular Phones and Wireless Data Systems. By A. W. Scott and R. Frobenius

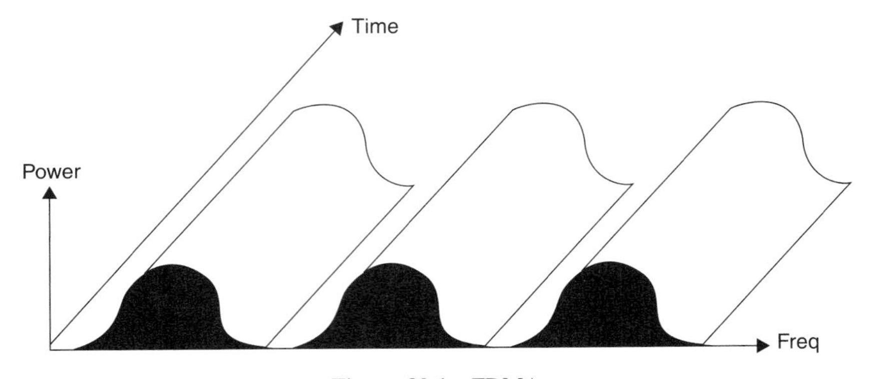

Figure 30.1 FDMA.

frequencies were 869–894 MHz, and the uplink frequencies were 824–849 MHz. Each FM modulated voice channel required an RF bandwidth of 30 kHz, so a total of 25 MHz/30 kHz = 832 voice channels were available. This is the same FDMA technique that was used for the original mobile phone system established around 1950. However, the cellular phone system had three improvements over the original mobile phone system. The major improvement was that the 832 channels were not permanently assigned to individual users; they were assigned to each user only when the user was making a call. If the user hung up and then made another call, a different available frequency was assigned for the second call and the process continued. Assuming that each user was calling only 1% of the time, this "frequency on demand" approach increased system capacity by 100 times.

The second difference was that half of the licensed bandwidth was assigned to AT&T, who was the major telephone company at the time. The other half was assigned to other telecom companies such as Motorola, GTE, and CellularOne to provide competition.

The third difference was to divide each coverage area into cells, each with a radius of about 6 miles to reduce the power required from the mobile phone transmitter.

Figure 30.2 shows the concept of a cellular phone system. Each mobile phone communicates to and from a base station in the cell in which the mobile is located. The base station in each cell is connected to an MSO. The MSO manages the assignment of frequencies to individual mobiles in the individual cells. The connection from the base station to the MSO is achieved by wire line or microwave relay, because both the base station and the MSO are fixed in location. The MSO changes the format of the mobile phone call to that of a wired telephone call and connects it to the PSTN. The MSO looks to the PSTN just like another PBX at a business organization. The phone messages from the MSO are routed by the PSTN to individual users using wire lines just like any wired phone system. The reply message is sent back to the MSO where its format is changed, and it is sent back to the appropriate base station for wireless transmission to the mobile user using a different set of downlink RF frequencies.

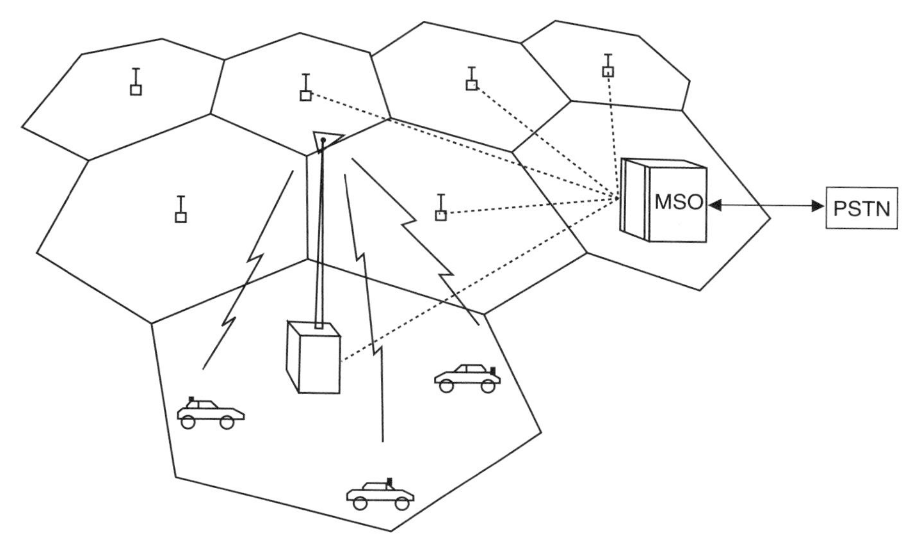

Figure 30.2 Cellular mobile phone.

In the U.S. cellular phone system just discussed, there are 832 channels each for transmission and reception. Unfortunately, only 60 of the 832 channels are available to use in each cell. The details of this limitation, which are called "frequency reuse," are illustrated in Figure 30.3. (A more appropriate name would be "frequency nonuse.") There are several reasons for it. The band is divided between two service providers, so that only half of the 30 kHz wide channels (416) are available to a given service provider. Figure 30.3 shows a common arrangement of seven hexagonal cells that form a cluster. Several other clusters of seven hexagonal cells are also shown. Only honeybees can make a perfect set of seven hexagonal cells. The cellular phone site designer is limited by topography, licensing, and obtaining property on which to erect a tower. However, the seven hexagonal pattern provides a simple drawing for explaining frequency reuse. As shown by the arrows at the bottom of the figure, if the diameters of the cells are 16 miles, the distance between one seven-cell cluster and the next is about 40 miles. Forty miles is around the curvature of the Earth, so the set of frequencies can be reused by the same or a different service provider to service a different set of users.

Referring to the seven-cell cluster at the top of the drawing in Figure 30.3, the star in cell 5 represents a mobile unit, which is assigned one of its 416 RF channels. Assume that the base station in cell 1 is using that same RF frequency for one of the mobile phones in its cell. Which of the RF signals is stronger in the mobile unit receiver that is in cell 5: the signal from the base station in cell 5 that is intended for it or the signal from the base station in cell 1 that is using the same frequency for another user? The obvious answer is the signal from the base station in cell 5, because it is closer to the mobile. This answer is statistically true 80% of the time. However, 20% of the time the RF signal from the base station in cell 1 will be larger, because of

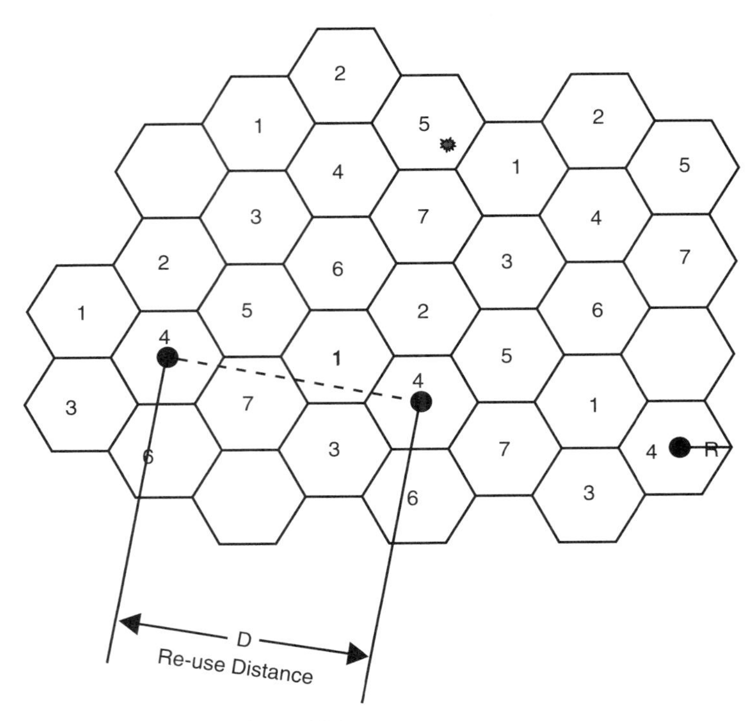

Figure 30.3 Frequency reuse.

the complex nature of cell phone propagation, as discussed in Chapter 2. Therefore, a particular cell phone frequency can never be reused in an adjacent channel. This effect reduces the number of usable channels in each cell to $416/7 = 60$.

30.2 TIME DIVISION MULTIPLE ACCESS (TDMA)

Figure 30.4 illustrates TDMA. TDMA uses the same set of RF frequency channels as FDMA. (Actually, the correct name for TDMA should be FDMA/TDMA, but this fine point is usually not respected.) With TDMA, each FDMA channel is shared in time. Figure 30.4a shows the use of three time slots. Figure 30.4b shows a general case of the use of N time slots. In order to use a different time slot for each different user, the analog signals from each user must first be digitized.

Referring to Figure 30.4a, the same frequency channels used for FDMA shown in Figure 30.3 are used. However, with TDMA, three users share the same frequency channel at three different times. For example, one user's data is transmitted in the channel for 6 ms, then another user's data is transmitted for 6 ms, and then a third user's data is transmitted for 6 ms.

Unfortunately, the use of TDMA does not increase system capacity, because the channel bit rate increases directly as the number of time slots used. For example, if

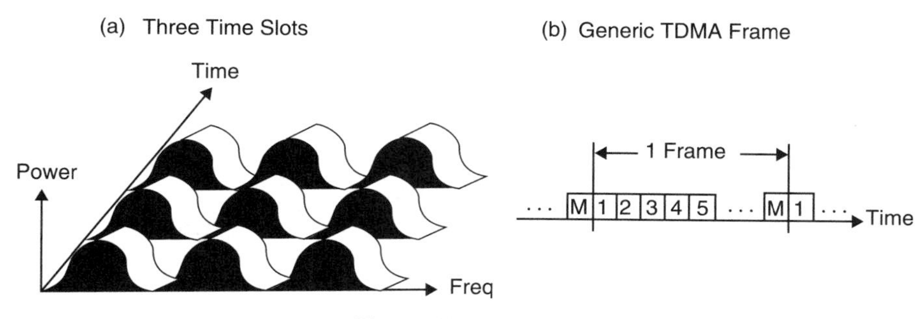

Figure 30.4 TDMA.

the bit rate for a single digitized voice channel is 16 kbps, as described in Chapter 29, and three time slotted channels are used, the RF channel must be designed to handle a bit rate of 3×16 kbps $= 48$ kbps.

Although TDMA does not provide greater capacity, it provides other system benefits. For example, three time slots are used in the IS-136 system, which is the digital upgrade to the original U.S. analog cellular phone system. As discussed in Section 30.1, the original analog signal used 30 kHz wide RF channels. Base stations were built and installed throughout the United States and soon reached their maximum user capacity. By digitizing and compressing the voice signal, the bandwidth of a single user channel could be reduced to 16 kbps. By using QPSK modulation, the RF channel with frequency guard bands, could be fitted into a 10 kHz wide bandwidth. Three of these 10 kHz wide digital channels could be fitted into the existing 30 kHz channel of the analog system. The expensive part of the base station equipment is the RF part, and the three times greater capacity digital system could be implemented without changing the RF components by using the three TDMA channel scheme.

When the European electronics community witnessed the market success of the U.S. cell phone systems, they developed GSM. It was designed from the start as a digital system, and it used eight time slots to simplify the base station. By using eight time slots in each RF system, the base station hardware could be reduced by eight times. There is a limit to the number of time slot channels that can be used in a TDMA system. As the number of time slotted channels is increased, the bit rate increases and the bit period decreases. When the bit period gets as short as the multipath distances in the transmission path, the bits can no longer be clearly defined. For the GSM system, this effect limits the time slots to eight.

Both the U.S. IS-136 and the GSM TDMA systems could further double their capacity by using digital speech interpolation (DSI). When digital processing chips first became available in the 1970s, the wired telephone industry realized that they could increase the capacity of their long distance lines by digitizing the phone signal. The average telephone user only talks 40% of the time. During the other 40% of the time he is listening, and for the remaining 20% of the time he is thinking about what to say next. Long distance wire line telephone calls use two separate channels for each user, one to talk and one to listen. However, each channel is used only 40% of the time. By having two users share the line from one city to another, capacity

TABLE 30.1 User Capacity of Different Multiple Access Techniques in a 12.5 MHz RF Bandwidth

Analog FDMA	60
TDMA	180
TDMA with DSI	360
GSM	144
CDMA	600

can be doubled. Of course, the two users do not talk and then listen at fixed time intervals, so each user's telephone conversation has to be digitized and transmitted in short groups of bits. This digital processing is called DSI. The DSI technique developed for wire line long distance communications was easy to apply to wireless cell phones, with the result of doubling cell phone capacity.

Using the cell phone RF frequency allocation of 12.5 MHz for each service provider, the user capacity for each multiple access technique would be as shown in Table 30.1. The lower capacity of the GSM system results from its use of more speech bits, more error correction and control bits, and less effective modulation (FSK vs. QPSK). The achievement of the CDMA capacity of 600 users is discussed in Section 30.3.

30.3 CODE DIVISION MULTIPLE ACCESS (CDMA)

As the IS-136 TDMA cellular phone system was being implemented in the United States and a similar TDMA system called Pacific Digital Cellular was being implemented in Japan, and GSM TDMA systems were being implemented in the rest of the world, Qualcomm was proposing an entirely new multiple access system for cellular phones called CDMA. This multiple access system was an outgrowth of the spread spectrum technology developed by the Qualcomm founders for secure military communications systems during World War II.

With CDMA, all individual users use the same RF frequency at the same time. Each user is assigned a unique 128 bit code by the base station to identify his digitized voice channel during a particular call. At the end of the call, the particular code is returned to a code pool at the base station. Each time the user makes another call, a different available code is assigned. Qualcomm's original calculations predicted that the use of CDMA would increase the cellular phone system's user capacity by 20 times compared to an analog phone system, which would be 1200 users under the conditions listed in Table 30.1.

Figures 30.5–30.7 give a simplified explanation of cellular phone CDMA. This simplified explanation is appropriate for this RF measurements textbook.

The upper three digital waveforms of Figure 30.5 are used at the transmitter site. For the IS-136 TDMA system, the bit rate of a single digitized telephone voice signal is 16 kbps (as explained in Chap. 29). Curve a shows two information bits from this digitized cell phone signal. Curve b shows one of the unique sets of CDMA coding bits. For ease of understanding, only 32 coding bits are shown during the period of a

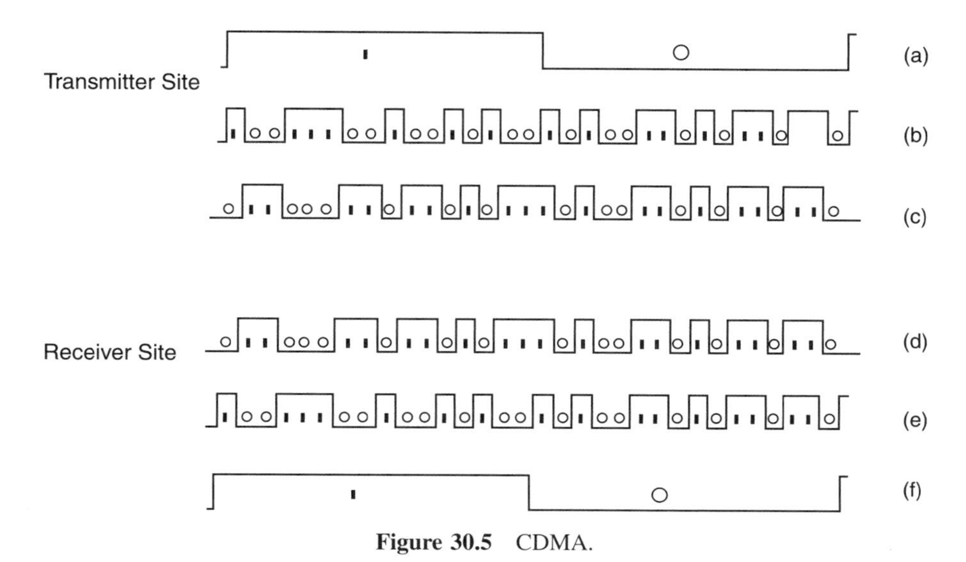

Figure 30.5 CDMA.

$$N = \frac{\text{CHIPS} \cdot \text{Ga} \cdot \text{Gv}}{\text{S/I} \cdot \text{M}}$$

N = Capacity of Cell

(1)	CHIPS = Processing Gain	128
(2)	S/I = Signal to Interference	5 (7 dB)
(3)	Ga = Gain by Using Sectors	3
(4)	Gv = Activity Factor	2.5
(5)	M = Interference from Adjacent Cells	1.6

$$N = \frac{128 \cdot 3 \cdot 2.5}{5 \cdot 1.6} = 120 \text{ Channels}$$

Number of FDMA Channels with CDMA

- Service Providers Licensed Bandwidth: 12.5 MHz
- CDMA Bandwidth: 1.25 MHz
- FDMA Channels: 12.5 MHz/1.25 MHz = 10
- Total Voice Channels: 10 × 120 = 1200

Figure 30.6 CDMA cellular phone capacity.

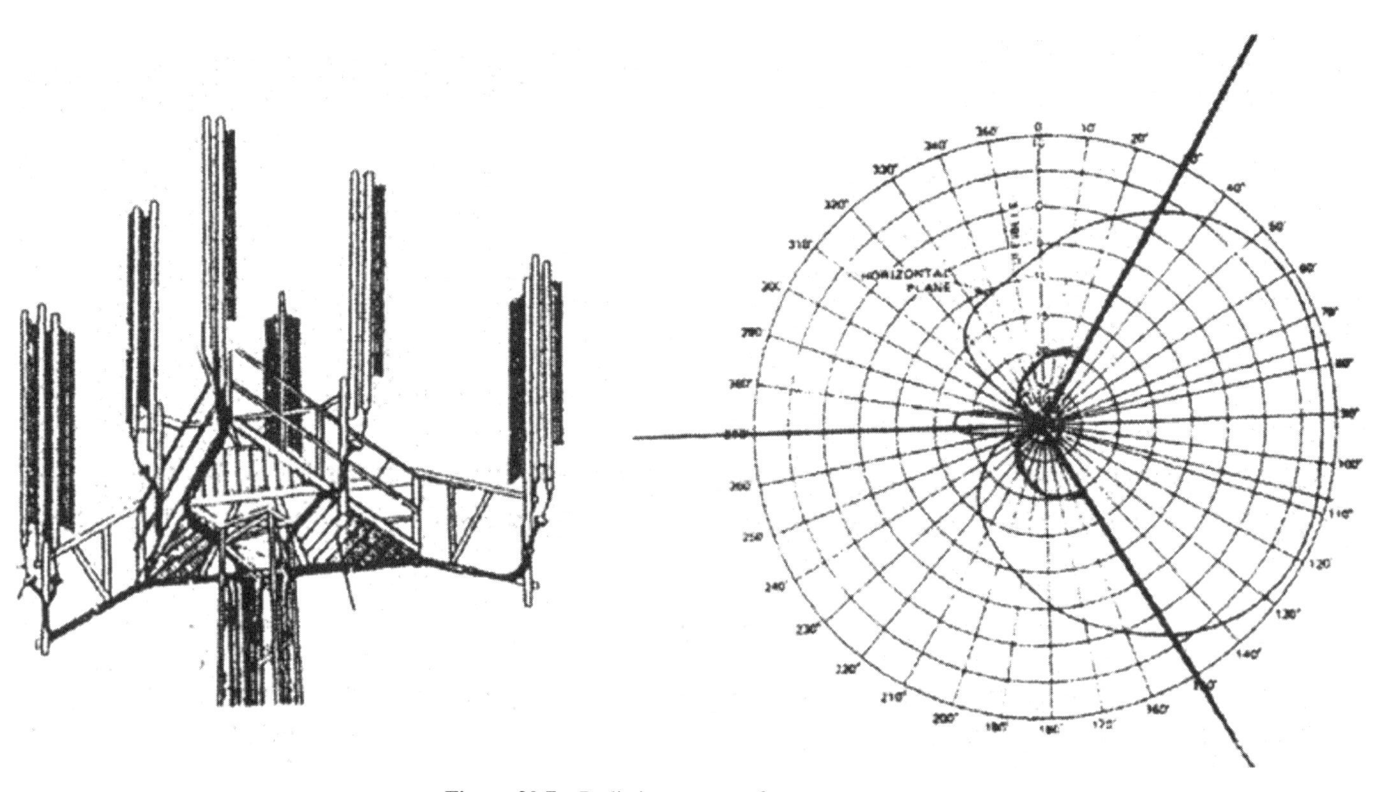

Figure 30.7 Radiation pattern of sector antenna.

single information bit instead of the actual 128 bits. To distinguish between information bits and coding bits, the coding bits are called "chips." This is somewhat confusing to RF test personnel who usually associate the word chip with a semiconductor electronic part. Nevertheless, the correct terminology of coding chips will be used throughout the rest of this book.

There are 2^{128} possible codes if 128 chip codes are used. The possible codes are analyzed by computer to select several thousand of the sequences that are the most different from each other. These "least correlated" sequences are recorded and given an identification number. Although they were generated as random numbers, the labeling process removes their purely random nature, so they are now called pseudorandom or PN codes.

During a particular cell phone call, the information bits are processed by the coding chips using an "exclusive nor" multiplication. This is the simplest multiplication process that can be implemented with digital semiconductor devices. If the bit and the chip being multiplied are the same, a digital 0 output is generated. If the bit and the chip being multiplied are different, a digital 1 is generated. Each information bit multiplied by each coding bit is shown in Figure 30.5c. The bit rate of the multiplied waveform is $16 \text{ kbps} \times 128 = 2.54 \text{ Mbps}$. This waveform is QPSK modulated onto the RF carrier and transmitted to the receiver site. Because QPSK modulation is used, the RF bandwidth is 1.25 MHz. This RF bandwidth is 64 times greater than the original bitstream. CDMA is defined as "spread spectrum." This spreading operation is called "direct sequence spread spectrum."

After amplification in the receiver, the received RF signal is demodulated. Figure 30.5 shows that curve d is identical to curve c at the transmitter if there are no errors in the transmission. When the signal is despread using the same PN code as used in the transmitter (Fig. 30.5e), the original 16 kbps information bitstream is obtained.

CDMA modulated signals from other users are also received in the subject receiver. However, the other signals are not despread by the assigned CDMA chip sequence. An interfering signal appears when spread with the wrong code as noise whose amplitude is $1/128$ (-21 dB) of the correctly coded signal. If 2 other signals that are not spread by the correct code are present, the noise is 3 dB higher (-18 dB of the correctly coded signal). If 10 other signals that are not spread by the correctly coded signal are present, the noise will be 10 dB higher (-11 dB compared to the correctly coded signal). If 25 other signals that are not spread by the correctly coded signal are present, the noise will be 14 dB higher (-7 dB compared to the correctly coded signal). Therefore, even if 25 interfering signals intended for other users are present, the CDMA coding process reduces their total signal to 7 dB below the desired signal, which will permit the required 10^{-3} BER to be obtained with QPSK modulation.

Qualcomm estimated that a total of 1200 voice channels could be obtained in a single cell using CDMA, compared to 60 voice channels for analog cell phone and 360 voice channels for digitized TDMA using DSI. These voice channel capacities were compared in Table 30.1.

Qualcomm's reasoning is shown in Figure 30.6. The formula shows that the voice channel capacity of the cell (N) depends on the number of coding chips (128) divided

SYSTEM	Peak Network Downlink Speed	Average User Downlink Downlink Speed	Data Type
cdma2000			
1×RTT	153 kbps	70 kbps	1
1×EV-DO	2.4 Mbps	700 kbps	2
1×EV-DO RevA	3.1 Mbps	900 kbps	3
GSM Evolution			
GPRS	115 kbps	40 kbps	1
EDGE	473 kbps	130 kbps	1
WCDMA	2 Mbps	320 kbps	1
HSDPA	14 Mbps	1100 kbps	2

Data Type: 1 = Data in voice channels
2 = Data only
3 = Data only with VoIP

Figure 30.8 Comparison of cell phone data systems.

by the S/N required for a BER of 10^{-3}, which for QPSK is 7 dB or 5 times. This ratio was explained in the previous paragraph.

The number of channels can be further increased by dividing the cell into three sectors and using a different PN code in each sector. Figure 30.8 shows a typical antenna setup used in a three sector cell along with the antenna pattern in a single sector. Note that the power is focused by the antenna into a 120° sector, but there is a large amount of radiation from this antenna into the other two sectors. Because of this side lobe radiation, the same frequency normally cannot be reused in these other two sectors. However, with CDMA, different codes can be used in each sector, with the result that the interfering side lobe signals are reduced by the CDMA coding ratio of 21 dB. Therefore, by using three sectors, each of which uses a different CDMA code set, the capacity in the cell can be increased (line 3, Fig. 30.8) by three times.

Because each user talks only 40% of the time, two users were possible on the same channel in TDMA by using the complex digital processing of DSI. The full advantage (line 5, Fig. 30.8) of $1/40\% = 2.5$ can be obtained with CDMA by simply turning a transmitter off if the user is not talking.

The interference from adjacent cells was estimated by Qualcomm (line 5, Fig. 30.8) to reduce overall capacity by 1.6 times if the same RF frequency is used, but with a different PN code set. Recall that with FDMA or TDMA, the same RF frequency could not be used in adjacent cells because of the possibility of interference, so the capacity reduction factor was 7.

All cell phone engineers agreed with Qualcomm's assumptions 1–4, but there was much controversy about assumption 5. Accepting assumption 5 predicted a user capacity of 120 channels in a 1.25 MHz bandwidth. The cellular phone provider's total bandwidth was 12.5 MHz, which made possible 10 CDMA 1.25 MHz channels, for a total capacity of 1200 users.

The only way to know the actual reduction in CDMA capacity attributable to adjacent channel interference was to make extensive field trials. This was an extremely large and expensive undertaking. Trials had to be made with several base stations and several thousand mobile phones. The cost was in the billions of dollars, and Qualcomm used investor money, expecting to recoup their investment by charging a royalty for the use of the CDMA principle. Qualcomm made arrangements with seven service providers in different parts of the U.S. and provided all of the mobile units and base station equipment to conduct the tests to prove experimentally the user capacity that CDMA provided. These field tests lasted into the early 1990s. The results showed that the capacity was about 600, which was only half of the theoretical predictions. However, the experimental results were almost twice as good as the best TDMA system. At that point everyone in the industry had to agree that a CDMA system provided the most user capacity in a cell phone application.

CDMA had other unique problems that many in the industry believed would make CDMA impractical, but these beliefs turned out to be incorrect. One problem was that of acquisition of a CDMA signal in the receiver. The PN spreading code had to be accurately aligned in time with the incoming bitstream. Because the receiver did not know how long it took for the CDMA signal to reach it, it used an acquisition procedure in which it moved the PN despreading code one chip at a time until time alignment was obtained, and the CDMA signal was despread. The PN despreading code timing was then slowly changed to match the timing of the incoming CDMA signal due to mobile motion.

CDMA RF signals reflected from different objects arrive at the receiver at different times. This is the multipath effect. CDMA uses a Rake receiver, which has two or more acquisition channels, so that these different incoming signals can be distinguished, and each multipath signal can be added together to increase the total received signal rather than allowing multipath fading. The use of a Rake receiver eliminates the problem of multipath fading.

The third problem of CDMA was controlling the received power at the base station. Propagation losses due to transmitter–receiver separation vary over a 70 dB range. The CDMA coding reduces interference by 21 dB between the desired and an interfering signal if both signals have the same power level at the receiver. Therefore, the received signal from all transmitters must be equal at the receiver. This is achieved in the mobile to base path by controlling the mobile power by an open group RF pilot transmitted from the base station and a closed loop instruction from the base station based on the measurement of the power received from the mobile unit.

The Qualcomm field trials proved the many advantages of CDMA, which were as follows:

- Large voice channel capacity: It has more than 10 times that of analog systems and about 2 times more than TDMA systems.
- Soft saturation: Because every user uses the same frequency, more than the 600 user capacity could be obtained during peak loading times, with just a small increase in S/N.

- Soft handoff as the mobile moves from cell to cell: Because every mobile uses the same frequency, it can connect with two base stations during handoff, until the connection with the new base station is firmly established.
- The Rake receiver reduces multipath fading and so reduces transmitter power requirements by 10 dB.

After the announcement of the CDMA field trials, about half of the U.S. cell phone operators shifted from TDMA to CDMA. At about the same time, the FCC announced the auction of RF bandwidth in the 1860–1990 MHz band for an additional cell phone band in the United States. It was called the PCS band. Unlike previous FCC licensing, the PCS licenses were auctioned off to qualified service providers rather than being given to whomever the FCC selected. The FCC expected to raise $3 billion for the U.S. Treasury. Actually, this auction raised $8 billion. This total amount represented $10 per person in each city. The multiple access techniques chosen by the service providers for the winners of this auction were as follows:

70% chose to build CDMA systems, because of their greater channel capacity

20% chose to build GSM TDMA systems because of delays in the production of the Qualcomm CDMA chips

15%, primarily AT&T (who was once again allowed to provide regional telephone service), chose to build an IS-136 TDMA system to be compatible with the similar 850 MHz cell phone band systems they had acquired from CellularOne

The service providers who selected CDMA had a two times more capacity advantage over those who chose TDMA.

As the cell phone Industry moved into the 21st century, most independent cellular phone licensees were bought or merged with the major service providers. Thus, by 2005 there were only three major cell phone service providers in the United States: Cingular, Verizon, and Sprint. Cingular used both IS-136 and GSM TDMA systems and in 2006 converted their older IS-136 systems to GSM. Verizon and Sprint used CDMA systems in both frequency bands.

30.4 3G CELL PHONES

By the mid-1990s, the worldwide cellular phone industry realized that voice cellular phones were approaching market saturation and that future growth required cellular phones that could handle voice, music, video, e-mail, Web access, and business data downloads. These cell phone systems were called 3G. Previous systems were defined at this time as 1G, which were analog voice systems, and 2G, which were digital voice systems.

An international goal list, but not a specification, was written for 3G, with the following goals:

- Clear, natural sounding telephone voice
- All wired telephone features: redial, caller ID, memory, and so forth
- Location-based services (911, navigation instructions)

- e-mail
- High data rate Internet access: 144 kbps mobile, 384 kbps pedestrian, and 2 Mbps indoor
- Video phone
- Easy migration from 2G systems
- Seamless worldwide service

The use of CDMA gave adopters a significant capacity advantage over GSM users in the United States and the rest of the world. However, because of the large amount of TDMA installed base station equipment, the GSM community could not immediately shift from TDMA to CDMA. The GSM community has therefore adopted a "GSM Evolution" strategy of upgrading their equipment from GSM to GPRS to EDGE to CDMA and finally to HSDPA in an evolutionary manner. The shift from GSM and EDGE TDMA systems to CDMA is a major retrofit step. Ultimately, by 2012 each family of cellular phones will have upgraded to high-speed packet data systems where all transmitted signals will be data based, and voice will be handled by VoIP. These high-speed cell phone packet data systems are discussed in the next section.

30.5 HIGH DATA RATE SYSTEMS FOR CELL PHONES

A comparison of high data rate systems for cell phones is provided in Figure 30.8. The peak network download is with only one mobile on the system. This is the only condition that can be accurately defined. The average user download is a more useful performance measure, but the exact value depends on the distribution of users within the cell. Type 1 systems simply dedicate one or more voice channels for data transfer. Type 2 systems are data only systems. Type 3 systems are data only systems that can also handle voice as VoIP.

Cdma2000 Systems

The cdma2000 systems are an upgrade of the original IS-95 CDMA systems. They are currently used throughout the United States and in Japan, Korea, and other countries. They are designed to use 12.5 MHz of RF frequency and are backward compatible with IS-95. A service provider can mix IS-95 and cdma2000 cells within its system.

Cdma2000 $1\times$RTT systems are simply CDMA voice systems using some or all of their voice channels for data. Designation $1\times$ signifies the 12.5 MHz RF bandwidth. Cdma2000 $1\times$EV-DO are data only systems. They are hardwired as data systems. If a mobile phone is to serve both as a voice and data phone, it must switch to another of the 10 CDMA 1.25 MHz bandwidths in a cell that is dedicated to voice.

CDMA voice channels use 128 or more coding chips on single digitized voice channels. Each channel has 16 kbps digitized voice signals. In contrast, data signals have bit rates up to 1 Mbps. If these high bit rate signals are coded with

128 coding chips, the bandwidth of each channel would far exceed the licensed frequency allocations. Therefore, CDMA data only systems use PN coding of only 16 chips or less. This limits the number of possible CDMA data channels. High user capacity is achieved by using packet data switching. CDMA data only systems for both cdma2000 and WCDMA achieve their high data rate performance by using combined CDMA coding and packet data.

As described later, a WCDMA system has the capability to change any or all of its RF channels in a cell from voice only to data only in 10 ms. Consequently, almost all channels can serve as voice channels for peak voice usage conditions, for example, during the 5 PM commute time, and then switch to data channels during the night hours from 2 to 5 AM for large batch data transfers.

Cdma2000 voice and data channels are hardwired, and they can only be used for one but not both purposes. Cdma2000 EV-DO will solve this problem by making all cells data only and handling voice calls by VoIP.

The technology for achieving high data rates is similar for EV-DO and HSDPA. This technology will be explained in a subsequent paragraph in this section on HSDPA.

In many parts of the world the cell phone frequency allocation is 50 MHz. Only three 12.5 MHz cdma2000 voice or data systems can be fit into 50 MHz because of guard band requirements. These cdma2000 systems are designated as $3\times$ instead of $1\times$. Consequently, WCDMA voice or data has a $4/3 = 1.33$ capacity advantage over cdma2000.

IS-95 CDMA systems have been operating in the United States since the early 1990s. Cdma2000 with EV-DO systems have been operating since 2005. Cdma2000 EV-DO has an installation lead of over 3 years compared to the WCDMA data system.

HSDPA High Data Rate Systems

The HSDPA high data rate system of the GSM Evolution family has the following features:

High-speed shared channels and short transmission time interval (TTI)
Fast scheduling and user diversity
Higher order modulation
Fast link adaptation

These features are explained in the following paragraphs.

High-Speed Shared Channels and Short TTI. HSDPA uses high-speed shared data channels called high-speed downlink shared channels (HS-DSCH). Up to 15 of these can operate in the 5 MHz WCDMA radio channel. Each uses a fixed spreading factor of 16. User transmissions are assigned to one or more of these channels for a short TTI of 2 ms, which is significantly less than the interval

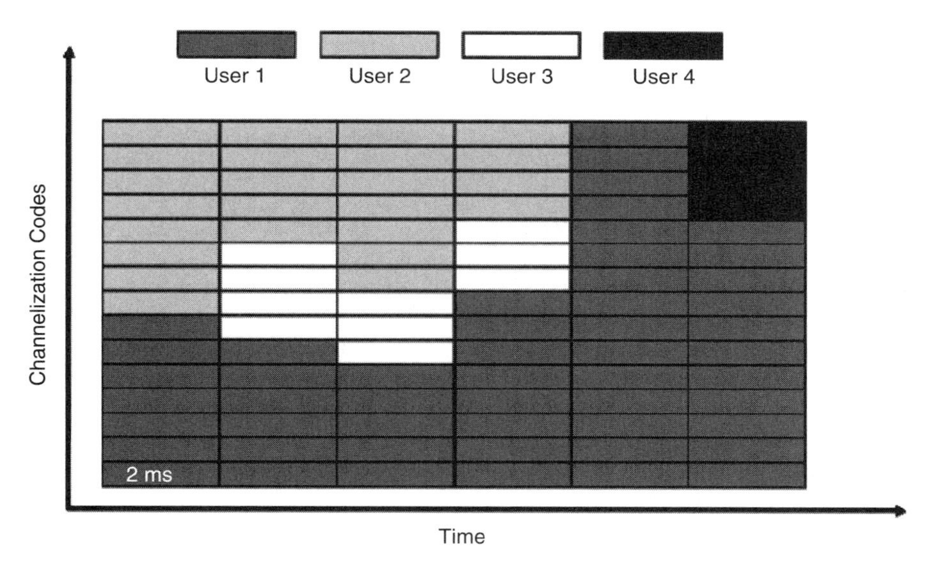

Figure 30.9 Example of HS-DSCHs.

of 20 ms used in WCDMA. The network can then adjust how users are assigned to different HS-DSCH channels every 2 ms. The result is that resources are assigned both in time (TTI) and code domains (HS-DSCHs). Figure 30.9 shows an example of three users with different amounts of data at different times and into different code channels.

Fast Scheduling and User Diversity. Fast scheduling exploits the short TTI by assigning channels to the users with the best instantaneous channel conditions, rather than in a round-robin fashion. Because channel conditions vary somewhat randomly across users, most users can be serviced using optimal radio conditions and can therefore obtain optimal data throughput. Figure 30.10 shows how a scheduler might choose between 2 users based on their varying radio conditions, mainly to emphasize the user with better instantaneous signal quality. With about 30 users active in a sector, the network achieves significant user diversity and significantly higher spectral efficiency. The system also insures that each user receives a minimum level of throughput. The result is referred to as "proportional fair scheduling."

Higher Order Modulation. HSDPA uses both QPSK modulation and, under good radio conditions, 16QAM. The benefit of 16QAM is that 4 bits of data are transmitted in each radio symbol, as opposed to 2 bits with QPSK. 16QAM increases data throughput, whereas QPSK is available under adverse propagation conditions.

Fast Link Adaptation. Depending on the condition of the radio channel, different levels of FEC can also be employed. For example, a three-quarter coding rate means that three-quarters of the bits transmitted are user bits and one-quarter are error

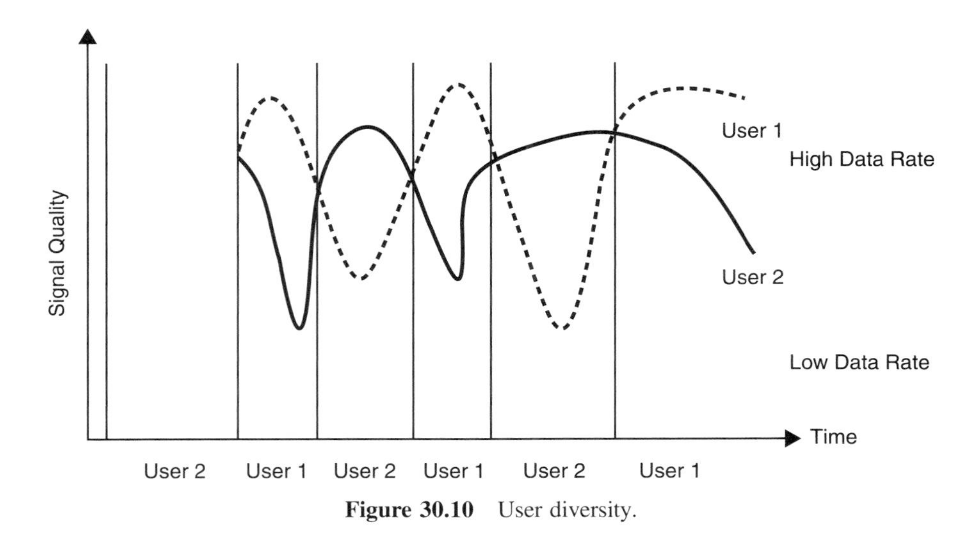

Figure 30.10 User diversity.

correcting bits. The process of selecting and quickly updating the optimal modulation and coding rate is referred to as fast link adaptation. This is done in close coordination with fast scheduling. Figure 30.11 shows throughput rates under various conditions of modulation, FEC, and the number of HS-DSCH codes that are used. Note that the throughput varies from 1.2 Mbps with QPSK, 1 : 4 FEC, and 5 codes to 14.4 Mbps with 16 QAM, 4 : 4 FEC, and 15 high-speed shared channel codes.

- The same radio carrier can service WCDMA voice and data users
- Uses high speed shared channels and short transmission time intervals
  - 15 HSSC can operate in 5 MHz WCDMA channel
  - Each channel uses a fixed spreading factor of 16
  - User transmissions are assigned to one or more channels for 2 ms
  - Fast scheduling assigns channels to users with the best channel conditions
- Uses either QPSK or 16QAM
- Can change FEC from 4 : 4 to 1 : 4
- HSDPA Throughput rates

Modulation	FEC coding	Throughput (Mbps)	
		5 codes	15 codes
QPSK	1.4	1.2	1.8
16QAM	4.4	2.4	14.4

- Field results: QPSK, 5 codes: >1 Mbps

Figure 30.11 HSDPA.

30.6 MEASUREMENT OF THE DISTORTION OF DIGITALLY MODULATED SIGNALS BY RF COMPONENTS

The complex digital processing of voice, video, and data signals described in this and the following chapter to obtain maximum transmission capacity in limited available bandwidths imposes severe restrictions on the performance of RF devices, which are analog in nature. The abrupt time variations of the digitally modulated carrier are difficult for RF components to reproduce. Performance measurements to insure that the RF components are reproducing the digital modulation without distortion are complex.

The basic RF instruments for measuring the characteristics of digitally modulated signals are the vector signal generator, spectrum analyzer, and VSA. Chapters 32–35 describe the use of these instruments to characterize digitally modulated RF signals in terms of their ACP; constellation, vector, and eye diagrams and EVM; CCDF; and BER, respectively. Chapter 36 demonstrates the use of these RF measurement instruments with special software to measure the key performance of the EDGE, WCDMA, and HSDPA family of GSM Evolution systems.

30.7 ANNOTATED BIBLIOGRAPHY

1. Data Capabilities: GPRS to HSDPA and Beyond, White Paper for 3G Americas, http://www.rysavy.com, Rysavy Research, Hood River, OR, 2005.
2. Rappaport, T., Multiple access techniques in wireless communications, In Wireless Communications Principles and Practice, Prentice Hall, Upper Saddle River, NJ, 1996.

Reference 1 is a thorough reference on TDMA and CDMA systems.

Reference 2 is a chapter covering multiple access techniques from a book that serves as a good reference for wireless systems.

CHAPTER 31

OFDM, OFDMA, AND WiMAX

Another multiple access scheme, Orthogonal Frequency Division Multiple (OFDM), was developed for short-range, high data rate wireless LANs. It provides data transfer up to 200 Mbps. This successful technology was then extended to WiMAX, which may be the next generation (4G) of cellular phones. Most of these systems are covered by IEEE 802.11 specifications. These specifications, OFDM principles, and WiMAX will be discussed in this chapter.

31.1 802.11 SPECIFICATIONS

IEEE has developed a series of specifications covering electronic data transmission. Perhaps the most widely used is 802.3, which covers Ethernet. IEEE, with the help of industry, has also developed the 802.11 series, which covers wireless data communication. A summary of the current 802.11 specifications is provided in Figure 31.1.

In the 1940s the FCC established the ISM frequency bands. These bands were originally for industrial and consumer microwave ovens, nuclear accelerators, and oncology equipment. In all of these original applications, the RF/microwave power was contained inside an enclosure, with strict regulations on allowable leakage power. In 1960 the FCC allowed the same frequencies to be used without a license for wireless communications. (Many other licensing agencies worldwide also made these frequencies available for unlicensed wireless communications.)

System	Frequency	Channels	Data Rates	Range
		Local Area Networks (LAN)		
802.11a	5 GHz	12	6,12,24...54 Mbps	100 ft
802.11b (Wi-Fi)	2.4–2.48 GHz	14	1,2, 5.5, 11 Mbps	300 ft
802.11g	2.4–2.48 GHz	14	up to 54 Mbps	300 ft
802.11n (MIMO)	5 GHz	6	up to 200 Mbps	100 ft
		Personal Area Networks (PAN)		
802.15 (Bluetooth)	2.4 – 2.48 GHz	1	720 kbps	30 ft
		Wide Area Networks (WAN)		
802.16 (WiMAX)	Various in 2.3 to 3.5 GHz range		75 Mbps	5-20 mi

Figure 31.1 IEEE specifications.

One of the first applications of the ISM frequencies for wireless communications was for the tracking of packages by package delivery companies like UPS and FedEx. All of the developed systems were proprietary, and none of the RF parts were interchangeable between systems.

To provide a specification for short-range, wireless networks, the IEEE 802 standards group began the development of an 802.11 wireless LAN specification. New modulation and multiple access techniques were proposed each year. After over 10 years of study, the 802.11 standards committee had not created a specification.

The wireless communication industry became frustrated by the IEEE's indecision. Several industry leaders including Intersil, Lucent, Nokia, 3Com, and Symbol formed a trade organization called WECA to prepare a specification to establish 802.11 interoperability. The resulting 802.11b specification was completed in early 2000. WECA also established product certification through a third party lab and a WiFi interoperability logo on all approved devices. The multiple access technique used was a spread spectrum technique. As shown in the second line of Figure 31.1, it achieved a range of 300 ft at data rates of 1, 2, 5.5, and 11 Mbps, depending on the number of users. The operating frequency was in the 2.40–2.48 GHz ISM band.

IEEE eventually completed their specification called 802.11a. It operates in an unlicensed government band at selected frequencies in the 5–6 GHz range that overlaps an ISM band. It uses a new multiple access technique called OFDM, which is described in Section 31.2. As shown in the first line of Figure 31.1, this new specification provided ranges up to 100 ft at data rates of 6, 12, 24, and up to 54 Mbps.

IEEE then developed a similar 802.11 g specification that operates in the 2.40–2.48 GHz band, with data rates up to 54 Gbps at ranges up to 300 ft. The higher range was achieved because of the better RF propagation characteristics at this lower frequency.

These three systems are usually packaged together to fit into a standard PCMCIA computer slot or packaged inside a computer.

A fourth system called multiple input–multiple output (MIMO) or 802.11n (line 4, Fig. 31.1) is an OFDM system that achieves data rates up to 200 Mbps by using two frequency channels and multiple antennas.

The fifth line of Figure 31.1 shows the performance of Bluetooth, which is specification 802.15. This unlicensed short-range wireless system is classified as a "personal area network" (PAN). It uses a frequency hopping multiple access technique and achieves a range up to 30 ft and data rates up to 720 kbps. It is widely used to achieve a wireless connection between a cell phone handset and a user microphone/speaker earpiece. It is also used for wireless data transmission between laptop/desktop computers and an Ethernet system.

The sixth line of Figure 31.1 shows Wide Area Networks (WANs) called WiMAX, using OFDM and OFDMA. Specifiation 802.16b is for line of site systems like wireless DSL from the service provider to an office or residence. Specification 802.16e is for long-range mobile phone access, which is expected to be the basis of 4G mobile phone systems. These WiMAX systems are described in Section 31.3.

31.2 OFDM MULTIPLE ACCESS PRINCIPLES

The OFDM signal is shown in Figure 31.2. OFDM stands for Orthogonal Frequency Multiple Access. The OFDM signal has the following features:

1. The RF signal is divided into 52 separate frequencies, each of which carry a portion of the data. These subcarriers are spaced 312 kHz apart.
2. Four pilot tones at different frequencies across the bandwidth continuously monitor transmission quality and provide a phase reference for demodulation.
3. Modulation can be adjusted between BPSK, QPSK, 16QAM, and 64QAM.
4. Various levels of FEC can be used, depending on signal transmission quality.
5. The modulation type and FEC can be changed with each transmission every 4 ms, based on transmission quality.
6. OFDM has less spreading of power into adjacent channels than CDMA.

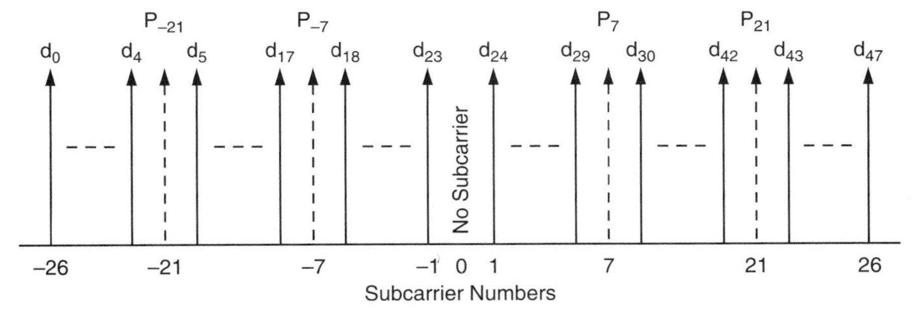

Figure 31.2 OFDM.

Data Rate	Mandatory Modulation	Optional Modulation	Coding Rate	Coded Bits per OFDM Subcarrier	Coded Bits per OFDM Symbol	Data Bits per OFDM Symbol
6 Mbps	BPSK		1/2	1	48	24
9 Mbps		BPSK	3/4	1	48	36
12 Mbps	QPSK		1/2	2	96	48
18 Mbps		QPSK	3/4	2	96	72
24 Mbps	16QAM		1/2	4	192	96
36 Mbps		16QAM	3/4	4	192	144
48 Mbps		64QAM	2/3	6	288	192
54 Mbps		64QAM	3/4	6	288	216

Figure 31.3 802.11a data rate summary.

"Orthogonal" means that a minimum of the modulated spectrum of each subcarrier is at the maximum of the modulated spectrum of the adjacent subcarriers, which minimizes subchannel interference.

Figure 31.3 shows the data rates that can be achieved with different conditions of modulation and coding rate (FEC). The coding defines the number of data bits divided by the total number of bits transmitted. The coding rate of one-third on the first line of the table means that for every 3 bits that are transmitted, 1 is a data bit, and 2 are error correction bits. In the last line of the table, 3 out of every 4 bits transmitted are data bits and 1 is an error correction bit. Note that as the modulation type gets more complex and less error bits are used, the data rate increases to 54 Mbps. Thermal noise is not a major consideration in the system S/N because the transmitter to receiver separation is so small. The major source of noise is interference with other users of the system.

Figure 31.4 is a drawing of a MIMO system covered by specification 802.11n. The term MIMO refers to the antenna system. By using multiple antennas and very

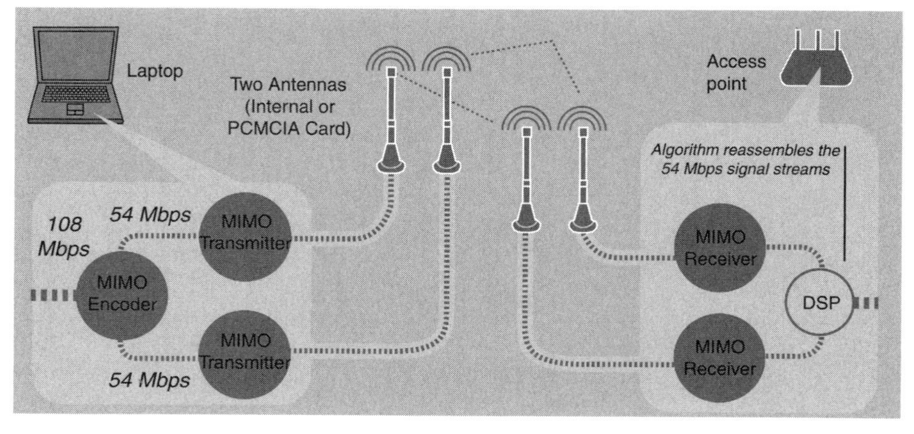

Figure 31.4 MIMO system.

complex signal processing, two independent signal paths are established at the same RF frequency, thereby doubling capacity. By using two 20 MHz RF channels instead of just one, capacity can be doubled again. Beginning with the 54 Mbps capability of the 802.11a OFDM technology and adding MIMO and two data channels, data rates of 200 Mbps can be achieved.

31.3 WiMAX

The advantage of WiMAX is that it provides greater data capacity in a given RF bandwidth. There are two WiMAX specifications:

1. 802.16d: stationary line of site systems
2. 802.16e: mobile cell phone systems

One example of a line of site system would be a wireless DSL connection between the service provider and a home or office. With no obstacles in the path, a transmitter to receiver distance of 20 miles could be obtained. If this line of site condition is not possible, the connection distance would be reduced to a typical cell phone range of about 6 miles.

WiMAX cell phone systems use OFDMA. OFDMA means Orthogonal Frequency Multiple Access. OFDMA is compared to conventional cell phone systems using TDMA or CDMA and to wireless data systems using OFDM in Figure 31.5. Figure 31.5a shows a conventional cell phone system where a dedicated RF channel using a single frequency is utilized for each user during the time of the call. With OFDM (Fig. 31.5b, described in Sec. 31.2) the carrier is divided into 48 (or more) subcarriers with pilot tones, and several bits of data are modulated onto each subcarrier. With OFDMA (Fig. 31.5c) the carrier is divided into subcarriers in the same way as it is with OFDM, but bits from several different users are assigned to the subcarriers.

Figure 31.6 shows another comparison of OFDM and OFDMA. The upper part of the figure shows OFDM. The user data does not overlap in time. The lower part of the figure shows OFDMA. The data bursts from different users overlap in time. This maximizes data capability to many users. It also adds support for handovers to allow mobility in a complex environment.

Figure 31.7 compares the spectral efficiency and the throughput in a cell sector for mobile WiMAX, HSDPA, and 3xEV-DO. (3xEV-DO means three 1.25 MHz EV-DO systems are fit into the 5 MHz bandwidth used by WiMax and HSDPA.) Note that mobile WiMAX has more than two times the spectral efficiency of the other systems and more than three times the sector throughput.

WiMAX cellular phone systems are already being installed in regions throughout the world that do not have the conventional GSM family of base station equipment installed, because of the greater bandwidth capability of WiMAX. It is expected that 4G cell phones will be WiMAX based. Possibly some service providers currently using GSM, GPRS, and EDGE will not upgrade to WCDMA and HSDPA, but will upgrade to WiMAX because of its greater bandwidth capacity.

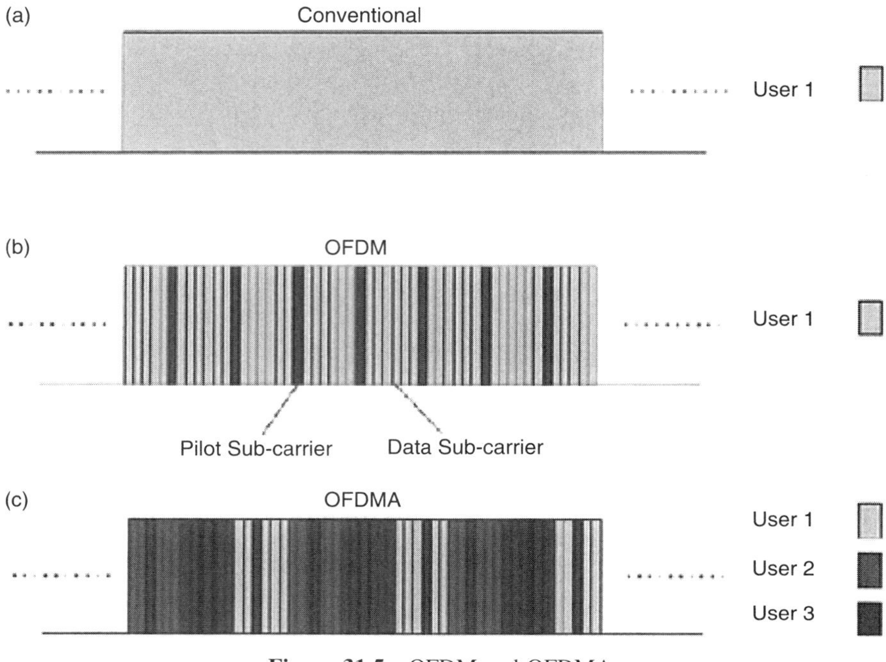

Figure 31.5 OFDM and OFDMA.

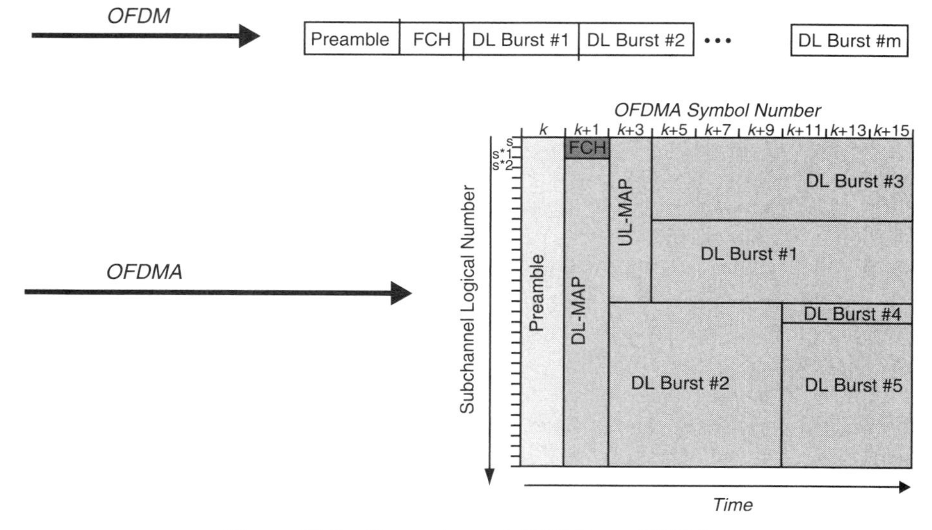

Figure 31.6 Comparison of OFDMA to OFDM.

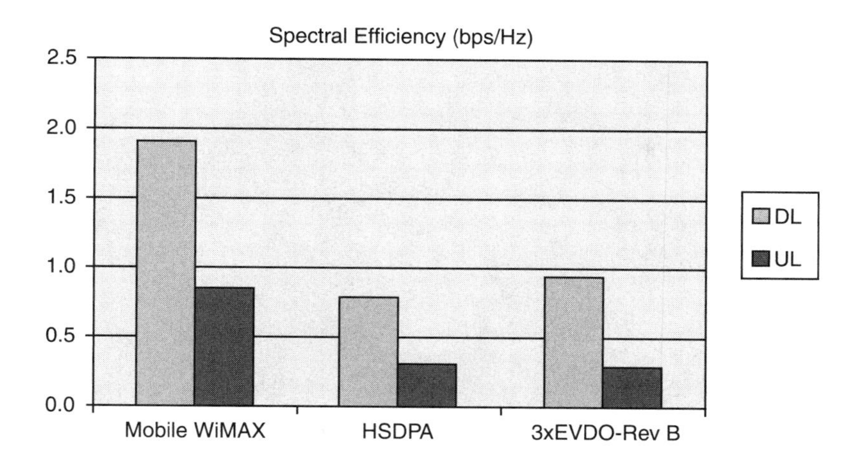

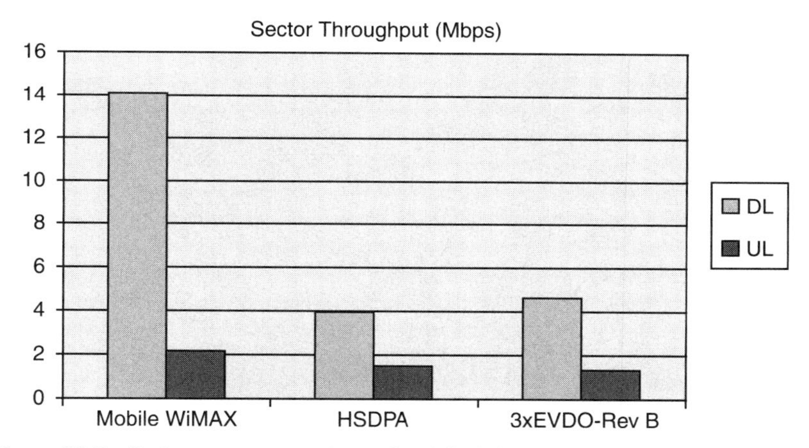

Figure 31.7 Performance comparison of mobile WiMAX, HSDPA, and 3XEV-DO.

31.4 ANNOTATED BIBLIOGRAPHY

1. Agilent Application Note 1380-1: RF Testing of Wireless LAN Products, Agilent Technologies, Santa Clara, CA.
2. Mobile WiMAX, Part 1 and 2, WiMAX Forum, http://www.wimaxforum.org/news/downloads.
3. Fixed and Mobile WiMAX Overview, http://fugitsu.com/wimax, Fujitsu Corporation, Tokyo.

Reference 1 describes 802.11 wireless LANs and their testing.

Reference 2 describes Mobile WiMax in more detail than this book.

Reference 3 describes fixed and mobile WiMAX in more detail than this book.

CHAPTER 32

ACP

32.1 ACP

Digitally modulated signals are carefully designed to occupy specific channel bandwidths as given by system specifications. A fundamental relationship between the time and frequency domains is that any abrupt change in one domain consumes a great deal of resources in the other domain. For example, a filter that has a sharp cutoff in the frequency domain also requires a great deal of time for settling to a steady state in the time domain. Conversely, a signal that has abrupt changes in amplitude or phase with respect to time will occupy a significant spectrum in the frequency domain. Thus, system designers filter the digitally modulated baseband waveform so that abrupt changes occur more gradually, which helps to limit the amount of spectrum that the signal would occupy. In order for this baseband filtering to work properly, the amplitude envelope and phase of the signal must be represented accurately. If the amplitude or phase is distorted by components in the system, then the baseband filtering will lose its effectiveness and the signal will occupy additional spectrum. Some of the signal energy beyond what was originally specified will spill into the adjacent channels. The measure of this problem is called ACP, which is reported in terms of decibels below the main channel (dBc). The typical measurement application will provide a table showing the integrated power in the main channel and the integrated power (dBm) and relative power (dBc) in the adjacent channels.

The ACP issue was discussed in Chapter 19 on RF power amplifiers. Measurements were described that were made at 75 MHz on a generic power amplifier operating below the RF band. This lower operating frequency allowed the modulated signal to be measured in the time domain as well as the frequency domain. In this chapter ACP measurements are shown with actual cellular phone modulation formats on the RF amplifier whose CW operating performance in the 2.40–2.50 GHz band was described in Chapter 19.

32.2 MEASURING ACP

Conceptually, measuring ACP requires integrating the power in the main channel and then comparing that to the integrated power in the neighboring channels. Note that we cannot simply place a marker and read the power level as we can with a CW signal because the modulation has spread the energy over a fixed bandwidth. Instead, most spectrum analyzers have a built-in application for measuring ACP, which performs integration over a specified channel bandwidth. In practice, a given specification may require measuring several offsets, each with different channel widths. Newer spectrum analyzers can handle these requirements through the measurement setup menus.

32.3 ACP FOR NORTH AMERICAN DIGITAL CELLULAR (NADC) VERSUS GSM MODULATION FORMATS

Comparing the ACP performance of NADC with GSM demonstrates the effect of amplitude distortion caused by amplifier nonlinearity. NADC uses $\pi/4$DQPSK modulation, which has phase transitions that have been filtered at baseband so as to occupy less bandwidth (the channel width is 30 kHz for three time slotted voice channels). The result of this filtering is that the signal's amplitude envelope varies with time as the signal changes phase states. Although this scheme uses the available spectrum efficiently, it relies on linear amplification to maintain the properties of the baseband filtering, so you cannot operate the amplifier at saturation. This means that the amplifier will not be operating at its most efficient power level, from the standpoint of converting battery power to RF signal power. Instead, the signal level must be "backed off" from saturation to allow linear operation.

GSM uses MSK modulation, an FM-based scheme that produces a signal with a constant amplitude envelope. Because the modulation is FM based, it occupies more spectrum (the channel width is 200 kHz for eight time slotted voice channels). However, because the amplitude envelope of the signal is constant the amplifier can run at saturation, which allows for greater power efficiency. This allows for handsets with greater talk times in a smaller form factor, because the battery power is used more efficiently. The trade-off in this design is that fewer users can be accommodated in the same spectrum, because the channel widths are larger.

32.4 BACKOFF

The most efficient operating point for power amplifiers is at saturation from the standpoint of converting battery power to RF signal power. Unfortunately, at this operating point, any variations in the signal amplitude envelope above the average power level will not be represented in the output because the amplifier is already at its maximum output power level. Depending on the amount of amplitude variation in the signal format, the power level needs to be reduced or backed off from the saturated power level until the output signal is no longer distorted.

In the linear region of operation, if we want to reduce the output power by 1 dB we can simply reduce the input power by 1 dB. However, as we increase the power from the 1 dB compression point all the way to saturation, we may need to increase the input power by several decibels in order to increase the output power by 1 dB. It is therefore helpful to construct a backoff curve for the amplifier based on its input/ output characteristics. Then if we want a certain amount of backoff, we can simply look up the required input signal level required.

For most signal formats, amplifiers can be operated at their 1 dB compression point, which represents a good compromise between output linearity and power efficiency. In some signals, such as multiuser CDMA signals, the amplitude varies so much that the operating point needs to be reduced below the 1 dB compression point. Other systems, such as GSM, have no amplitude variation and can be operated at saturation.

32.5 ACP MEASUREMENT RESULTS FOR NADC AND GSM

Figure 32.1 shows the ACP of a $\pi/4$DQPSK signal after running through a power amplifier with various amounts of backoff from saturation. The measurement application is set up to measure the adjacent channel on either side of the main channel with no guard bands. The results for the signal coming from the test source are that the ACP is -28 dBc. The specifications require the ACP level to be -26 dBc or better. For the power amplifier under test in this example, it fails for signals greater than 3 dB backed off from saturation. The traces in the figure show that the amount of energy in the neighboring channels is steadily increasing as the input signal is increased.

Figure 32.2 shows the ACP of a GMSK signal for the test signal from the signal generator as well as after going through the power amplifier at its saturation power level. The results are the same in both circumstances: the ACP is -16 dBc. Because the amplitude envelope for GMSK is constant, the amplitude distortions caused by compression in the power amplifier do not affect the signal. This makes the power amplifier design much more straightforward. The amplifier designer can focus on making the amplifier highly efficient without worrying about linearity considerations. As GSM is phased out to accommodate new signal formats that can achieve higher data rates, amplifier designers struggle to maintain the highest

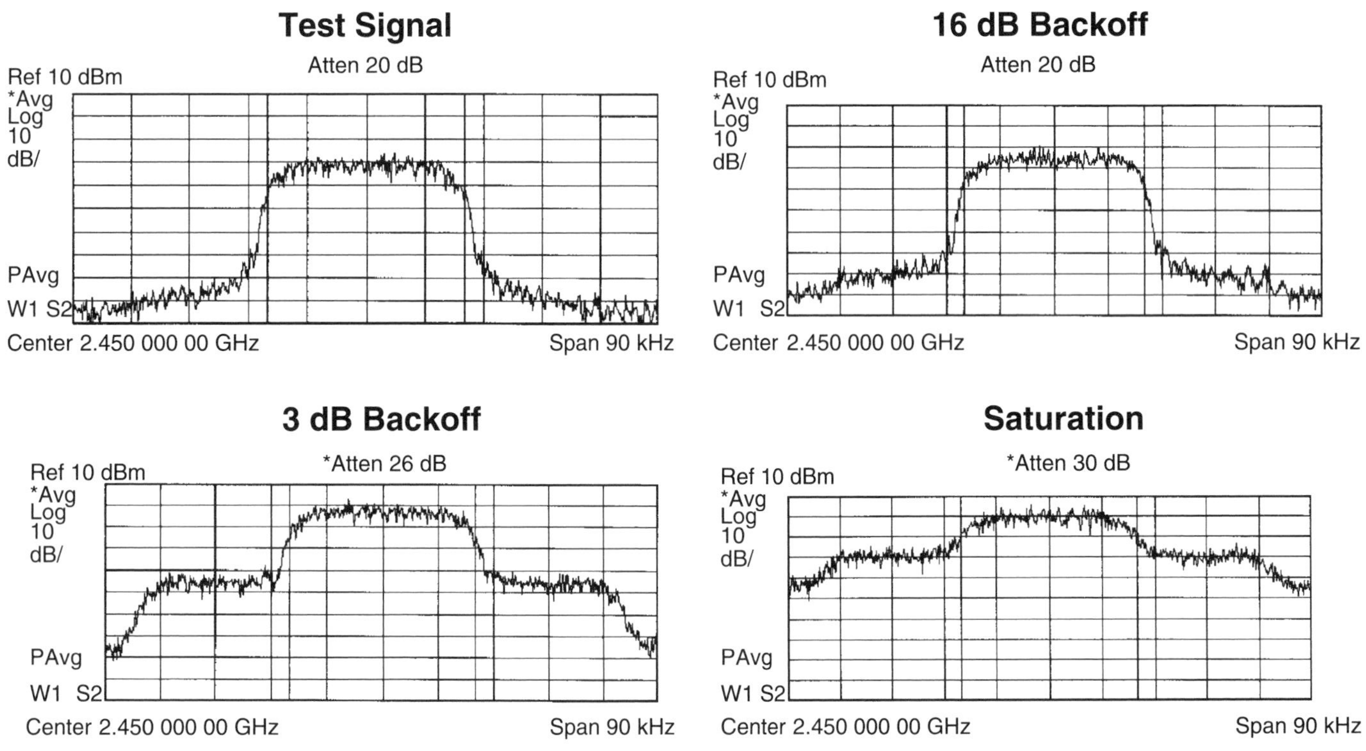

Figure 32.1 ACP of $\pi/4$DQPSK signal for various amounts of backoff.

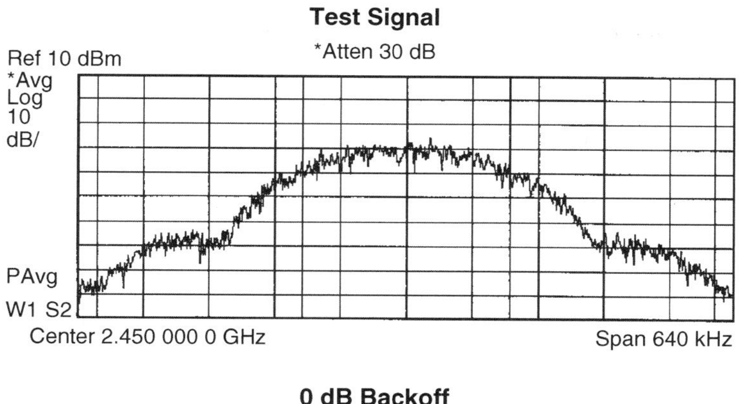

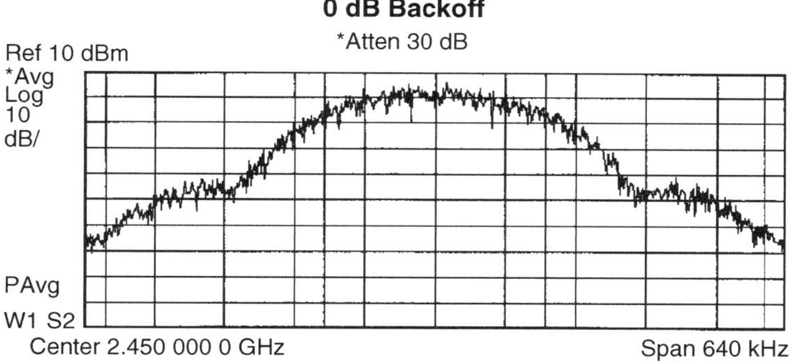

Figure 32.2 ACP of GMSK signal for various amounts of backoff.

power efficiency possible while achieving greater degrees of linearity to avoid causing unwanted degradation of the ACP performance of the system.

32.6 ZERO SPAN

Most ACP measurements use integration over the specified channel bandwidth to measure channel power in the main channel and compare it to the neighboring channels. Zero span is another technique that can be employed on the spectrum analyzer to measure power in a specified bandwidth, limited by available resolution bandwidth settings. If the user sets a span of 0 Hz on the spectrum analyzer, the analyzer will stop sweeping frequency and instead measure the amplitude envelope of the signal passing through the resolution bandwidth filter versus time. This can be useful for examining pulsed signals such as those found in TDMA systems. If the resolution bandwidth filter is set to the same width as the desired channel measurement, then the power level shown on the display will be the channel power. Some spectrum analyzer ACP measurement applications use this technique instead of integrating over the channel.

32.7 ANNOTATED BIBLIOGRAPHY

1. Application Note 1303: Spectrum Analyzer Measurements and Noise. Part II: Measurement of Noise-Like Signals, Agilent Technologies, Santa Clara, CA.
2. Product Note: Comparing Power Measurements on Digitally Modulated Signals, Document 8560E/8590E, Agilent Technologies, Santa Clara, CA.

References 1 and 2 provide additional background on ACP measurements.

CHAPTER 33

CONSTELLATION, VECTOR, AND EYE DIAGRAM AND EVM

Chapter 10 described the VSA and showed examples of the constellation, vector, and eye diagrams that it could display. The diagrams give insight into the causes of bit errors in the wireless data transmission system. The concept of EVM was also described, which can give a quantitative measurement of signal distortion and can be used for pass/fail decisions. Note that this troubleshooting is possible using the output signal from the transceiver system. No probe point at the IQ modulator is necessary. This is an important advantage of using the VSA for troubleshooting, especially with the highly integrated systems found in today's commercial applications.

This chapter will discuss measurements with the VSA in greater detail, showing specific measurement examples. The chapter is organized into five sections. Section 33.1 will show an example of power amplifier compression, which was discussed in Chapter 19. Power amplifier compression is a major source of distortion of phase modulation or QAM modulation in wireless communication systems. Amplifier distortion can be reduced or eliminated by operating the amplifier below saturation. This is called backoff. Backoff is controlled by reducing the RF input power, and the trade-off is described.

Section 33.2 discusses constellation, vector, and eye diagrams and what they tell about the problems in the RF system. The relation of the eye diagram to the vector diagram is also explained. Section 33.3 discusses EVM measurements.

Section 33.4 shows combined measurements of constellation, vector, and eye diagrams and EVM on an RF amplifier at various levels of compression and shows

RF Measurements for Cellular Phones and Wireless Data Systems. By A. W. Scott and R. Frobenius

the effect of IF filter group delay on a modulated RF signal. Section 33.5 shows a trouble shooting tree to aid in the analysis of a problem with a poorly performing wireless communication system, based on the constellation, vector, and eye diagrams and EVM measurements that are used to diagnose the problems.

33.1 POWER AMPLIFIER BACKOFF

Power amplifier saturation is a major source of modulation distortion in wireless communication systems. At low signal levels, the RF power output of an amplifier is directly proportional to the RF power input, as discussed in Chapter 19. This is called the "linear" range of operation. Every RF amplifier becomes nonlinear as it approaches its maximum RF output power level, therefore distorting any phase or QAM modulated signal.

Figure 33.1 shows the RF power output as a function of RF power input for the amplifier discussed in Chapter 19. This curve can be measured at a single frequency on a point-by-point basis with a signal generator and a power meter or a spectrum analyzer. It is most easily measured with a VNA in its power sweep mode. From the power output versus power input curve, the amplifier's backoff can be indicated as shown in Figure 33.1. Note that the output power reaches saturation of 19 dBm at an input power of 13 dBm. When the input RF power is adjusted to 6 dBm, the output power is 3 dB below saturation. The RF output power can be controlled by adjusting the RF input power.

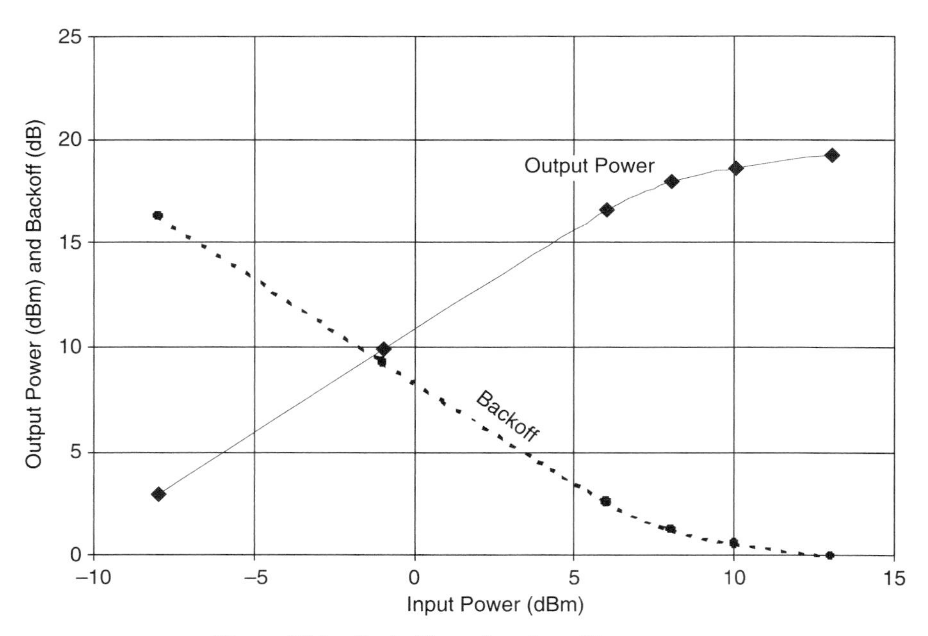

Figure 33.1 Backoff as a function of input power.

33.2 CONSTELLATION, VECTOR, AND EYE DIAGRAMS

The basis of the constellation, vector, and eye diagrams is the demodulated I/Q baseband signal from the VSA. Setting up a digital demodulation measurement requires providing the analyzer with all of the detailed parameters of the modulation format. These include symbol rate, constellation format (QPSK, QAM, etc.), baseband filtering information, and whether the signal is a pulsed format. Fortunately, many widely used standards are already preloaded into the analyzer as sets of demodulation parameters that can be called up by selecting them from a menu. Once the analyzer has been switched to the digital demodulation mode and the demodulation format information has been provided, the constellation, I/Q, and eye diagrams can be set up by activating a trace with "measured IQ" as its data and choosing one of the three formats as the "trace format."

Constellation Diagram

As discussed in Chapter 10, which described the VSA, the constellation measurement shows the ideal states for the modulation format that has been selected, either as crosshairs or circles. The size of the crosshairs or circles can be changed to show a specific size of EVM, for example, 5%. We could then see whether the sample points, shown as dots, fall within a specific EVM circle (like 5%) for each measurement.

Vector Diagram

The vector diagram measurement adds traces of the signal between sample times to the constellation measurement. The traces show the amplitude envelope and phase information of the RF carrier as it transitions from one point on the constellation to another. The number of points per symbol can be adjusted to show more or less detail for this trace, at the expense of requiring a longer result length and slower refresh times for the display.

Eye Diagram

Unlike baseband digital signals, RF signals have an in phase I component as well as a quadrature phase Q component. Consequently, RF signals have two eye diagrams: one each for the I and Q components of the signal. The "I-eye" shows signal transitions with respect to the horizontal or in phase axis of the constellation versus time. The "Q-eye" shows signal transitions with respect to the vertical or quadrature phase axis versus time. Because the x axis of the eye diagram is time, this measurement can be useful for diagnosing timing issues in either the I or Q channel of your modulator or digital signal processing hardware.

To better understand how the eye diagram is constructed, it is helpful to examine a hypothetical signal as it transitions through the constellation and the corresponding eye diagram that would be measured for that signal. To simplify the illustration,

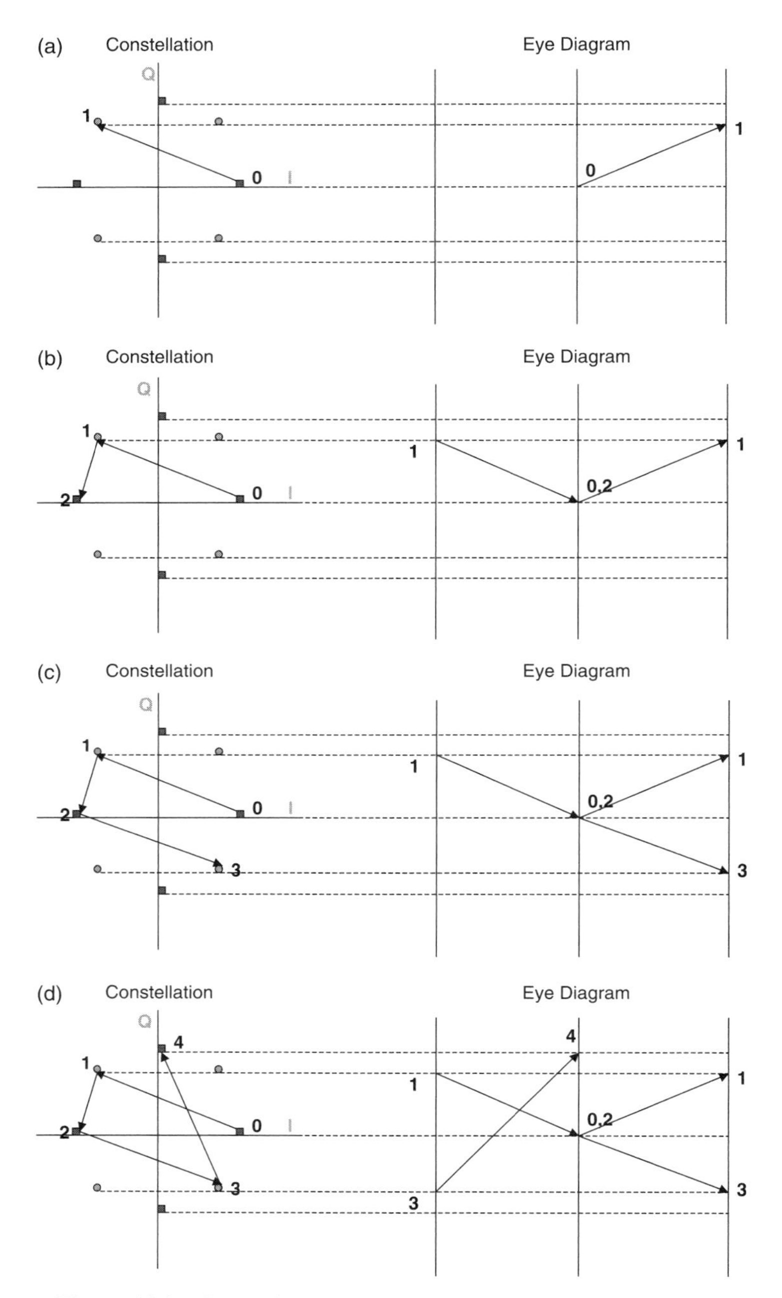

Figure 33.2 Generation of eye diagram from constellation diagram.

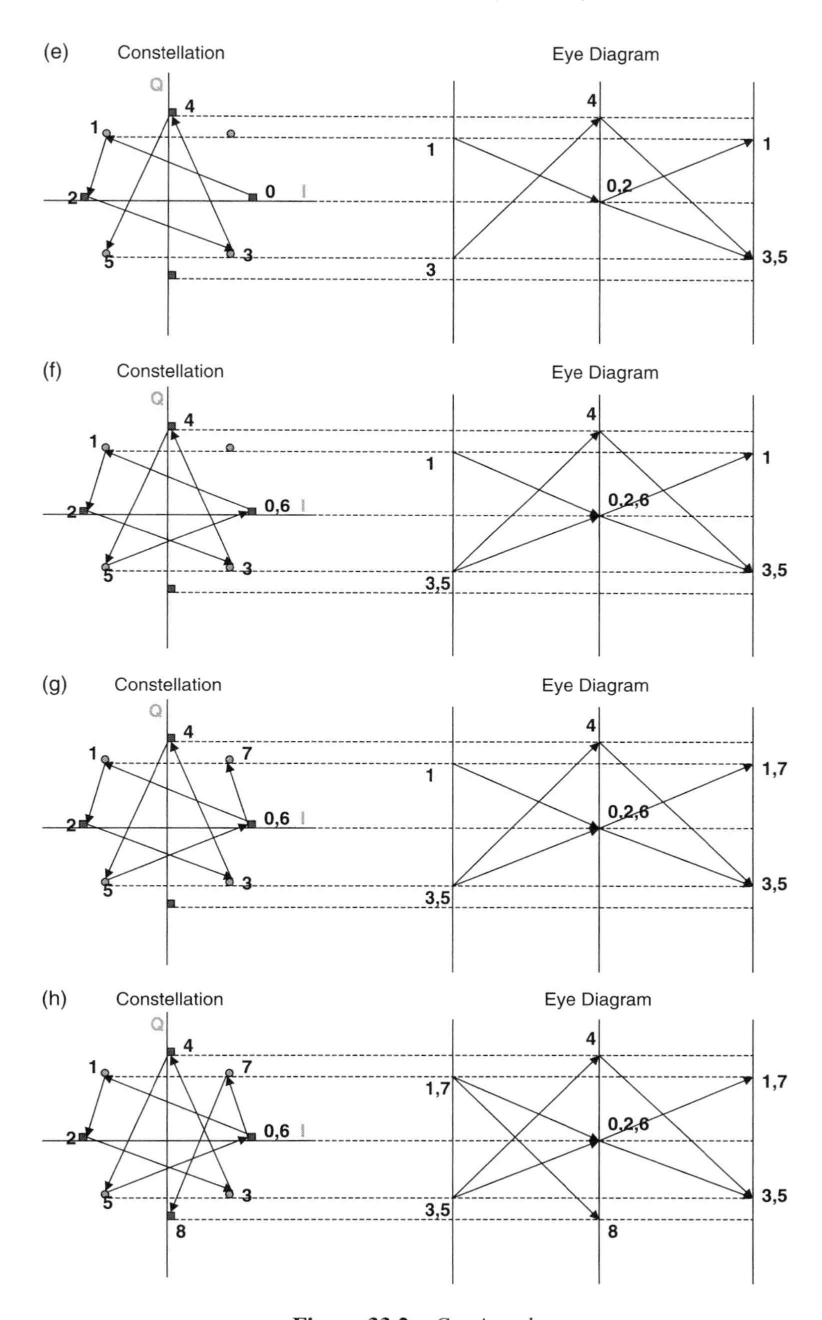

Figure 33.2 *Continued.*

we choose to draw the Q-eye, which shows travel along the vertical axis versus time. Figure 33.2 shows the signal for a $\pi/4$DQPSK signal as it transitions from one state on the constellation to another. The signal format consists of two QPSK constellations, one of which has been rotated by 45°, shown in the figure as the squares at 0°, 90°, 180°, and 270°. For each successive symbol, the signal alternates between using the rotated and unrotated constellation. This configuration provides the benefit that the signal amplitude envelope avoids the center of the constellation during transitions between states. High efficiency power amplifiers exhibit switching "glitches" when the amplitude falls to zero, so by avoiding zero amplitude transitions this format allows high efficiency amplifiers to be employed without suffering spectral growth that would be caused by any switching problems.

As the signal travels from starting point 0 to point 1 on the constellation (Fig. 33.2), its corresponding travel along the vertical axis is drawn from left to right on the eye diagram. When the right margin is reached, the "pen" is lifted and moved to the same level on the left margin, so that when the signal moves from position 1 to position 2 the transition is drawn on the eye diagram starting from the left margin and returning to the middle level vertically. Note that the 0° and 180° states share the same position vertically. Likewise, the 45° and 135° states share the same vertical position for the unrotated constellation, as do the 225° and 315° states.

As the signal transitions to the next state, labeled 3 (Fig. 33.2c), the trace is drawn to the lower level on the eye diagram. Because the right margin is reached, the trace resumes at the same level on the left margin. Now the example signal travels to the top of the rotated constellation. This transition along the vertical axis is drawn on the eye diagram. The process is repeated for transitions 5, 6, 7, and 8 (Fig. 33.2e–h).

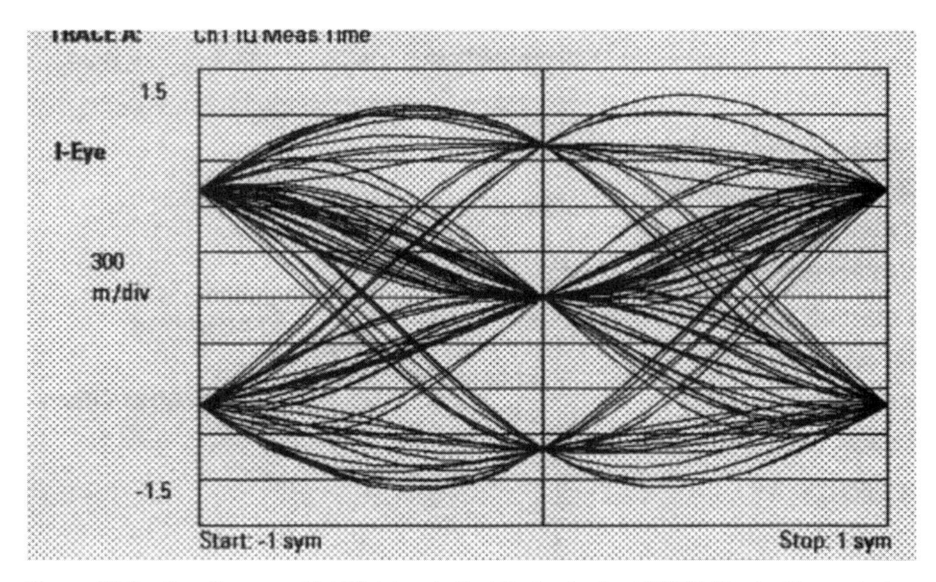

Figure 33.3 Eye diagram with 157 states. Agilent Technologies © 2008. Used with permission.

Note that because of the alternating use of the rotated and unrotated constellation, the eye diagram also alternates between having two vertical levels at the left and right edges and three possible vertical levels in the middle. This is seen more clearly in the measured eye diagram shown in Figure 33.3, which shows 157 symbols worth of data, instead of just 8.

33.3 EVM

The basis of the demodulation measurement on the VSA is the comparison of a measured signal with a "perfect" reference signal with respect to both magnitude and phase. For each batch of symbols to be analyzed, the demodulated bits of the measured signal are used as the input for creating a perfect reference signal within the analyzer. Note that the error vector can be calculated during the transition between constellation points as well as at the sample times when the symbols are determined.

EVM Troubleshooting

Although EVM is a convenient quantity for specifying performance on a data sheet, analyzing the error vector information is very useful for troubleshooting system performance. The error vector display and the tabulated error vector data on the VSA provide useful information for isolating the cause of system impairments.

EVM Versus Time

The EVM versus time display can provide useful information, especially in pulsed multiple access or TDMA systems. For example, if the EVM is significantly greater at the start or end of the pulse, this can indicate timing problems with amplifiers switching on or off too early or too late. High EVM values between symbol times can indicate improper baseband filtering being applied to the signal.

EVM Spectrum

If we apply a fast Fourier transform to the EVM versus time display, then we will see the distribution of error "energy" with respect to frequency. This can be very useful for revealing the presence of hidden spurs within our signal. Figure 33.4 shows the spectrum of a QPSK signal in the presence of a spur within the bandwidth of the modulation. The display shows how the signal would appear on a spectrum analyzer. The spurious tone is completely hidden within the signal. Figure 33.5 provides the error vector spectrum for the same signal. The error energy is contained mainly within the signal bandwidth. However, a very sharp concentration of error energy in the EVM spectrum reveals the location of the spurious signal, even though it is "buried" within the desired signal itself.

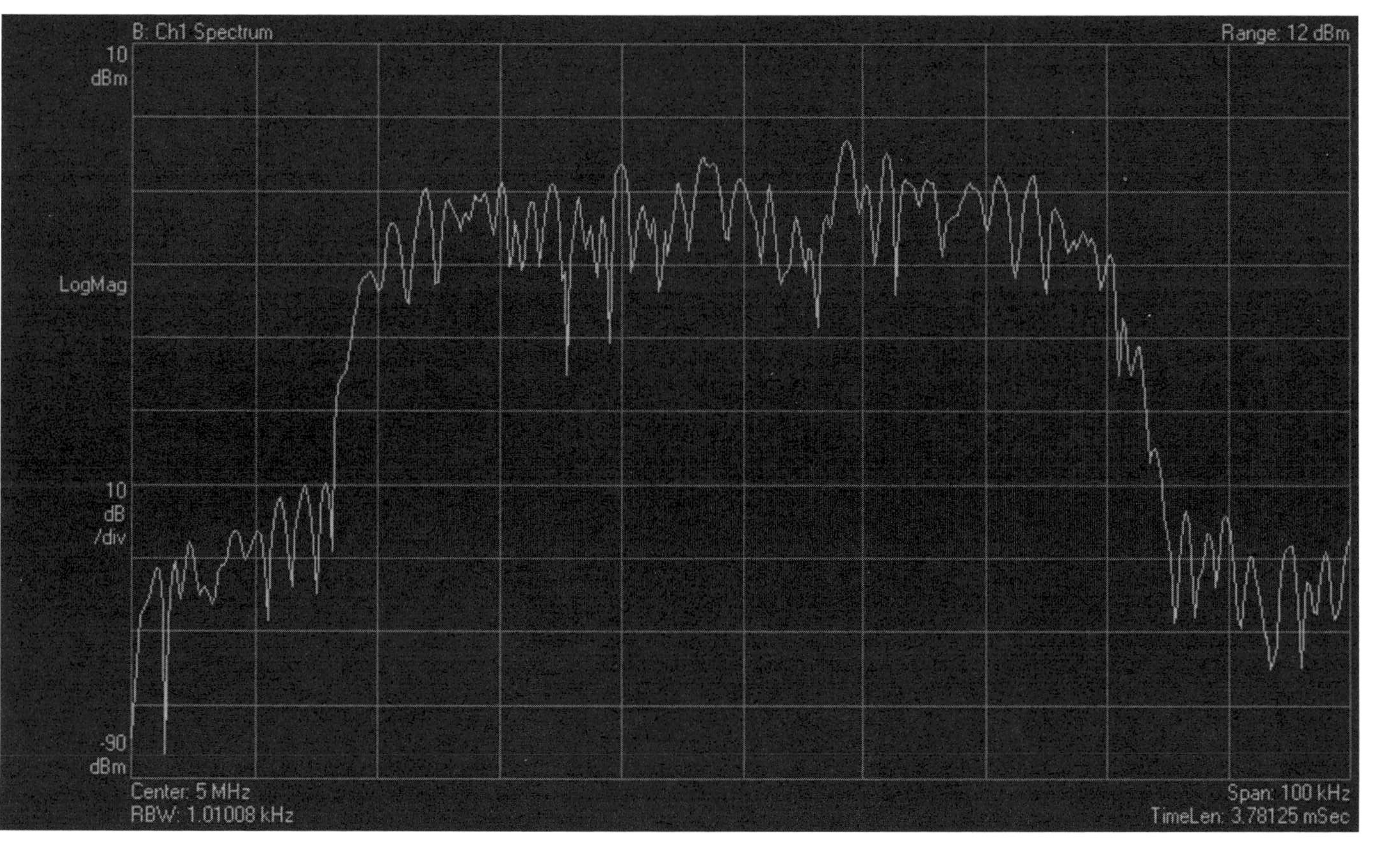

Figure 33.4 Spectrum analysis of modulated signal.

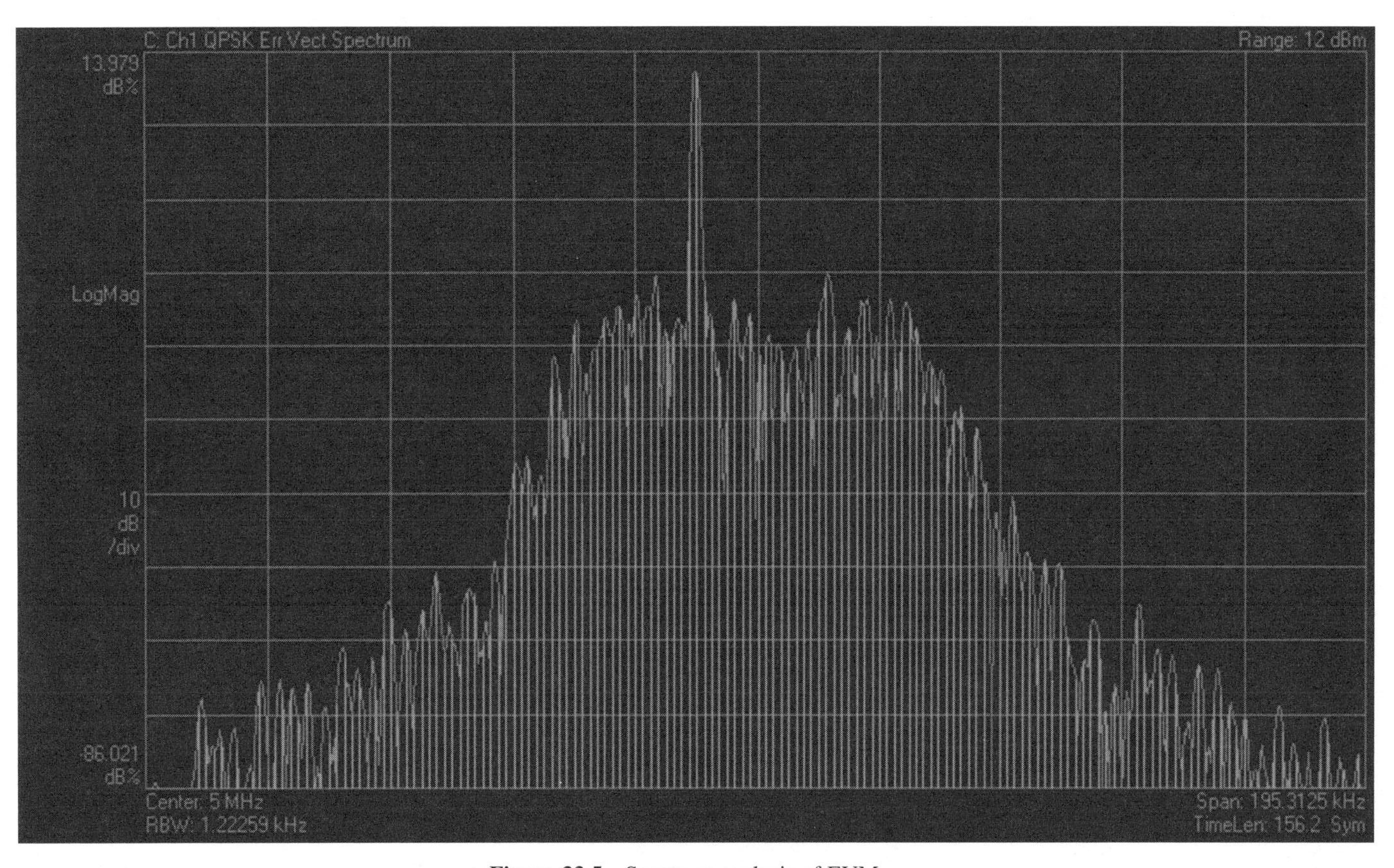

Figure 33.5 Spectrum analysis of EVM.

IQ Modulator Impairments

The EVM parameters table lists metrics for the performance of the IQ modulator. Two important design and implementation tasks for the IQ modulator are that the I and Q channels should have symmetric gain and that the two channels should have a 90° phase offset between them. The gain imbalance parameter in the table shows the difference in gain between the I and Q channels. Figure 33.6 shows a QPSK signal with 1 dB of gain imbalance. The constellation points appear as a rectangle instead of a square. The quadrature error parameter shows the number of degrees deviation from 90° offset for the I and Q channels. Figure 33.7 shows a QPSK signal with 2° of quadrature error. The constellation appears as a "tilted" parallelogram instead of a square.

33.4 MEASUREMENTS OF CONSTELLATION, VECTOR, AND EYE DIAGRAMS AND EVM ON AN RF POWER AMPLIFIER AND ON AN IF FILTER

Figures 33.8–33.12 show VSA measurements made on the RF power amplifier described in Chapter 19 as a function of backoff from saturation. The graph of Figure 33.1 was used to determine the RF input to the amplifier to achieve the different amount of backoff. The VSA was set to show four measurements on its display:

Upper left: vector diagram
Upper right: eye diagram
Bottom left: EVM
Bottom right: constellation diagram

Figure 33.8 is the test signal. The RMS EVM is less than 1%. (Note that the vertical scale divisions are 2%.) The constellation points are at the crosshairs.

Figure 33.9 shows the same measurement in the linear range of the amplifier with 9 dB of backoff. The performance is about the same as was measured with the test signal.

Figure 33.10 provides the same measurement with the amplifier backed off 2 dB from saturation. The RMS EVM is about 4%, and the constellation points are spreading around the crosshairs.

Figure 33.11 shows the amplifier at 1 dB backoff. The RMS EVM is about 10%. The constellation points are spread more, and the vector diagram and eye diagram show more spreading.

Figure 33.12 presents the amplifier performance at saturation with no backoff. The EVM is about 14%. The constellation points are widely spread, more in angle than in radius because of AM to PM (the phase change with increasing input power). The cross-over points in the time axis of the eye diagram are spreading in time.

The overall cell phone system specifications of the IS-136 system in which the amplifier is to be used requires the RMS EVM to be less than 16%. When the

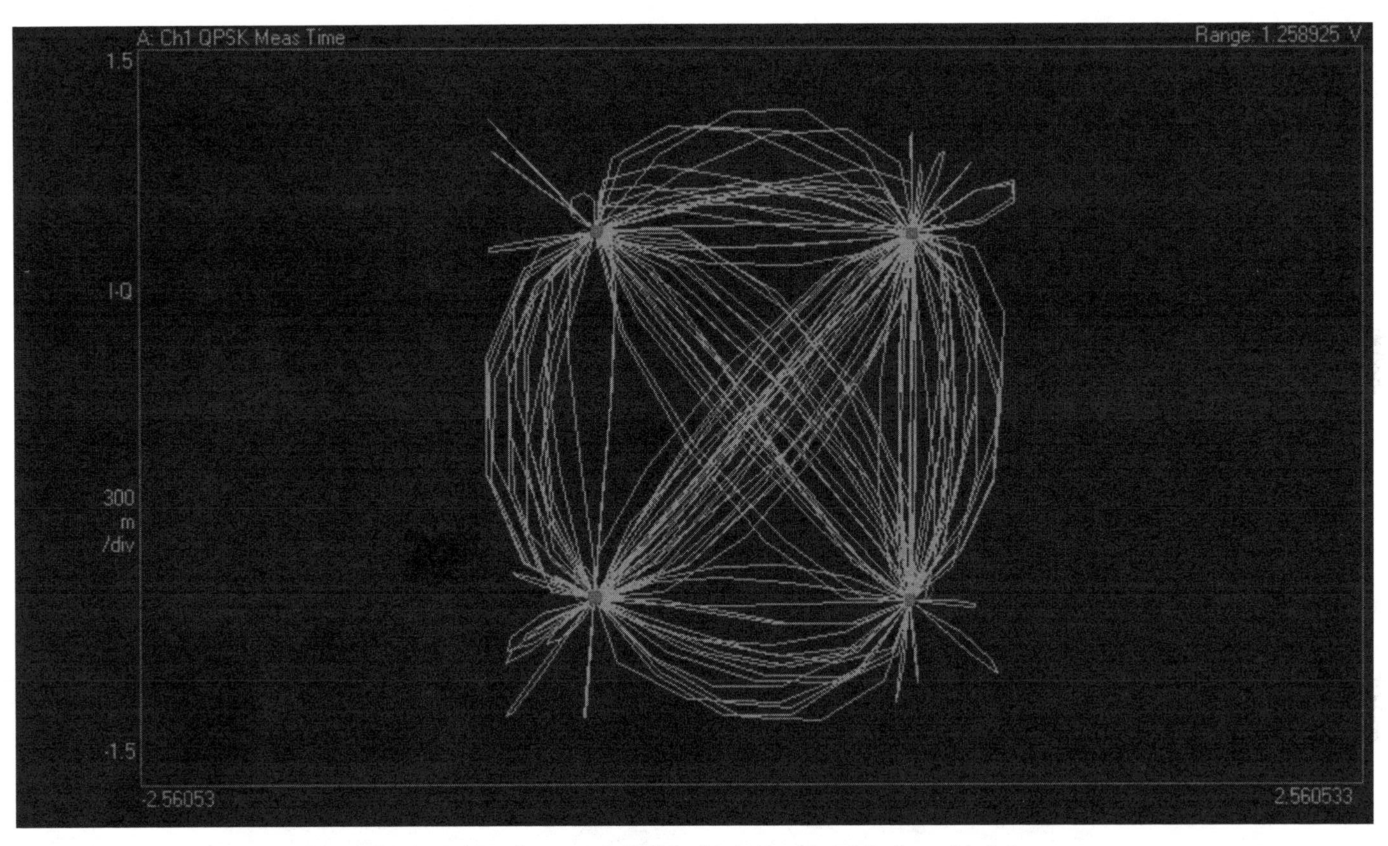

Figure 33.6 Vector diagram of QPSK with 1 dB of I and Q channel imbalance.

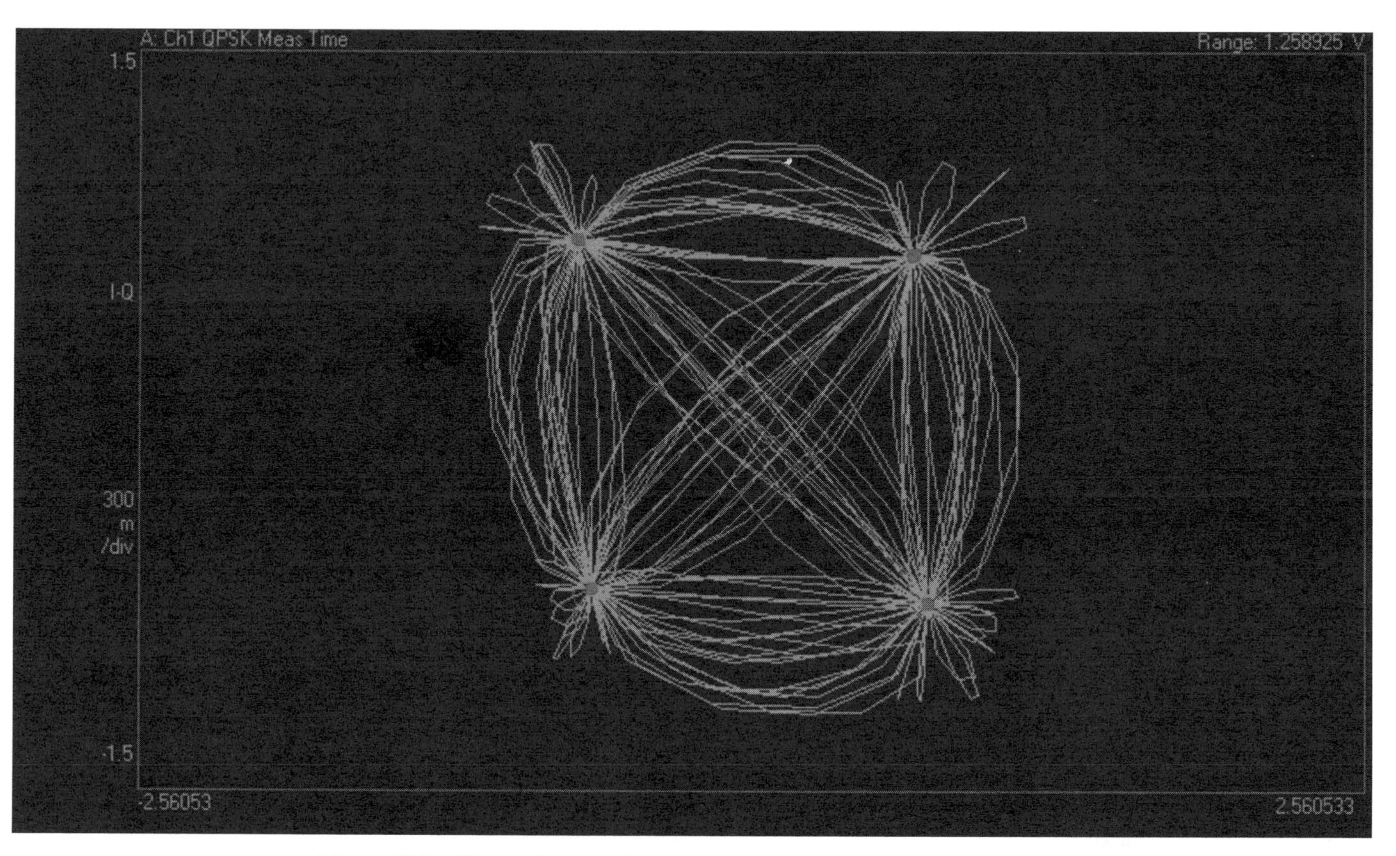

Figure 33.7 Vector diagram of QPSK with 2° of I and Q phase imbalance.

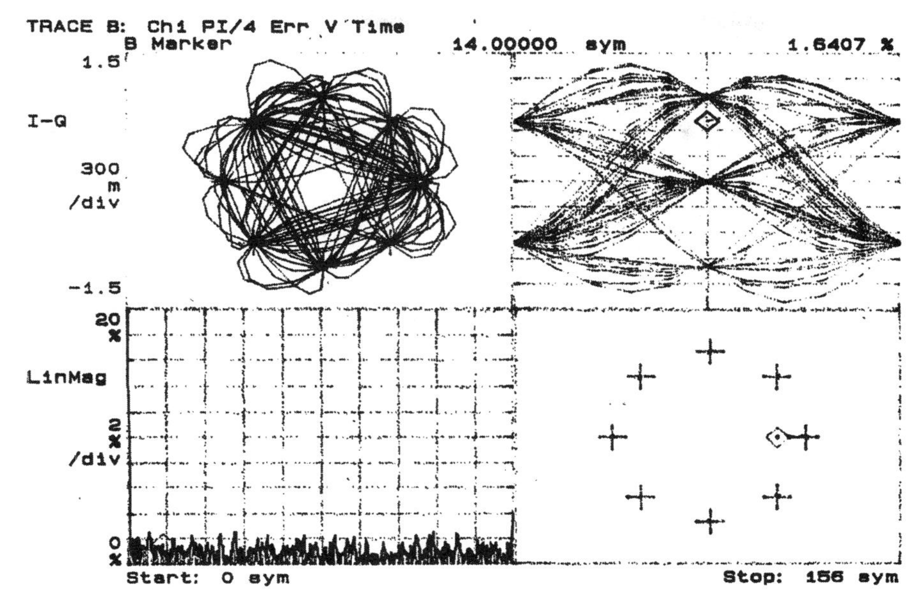

Figure 33.8 VNA analysis of $\pi/4$DQPSK test signal.

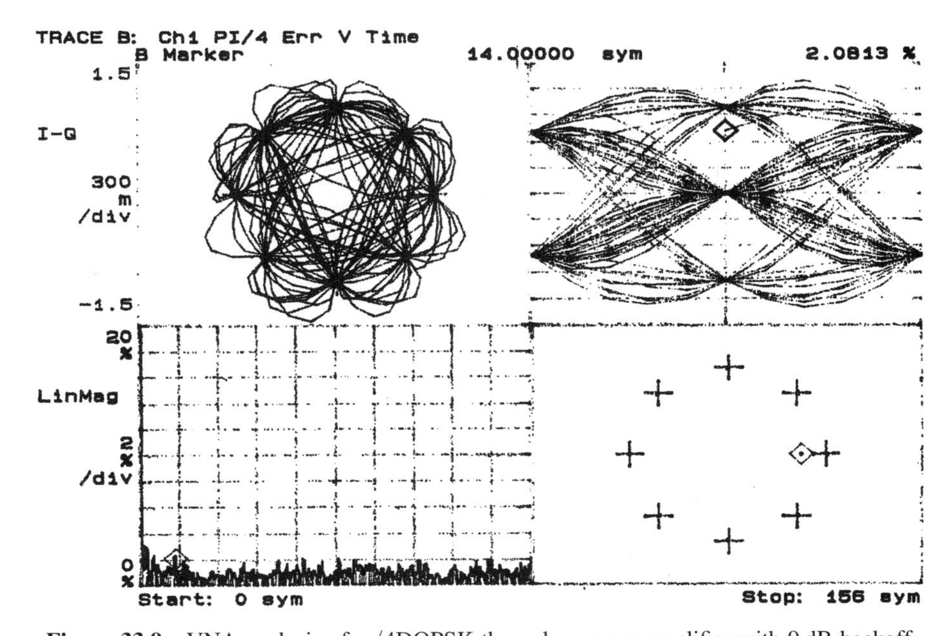

Figure 33.9 VNA analysis of $\pi/4$DQPSK through a power amplifier with 9 dB backoff.

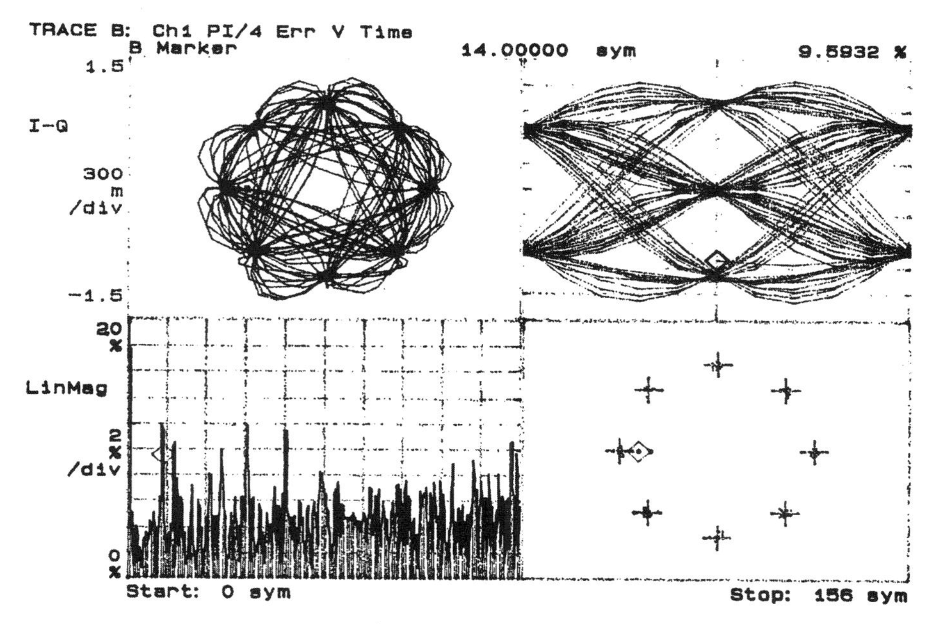

Figure 33.10 VNA analysis of $\pi/4$DQPSK through a power amplifier with 2 dB backoff.

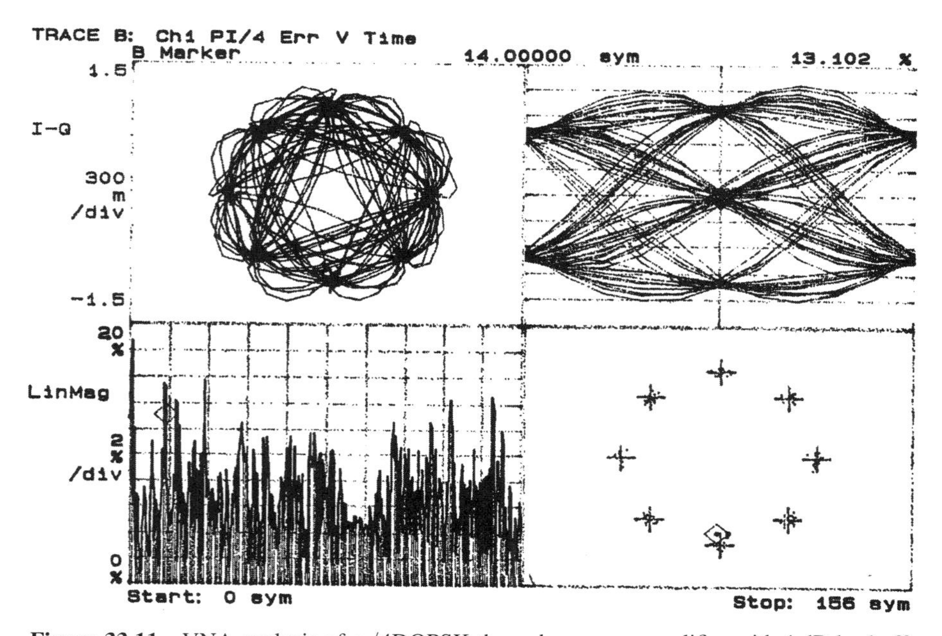

Figure 33.11 VNA analysis of $\pi/4$DQPSK through a power amplifier with 1 dB backoff.

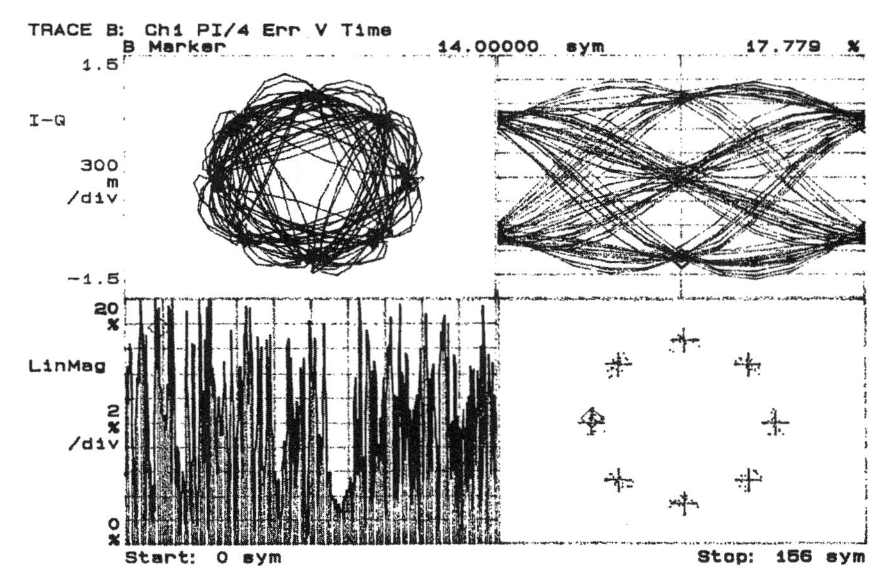

Figure 33.12 VNA analysis of $\pi/4$DQPSK through power amp with 0 dB backoff.

amplifier is driven to saturation, it uses up almost all of the specifications margin, which must also include the effects of IF group delay and PLO phase noise.

Figure 33.13 summarizes the results of Figures 33.8–33.12, showing EVM as a function of backoff. The amplifier must be backed off to 3 dB below saturation, which is its 1 dB compression point, to keep the EVM below 6%.

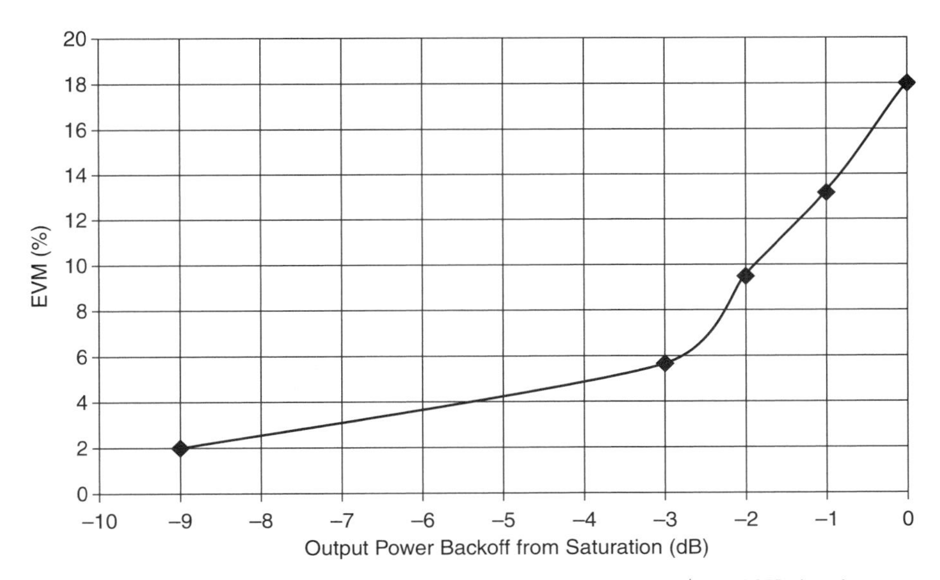

Figure 33.13 EVM versus power amplifier backoff with $\pi/4$DQPSK signal.

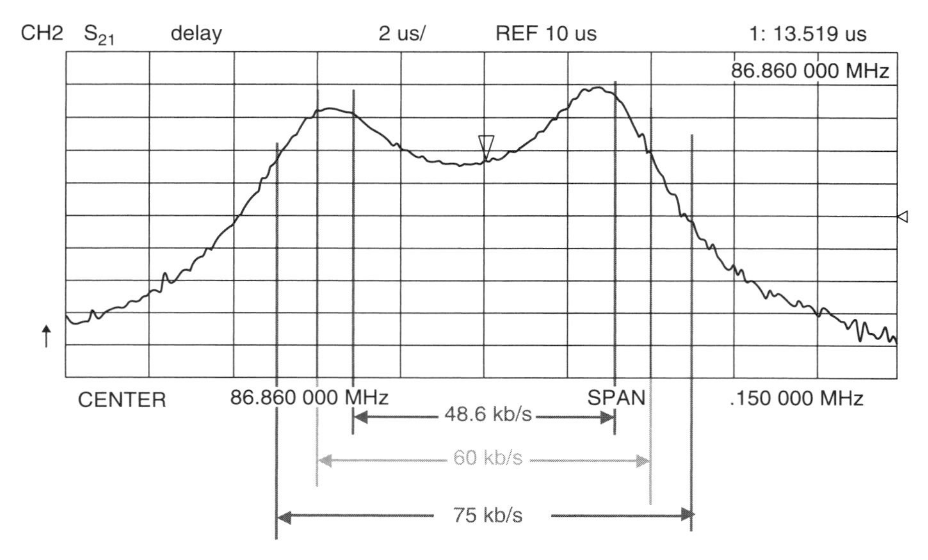

Figure 33.14 Group delay of IF filter.

Figure 33.14 shows the group delay measurement of the IF filter discussed in Chapter 22. Added to the figure are the approximate bandwidths of RF signals with data rates of 48.6, 60, and 75 kbps. The filter is designed for use in the 48.6 kbps IS-136 system.

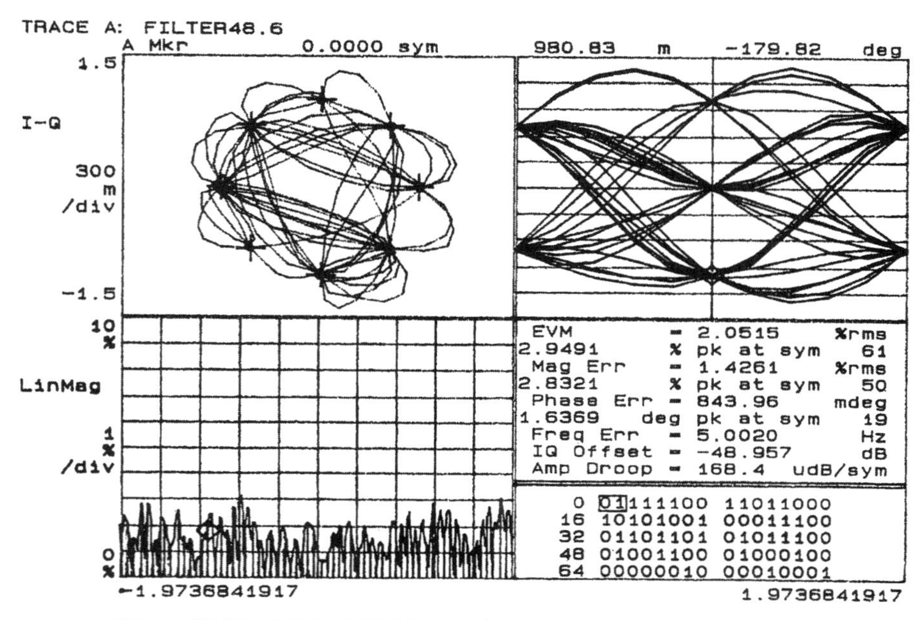

Figure 33.15 EVM of 48.6 kbps $\pi/4$DQPSK signal through IF filter.

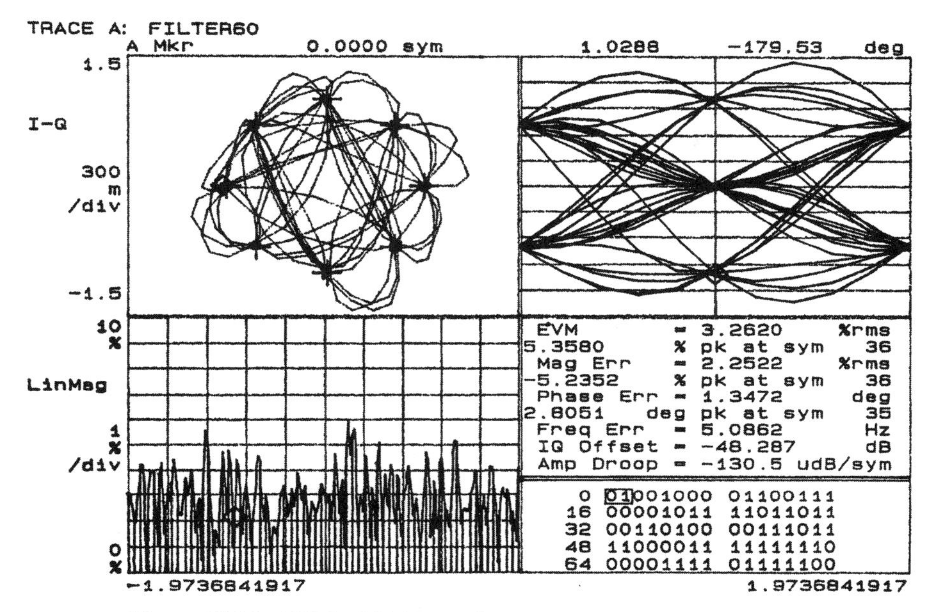

Figure 33.16 EVM of 60 kbps $\pi/4$DQPSK signal through IF filter.

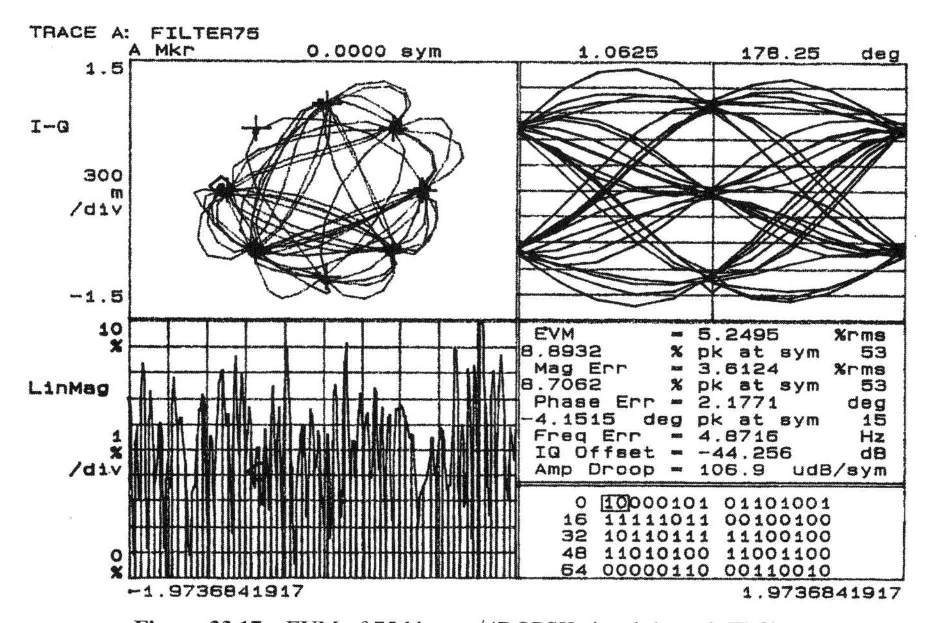

Figure 33.17 EVM of 75 kbps $\pi/4$DQPSK signal through IF filter.

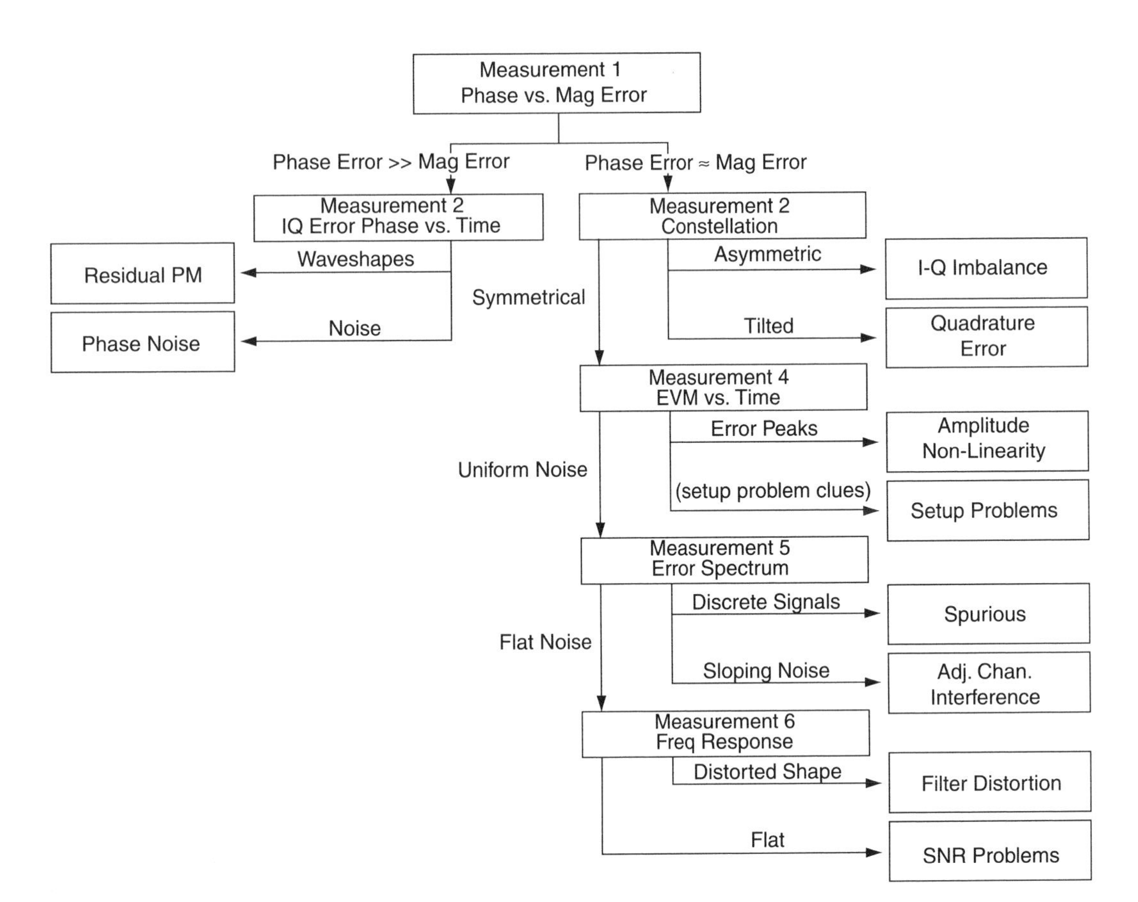

Figure 33.18 EVM trouble shooting tree.

Figures 33.15–33.17 illustrate VSA measurements of the vector diagram, eye diagram, EVM, and EVM table at the three data rates. Note from Figure 33.15 that the EVM of the 48.6 kbps signal after passing through the filter is 2% RMS, which is due to the filter group delay. When the data rate is raised to 60 kbps, the EVM increases to 3.26%. When the data rate is raised to 75 kbps, the EVM increases to 5.25 RMS and the constellation points (as shown in the vector diagram) and the eye diagram spread. These effects are caused by the group delay of the IF filter.

33.5 EVM TROUBLE SHOOTING TREE

When properly applied, measurements of EVM and related quantities can provide insight into the quality of a digitally modulated signal. They can also pinpoint the causes of any problems uncovered during the testing process.

Figure 33.18 shows an EVM trouble shooting tree. Measurement 1 is to determine if phase errors or magnitude errors are the major cause of problems. This can be determined by comparing each source of EVM, for example, as shown in the lower left-hand display of Figure 33.17. In this measurement, the magnitude and phase EVM errors are about the same, so phase is not a problem. However, if the phase error was significantly greater, we should make measurement 2, which is a measurement of IQ phase versus time, and look for residual phase modulation and phase noise.

If the phase and amplitude errors are approximately equal, measurement 3 should be made, which is an analysis of the constellation diagram to see if IQ imbalance or quadrature error is the problem. These problems were shown in Figures 33.6 and 33.7, respectively.

If there are no problems evident in the constellation diagrams, proceed to measurement 4, showing EVM versus time. If error peaks are present, the problem could be amplitude nonlinearity or setup problems.

Next proceed to measurement 5: error spectrum. Discrete signals such as shown in Figure 33.5 may be caused by power supply switching transients. A sloping noise spectrum suggests adjacent channel interference.

Measurement 6 will show filter distortion if the frequency spectrum is distorted or S/N problems if the spectrum is flat.

33.6 ANNOTATED BIBLIOGRAPHY

1. 89600 Series Vector Signal Analyses Software, http://www.agilent.com/find/89600, Agilent Technologies, Santa Clara, CA, 2006.
2. Product Note 89400-14A: 10 Steps to a Perfect Digital Demodulation Measurement, Agilent Technologies, Santa Clara, CA.
3. Product Note 8944-14: Using Error Vector Magnitude Measurements to Analyze and Troubleshoot Vector-Modulated Signals, Agilent Technologies, Santa Clara, CA.

Reference 1 is an Agilent CD that allows the demodulated signal from the spectrum analyzer to be analyzed on a laptop computer. The CD includes an extensive tutorial and help file and recorded test signals of various modulation types with various distortions. The CD costs around $20,000 for continuous use with demodulated signals from actual equipment. A free copy can be obtained from Agilent that is fully functional for 2 weeks. Thereafter it can be used indefinitely with its recorded signals only. Even in this mode it is still an extremely valuable reference.

References 2 and 3 explain the use of the EVM trouble shooting tree in more detail than this book does.

CHAPTER 34

CCDF

The need for CCDF (Complementary Cumulative Distribution Function) arises primarily when dealing with digitally modulated signals in spread spectrum systems such as IS-95, cdma2000, and WCDMA. Because these types of signals are noiselike, CCDF curves provide a useful characterization of the signal power peaks. CCDF is a statistical method that shows the amount of time the signal spends above any given power level. The mathematical origins of CCDF curves are the familiar probability density function (PDF) and CDF curves explained in an introductory probability and statistics course.

The CCDF measurement is an excellent way to fully characterize the power statistics of a digitally modulated signal. Modulation formats can be compared via CCDF in terms of how stressful a signal is on a component such as an amplifier. The CCDF curve can also be used to determine the impact of filtering on a signal. From the CCDF curves of multitone signals, amplifier designers know exactly how much headroom to allow for the peak power excursions to avoid compression. CCDF curves are a powerful way to view and characterize how various factors affect the peak amplitude excursions of a digitally modulated signal.

The number of active codes in a CDMA signal significantly affects the power statistics. Furthermore, different combinations of active codes yield different power CCDF curves, because of orthogonal coding effects. Multicarrier signals also cause a significant change in CCDF curves, similar to the effect of multitone signals. CCDF is becoming a necessary design and testing tool in 3G communications systems.

RF Measurements for Cellular Phones and Wireless Data Systems. By A. W. Scott and R. Frobenius

Perhaps the most important contribution of CCDF curves is in setting the signal power specifications for mixers, filters, amplifiers, and other components. The CCDF measurement can help determine the optimum operation point for components.

34.1 CCDF CURVES

Figure 34.1a shows a power versus time plot of a nine-channel IS-95 signal. The signal in the form shown in Figure 34.1a is difficult to quantify because of its inherent randomness. In order to extract useful information from this noiselike signal, we

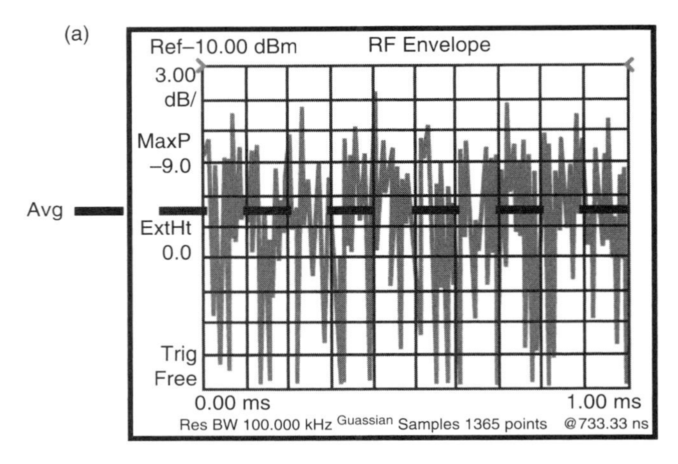

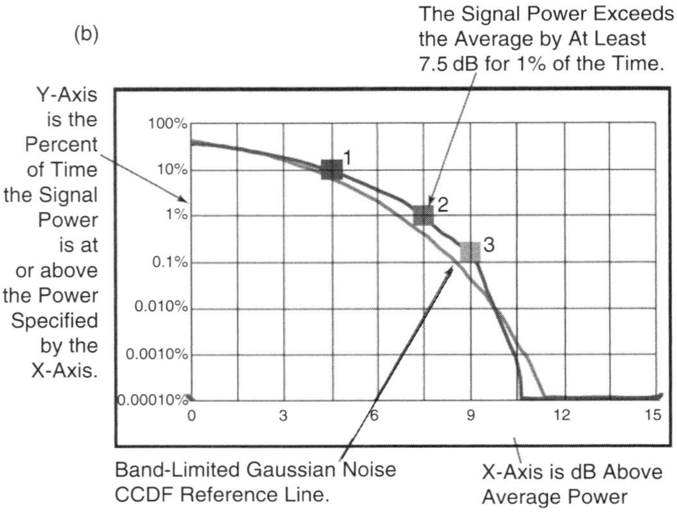

Figure 34.1 CCDF.

need a statistical description of the power levels in this signal, and a CCDF curve gives just that.

A CCDF curve shows how much time the signal spends at or above a given power level. The power level is expressed in dB relative to the average power. For example, each of the lines across the waveform shown in Figure 34.1a represents a specific power level above the average. The percentage of time the signal spends at or above each line defines the probability for that particular power level. A CCDF curve, as shown in Figure 34.1b, is a plot of relative power levels versus probability.

Figure 34.1b displays the CCDF curve of the same nine-channel IS-95 signal. Here the x axis is scaled to decibels above the average signal power, which means the peak to average ratios are being measured as opposed to absolute power levels. The y axis is the percentage of time the signal spends at or above the power level specified by the x axis. For example, at $t = 1\%$ on the y axis, the corresponding peak/average ratio is 7.5 dB on the x axis. This means that the signal power exceeds the average by at least 7.5 dB for 1% of the time. The position of the CCDF curve indicates the degree of peak/average deviation, with more stressful signals further to the right.

34.2 DERIVATION OF CCDF CURVES

The mathematics of CCDF curves is shown in Figure 34.2 and is as follows: the PDF is derived from a signal level versus time graph as shown in the upper two charts. To obtain the CDF, the integral of the PDF is computed as shown in the middle two charts. Finally, subtracting the CDF from 1 results in the CCDF. Thus, the CCDF is the complement of the CDF (CCDF = 1 − CDF).

To generate the CCDF curve in the form shown in Figure 34.1b, the y axis is converted to logarithmic form and the x axis begins at 0 dB. CCDF emphasizes peak amplitude excursions, whereas CDF emphasizes minimum values.

34.3 COMPARISON OF VECTOR DIAGRAMS AND CCDF

The modulation format of a signal affects its power characteristics. Using CCDF curves, the power statistics of different modulation formats can be fully characterized.

In Figure 34.3a and b, the vector plot of a QPSK signal is compared to that of a 16QAM signal. A QPSK signal transfers only 1 bit/symbol, whereas a 16QAM signal transfers 4 bits/symbol. Therefore, the bit rate of a 16QAM signal is twice that of a QPSK signal for a given symbol rate. The 16QAM signal appears to have higher peak/average power ratios, but it is difficult to make quantitative observations from these plots.

Figure 34.3c shows the CCDF curves of both the 16QAM and QPSK signals. The 16QAM signal has a more stressful CCDF curve than that of the QPSK signal. Although 16QAM is capable of transmitting more bits per state than QPSK for a given symbol rate, it also produces greater peak/average ratios than QPSK.

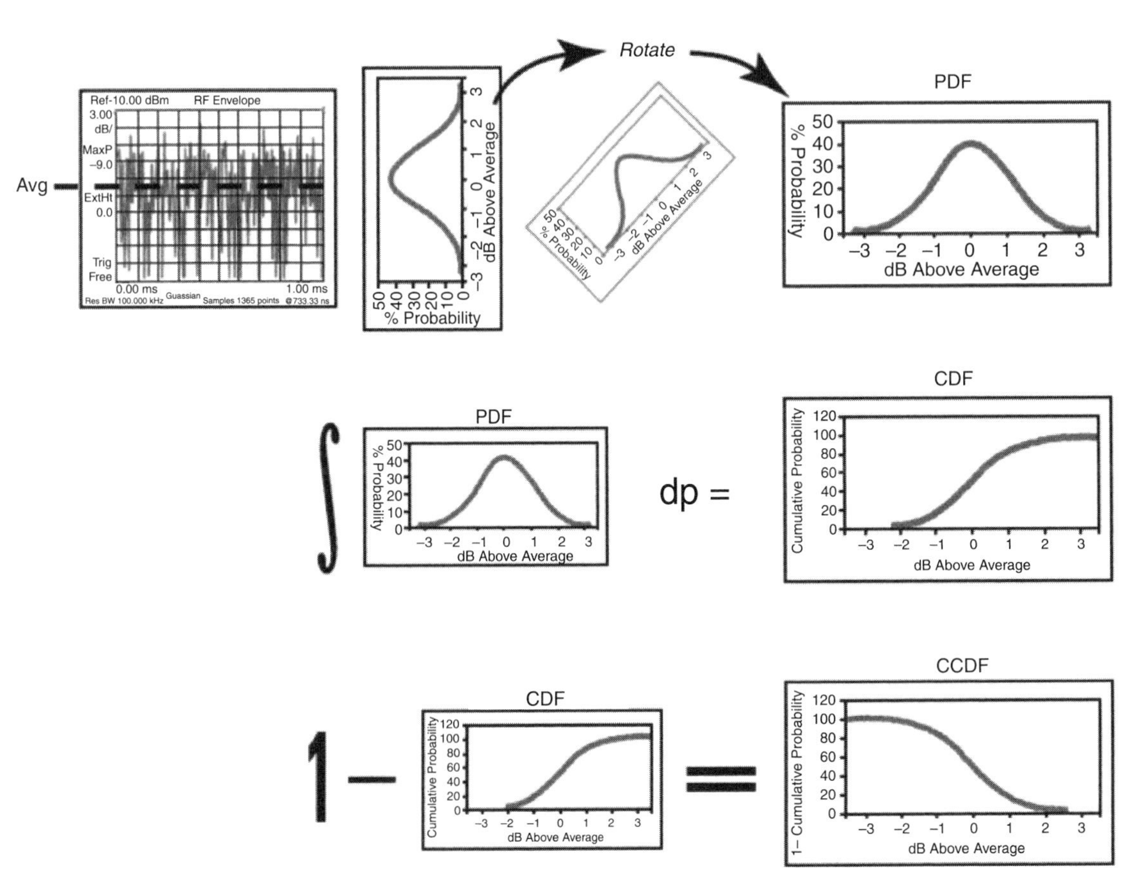

Figure 34.2 Derivation of CCDF. Agilent Technologies © 2008. Used with permission.

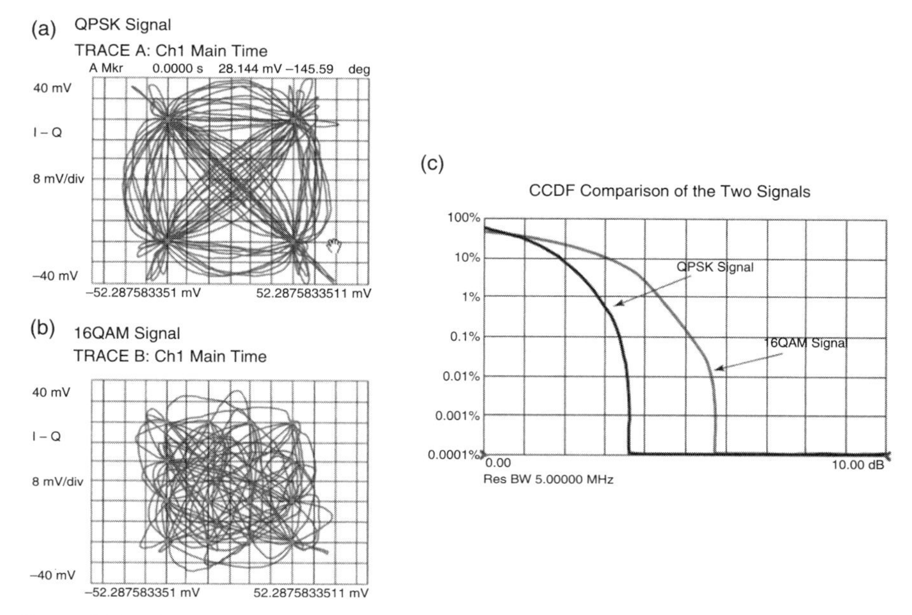

Figure 34.3 CCDF comparison for QPSK and 16QAM. Agilent Technologies © 2008. Used with permission.

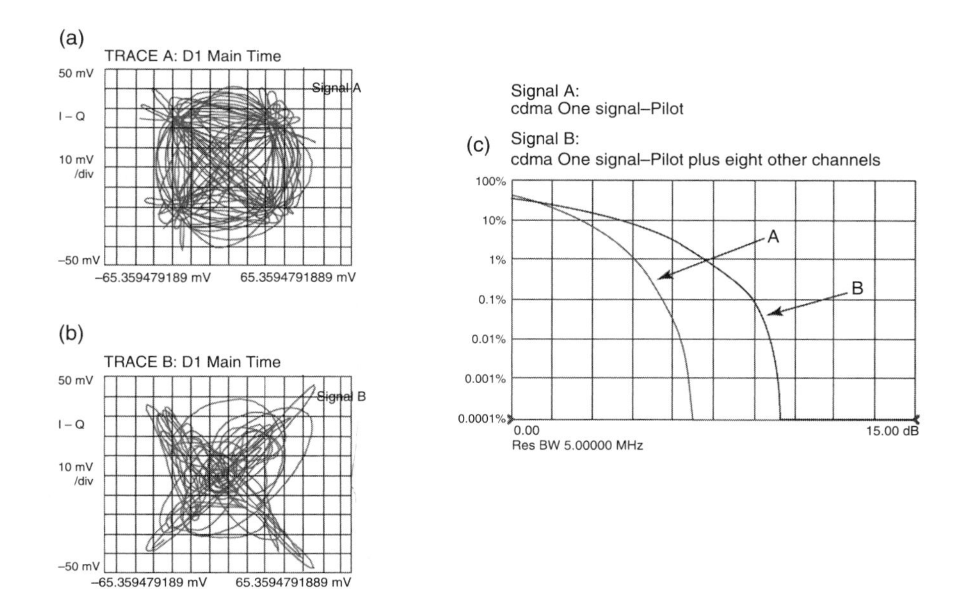

Figure 34.4 CCDF comparison for single channel and nine-channel CDMA. Agilent Technologies © 2008. Used with permission.

34.4 THE EFFECT OF THE NUMBER OF ACTIVE SPREAD SPECTRUM CODES

The number of active codes in a spread spectrum system affects the power statistics of the signal. Increasing the number of channels in a spread spectrum system increases the power peaks, thus putting higher stress on system components.

Figure 34.4 presents vector diagrams and CCDF curves of two IS-95 signals. Signal A with just the pilot channel has a significantly lower statistical power ratio than signal B that has the pilot plus eight other randomly chosen channels, as indicated by the CCDF curves. As expected, the signal with higher channel usage has a more stressful CCDF curve.

34.5 CCDF IN COMPONENT DESIGN

Compression of signals occurs in nonlinear components. For instance, an amplifier compresses a signal when the signal exceeds the amplifier power limitations. To avoid compression, the optimum input power level for the amplifier needs to be known. CCDF curves can be used to determine input power levels and serve as a guide for RF power amplifier designers and systems engineers.

It is difficult to see compression effects in the time domain measurements shown in Figure 34.5. Some clipping is evident, but there is no convenient way of describing the compression quantitatively. However, compression of the output signal is clearly evident by comparing the power CCDF curves of the input signal and the amplified output signal.

As seen in Figure 34.6, CCDF curves are clearly an excellent tool for displaying compression effects. By definition, CCDF curves measure how far and how often a signal exceeds the average power. If a signal passing through an amplifier were perfectly (linearly) amplified, the output waveform of the signal in the time domain would perfectly resemble the input waveform, with a gain in power. Both the average and envelope power of the amplified signal would increase by a common factor. Therefore, the peak/average power ratio would not change, and the two CCDF curves would appear identical. However, when the output signal exceeds the power limitations of the amplifier, clipping occurs; the output waveform no longer resembles an amplified version of the input waveform, and the peak/average ratios change. The CCDF curve of the clipped signal decreases and no longer matches the original input signal. This effect makes CCDF a good indicator of compression.

Figures 34.7 and 34.8 show CCDF measurements of an IS-95 CDMA signal passing through the RF amplifier whose analog performance was measured in Chapter 19. These measurements were made with a special CCDF processing option of the PSA spectrum analyzer described in Chapter 9.

Figure 34.7 shows the CCDF curve when this amplifier is operated in the linear range, and it is identical to the CCDF curve of the input signal itself. In contrast, Figure 34.8 shows the severe compression of the CCDF curve when the same amplifier is operated is at saturation.

(a) Amplifier Input Signal Versus Time

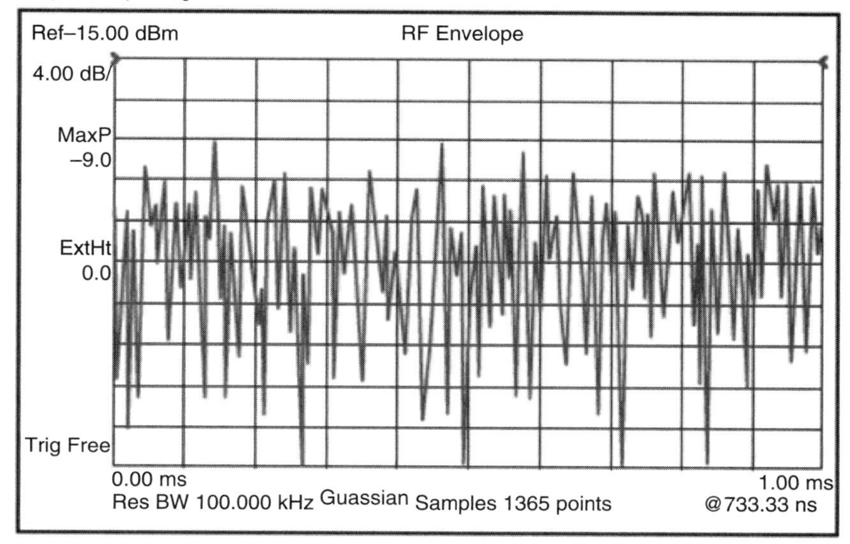

(b) Amplifier Output Signal Versus Time

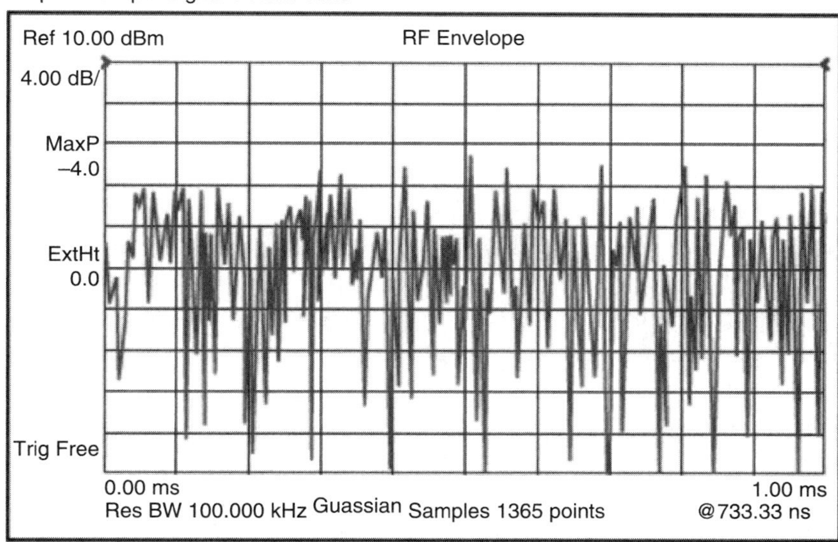

Figure 34.5 Amplifier input and output versus time. Agilent Technologies © 2008. Used with permission.

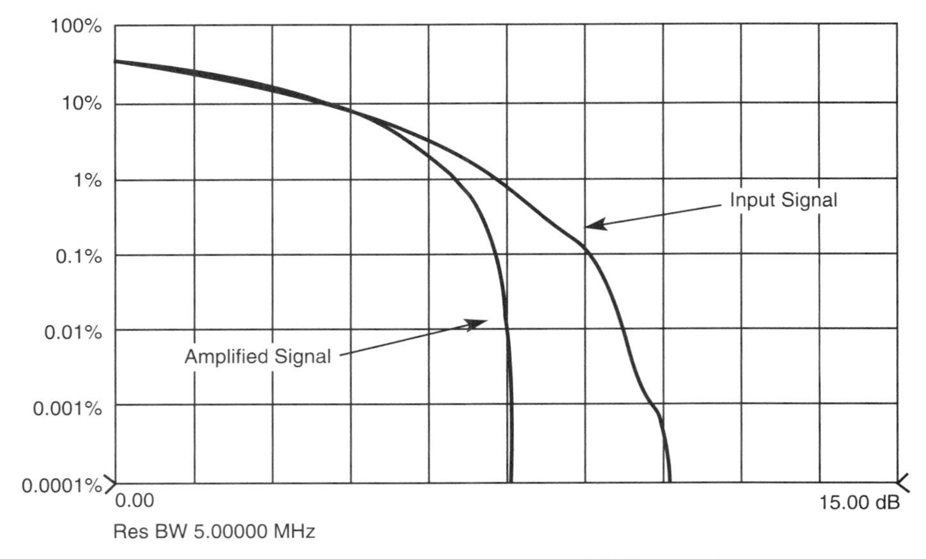

Figure 34.6 Amplifier input and output CCDF comparison.

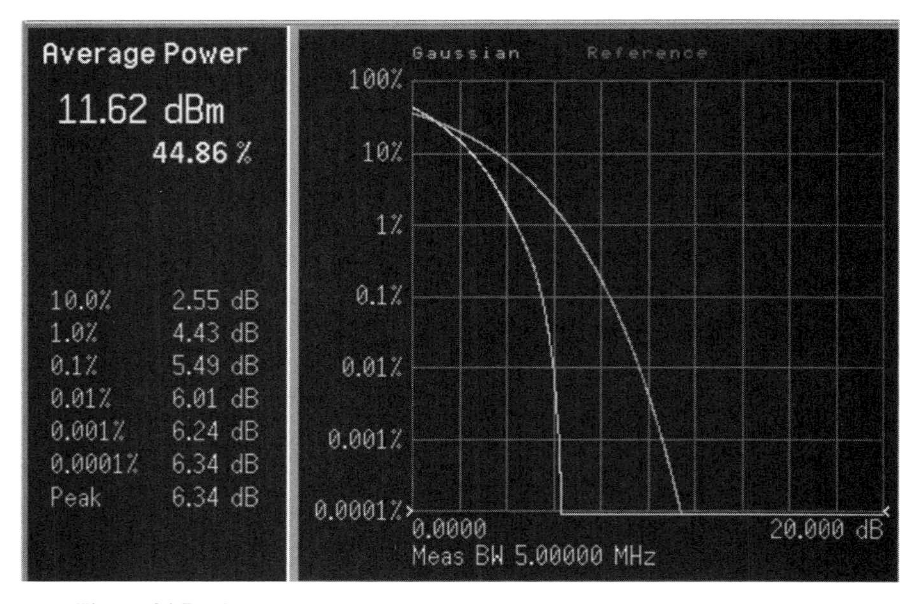

Figure 34.7 CCDF measurement of CDMA signal with AMP in linear range.

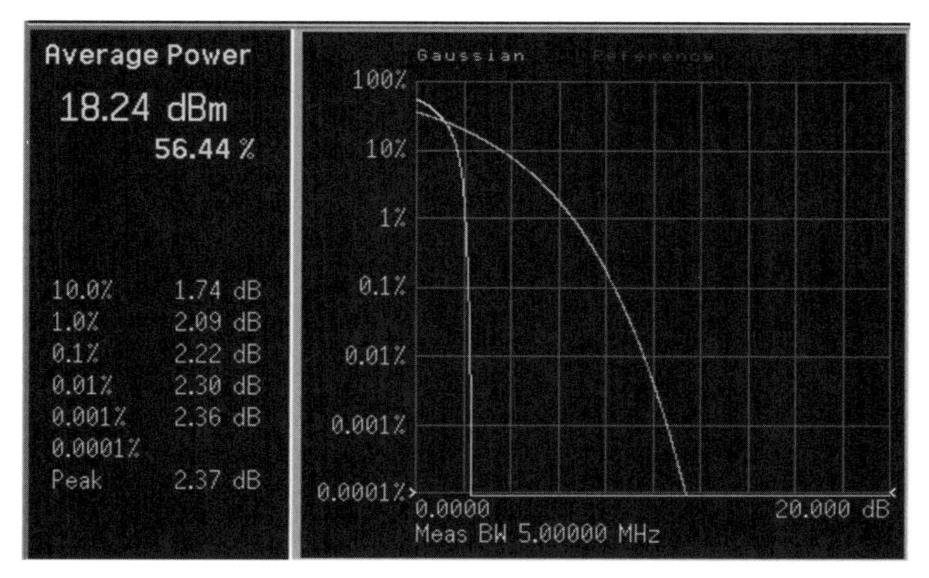

Figure 34.8 CCDF measurement of CDMA signal with AMP at saturation.

34.6 ANNOTATED BIBLIOGRAPHY

1. Application Note: Characterizing Digitally Modulated Signals with CCDF Curves, Document 5968-6875E, Agilent Technologies, Santa Clara, CA.

Reference 1 presents additional information on CCDF testing that is not presented in this book.

CHAPTER 35

BER

35.1 BER (BIT ERROR RATE) TESTING

BER testing compares a received bitstream to a transmitted bitstream to determine the BER of a digital communication system. Because input and output bits are being compared, the BER must be measured on a complete system including a modulator and demodulator, rather than on individual components like an RF amplifier.

BER testers (BERT) give no information about the cause of the bit errors, but give only quantitative pass/fail information. The BERT may be a stand-alone system or can be part of another instrument. The vector signal generator described in Chapter 5 can be procured with a BERT option in addition to its signal generation capabilities.

Figure 35.1 provides a BERT block diagram. All components except the cell phone under test and the antennas are included in the vector signal generator. The pattern generator shown in the lower left-hand box generates the digital test signal. The test signal is called PRBS, which has been recorded and given an identifying number, although it has a series of random bits. It also has a special known bit sequence at its start. The use of the special bit pattern at the start simplifies the synchronization of the transmitted and received bit patterns in time. PN9 and PN15 are common PRBS signals. A small sample of the bit pattern is shown along the bottom of the figure.

The PRBS pattern is encoded into the BERT system's modulation and modulated onto an RF carrier generated by the VSA. The modulated signal is wirelessly transmitted to the receiver of the UUT. The RF signal is amplified and detected in

RF Measurements for Cellular Phones and Wireless Data Systems. By A. W. Scott and R. Frobenius

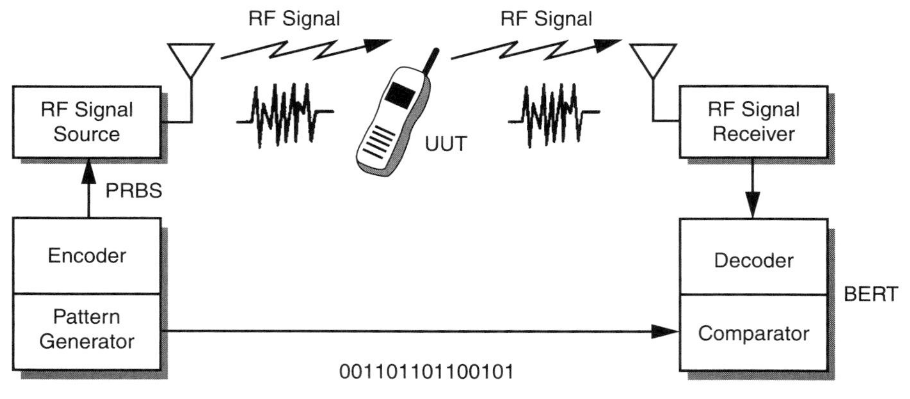

Figure 35.1 BER test setup.

the UUT receiver and now includes any bit errors created in the receiver. The received bitstream is now connected to the transmitter portion of the UUT terminals in the bottom of the cell phone. The transmitted signal, modulated with the bitstream from the UUT receiver, is then transmitted to the BERT RF receiver and demodulated. If there were no bit errors, this received bitstream would be identical to the original bitstream from the BERT. The transmitted and received signals are compared in a comparator, and the BER is determined.

Figure 35.2 shows the display of the vector signal generator when it is operating in the BERT mode. The RF carrier frequency in this example is 4 GHz, and it is accurate to all of the significant figures that are shown. Its amplitude is −100 dBm, which is typical of the received RF signal when a mobile is at the edge of a cell. As the

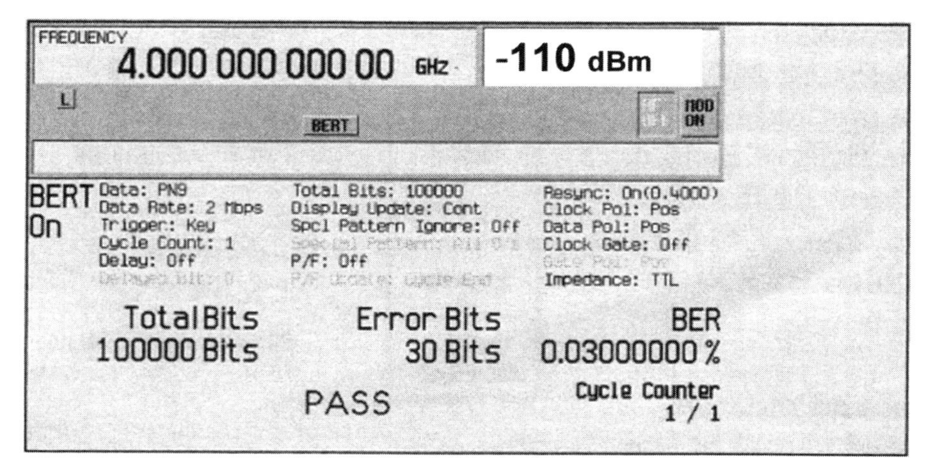

Figure 35.2 BER display on an ESG-D signal generator. Agilent Technologies © 2008. Used with permission.

display shows, a total of 100,000 bits have been analyzed, and the number of error bits is 30. The BER is 0.030% = 0.0003. The BER requirement for the particular cell phone under test is 0.001, so the tester indicated a "pass."

35.2 ANNOTATED BIBLIOGRAPHY

1. Product Note: Option UN7 for E4438C, Agilent Technologies, Santa Clara, CA.

Reference 1 describes the BERT option for a VSA.

CHAPTER 36

MEASUREMENT OF GSM EVOLUTION COMPONENTS

The VSA displays constellation, vector, and eye diagrams and measures EVM. However, it can also be supplied with software that will perform tests to specific system requirements, including making pass/fail measurements.

This type of measurement to system specifications is illustrated in this chapter. Measurements are shown for the GSM Evolution systems described in Chapter 30, including EDGE, WCDMA, and HSDPA. Special Agilent software is used to control the VSA. The test signals are provided by the vector signal generator described in Chapter 5, using special software to generate the EDGE, WCDMA, and HSDPA modulated RF signals. Measurements are made in the 1.9 GHz PCS frequency band.

Various RF components distort the RF signal that is carrying the digital information in its modulation. One of the most critical components is the power amplifier that was described in Chapter 19. Its distortion of modulated signals was shown in Chapters 32, 33, and 34 to illustrate ACP, constellation diagrams/EVM, and CCDF, respectively.

Similar software packages are available for evaluating the 802.11a, b, g, and n short-range, high data rate systems, as well as 802.16 WiMAX systems.

The gain and phase of a power amplifier was measured as a function of RF input power in Chapter 19. These measurement results are shown again in Figure 36.1. Measurements of the distortion of EDGE, WCDMA, and HSDPA signals by this amplifier operating in its linear range, at its 1 dB compression point, and at saturation will be described in this chapter.

RF Measurements for Cellular Phones and Wireless Data Systems. By A. W. Scott and R. Frobenius

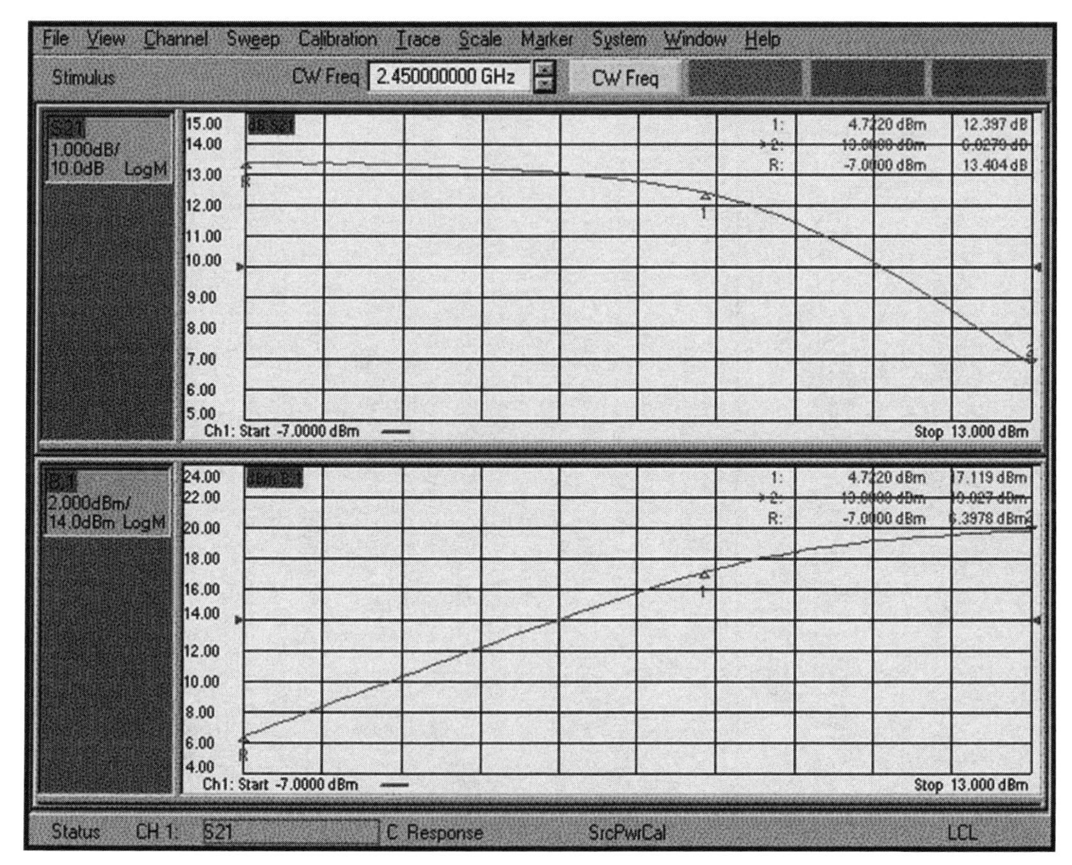

Figure 36.1 Power amplifier swept gain and output power versus input power.

36.1 MEASUREMENT OF EDGE SIGNAL DISTORTION

The Agilent software package allows distortions of GSM, GPRS, and EDGE signals to be analyzed. Because GSM and GPRS use digital frequency modulation, their waveforms are not distorted by power amplifier nonlinearity. Consequently, these modulations will not be discussed here.

Figure 36.2 shows the measurement of an EDGE polar vector test signal. The $3/8\pi 8$PSK modulation pattern is shown on the right. The tabular information on the left shows that the RMS EVM is only 0.17% and the peak EVM is only 0.44%. As indicated in the upper right-hand corner of the display, this signal passes the specification. This same display is obtained when the EDGE signal is amplified by the power amplifier operating in its linear range. Figure 36.3 shows the EDGE polar vector measurement when the amplifier is operating at saturation. The vector diagram is badly distorted. The RMS EVM has increased to 17.44% and the peak EVM has increased to 41.82%; thus, the signal fails the specification.

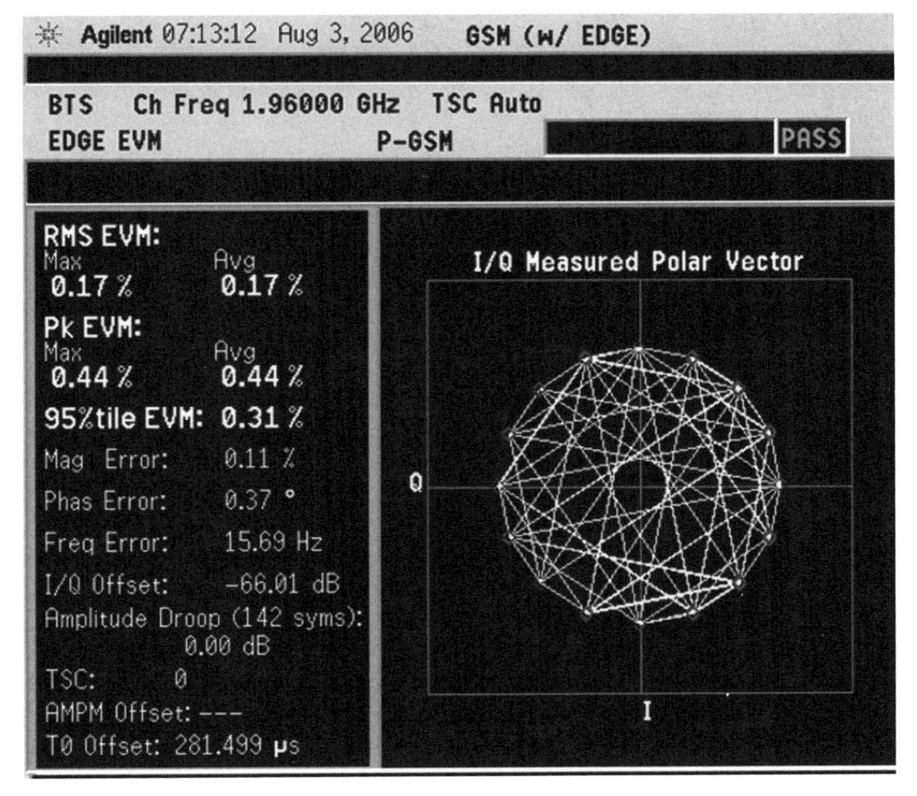

Figure 36.2 EDGE polar vector.

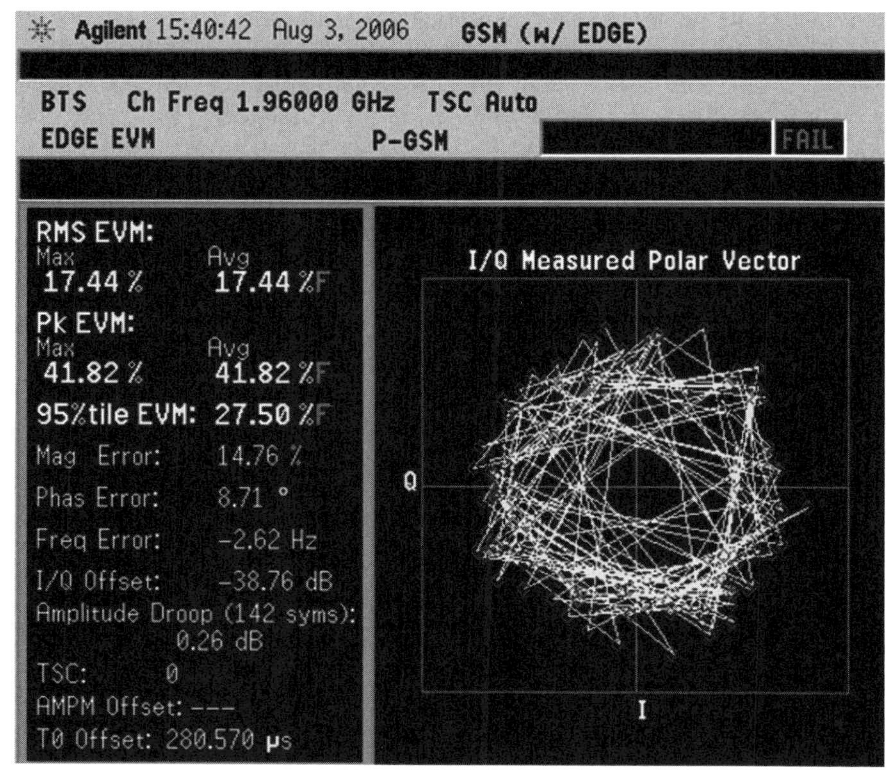

Figure 36.3 EDGE polar vector at amplifier saturation.

36.2 MEASUREMENT OF WCDMA AND HSDPA DISTORTIONS

Figure 36.4 shows adjacent power spreading of a WCDMA signal. The spreading was measured on four channels below and four channels above the main channel. All channels are 5 MHz wide. For the case of the test signal only, and the test signal amplified by the amplifier operating in its linear range (top chart), the ACP is −65 dBc in all adjacent channels. However, when the amplifier is operated at saturation, the ACP in the nearest 5 MHz channels is only −25 dBc.

Figure 36.5 shows HSDPA code domain measurements. The upper left-hand diagram shows the channels with different PN codes horizontally. The thick bars are the channels in use. The left-hand bar is a channel that is also in use and that has been selected for detailed analysis. The upper right-hand diagram shows symbol power over a 7.36 K symbol range.

The lower left-hand chart in Figure 36.5 shows the vector diagram of the 16QAM modulated signal in the channel selected for detailed analysis. The lower right-hand chart shows tabular data for this selected channel. Because this is the test signal, the EVM is only 1.18% RMS and 2.51% peak.

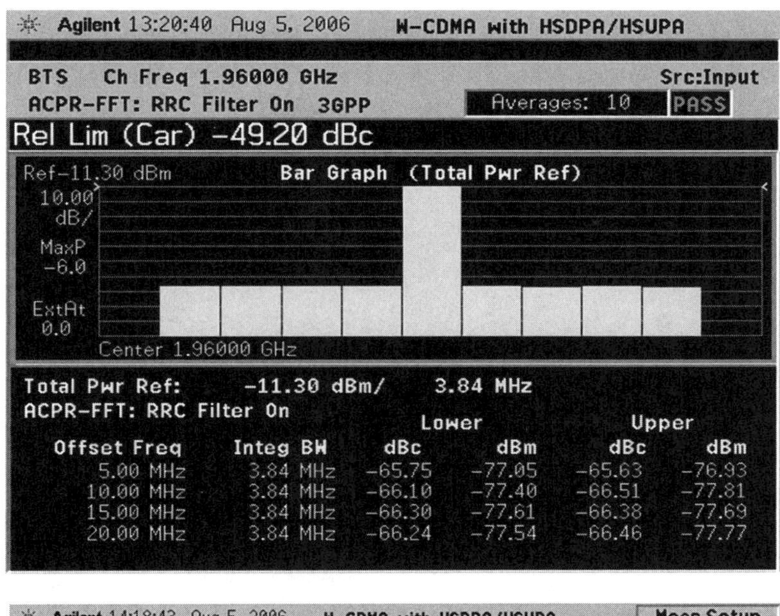

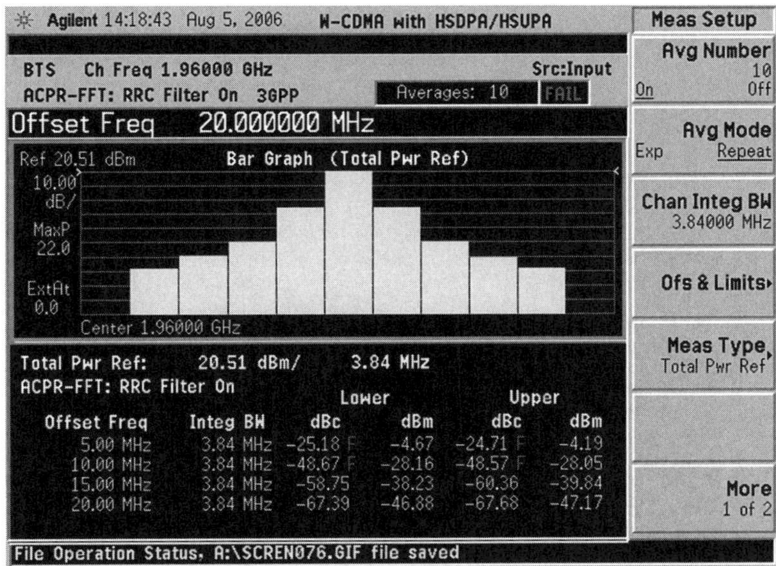

Figure 36.4 ACP of WCDMA/HSDPA signal.

Figures 36.6, 36.7, and 36.8 show the HSDPA polar vector diagram when the amplifier is operating in its linear region, at its 1 dB compression point, and at saturation, respectively. Note that the signal passes when the amplifier is operating in its linear region or at its 1 dB compression point, but it fails at saturation.

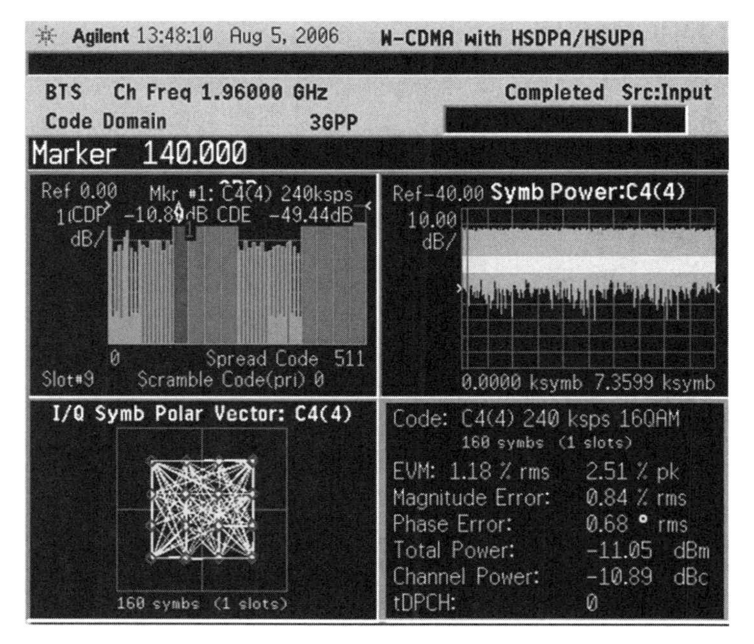

Figure 36.5 HSDPA code domain measurements.

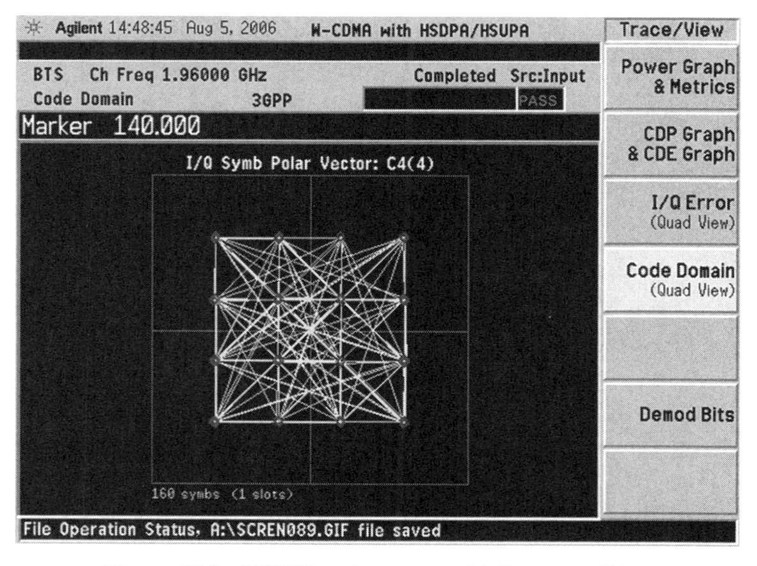

Figure 36.6 HSDPA polar vector with linear amplifier.

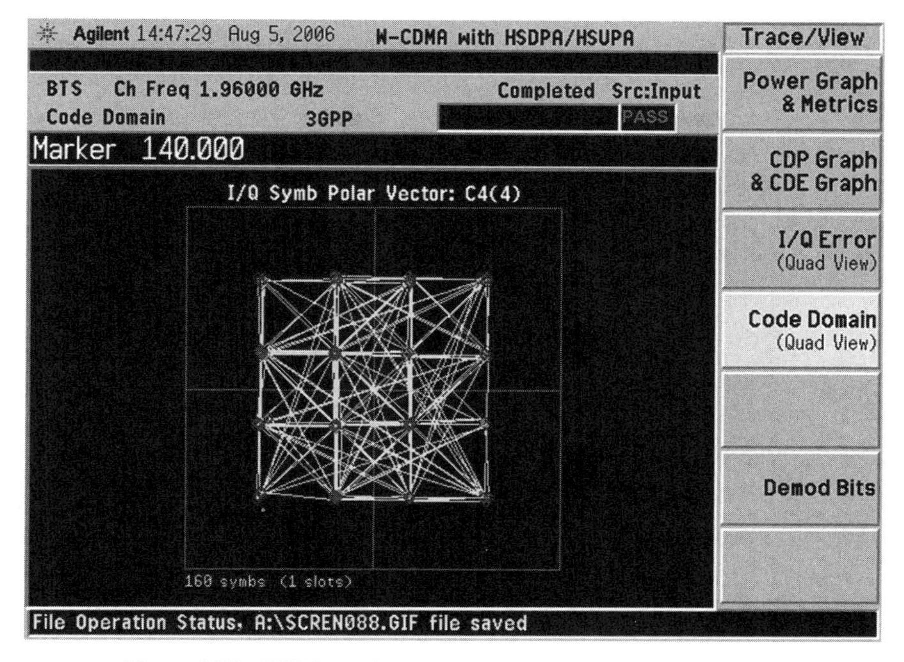

Figure 36.7 HSDPA polar vector with AMP at 1 dB compression.

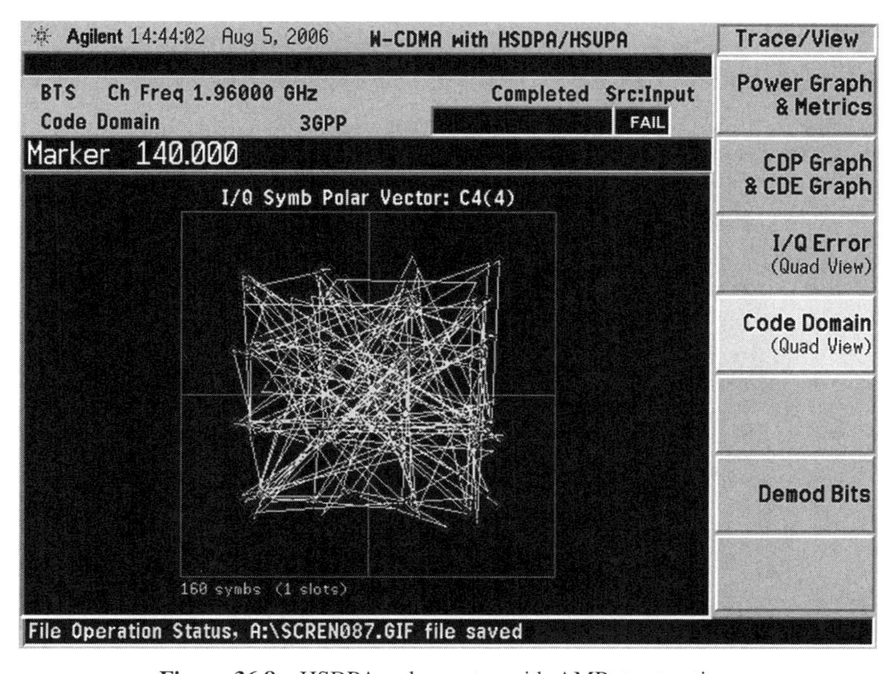

Figure 36.8 HSDPA polar vector with AMP at saturation.

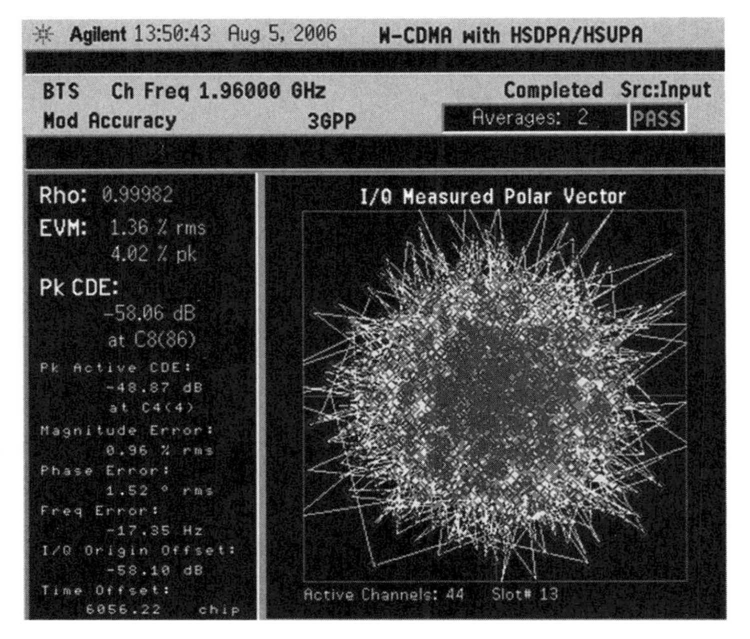

Figure 36.9 HSDPA composite EVM with AMP in linear range.

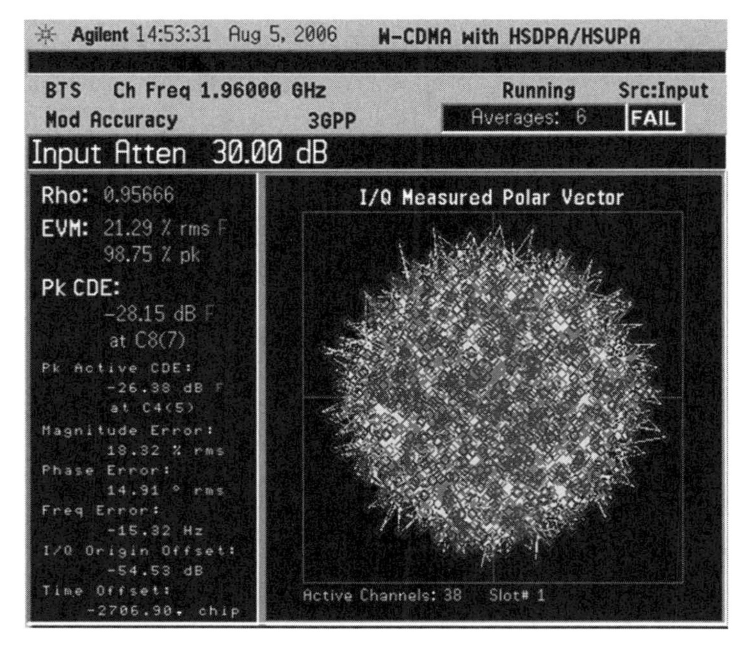

Figure 36.10 HSDPA composite EVM with saturated amplifier.

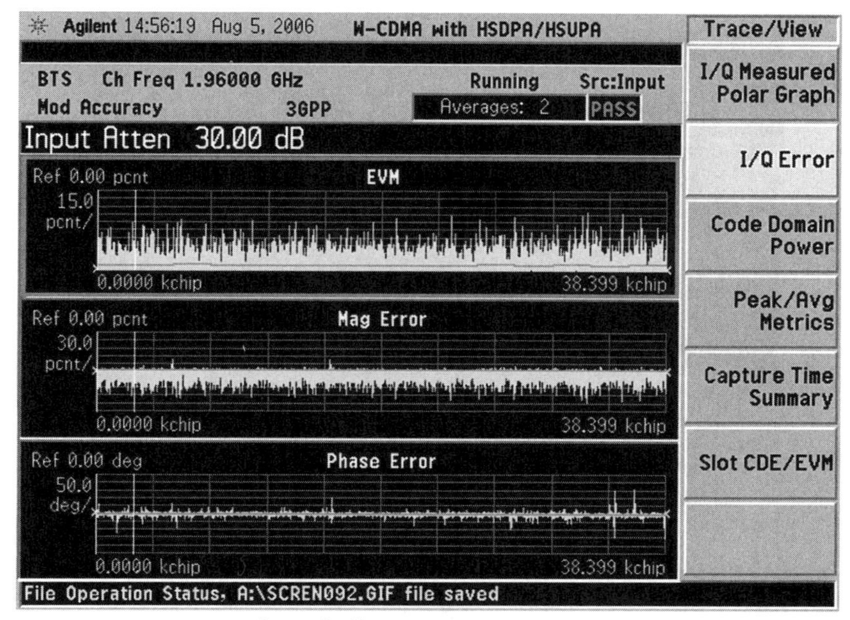

Amp in linear range

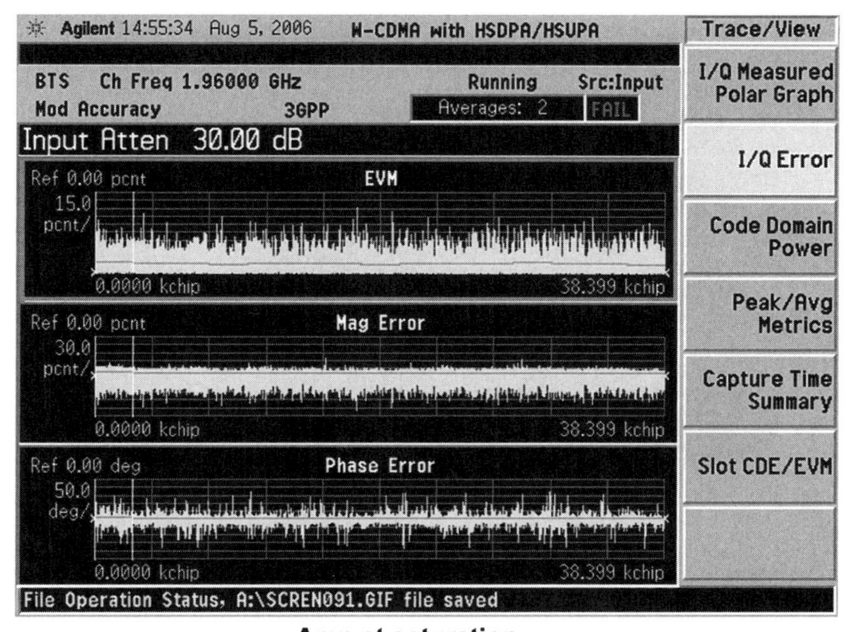

Amp at saturation

Figure 36.11 I/Q error of HSDPA signal.

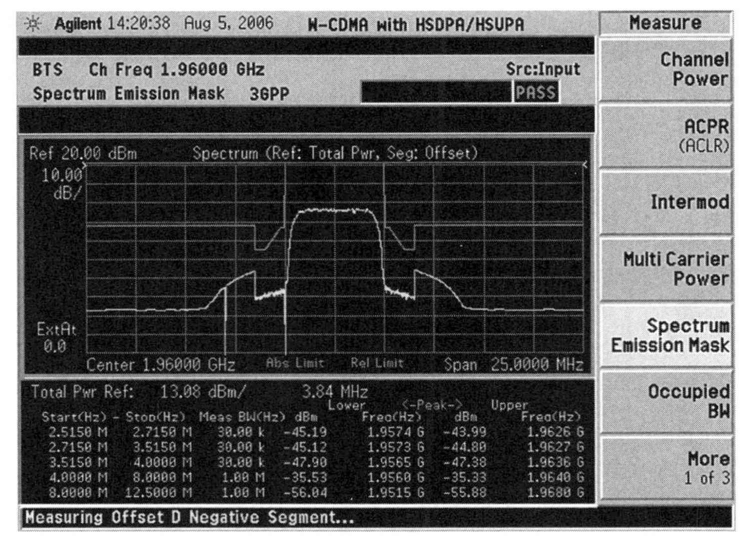

Amp in linear range

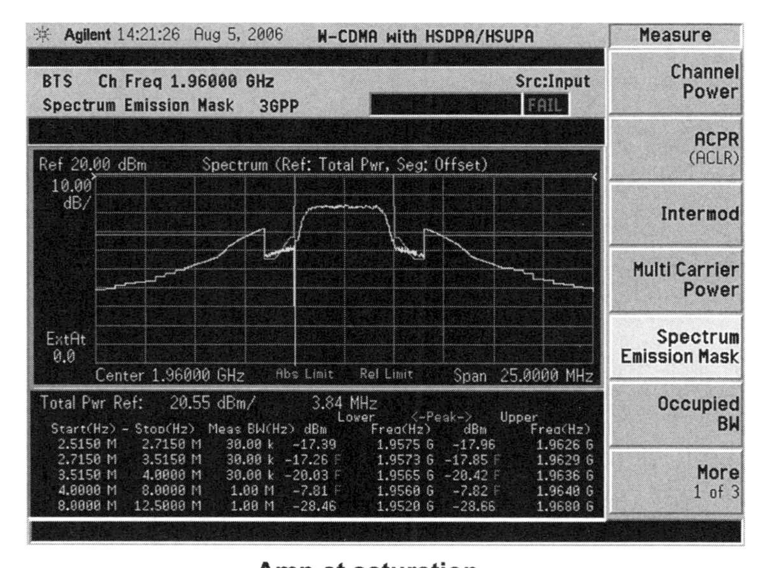

Amp at saturation

Figure 36.12 Spectrum emission mask.

Figures 36.9 and 36.10 show the HSDPA composite EVM with the amplifier operated in its linear range and at saturation, respectively. The polar vector shows undistorted transitions between the phase points in Figure 36.9 but severe compression of the transitions in Figure 36.10. The EVM when the amplifier is operated in its linear

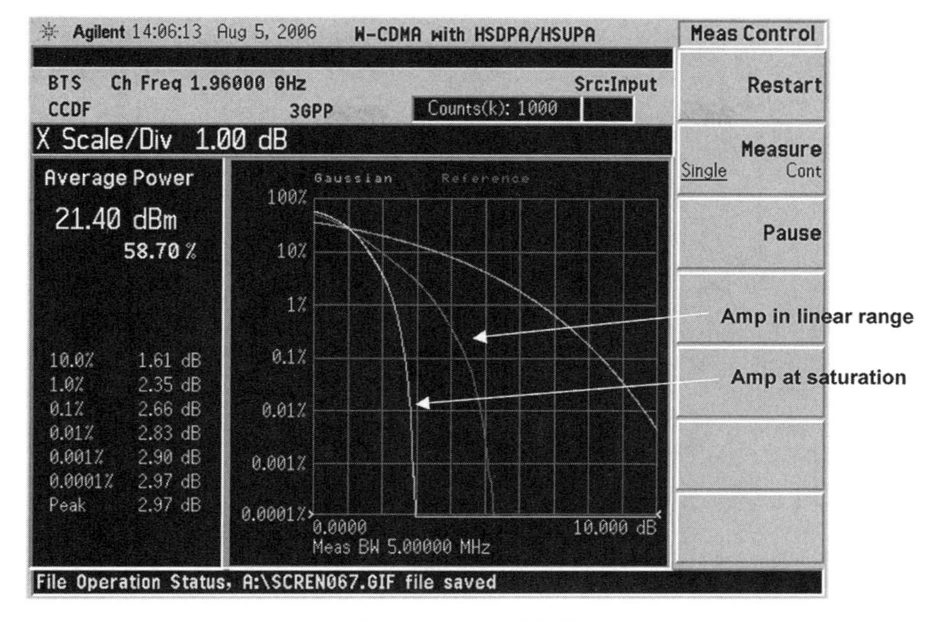

Figure 36.13 CCDF.

range is 1.36% RMS and 4.02% peak, and the amplifier passes. The EVM when the amplifier is operated at saturation is 21.29% RMS and 98.75% peak, and the amplifier fails the specification.

Figure 36.11 shows the I/Q error of the HSDPA signal when the amplifier is operated in the linear region and at saturation.

Figure 36.12 provides a spectrum emission mask. Note that when the amplifier is operated its linear range, the spectrum is within the spectrum mask limits and the system passes. When the amplifier is operated at saturation, the spectrum is outside of the limits and the system fails.

Figure 36.13 shows the CCDF of the amplifier in the linear range and at saturation. Note the extreme compression of the signal when the amplifier is operated at saturation.

36.3 ANNOTATED BIBLIOGRAPHY

1. Product Note: Technical Overview with Self Guided Demonstration. Option 202: GSM with EDGE Measurement Personality, Agilent Technologies, Santa Clara, CA.
2. Product Note: Technical Overview with Self Guided Demonstration. Options BAF and 210: W-CDMA and HSDPA/HSUPA Measurement Personalities, Agilent Technologies, Santa Clara, CA.

References 1 and 2 are Agilent brochures describing how to make measurements of the type described in this chapter, using the software options listed.

TERMINOLOGY

ACP	adjacent channel power (1)
ADPCM	adaptive differential PCM (1)
AFC	automatic frequency control (28)
AlN	aluminum nitride (22)
APC-7	Amphenol 7 mm precision connector (12)
ARQ	automatic repeat request (1)
BER	bit error rate (1)
BERT	bit error rate tester (35)
BJT	bipolar junction transistor (19)
BNC	bayonet Navy connector (12)
BPSK	binary phase shift keying (1)
BTS	base transceiver station (15)
CCDF	complementary cumulative distribution function (1)
CDMA	code division multiple access (1)
CMOS	N-doped and P-doped semiconductor MOSFET (19)
CR	carriage return (29)
CW	continuous wave (9)
DAC	digital to analog converter (28)
DBPSK	differential binary phase shift keying (4)
DCS	Digital Cellular System (28)
DPSK	differential phase shift keying (1)
DQPSK	differential quadrature phase shift keying (1)

RF Measurements for Cellular Phones and Wireless Data Systems. By A. W. Scott and R. Frobenius

DSI	digital speech interpolation (30)
DUT	device under test (1)
EDGE	Enhanced Data Rates for GSM Evolution (1)
EHF	extremely high frequency (2)
ENR	excess noise ratio (1)
EVM	error vector magnitude (1)
FBAR	film bulk acoustic resonator (22)
FDMA	frequency division multiple access (1)
FEC	forward error correction (1)
FET	field effect transistor (1)
FSK	frequency shift keying (1)
1G	first generation (1)
2G	second generation (1)
3G	third generation (1)
4G	fourth generation (1)
GaAs	gallium arsenide (1)
GaN	gallium nitride (19)
GMSK	Gaussian minimum shift keying (4)
GPRS	General Packet Radio Service (28)
GSM	Global System for Mobile Communications (1)
HBT	heterojunction bipolar transistor (19)
HDTV	high definition TV (29)
HEMT	high electron mobility transistor (19)
HET	hot electron transistor (19)
HSDPA	High-Speed Downlink Packet Access (1)
HS-DSCH	high-speed downlink shared channel (30)
IC	integrated circuit (1)
IF	intermediate frequency (1)
IIP3	input IP3 (1)
IP3	third-order intermodulation product (1)
ISM	Industrial, Scientific, and Medical (frequency bands) (28)
LAN	local area network (1)
LDMOS	laterally defused MOSFET (19)
LNA	low noise amplifier (1)
LO	local oscillator (1)
LRM	load, reflection, and match model (14)
LTCC	low temperature cofired ceramic (17)
MDS	minimum detectable signal (27)
MES	metal–semiconductor (19)
MESFET	metal–semiconductor field effect transistor (19)
MF	medium frequency (2)
MIMO	multiple input–multiple output (31)
MOS	metal–oxide–semiconductor (19)
MOSFET	metal–oxide–semiconductor field effect transistor (19)

MSO	Mobile Switching Office (15)
MSK	minimum shift keying (4)
NADC	North American digital cellular (32)
NMOS	N doped semiconductor MOSFET (19)
OFDM	orthogonal frequency division multiplexing (1)
OFDMA	orthogonal frequency division multiple access (1)
OIM3	output third-order intermodulation product (27)
OIP3	output IP3 (1)
OOK	on–off keying (1)
ORR3	ratio of output power to OIM3 (27)
PAE	power-added efficiency (28)
PAN	Personal Area Network (31)
PBX	Private Branch Exchange (15)
PCM	pulse code modulation (1)
PCS	Personal Communications System (1)
PDF	probability density function (34)
PHEMT	pseudomorphic high electron mobility transistor (19)
PLL	phase locked loop (1)
PLO	phase locked oscillator (1)
PMOS	P-doped semiconductor MOSFET (19)
PN	pseudorandom number (1)
PRBS	pseudorandom bit sequence (1); PN9, PN15, common PRBS signals
PSK	phase shift keying (1)
PSTN	Public Switched Telephone Network (15)
QAM	quadrature amplitude modulation (1)
16QAM	16-quadrature amplitude modulation (1)
64QAM	64-quadrature amplitude modulation (1)
QPSK	quadrature phase shift keying (1)
RFIC	RF integrated circuit (1)
RSS	root sum of squares (13)
S-parameter	scattering parameter (1)
S_{11}	return loss
S_{21}	insertion gain/loss
SAW	surface acoustic wave (22)
S/N	signal/noise ratio (1)
SHF	superhigh frequency (2)
SiC	silicon carbide (19)
SiGe	silicon germanium (1)
SMA	subminiature version A (coaxial RF connectors) (8)
SOC	system on a chip (1)
SOLT	short, open, load, and through (1)
SONET	synchronous optical network (4)
SWR	standing wave ratio (1)

TDD	time division duplexing (28)
TDMA	time division multiple access (1)
TNC	threaded Navy connector (12)
TOI	third-order intercept (19)
TRL	through, reflection, and load (14)
TRM	through, reflection, and match (14)
TTI	transmission time interval (30)
UHF	ultrahigh frequency (2)
UMTS	Universal Mobile Telecommunications System (19)
UUT	unit under test (1)
VCO	voltage controlled oscillator (1)
VHF	very high frequency (2)
VNA	vector network analyzer (1)
VoIP	Voice over Internet Protocol (29)
VSA	vector signal analyzer (1)
WCDMA	Wideband Code Division Multiple Access (1)
WiMAX	Wireless Interoperability for Microwave Access (29)
ZIF	zero IF (1)

Units of Measure

A	amps
mA	milliamps
bps	bits per second
kpbs	kilobits per second
Mbps	megabits per second
dB	decibel (1)
dBc	decibels below the carrier (17) or channel (32)
dBi	decibels relative to isotropic (1)
dBm	decibels relative to 1 mW (1)
ft	foot/feet
Hz	Hertz
kHz	kilohertz
MHz	megahertz
GHz	gigahertz
THz	terahertz
in.	inch(es)
m	meter
mm	millimeter
μm	micron (micrometer) (1)
nm	nanometer
s	second
ms	millisecond
μs	microsecond
ns	nanosecond (10^{-9})

ps	picosecond
W	watt
mW	milliwatt
MW	megawatt
Ω	ohm
V	volt

INDEX

RF Measurements for Cellular Phones and Wireless Data Systems. By A. W. Scott and R. Frobenius